Impact Studies

Series Editor: Christian Koeberl

Christian Koeberl
Herbert Henkel
(Editors)

Impact Tectonics

With 187 Figures

ESF IMPACT

Springer

EDITORS

PROFESSOR CHRISTIAN KOEBERL
Department of Geological Sciences
University of Vienna
Althanstrasse 14
1090 Vienna
Austria
Email: christian.koeberl@univie.ac.at

DOCENT HERBERT HENKEL
Department of Land and Water Resources
Engineering
Royal Institute of Technology
10044 Stockholm
Sweden
Email: herbert@kth.se

ISSN 1612-8338
ISBN 3-540-24181-7 **Springer Berlin Heidelberg New York**

Library of Congress Control Number: 2004117859

Springer is a partof Springer Science+Business Media
springeronline.com

Printed in The Netherlands

Cover design: E. Kirchner, Heidelberg
Production: A. Oelschläger
Typesetting: Camera-ready by the Editors
Printing: Krips, Meppel
Binding: Litges+Dopf, Heppenheim

Printed on acid-free paper 32/2132/AO 5 4 3 2 1 0

Preface

This volume is the latest of eight books as a result of the activities of the scientific program on "The Response of the Earth System to Impact Processes" (IMPACT) of the European Science Foundation (ESF). The ESF is the European association of national funding organisations of fundamental research, with more than 60 member organisations from more than 20 countries. One of the main goals of ESF is to bring European scientists together to work on topics of common interest. The ESF IMPACT program deals with all aspect of impact research, mainly through the organisation of workshops, exchange program, short course, and related activities. An important aspect of the program is to stimulate interdisciplinary and international research.
Impacts of asteroids or comets on the earth surface have played an important role in the evolution of the planet. The ESF IMPACT program is an interdisciplinary program aimed at understanding impact processes and their effect on the earth system, including environmental, biological, and geological changes, and consequences for the biodiversity of ecosystems. The key goal of the program, to comprehend the processes that are responsible for these interactions, has been met with by eight workshops and published proceedings from these.

The workshop from which this volume resulted, was devoted to impact tectonic studies. It was located in the small town of Mora in the western part of the ring basin of the very large Siljan impact structure in Sweden. Several excursions were arranged to localities were impact tectonic processes can be studied in the 30 km diameter central uplift of crystalline rocks and in numerous quarries that have been mined in the Paleozoic limestones that collapsed into the excavated crater forming the ring basin.

A selection of 20 papers resulted from this IMPACT workshop, which was held in mid 2002. Each of the manuscripts was reviewed by at least two referees and was considered for publication on the basis of originality and proximity to the topics discussed at the workshop.

The contributions that have been collected in this volume cover the following themes:

- General topics related to impact tectonics
- Structural and tectonic aspects of impact craters
- Impact studies by numerical and experimental methods
- Economic aspects of impact craters.

The conundrum of *pseudotachylite* formation is extensively presented. Their formation is associated both with large impact structures and tectonic shear zones. The common factor is a highly mobile infra-crustal environment and the melting of involved rocks along shear surfaces. The resulting pseudotachylic breccia fills irregular fractures and cools rapidly. The volumes of such breccias increases with increased crater size and they are generally located in the crystalline basement of the crater structures.

Shatter cone formation is another tectonic feature characteristic for impact structures. These are compared to seemingly similar cone-in-cone (cic) structures found in some sediments. The shatter cones represent finely spaced sets of motion surfaces arranged in a cone-like pattern that formed in connection with the passage of a shock wave and the subsequent displacement flow, whereas the cic structures seem to represent a silicification of original sedimentary layering.

Dykes that contain *exotic fragments* (i.e., not from nearby sources) of rocks is another characteristic feature of impact structures. Both during the initial compression flow and the subsequent collapse flow in large craters, pre-impact lithologies are redistributed and may be found within the crater basement of the central uplift and within displaced parts of the original crater wall.

Impacts cause a considerable *fracturing* of the target area, both at the initial stage when shock- and rarefraction waves pass through the subsurface, and in connection with the collapse of large craters. It has been difficult to discriminate fracturing events that pre-date an impact, that are added by the impact and that have occurred after the impact. A pre-existing fracture pattern may influence the shape of a crater, together with the incidence angle. In several papers these issues are further developed by studying examples of polygonal craters on mars, by modelling of possible incidence angles, by field studies of actual fracture systems, by analysis of terrain patterns and geophysical indications based on electromagnetic measurements.

Studies of several more remotely located impact structures are presented in this volume, comprising two impact structures found in the Libyan desert (the *Oasis* and *BP* craters), the *Mjölnir* crater located off-shore in the Barents sea, and the *Kentland* structure in USA.

A compilation is presented in the paper by Puura and Plado of impact structures occurring within the *Baltic shield* and the nearest *platform regions*. Several new impact structures are suggested and presented in the volume. It seems that the Baltic Shield continues to be a region with a rather high crater density with over 20 structures including some that are suspected craters due to the occurrence of several typical geological, morphological and geophysical features. Two of these structures, Lockne and Duobblon, occur at the Caledonian thrust front and are partially covered by younger thrusted units.

The *Lockne* structure has been intensively studied for several years by a research team from the Stockholm University . In this volume three papers deal with the Lockne structure, presenting a summary of the research efforts, a numerical mechanical model of the impact event, and the character of the associated brecciation of the crater basement.

A review is given of the *Siljan* structure, one of the largest impact structures in Europe, which once was the target for a-biogenic hydrocarbon investigations. This very large structure has remarkably small geophysical signatures despite an almost oversized central crystalline core with 30 km diameter. The sedimentary ring of the structure contains pre-impact cover rock sequences that were down-warped by the crater collapse flow, in some cases in the form of very large, km-sized, xenoliths within the granitic basement. The stratigraphic succession within individual cover rock fragments is almost undisturbed.

The *economic aspects* of impact craters have occasionally been discussed. In this volume an extensive review is made of this issue. There seems to be two main types of phenomena that can result in economic accumulations of natural resources. The intense brecciation of the crater basement can act as a reservoir structure for oil, gas, and water on the one hand. It can also be subjected to intense hydrothermal activity that redistributes and accumulates dissolved metals along the intense thermal gradients that reside in the target area for a rather long period after large impacts.

Herbert Henkel
Royal Institute of Technology
Stockholm, Sweden

Christian Koeberl
University of Vienna
Vienna, Austria

October 2004

Acknowledgements

The editors are grateful to the ESF IMPACT program for organizing the workshop and providing financial support for the preparation of this proceedings volume. We particularly appreciate the help, patience and cooperation of the many reviewers and authors who have contributed their expertise to this project. We especially thank docent Joanne Fernlund (Royal Institute of Technology Stockholm) for her efforts to bring all the manuscripts into a presentable shape.

Contents

Impact Tectonics – General Aspects

Structural and Tectonic Aspects of Impact Craters

Impact Studies by Numerical and Experimental Methods

Economic Aspects of Impact Structures

List of contributors

Sven Aaro
Geological Survey of Sweden
Division of Geophysics
Box 670
S-751 28 Uppsala, Sweden
(sven.aaro@sgu.se)

Marko Aittola
Planetology Group
Division of Astronomy
Department of Physical Sciences
University of Oulu
P.O. Box 3000
FIN-90014, Oulu, Finland

Ann Bäckström
Department of Land and Water Resources Engineering
Royal Institute of Technology
S-100 44 Stockholm, Sweden
(annb@kth.se)

Ilka von Dalwigk
Department of Geology and Geochemistry
Stockholm University
S-10691 Stockholm, Sweden
(ilka@geo.su.se)

Raymond A. Donelick
Apatite to Zircon Inc.
1075 Matson Road
Viola, ID 83872-9709, USA

Henk Brinkhuis
Laboratory of Palaeobotany and Palynology
Utrecht University
Budapestlaan 4
NL-3584 CD Utrecht, The Netherlands

Joanne Fernlund
Department of Land and Water Resources Engineering
Royal Institute of Technology
S-10044 Stockholm, Sweden
(Joanne@kth.se)

Tom Flodén
Department of Geology and Geochemistry
Stockholm University
S-10691 Stockholm, Sweden
(tom@geo.su.se)

Roger L. Gibson
Impact Cratering Research Group
School of Geosciences
University of the Witwatersrand
Private Bag 3, P.O. WITS 2050
Johannesburg, South Africa
(gibsonr@geosciences.wits.ac.za)

Nicole Heller
Department of Geology
Grand Valley State University
1 Campus Drive
Allendale, MI 49401, USA

Herbert Henkel
Department of Land and Water Resources Engineering
Royal Institute of Technology
S-10044 Stockholm, Sweden
(herbert@kth.se)

Jan-Martin Hertzsch
School of Mathematics
University of Bristol
BS8 1TW Bristol, United Kingdom
(Jan-Martin.Hertzsch@bristol.ac.uk)

Boris A. Ivanov
Institute for Dynamics of Geospheres
Russian Academy of Sciences
Leninsky Prospekt, 38-6
ROS-117939 Moscow, Russia
(baivanov@idg.chph.ras.ru)

Thomas Kenkmann
Institut für Mineralogie, Museum für Naturkunde
Humboldt-Universität
D-10099 Berlin, Germany
(thomas.kenkmann@museum.hu-berlin.de

Ü. Kestlane
Institute of Geology
Tallinn Technical University
7 Estonia Avenue
EE-10143 Tallinn, Estonia

Juho Kirs
Institute of Geology
University of Tartu
Vanemuise 46
EE-51014 Tartu, Estonia

Christian Koeberl
Department of Geological Sciences
University of Vienna
Althanstrasse 14
A-1090 Vienna, Austria
(christian.koeberl@univie.ac.at)

Mare Konsa
Institute of Geology
Tallinn Technical University
Estonia pst.7
EE-10143 Tallinn, Estonia
(mare@gi.ee)

Veli-Petri Kostama
Planetology Group
Division of Astronomy
Department of Physical Sciences
University of Oulu
P.O. Box 3000
FIN-90014 Oulu, Finland

A. Kraut
National Heritage Board
18 Uus St.
EE-10111 Tallinn, Estonia

J.W. de Leeuw
Department of Geochemistry
Institute of Earth Sciences
Utrecht University
PO Box 80.021
NL-3508 TA Utrecht, The Netherlands

Robert Lilljequist
Ecominas
Calle Horno 9
Estepuma, Spain
(robertlilljequist@yahoo.se)

Maurits Lindström
Department of Geology and Geochemistry
Stockholm University
S-10691 Stockholm, Sweden
(maurits.lindstrom@geo.su.se)

C.V. Looy
Laboratory of Palaeobotany and Palynology
Utrecht University
Budapestlaan 4
NL-3584 CD Utrecht, The Netherlands

Stefano Lugli
Dipartimento di Scienze della Terra
Università degli Studi Modena e Reggio Emilia
Largo S. Eufemia 19
I-41100 Modena, Italy
(lugli.stefano@unimire.it)

Victor L. Masaitis
Karpinsky Geological Institute
Sredny prospekt 74
ROS-199106 St.-Petersburg , Russia
(vicmas@vsegei.sp.ru)

Mikhail S. Mashchak
Karpinsky Geological Institute
Sredny prospect 74
ROS-199106 St. Petersburg, Russia
(mvn@mail.wplus.net)

H. Jay Melosh
Lunar and Planetary Laboratory
University of Arizona
Tucson, AZ 85721, USA
(jmelosh@lpl.arizona.edu)

T. Moora
Institute of History
6 Rüütli St.
EE-10130 Tallinn, Estonia

Mikhail V. Naumov
Karpinsky Geological Institute
Sredny prospect 74
ROS-199106 St. Petersburg, Russia
(mvn@mail.wplus.net)

Teemu Öhman,
Institute of Geoscience
Department of Geology
University of Oulu
P.O. Box 3000
FIN-90014, Finland
(teemu.ohman@oulu.fi)

Jens Ormö
Centro de Astrobiología (CSIC/INTA)
Instituto Nacional de Técnica Aeroespacial
Ctra de Torrejón a Ajalvir, km 4
E-28850 Torrejón de Ardoz, Madrid, Spain
(ormo@inta.es)

Jüri Plado
Institute of Geology
University of Tartu
Vanemuise 46
EE-51014 Tartu, Estonia

Jeffrey Plescia
MP3-E163, Applied Physics Laboratory
Johns Hopkins University
11100 Johns Hopkins Road
Laurel, MD 20723-6099, USA
(jeffrey.plescia@jhuapl.edu)

Michael C. Pope
Department of Geology
Washington State University
Webster Physical Science Building 1228
Pullman, WA 99164-2812, USA

Christina Poulos
Department of Geology
Grand Valley State University
1 Campus Drive
Allendale, MI 49401, USA

Ulla Preeden
Institute of Geology
University of Tartu
Vanemuise 46
EE-51014 Tartu, Estonia
(ullap@ut.ee)

J.-M. Punning
Institute of Ecology
Tallinn Pedagogical University
2 Kevade St.
EE-10137 Tallinn, Estonia

Väino Puura
Institute of Geology
University of Tartu
Vanemuise 46
EE-51014 Tartu, Estonia
(puura@ut.ee)

Jouko Raitala
Planetology Group, Division of Astronomy
Department of Physical Sciences
University of Oulu
P.O. Box 3000
FIN-90014, Finland

Anto Raukas
Institute of Geology
Tallinn Technical University
7 Estonia Avenue
EE-10143 Tallinn, Estonia
(raukas@gi.ee)

Wolf Uwe Reimold
Impact Cratering Research Group, School of Geosciences
University of the Witwatersrand
Private Bag 3, P.O. WITS 2050
Johannesburg, South Africa
(reimoldw@geosciences.wits.ac.za)

Mark A. Sephton
Planetary and Space Sciences Research Institute
The Open University
Walton Hall
Milton Keynes, MK7 6AA, United Kingdom
(m.a.sephton@open.ac.uk)

Valery Shuvalov
Institute for Dynamics of Geospheres
Russian Academy of Sciences
38 Leninsky pr.
ROS-119334 Moscow, Russia
(shuvalov@idg.chph.ras.ru)

Erik Sturkell
The Icelandic Meteorological Office
Bustadarvegur 9
IS-150 Reykjavik, Iceland
(erik@vdur.is)

Filippos Tsikalas
Department of Geosciences
University of Oslo
P.O. Box 1047 Blindern
N-0316 Oslo, Norway
(filippos.tsikalas@geo.uio.no)

H. Visscher
Laboratory of Palaeobotany and Palynology
Utrecht University
Budapestlaan 4
NL-3584 CD Utrecht, The Netherlands

John C. Weber
Department of Geology
Grand Valley State University
1 Campus Drive
Allendale, MI 49401, USA

"Pseudotachylites" in Large Impact Structures

Wolf U. Reimold and Roger L. Gibson

Impact Cratering Research Group, School of Geosciences, University of the Witwatersrand, Private Bag 3, P.O. Wits 2050, Johannesburg, South Africa

(reimoldw@geosciences.wits.ac.za; gibsonr@geosciences.wits.ac.za)

Abstract. The catastrophic processes associated with meteorite impact are capable of producing dark-matrix breccias by a variety of mechanisms during both the shock compression and crater modification phases. Distinguishing the origins of these breccias provides vital clues to understanding the cratering process. Unfortunately, the criteria by which these distinctions can be made are not always unequivocal. This has led to unfettered use of terminology, which must change if the science is to move forward. Principal among this is the problematic relating to the origin of so-called *"pseudotachylite / pseudotachylite-like / pseudotachylitic breccia"* in impact crater settings. The purpose of this discussion is to emphasize the problematics concerning the definition and recognition of such material. Ill-defined nomenclature coupled with insufficient field and laboratory investigations have led to confusion with respect to recognition of *bona fide* pseudotachylite = friction melt *(according to its conventional geological terminology, adhered to by these authors)*, and separation from shock melt and other impactogenic or tectonically produced breccia types. It is essential to deal with these materials as carefully as possible, as they do have the potential to contribute to the understanding of the impact cratering process and have implications regarding the recognition of impact structures and understanding of their formation. It is categorically rejected to apply blanket terminology – such as indiscriminate usage of the term "pseudotachylite" – as label for (ultra) cataclasites, (ultra) mylonites, or dike (veinlet) breccia in impact crater floors of unknown or unrecognizable (e.g., because of thorough metamorphism or alteration of breccia matrix) genus, or the term "tagamite" for dense, black, finest-grained or aphanitic melt rock that could represent impact melt rock or *bona fide* pseudotachylite, or any of the other breccia types listed herein. Also, the use of simplifying terms such as A- or B-type, "S (shock)-type" or "E (endogenic)-type pseudotachylite", without discriminating by rock nature or distinctive criteria, in order to contribute to fully understand the genesis of such materials, is discouraged. Recent work at Vredefort has shown that much, if not all of what has conventionally been termed "Vredefort pseudotachylite" could represent shock melt, in contrast to previous thinking where much of the so-called Vredefort "pseudotachylite" was considered to have formed relatively late (in the impact cratering process) during decompression or fault-related melting related to the modification phase of cratering. It is discussed that current A, B or S, E "pseudotachylite" classifications are insufficient and unsatisfactory.

1 Introduction

Rocks of the Earth's crust may experience a variety of catastrophic processes that can lead to the formation of new rocks characterized by fragments of the original rock(s) hosted in a much finer-grained, aphanitic, or even glassy matrix. Such *breccias* may owe their origin, *inter alia*, to violent, often volatile-assisted, stopping of wallrocks adjacent to magma chambers, more passive intrusion of melts into country rocks in magmatic and high-temperature metamorphic environments, post-tectonic or volcanic impregnation of cataclasites with deposits from hydrothermal solutions, slip and frictional sliding along fault or shear zones, or the catastrophic deformation processes associated with impact cratering.

In essence, shock compression is capable of melting minerals and whole rocks, at shock pressures above circa 40 GPa. This process results in strongly heterogeneous distribution of shock pressure and post-shock temperature (potentially, as at Vredefort, coupled with crustal heat inherent to the target rock), whereby mixtures of shock melted and crystalline remnants a breccia is generated. These impact melt breccias may be sharply separated from crystalline crater basement, at least partially resulting in a vein pattern, or grade into more and more massive crystalline material. Injection of shock melts into less shocked basement is also possible – resulting also in a distinct vein pattern that may closely resemble injection veins of friction melt (pseudotachylite) from tectonic settings. Dark-matrix breccia veining can also result from injection of impact melt breccia into the crater floor – as described from Puchezh-Katunki, where such impact melt rock injections have been interpreted to depths of >4000 m.

It is actually necessary to quantify the term "dark-matrix breccia" further. This term has been elected in this context, as the vast majority of veins and dikes that have been termed "pseudotachylite" in the impact cratering literature are either blackish or dark-grey; however, a number of descriptions of "pseudotachylitic breccia" have been reported (e.g., Fletcher and Reimold 1989; Killick and Reimold 1990; Reimold 1991; Reimold and Colliston 1994) of different color breccias from the Witwatersrand Basin. These workers have discussed that composition/nature of precursor material may determine the actual color of the breccia. In addition, secondary alteration can also strongly

Fig. 1. Oblique view of the 110 km diameter lunar impact crater Theophilus (Apollo 16, NASA AS16-0692). The image displays a complex central peak area and a series of distinct terraces along the inner crater wall.

change the color of such material – with many altered pseudotachyltiic breccias in the Vredefort Dome being of light-grey or beige color.

To complicate the options for pseudotachylite-like vein formation further, increasing recognition has been given to the possibility that friction melting and shock melting could actually go hand-in-hand (Kenkmann et al. 2000a; Langenhorst et al. 2002). Friction melting can further take place during the crater modification stage – involving uplift formation and collapse, first under the extreme high-strain rate uplift conditions, and then due to extensional collapse involving sliding and thrusting of rock masses against and over each other, and also by mass sliding along large listric faults at the crater margin, which would result in the distinct terracing observed along the inner margin of large impact structures (Fig. 1). In this setting, shock pressure decay has been completed by the time this process evolves, so that pure friction melting would take place.

Despite their superficial – and maybe even microscopic – similarities, such breccias signify markedly different geological environments and processes, and the question is whether they have distinctive features that can aid in elucidation of their origin(s). This problem is typified by ”pseudotachylite” – a term that was originally introduced to describe breccias in an impact setting (at Vredefort - although the setting was unknown at the time), but which was subsequently extended to morphologically similar breccias found in non-impact settings and which has, consequently, taken on a new meaning; but which is still used indiscriminately by impact workers. This review aims to explain why the present indiscriminate use of this term hampers, rather than aids, understanding of impact processes on Earth and why alternative terminology is needed for impact settings.

2 Historical Overview

Shand (1916) introduced the term "pseudotachylyte" (modern spelling "pseudotachylite") to describe the voluminous breccias (Figs 2a, b) occurring in the granitoid core of the Vredefort Dome in South Africa. He compared these breccias with "trap-shotten gneiss" (a term coined initially for magmatic breccias from India, where clasts of country rock were noted to occur in a matrix then thought to be of igneous origin) and with so-called “flinty crush-rock” from Scotland, occurring as dark-matrix breccia veins that, at the time,

Fig. 2. Prominent exposures of pseudotachylitic breccia in the Vredefort Dome: **Fig. 2a** (top) Part of the massive exposure in the main quarry at Leeukop Hill. Face shown is ca. 10 m wide. **Fig. 2b** (bottom) An exploratory cutting on the eastern side of the Leeukop Hill. When comparing these two photographs, note the distinct difference in shape of large clasts – generally rounded in **Fig. 2a** which shows part of a several m thick, at least 50 m wide, occurrence, and mostly angular blocks in **Fig. 2b** that shows a large block broken off the massive granite on the right side of this view. However, narrow cracks in this broken and only locally brecciated material have been invaded by melt, providing some support for the suggestion that melt may be produced at generation planes and then intrudes into open spaces.

Fig. 2c. Strongly brecciated and intricately fractured hinge zone of the large (hundreds of meters) fold structure just south of Parsons Rest farm in the northwestern sector of the collar. The entire hinge zone is invaded with up to decimeter thick veins of pseudotachylitic breccia. The numerous quartzite clasts are generally angular. Not much evidence for movement beyond the centimeter to decimeter scale has been observed. Person in upper half of image for scale.

were also believed to be of igneous origin. He concluded that a catastrophic process (Shand actually used the term "shock") was responsible for the abundant brecciation observed in the dome. Other early workers, such as Hall and Molengraaff (1925), emphasized the remarkable abundance of this breccia type in the Vredefort Dome, but, like their contemporaries, could only speculate on possible endogenic processes responsible for this unique brecciation phenomenon.

Whilst the term "pseudotachylite" has been used to describe breccias from other mid- to large-sized impact structures (for example, the Ries Crater [Germany, 24 km diameter], Rochechouart [France, 23 km diameter], or Manicouagan [Canada, 100 km diameter] – for a comprehensive review, cf. Dressler and Reimold 2004), there is only one other known terrestrial impact structure where such material occurs in similar abundance and volume - the Sudbury Structure in Ontario, Canada (Dressler and Reimold 2004, and references therein).

By the 1960s, structural geologists working in brittle fault and mylonitic shear zones discovered dark-matrix breccias (e.g., Park 1961; Philpotts 1964). Early debate centered on whether the matrices of these breccias were cataclastic or melt-derived. At the time, neither Vredefort nor Sudbury was seriously considered to be of impact origin and the term pseudotachylite was applied to these breccias in fault and shear zones that resembled the material initially described by Shand (1916). Given the relative scarcity of terrestrial impact structures, and the limited number of confirmed impact structures listed around that time, it is unsurprising that the usage of the term in the geological literature rapidly assumed tectonic connotations. It was only in the 1970s that impact workers started to try to claim back the term. However, they were hampered not only by the lingering uncertainty concerning the origin of the breccias in the Vredefort type locality, but also by the fact that the Vredefort breccias dwarfed all other occurrences except Sudbury, besides which the origin of Vredefort by impact or internal processes was still highly controversial.

2.1 The "Sense" of "Pseudotachylite"

Modern textbook definitions of "pseudotachylite" are still basically descriptive but do sometimes recognize that such breccia appears to occur in two different geological settings - tectonic fault and shear zones on the one hand, and impact structures on the other. A typical example is found in Bates and Jackson (1987), who define *pseudotachylite as* "[a] dense rock produced in the compression and shear associated with intense fault movements, involving extreme mylonitization and/or partial melting. Similar rocks, such as the Sudbury Breccias, contain shock metamorphic effects and may be injection breccias emplaced in fractures formed during meteoric [sic] impact ... material carries fragmental enclosures and shows evidence of having been at high temperature...". This definition already highlights a specific problem concerning "pseudotachylites", namely the fairly commonly observed relationship between (ultra) mylonites, and/or (ultra) cataclasites and a melt phase (see Reimold 1995, 1998).

Another example, from Passchier et al. (1990), states that "Pseudotachylites are formed by fault-associated melting of the host rock in response to the heat produced by major seismic slip events, or possibly in some cases by intense cataclasis." Clearly, these authors have placed the emphasis only on fault/shear related processes. *The tectonic worker consequently associates "pseudotachylite" with friction melting triggered by faulting/shearing.* Much of the existing literature on pseudotachylites, accordingly, is dominated by discussion of tectonic breccias, but some groups have, in the recent past, worked extensively on impact-related, pseudotachylite-like breccias (Dressler and Reimold 2004, for an extensive review of this literature). Reimold (1995, 1998) and this work recommend that the term "pseudotachylite" should only be used as a synonym for *bona fide* friction melt, and that all other breccias of pseudotachylite-like appearance should be assigned the preliminary classification "pseudotachylitic breccia" until such time as the nature and origin of the material is identified, at which point other genetic nomenclature can be applied.

Tectonic pseudotachylite occurrences are generally limited to breccia veins with centimeter (rarely up to decimeter) width. One of us (WUR), recently, had opportunity to observe pseudotachylite occurrences typical for the

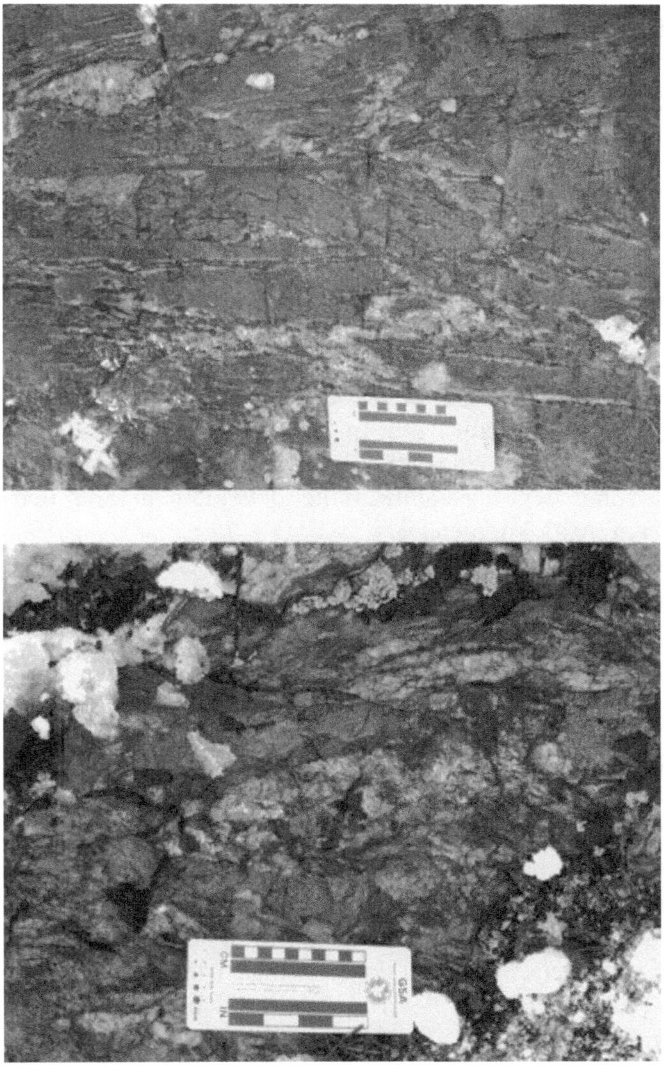

Fig. 3 a (upper) and **b** (lower). Two impressions of tectonically produced breccia veinlets from the Homestake Lake fault zone in Colorado (USA). The dark-matrix breccia either occurs in the form of centimeter (or less) wide veinlets and stringers (**a**), generally following the fabric trend in the fault zone, but occasionally oriented oblique to it. Clasts in and along such occurrences have both roundish or angular shapes. As shown in (**b**), occasionally wider developments of breccia can be seen, with some well-rounded clasts attesting to the highly dynamic nature of this system. However, despite the fact that this fault zone is of a kilometer scale, the breccia occurrences are seemingly sparse and cannot be compared in extent to the massive occurrences at Vredefort or Sudbury.

prominent and extensive Homestake Lake Fault Zone in Colorado (Fig. 3a and b). In comparison to the voluminous breccias at Vredefort, the generally several millimeter-wide occurrences of dark-matrix breccia that occur over a width of centimeters to perhaps several decimeters did not appear very impressive. However, one really significant tectonic pseudotachylite-bearing fault zone, up to 1000 meters wide and containing some 4% of pseudotachylite veining, has been described by Camacho et al. (1995) from the Woodroffe Thrust in the Musgrave Block of the interior of Australia. Up to 2-3 m wide extensional deformation zones with tectonically produced "pseudotachylyte" forming up to 30% of the total rock volume are also known from the Insubric Line bordering the Ivrea Zone of northern Italy (e.g., Techmer et al. 1992). Here, pseudotachylite formation has been related to transpressional strike-slip movements along the fault zone. In both cases, these pseudotachylites occur within crustal-scale fault zones that are several orders of magnitude larger than the actual zones with pseudotachylite development and that involve slip magnitudes of many kilometers.

Endogenic pseudotachylites are not rare. Numerous investigations of recent years have provided evidence that tectonically produced pseudotachylites are quite common in the brittle-to-ductile transition zone of the Earth's crust. Such occurrences occur in a variety of tectonic settings and at different crustal depths, including strike-slip fault zones (e.g., Insubric Line – Techmer et al. 1992), on thrust planes (e.g., Silvretta, Alps – Koch and Masch 1991), subduction zones at eclogite facies conditions (Norway - Austrheim and Boundy 1994), under granulite facies metamorphic conditions (Antarctica - Passchier et al. 1991; Clarke and Norman, 1993), and in landslides (Silvretta - Alps, Himalaya - e.g., Masch et al. 1985; Heuberger et al. 1984; Köfels – e.g., Preuss, 1971; Erismann and Abele 2001; Arequipa, Peru – Legros et al. 2000). Explanations for these occurrences range from stick-slip faulting processes to catastrophic volume reduction. Early-formed pseudotachylites within a fault zone may be obliterated or deformed during subsequent shearing or faulting events (e.g., Passchier et al. 1991).

The volumetrically most important occurrences of pseudotachylite-like breccias are, as stated, the Vredefort Structure in South Africa and the Sudbury Structure in Canada. After lengthy controversies, both structures are now widely accepted as the world's largest known impact structures (e.g., Gibson and Reimold 2001a; Grieve and Therriault 2000). In the Sudbury Structure, breccia dikes of up to 42 km length and, at least, 400 m width have been described, whereas at Vredefort several occurrences of kilometer length and up to a hundred meter width have been mapped. Dressler and Reimold (2004) report a dike-like occurrence of network breccia that could be mapped for 2.6 km at a width of up to 100 m. Smaller impact structures do not appear to exhibit either voluminous or extensive pseudotachylite development (although this statement is relative and its validity depends entirely on what is considered "pseudotachylite"). In the following, this issue and other problematics concerning impact-related "pseudotachylite-like" breccias are discussed. The discussion focuses largely on the so-called "pseudotachylite" in the Vredefort Structure, not only because this is the type locality for this material, which, itself, is a major reason for the difficulty regarding nomenclature and application of this term in relation to impact structures, but also because the excellent preservation of breccias in a range of impact-related environments allows evaluation of the relative contributions of shock and slip processes to breccia formation.

3
What Should Be Called "Pseudotachylite?"

Allaby and Allaby (1996), in their Oxford Concise Dictionary of Earth Sciences, are succinct: "Pseudotachylite [is a] rare glassy rock produced by frictional melting during extreme dynamic metamorphism in a fault and thrust zone". This definition is different from that provided by Bates and Jackson (1987). These contrasting nomenclatures and the generally loose usage of the term "pseudotachylite" are the reasons for widespread misunderstandings and neverending discussion. In the context of this discussion of "pseudotachylites" in impact structures, these definitions are entirely ill-suited, for example, for the case of the small (2.5 km diameter) Roter Kamm impact crater, the rocks of which are only poorly exposed in the dune sands of the southern Namib Desert. A major occurrence of pseudotachylite-like breccia in an exposed area of the crater rim was described by Reimold and Miller (1989). Admittedly, evidence for melting in this material was rare, but glassy patches were described, and – in the absence of definite evidence of shock metamorphism - it was concluded that this material represented pseudotachylite. Later work (Degenhardt et al. 1996) established beyond doubt, however, that this material represents an extremely fine-grained, partially cataclastic schist of the target geology. The absence of evidence for faulting (impact-induced or not) at this locality makes the most probable cause of this breccia impact-induced cataclasis involving mainly brittle fracture, but possibly with a component of shock melting caused by local shock pressure excursions.

Several problems exist with the definition of Allaby and Allaby (1996). For one, it is apparent that pseudotachylitic breccias are not a rare phenomenon: That it is inappropriate to consider tectonically formed pseudotachylite a "rare" phenomenon has already been discussed above, and the breccias at the Vredefort type locality and in the Sudbury Structure are certainly not rare. It is also clear from the literature that most workers do not regard a glassy, or even aphanitic, matrix as a prerequisite either (many pseudotachylites are at least partially crystallized, and microlites are frequently observed; just a few examples: Clarke (1990); Austrheim and Boundy (1994); Hetzel et al. (1996); Müller et al. (2002). Many of the more voluminous breccias in the Vredefort Structure have well-crystallized igneous textures or, at least, microlites in a glassy or cryptocrystalline mesostasis (e.g., Reimold and Colliston 1994). Second, this statement "produced by frictional melting": this is the view of a tectonic worker who is not concerned with impact-produced breccias. However, the type locality for "pseudotachylite" after Shand (1916) is the world's currently largest known impact structure (Vredefort) and the literature is awash with descriptions/reports of such breccias from other impact structures. In addition, the possibly planetary importance of the high-temperature processes that may lead to the formation of large-scale breccias of this/these type(s) has also been speculated about (Fiske et al. 1995a). One benefit one can take from the Allaby and Allaby (1996) definition is that it makes it very clear what tectonic workers/structural geologists consider as "pseudotachylite": *only bona fide friction melt!*

The definition then refers to "dynamic metamorphism" as the environment in which such breccias form. This statement does not refer to any specific metamorphic regimes in terms of pressure and temperature, as well as strain rate. To refer to "dynamic metamorphism" implies that there is no external heat source to cause changes which are driven by differential stress. Thus, this statement does not recognize the impact process/shock metamorphism. Impact breccias form at much higher strain rates than those attained under "normal" tectonic conditions that have been termed "pseudotachylites". The review by Spray (1998) has demonstrated this in detail. And the tectonic literature provides ample documentation that pseudotachylites have been formed over the entire lithospheric pressure-temperature regime, from lower greenschist to upper granulite and even eclogite facies conditions.

Fig. 4. (**a** top) Width of the image is ca. 1 m. Typical only a cm wide, veinlet of pseudotachylitic breccia in Hospital Hill quartzite on farm Parsons Rest, NW collar of the Vredefort Dome. Note the subparallel, NNW trending joint set that has displaced the veinlet locally. (**b** bottom) The pen used for scale is ca 13 cm long. Two veins of pseudotachylitic breccia in Hospital Hill quartzite near the Smilin Thru Resort in the northern collar of the Vredefort Dome. The ca. 7 cm wide, vertical vein is seemingly cut by the thinner, NW-SE trending vein. Pen for scale: ca. 12 cm long. Displacements, however, are not observed. Without detailed analysis of the wider outcrop, it is not possible to ascertain what the actual relationship between these two veins could be: distinct generations emplaced consecutively, or both veins being part of a complex vein system that may not necessarily reflect temporally separate deformation events.

Fig. 4 c. En echelon veinlets of pseudotachylitic breccia, also from the Smilin Thru quartzite exposure. The width of this image is about 1.2 m. These veins do not indicate motion direction, but it is clear that they were formed under tension.

Finally, to restrict "pseudotachylite" formation to "thrust and fault zone" environments is not correct either. Contrary to the tectonic occurrences of friction melt, pseudotachylite-like breccias in impact settings may or may not be related to such structural elements. Again, the Vredefort impact structure provides a case in point: In the Vredefort Dome, the deeply eroded root of the central uplift of this impact structure, apparently fault-related brecciation is observed at many sites in the collar of supracrustal strata, but this is not the rule. Inevitably, these fault-related (i.e., displacements along a linear disruption of the host rock fabric are noticeable) occurrences are thin (examples of pseudotachylitic breccia development in collar strata are shown in Fig. 4a-c) and generally (with just a handful of known exceptions – mostly in the hinge zones of large-scale, i.e., tens or even hundreds of meters, fold structures and, in rare cases, at contacts between mafic intrusions and shale or quartzite) not wider than 1-5 cm. This applies, in comparison to the absolute bulk of breccia observed in the granitoids of the core, to a relatively minor breccia proportion. The overwhelming amount of pseudotachylitic breccia observed in the dome occurs in bodies of generally irregular geometries that - even if they are up to 2.6 km long - do not exhibit a distinct alignment to a recognizable structural element. At a handful of places in the core granitoids,apparent movement on possible faults has also been observed. For example, B.O. Dressler noted a dike-like occurrence of pseudotachylitic breccia in the Kudu Quarry immediately north of the town of Parys, which occurs parallel to a mafic dike or transposed large inclusion of mafic rock that shows signs of tectonic movement.

In contrast, in the Witwatersrand Basin in the environs of the Vredefort Dome, the bulk of breccia observed occurs along bedding-plane parallel and normal fault zones. All three types of occurrences have been linked through detailed field observation (reviewed by Reimold and Colliston 1994 and Dressler and Reimold 2004) as well as absolute chronological data (Trieloff et al. 1994; Spray et al. 1995; Kamo et al. 1996; Moser 1997; Friese et al. 2003) to the Vredefort impact event.

In the discussion of apparent relationships between breccia occurrences and apparently linear structures, the work by Brandl and Reimold (1990) and subsequent field work by these authors on pseudotachylite (friction melt) from the Sand River Gneiss, Limpopo Province of South Africa, is relevant. These workers have shown that veins that may on surface appear perfectly straight, even planar, can just below surface turn – as determined by drilling of such veins - by any angle oblique to the surface exposure, making a mockery of strike trend measurements. Similar observations have been made by us on apparently bedding-plane parallel (Fig. 5a), impact-produced veins of breccia in the collar strata of the Vredefort Dome (also F. Wieland, Univ. of the Witwatersrand, Johannesburg, pers. commun.). Dressler and Reimold (2004) report this effect from the Esperanza Quarry in the southwestern part of the Dome, in core granitoid.

Breccias of pseudotachylite-like appearance from diverse geological (fault/shear or impact) settings may macroscopically and microscopically closely resemble each other (see also below). Two examples shall suffice to emphasize this point: Massive dark-matrix breccia has recently been mapped on the northern anticlinal margin of the Witwatersrand Basin. First field observations resulted in emphasis of close resemblance to Vredefort breccia, with a dark, microcrystalline to aphanitic appearing matrix and sub-millimeter to centimeter size inclusions of granitic country rock. In contrast, subsequent microscopic analysis revealed that the matrix is neither glassy nor aphanitic, and instead represents fine-grained cataclastic material with ample product of secondary alteration. What is more, this breccia closely resembles the so-called Gardnos Breccia from the Gardnos impact structure in Norway (French et al. 1997; Gilmour et al. 2003). It has been demonstrated by French et al. (1997) that Gardnos Breccia represents definite impact breccia of both clastic (lithic impact breccia) and melt-bearing varieties. Macroscopically, it is not possible to distinguish the cataclasite from the Witwatersrand Basin from the Gardnos Brrccia. The breccia from the northern margin of the Witwatersrand Basin could have been generated as a consequence of the Vredefort impact event or, alternatively, in one of the tectonic events that have affected this region over nearly 3 billion years.

Clearly, the descriptive criteria for pseudotachylite are insufficient to separate breccias of such appearance but of different origin. Thus, the normally applied definitions are insufficient. In view of the overwhelming number of pseudotachylite occurrences being located in tectonic – and not in impact – settings, and as the formation of pseudotachylite (friction melt) by cataclasis and frictional melting has been demonstrated by Spray (1992, 1995), we propose *that the term "pseudotachylite" be reserved for bona fide friction melt breccia.*

4
The Early (1970s-end 1980s) Controversy

One persistent aspect of the earlier pseudotachylite literature was the question amongst tectonic researchers whether "pseudotachylite" was formed by cataclasis only or whether it involved melting (e.g., Wenk 1978). This issue was laid to rest through the friction experiments performed by Spray (1992, 1995). He showed that melting is greatly enhanced by early cataclasis through the large increase of surface area of the comminuted material, which, in turn, facilitates frictional heating. These experiments demonstrated why cataclastic breccia and material resulting from a melt phase commonly occur in intimate association in tectonic fault zones. Spray's experimental results also provided a physical basis for the explanation of the melting process, whereby he was able to relate fracture toughness and compositional (elemental composition and presence of volatiles in crystal structures) differences of minerals to the ease with which they would be able to enter the melt phase. Spray observed that mafic hydrous minerals, such as biotite or amphibole, were the first phases in a rock system to begin to melt selectively, followed by feldspar minerals, and only then by the more refractory quartz. These findings correlate very well with the compositional observations by various workers on natural pseudotachylite and pseudotachylitic breccia occurrences from both tectonic and impact settings (e.g., review by Reimold 1991).

5T
The Mid-1990s "Problematics" - A, B, S, E types, Tagamite and Shock Veins

Right through to the beginning of the previous decade, many workers – especially in South Africa – had remained reluctant to accept the long proposed (e.g., Dietz 1960) impact origin of the Vredefort Dome. Within this context, the origin of pseudotachylite in Shand's type locality was not considered significant. It took the confirmation of *bona fide* shock deformation in quartz (Leroux et al. 1994) and zircon (Kamo et al. 1996), and the identification of a meteoritic component in the Vredefort Granophyre (Koeberl et al. 1996) to change the conviction of, by far, most of these sceptics. Several geochronological and field studies showed that most, if not all, of the so-called "pseudotachylite" from Vredefort had to be impact-related (Trieloff et al. 1994; Spray et al. 1995; Kamo et al. 1996). The early 1990s then also saw the link made between the Vredefort Dome and the surrounding Witwatersrand Basin, and between the rock deformation (particularly large-scale brecciation) of the Dome and that in the Basin (e.g., Fletcher and Reimold 1989; Killick and Reimold 1990; Reimold and Colliston 1994). Therriault et al. (1997) and Henkel and Reimold (1998) showed that the Vredefort Structure encompasses the entire region known as the Witwatersrand Basin and indicated that the Vredefort impact structure could originally have been as large as 300 km in diameter (or even more – e.g., Phillips et al. 1999).

Martini (1978) demonstrated the existence of the high-pressure polymorphs of quartz, coesite and stishovite, within and in the immediate vicinity of narrow (millimeter-wide) veinlets of "pseudotachylite" (Fig. 7) from the collar of the Vredefort Dome (see also White 1993). Martini followed this up with a paper (Martini 1991), in which he proposed that these narrow veinlets represented a melt produced during the early compression phase of the impact cratering event. He termed this material "A-Type pseudotachylite", in contrast to the remaining occurrences of largely massive breccia (such as those described by Shand 1916), which he considered the result of "decompression melting" during the later modification stage of cratering and termed "B-Type pseudotachylite". Reimold et al. (1992) contested this classification into "A- and B-types" on the basis of lack of definitive recognition/separation criteria for such types.

A decade later Lambert (1981) proposed a classification scheme for breccias in impact structures, and he had also employed an A- and B-type terminology for dike breccias in impact structures, in fact further subdividing into A1, A2, B1, and B2 types. Lambert stated that A-type breccias formed under shock compression and, thus, would include Martini's A type; however, it is not clear which one of Lambert's sub-types would correspond to "pseudotachylite". B-type dikes would form during or after pressure release, with B1 dikes represented injections of material from the interior of the crater into openings in the crater floor, and B2 dikes formed "after, by, or during displacement of blocks" during crater modification. As it can be debated whether all "pseudotachylite" (i.e., all veins that have been termed such) has been formed *in situ* like tectonic friction melt would, it is not clear whether all that has been termed "B-type pseudotachylite" in impact craters could be categorized into the B1 or B2 types. In essence, we are not comfortable trying to make Martini's (1991) and Lambert's (1981) classifications compatible.

Detailed review of the breccia literature, specific to terminology employed by impact workers, by Reimold (1995, 1998) illustrated that a range of different breccia types has been indiscriminately classified as "pseudotachylite". This includes ultracataclasites and cataclasites, ultramylonites and mylonites, friction melt *sensu stricto*, impact melt rock, breccias of pre-, syn- or post-impact age in crater settings and involving both impact breccias and tectonic breccias, as well as the so-called "shock veins" of meteorites.

A good example of the confusion between impact melt rock and "pseudotachylite" is provided by the case of the term "tagamite". In the Russian literature, this geographic term originally referred to impact melt rock from the Popigai Structure, but then was applied to impact melt rock from a wide range of impact structures. However, it was also used for the description of narrow veinlets occurring in crater floors; for example those intersected to a depth of 4809 m (A. Deutsch, pers. commun.) in the deep VDW borehole into the central uplift of the Puchezh-Katunki impact structure (Russia) – irrespective of what these veinlets might actually represent. One of us (WUR) has personally seen thin sections of some such veinlets (kindly provided by Dr. Fel'dman, Moscow State University) that are nothing else than totally epidotized breccia veinlets that do not allow any speculations on the actual type of breccia that was present. Other veinlets from this impact structure studied by us and that were termed tagamite represent cataclasite or mylonite.

"Pseudotachylites" have also been described from the crater floor of the large Morokweng impact structure in northwest South Africa (e.g., Hart et al. 1997), but these veinlets were shown by Reimold et al. (1999a) and Koeberl and Reimold (2003) to also comprise a range of different breccias including impact melt injections and cataclasite, besides altered breccia of uncertain genesis. Most recently, Macdonald et al. (2003) described "pseudotachylite" from the old and deeply eroded Yarrabubba impact structure in the Archean Pilbara craton of western Australia, but, interestingly, these authors admitted themselves that at least some of these occurrences would be better termed mylonites.

So-called "shock veins" are known from numerous meteorites, especially from ordinary chondrites, and the fact that they often occur together with high-pressure polymorphs (ringwoodite, majorite, and others) of various rock-forming minerals (e.g., Langenhorst and Poirier 2000; Chen et al. 1996, 2003) has led many to believe that these breccia veinlets are the result of shock

Fig. 5. (**a** upper) Part of a vein system at Smilin Thru Resort, Vredfort Dome, that has a predominant subvertical vein that can be followed for several m and is visible here in the upper, middle part of figure (**a**). However, this image also demonstrates that prominent veins occur. It is, thus, basically impossible to determine whether a possible generation fault could have played a role, or whether these individual veins of highly irregular attitude are either the result of injection or of local generation. Compass for scale: ca. 9 cm sidelength. (**b** lower) A pseudotachylitic breccia exposure in medium-grained granitic gneiss of the core of the Vredefort Dome, indicating that there are many occurrences that show no evidence as to how these "pods" were formed or emplaced. Note the apparent flow structures that could provide a hint that flow took place in the third dimension. Knife for scale, 9 cm long. (**c**) Two more illustrations of breccia occurrences in granitoid of the core of the Vredefort Dome.

Fig. 5. (**c** upper) Several apparently irregularly shaped veins occur in close proximity to a decimeter-wide vein (at bottom of image). Whether there is a link between these two occurrences is not clear. (**d** lower) This seemingly irregular breccia occurrence is, upon closer inspection, revealed to be linked to a N-S trending, millimeter-wide veinlet of variable – subvertical to near-horizontal - attitude. In both (**c**) and (**d**) knife for scale is 9 cm long.

compression. A comparison between naturally occurring shock veins in meteorites and the results of friction experimentation on meteorites by van der Bogert and Schultz (1997) and van der Bogert et al. (2000) resulted in the conclusion that a combination of shearing and shock, or either process could be responsible for these veins in meteorites. Our own recent study of the H4/5 meteorite Thuathe (Reimold et al. 2003a) also concluded that the narrow black veinlets that occur ubiquitously in specimens of this meteorite are the result of friction melting, without any indication that locally enhanced shock pressure could have contributed to their formation.

In addition to "tagamite", there are other local geographic names used for impact melt rock/breccia in the literature, for example "kärnäite" for the impact melt rock occurring on the peninsula Kärnänsaari in the Lappajärvi meteorite crater in Finland, or "dellenite" for the impact melt rock of the Dellen Structure in Sweden. It appears to us that the generalized use of such local geographic nomenclature ought to be avoided in the interest of simplification and introduction of impact literature into the general geoscientific community. What is wrong with the general term "impact melt rock" that provides a clear genetic label? Reimold (1995, 1998) provided extensive evidence that this lack of discrimination continues to date! On the one hand, some workers have gone as far as to say that pseudotachylite can be used as a recognition criterion for impact structures (Fiske et al. 1995b) – disregarding the wealth of information about tectonic friction melts. And on the other, even the proposed IUGS Impactite Nomenclature (Stöffler and Grieve 1994) lists "pseudotachylite" as an impactite. Macdonald et al. (2003) recently cited "pseudotachylite" as a "shock metamorphic effect", in the face of a plethora of tectonic occurrences worldwide! Reimold (1995, 1998) acknowledged that it would be difficult to change long-time habits, but he recommended that anything not clearly identified as pseudotachylite (*bona fide* friction melt) should be separated and, at least, referred to as "pseudotachylite-like" or "pseudotachylitic breccia". It must be admitted that it is difficult to introduce conscientious application of this distinction even to close collaborators – but we believe that the reward of contributing to the understanding of the cratering process should be sufficient incentive to give it a try!

Much thought has gone into finding a term that would be completely independent from the association with "pseudotachylite", and many colleagues have also proposed alternative terms (including "frictionite" – which is, however, already used as a synonym for pseudotachylite; or "otavite" with regard to one of the type localities in the Vredefort Dome, the Otavi quarry northeast of Parys, but this term is already reserved for a Co mineral – apart from the general undesirable use of geographic names). One could have thought of another alphabetical term – such as the "X-breccia" – but as it is already impossible to properly distinguish the A, B, S and E breccias, and as a "H breccia" has recently been proposed by Ernstson et al. (2002) for a specific geometric development of dike breccias as observed by these authors in the Azuara structure of disputed impact origin, we refrain from introducing an X-, Y-, or Z-breccia concept.

One more example shall be given to illustrate the problems of indiscriminate use of the term "pseudotachylite": Fiske et al. (1995b), in an abstract, state: (1) "Pseudotachylite dikes and veins .... are important evidence for meteorite impact". This point has already been raised. Pseudotachylite by itself is not evidence for impact, as it occurs worldwide in tectonic deformation zones. Should Fiske et al. have referred to some "shock veins" that are widely considered as evidence of shock metamorphism in meteorites (Stöffler et al. 1991), this would have been a different matter, and yet – even the origin of shock veins by "simple" shock compression or a combination of shock and shear is now a matter of debate (as outlined above). However, even if found together with other diagnostic impact indicators (such as *bona fide* shock metamorphic effects), "pseudotachylite" in impact settings can not be considered diagnostic, and it must be proven whether this material indeed represents pseudotachylite (= friction melt) or another type of breccia, or whether it perhaps pre- or post-dates an impact event. (2) "... small pseudotachylite veins (Type A of Martini...) lacking any association with faults are also found. Commonly defining shatter cone surfaces". There are a number of problems with this statement: (a) Small pseudotachylite veins do not need to be of Martini's (1991) A-type (produced during shock compression) and, as, for example, discussed by Dressler and Reimold (2004), very large occurrences of breccia may also correspond to shock melt. Furthermore, shock experimentation by Kenkmann et al. (2000a) and Langenhorst et al. (2002) has shown that rock discontinuities such as lithological boundaries may provide appropriate locations for shock pressure excursions to high values that could lead to shock melting, or shock-cum-friction melting, at such discontinuities. Field evidence from the Vredefort Dome shows that existing zones of weakness, such as mylonitized fault zones, are preferred locations for pseudotachylitic breccia development. Thus, to differentiate a so-called "Type A pseudotachylite" from fault-association is wrong. Van der Bogert and Schultz (1997) and van der Bogert et al. (2000) also have shown that at least some so-called "shock veins" in meteorites may be associated with faulting/shearing as evidenced by displacements along them.

Finally, there is indeed evidence of thin melt films coating some shatter cone surfaces. Martini (1991) referred to this, and the work by both Gibson and Spray (1999) and Nicolaysen and Reimold (1999) has confirmed this. However, to make the assertion that such melt films correspond to the so-called "Type A pseudotachylite" is ill-advised. Temporal relationships between shatter cones and early formed pseudotachylitic breccia are not understood yet. Reimold and Colliston (1994), for example, showed that there is pseudotachylitic breccia in the Vredefort Dome that is overprinted by shatter cones, and Simpson (1981) recorded shatter cones overprinting a fault gouge on a presumed Vredefort-age fault. In addition, the process by which melt films on shatter cone surfaces are produced is not understood yet. Why would shock melting occur preferentially on shatter cone surfaces and, rather rarely, at intersections of Nicolaysen and Reimold's MSJS (multipli-striated joint surfaces that these authors linked unequivocally to the shatter cone phenomenon) in the interior of such a fractured specimen? One would expect that many other defect sites throughout a rock volume would render material susceptible to melting under strongly enhanced shock pressure. Nicolaysen and Reimold (1999) have shown that there are small displacements along MSJS throughout a rock specimen. Why could such a melt film on a shatter cone or MSJ surface not represent *bona fide* friction melt? What about the possibility that shatter cone surfaces represent extensional sites (extensional conditions developed after shock wave transition), in which case shock pressure would be low?

Fiske et al. (1995b) also claimed: (3) "High P polymorphs of quartz suggest that Type A pseudotachylite forms during compression and crater excavation". Apart from the controversial Type A classification (Reimold et al., 1992; Reimold 1995, 1998), there is no problem with shock veins containing high P polymorphs forming very early in the cratering process. The fact that some workers (Dressler and Sharpton 1997; also Dressler and Reimold 2004) have observed clasts in impact breccia that contain breccia veinlets resembling pseudotachylitic breccia is strong evidence that such a brecciation phase must occur early, likely during shock compression. And (4): "We have produced pseudotachylite-like material in shock experiments on quartz that may be analogous to Type A pseudotachylites". This statement may well be correct in that these authors produced similar melt to that experimentally generated by Kenkmann et al. (2000a) and Langenhorst et al. (2002). Whether they produced friction melt (pseudotachylite) or a shock melt, however, still remains to be proven.

6 The Setting of Pseudotachylitic Breccia in Impact Craters

Veins of dark-matrix breccia resembling tectonic pseudotachylite have been described from many impact structures - generally from crater floor rocks, but occasionally from crater rim settings. Such reports are very rare for small, simple bowl-shape craters, and it must be questioned whether any *bona fide* pseudotachylite (friction melt) has ever been described from these. For example, the only breccia ever recovered from the crater rim of the Tswaing (Pretoria Saltpan) meteorite crater that macroscopically resembled a pseudotachylite turned out to be a manganese/iron-oxide cemented monomict granitic cataclasite. The case of the Roter Kamm breccia has already been discussed above. Furthermore, narrow veinlets of pseudotachylitic breccia have also been described from clasts in impact melt breccia (Dressler and Sharpton 1997).

As reviewed in detail by Dressler and Reimold (2004), significant amounts of pseudotachylitic breccia have been observed only in very large impact structures. Even medium-sized structures like the Ries Crater (24 km diameter), Rochechouart (23 km), and Lappajärvi (17 km diameter) have only yielded centimeter-wide pseudotachylitic breccia veins. Even from the 100 km diameter Manicouagan structure in Canada, only centimeter-wide veinlets of pseudotachylitic breccia have been described to date. However, in such comparisons it is important to consider that these structures are exposed at different erosion levels and that the formation of pseudotachylitic breccia in the floors of impact structures may be mostly a feature of the uppermost floor level. The apparent increase in abundance of such breccia from the Vredefort Structure to the Sudbury Structure is likely a result of the relatively much increased degree of erosion in both case (e.g., Grieve and Therriault 2000; Gibson and Reimold 2001a).

Fig. 6. Two images of the recently identified massive breccia at De Pan, on the Rand Anticline at the northern margin of the Witwatersrand Basin. Leucogranite clasts of mostly – but not exclusively – angular to subangular shapes occur in a dark-grey matrix. Clasts are nearly all derived from the adjacent country rock (leucogranite), and the few apparent exotics may well be derived from mafic xenoliths or bands in the granitic country rock. Knife for scale, ca. 9 cm long.

Fig. 7. (**a** - upper, **b** – lower, **c** - right) Photographs show millimeter to sub-millimeter wide veinlets of pseudotachylitic breccia that would represent the so-called "A-Type pseudotachylite" of Martini (1991). Note the undulating geometry of these veins. In (**b**), a network of such veinlets is actually observed. Country rock: Government Reef quartzite, from northwest of the Smilin Thru Resort in the northern collar of the Vredefort Dome. Hammer for scale: ca. 25 cm long. (**c**) Drill core specimen from the Inlandsee borehole ca. 4 km south of the geographic center of the Vredefort Dome. This sample consists of pseudotachylitic breccia, of mottled appearance, that does not display a distinct boundary to the gneissic component in this image (at right and left sides). Rather the dark breccia appears to grade into the country rock. The latter only locally displays the originally strong gneiss fabric. Clearly the entire assembly of dark breccia and lighter gneiss seems to have been plasticised. Length of drill core segment ca. 15 cm.

The setting of the pseudotachylitic breccias in Vredefort has already been discussed above. It shall only be added here that some very narrow veinlets of strongly altered and thus not classifiable breccia were described (Fletcher and Reimold 1989) from the northeastern edge of the Witwatersrand Basin (Northcliff Koppie, Johannesburg; Zwartkops inlier of Witwatersrand strata in the western Johannesburg Dome). Whether these occurrences are the result of faulting due to the Vredefort impact event, can be speculated on, but to date, no time marker has been identified for this "tectonic" event. In addition, an occurrence of massive breccia in Archean granite was recently discovered by WUR, at a site called De Pan on the Rand Anticline (Fig. 6); as discussed above, this breccia represents a cataclasite, the matrix of which is strongly altered. It is also not clear whether this is a Vredefort-related breccia, although the massive formation is suggestive of this association.

The massive Sudbury Breccia occurs, apparently somewhat enriched in annular zones (Thompson and Spray 1994; Spray and Thompson 1995), within the Archean gneiss of the basement to the Sudbury Igneous Complex. Preferential enrichment of pseudotachylitic breccia around the Vredefort Dome has also been discussed, but is likely a result of limited knowledge of spatial distribution of such breccia in the region of the Witwatersrand Basin – due to the fact that most relevant information has come from the gold-mining districts that straddle a semi-annular zone around the Vredefort Dome where mining is possible at acceptable depths and where, thus, intersections of major, basin-wide fault zones (Fletcher and Reimold 1989; Killick and Reimold 1990; Killick et al. 1988; Killick 1993) could be investigated.

Hilke (1991), in his detailed discussion of breccia dikes from the 39 km diameter Carswell impact structure in Saskatchewan (Canada), chose to introduce still another term - "Reibungsschmelzbreccia " (which translates to "friction melt breccia"), instead of subscribing to "pseudotachylite", that he explicitly defines as a *genetic* term. This author stated that he feels that the term "pseudotachylite" is applied to "both impact-induced as well as impact-independent, tectonically formed dikes and an exact definition has not been fixed" (Hilke 1991). He then distinguished, allegedly on textural grounds and discordant contacts, between "early" and "late" veins. "Early veins" were ~200-300 μm wide and composed of glass and its alteration products, in "network-like geometries." Note that this geometric observation does not correspond to Martini's (1991) definition of early "Type A pseudotachylite". "Late" dikes were macroscopically distinguishable because of their blackness, they could be up to several centimeters wide, and transitions from crystalline to clastic matrix could be observed (according to Hilke). Significantly, this author emphasized that "the matrix can not be resolved by optical microscopy and consists of secondary phyllosilicates". This statement represents a clear contradiction to Hilke's claim that crystalline and clastic matrix types could both be observed at Carswell. It also leads to the serious question whether these veins can be classified as "friction melt breccia" (a term that we would consider equivalent to pseudotachylite = friction melt), if it is not possible to positively resolve the nature of matrix. It is also interesting that Hilke (1991) observed very small ("gering") and significant displacements along "early" and "late" veins, respectively. He declares that both "generations" of "Reibungsschmelzbreccie" were formed during the compression to early excavation phase of cratering; thus, Hilke's "late" generation can not be equated with Martini's (1991) "B-Type pseudotachylite" either, which Martini linked to the later modification stage of cratering.

From the 30 km diameter Azuara structure (Spain) of still controversial origin sub-millimeter veinlets of "glass-bearing dike breccia" were classified as pseudotachylite (Ernstson and Fiebag 1992). In view of the debate about the origin of Azuara, it is certainly premature to attempt a genetic discussion of this occurrence, and, anyway, information presented on this alleged breccia occurrence is insufficient to claim an origin of this material as friction melt.

Besides the studies mentioned above, not much very detailed work has been carried out on pseudotachylitic breccias. Veins are known from a number of other structures, such as Morokweng and Puchezh-Katunki, where they have been described from drill core only – thus not providing any information on structural controls that perhaps could be used to investigate the time during the cratering event when these veins were formed. The drill core Hättberg BH5 from the Siljan impact structure is also said to contain numerous pseudotachylites; however, the true nature of these occurrences has not been published yet (T. Kenkmann, pers. commun., 2003). Kenkmann also reports that a number of outcrops in the Siljan structure, such as the Trollberget location, showed such breccia occurrences.

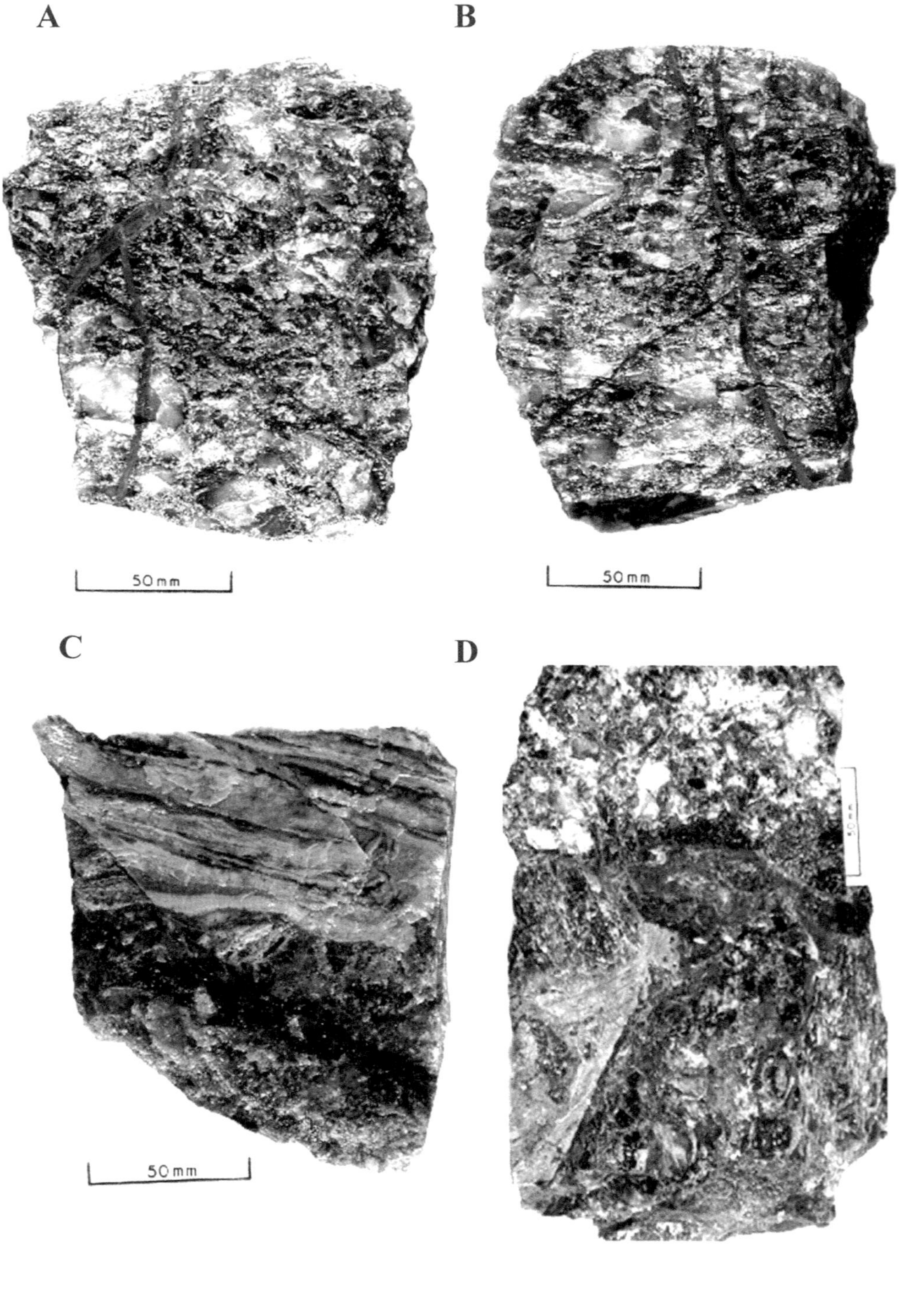
A
B
50mm
50mm
C
D
50mm

Fig. 8. (Previous page) Four examples of pseudotachylitic breccia in, or associated with, Ventersdorp Contact reef (small-pebble conglomerate) from Elandsrand Gold Mine, ca. 70 km north-northwest of the center of the Vredefort impact structure. (**a**- upper left) Two cross-cutting veinlets of dark-matrix breccia that do not display obvious displacement. (**b** upper right) Several veinlets of pseudotachylitic breccia, whereby the NE-SW trending veinlet has displaced the longer, N-S trending one by several millimeters. (**c** lower left) Two generations of breccia: the lighter-colored mylonite in the top half of the specimen has been transected by an apparently younger (younger by how much – microseconds or million years – is of course not known!) and very thin dark-matrix breccia, and in the upper right corner of the sample, it is obvious that the banded mylonite has been displaced by at least 3 cm. (**d** lower right) A complex sample with two – lighter and dark colored – breccia types that occur in intimate association with each other. Nevertheless, this sample gives the impression that the darker phase post-dates the lighter material, which is found in the form of very small inclusions in the swirl of dark breccia in the central part of the image. Note that the sample from Elandsrand discussed by Killick and Reimold (1990) comprised three distinct generations of breccia – based on cross-cutting relationships and inclusion of clasts of an older in relatively younger breccia. Scales on all four images: 50 mm wide.

In the Champagnac Quarry in the Rochechouart impact structure of France, several veins of up to 10 centimeter width have been observed, all at intermediate to high angles to the quarry floor and dipping towards the crater center. One could argue that such geometry could indicate formation of friction melt during collapse of the central uplift – however, we do not even know whether the Rochechouart breccias are the result of friction melting or shock melting. Indeed, some of the observations of such melt breccia by Reimold et al. (1987) and Bischoff and Oskierski (1987) may be suggestive of an origin by localized shock melting. These reports include observations of anastomozing veinlets and microscopic melt "pods". Also, these occurrences are closely related to hydrothermal quartz veins and impregnations with secondary quartz, which may suggest melt formation under extensional conditions – such as at crater modification. Accordingly, these authors concluded that melt formation took place late with regard to crater evolution, due to friction melting as a consequence of block movement in the crater floor during the modification stage of cratering. Kenkmann et al. (2000b) reported pseudotachylite at Champagnac that occured on fault surfaces belonging to a low-angle normal fault system that they interpreted to have accommodated inward movement during crater collapse.

The Champagnac Quarry represents only a small window into the Rochechouart crater floor. Far more extensive information is required before a comprehensive model for the 3D occurrence, distribution, and orientation of pseudotachylitic breccias can be established.

One of the latest additions to the terrestrial impact crater record, the Woodleigh Structure of Australia, was said to have a "penetrative network of pseudotachylite", intersected in a drill core (Mory et al. 2000a). Diagnostic shock deformation has been described (Mory et al., 2000b; Koeberl et al. 2001; Reimold et al. 2003b) from drill core, confirming Woodleigh's origin by impact. However, authors are divided with regard to the diameter of this structure, for which values of 120 km (Mory et al., 2000a,b; Uysal et al. 2001, 2002) or 60 km (Reimold and Koeberl 2000; Reimold et al. 2003b; Renne et al. 2002) have been favored. Regarding the alleged "pseudotachylite", Reimold et al. (2003b) have demonstrated that individual narrow (not wider than 1 mm) veinlets with a dark matrix are indeed found in a few drill core samples. They concluded that to use the term "penetrative" represents exaggeration. As to the nature of this material, all veins studied by Reimold et al. were found to be thoroughly altered and do not allow identification of a *bona fide* pseudotachylite. A range of other breccia types (such as cataclasite – of tectonic or impact origin, mylonite of pre-impact origin) could be considered as candidates as well.

This scenario for Woodleigh is very similar to what is known from Morokweng, a 70-80 km wide impact structure in the North West Province of South Africa (Reimold et al. 1999a, 2002). Early workers (Andreoli et al. 1995; Hart et al. 1997) reported "pseudotachylite" development in the crater floor as seen in drill core, but Reimold et al. (1999a) and Koeberl and Reimold (2003) have shown that there are occurrences of both impact melt injections and cataclasite in this drill core section, as well as some other breccia veins, the filling of which is so thoroughly altered that assignation of a breccia type is not possible anymore. As at Woodleigh, dark-matrix breccia, similar to pseudotachylite = friction melt, occurs in the crater floor of Morokweng, but – as discussed above - several types of breccia are represented. Both Woodleigh and Morokweng demonstrate that caution is advised to avoid such indiscriminate categorization of dark-matrix breccias as "pseudotachylite".

7
"Pseudotachylite" or "Pseudotachylitic Breccia", Comparison of Vredefort-Witwatersrand Breccias and Tectonic Friction Melt Rock

In this section, a number of aspects of breccias occurring in the Vredefort Dome-Witwatersrand Basin (VWB) and in tectonic fault/shear zone (TFS) settings are compared. Pseudotachylitic breccias in the Vredefort Dome – Witwatersrand Basin can be divided into two groups:

1. the breccias in the goldfields >80 km from the center of the structure and well outside the zone of shock metamorphism;
2. the breccias in the Vredefort dome, where shock-induced features like shatter cones, PDFs and coesite and stishovite indicate the elevated shock conditions that were reached in the rocks of the central uplift.

In the goldfields, the breccias have been dated as syn-Vredefort (Trieloff et al., 1994; see also Friese et al. 2003), and they occupy cataclastic fault zones that display a consistent dip inwards towards the center of the Vredefort Structure and that have consistent normal dip-slip sense of movement (Killick, 1993). The veins are large by fault standards (up to 1 m wide, etc.) but they are still subordinate within fault zones that are considerably wider, like tectonic pseudotachylites. Evidence has been presented that indicates that the Witwatersrand fault breccias include cataclasites and ultramylonites, but at least locally melting has taken place, as deduced from petrographic observations reviewed by, e.g., Reimold and Colliston (1994; see also Reimold et al. 1999b). We conclude that these breccias are likely the result of the collapse of the Vredefort impact crater during the crater modification phase; they comprise, in part, true pseudotachylites, besides much cataclastic and ultramylonitic material. A number of specific aspects, **A** – **F**, can be compared:

A. Many breccias from the central part, the Archean basement complex, of the Vredefort Dome - the central uplift of the large Vredefort impact structure, that have been described as "pseudotachylite" in the past are indeed *bona fide* melt breccia, though of still unconfirmed shock, friction, or shock-friction origin. All pseudotachylitic breccias of the Archean basement core and of the supracrustal collar of the Dome are invariably recrystallised due to post-impact regional metamorphism (Gibson and Reimold 2001a). They were formed in quartzite host rock in the collar of the structure. The matrix represents a very fine-grained, cherty groundmass of silica ± Fe,Mn oxides. And in the innermost part of the core, breccias described by Stepto (1979) and, recently, by Gibson and Reimold (2001b) and Gibson et al. (2002) are also mostly annealed. In these cases it is not possible to state with conviction what kind of breccia these occurrences may have represented. But in the case of the breccias from the center of the Dome, evidence is mounting that they could well be impact (shock) melt breccia instead of pseudotachylite = friction melt (Gibson et al. 2002; Gibson and Reimold 2004).

The Vredefort Dome exposes a wide range of rock types, but can be broadly divided into two zones on the basis of lithology – the collar of the dome is dominated by supracrustal strata and a well-layered structure caused by bedding on a mm to km scale; and the core of the dome is characterized by massive, crystalline gneissic basement. The average shock pressure increases from <10 GPa in the collar to as much as 45-60 GPa in the core (Gibson and Reimold 2004). The lithological differences have a clear effect on the morphology of the breccias – in the collar, veins tend to be more planar than in the crystalline rocks of the core, although they are still highly variable in orientation. A similar effect is seen in the alkali granites in the collar, where the breccias define anastomosing vein arrays. However, another factor is shock pressure.

Gibson et al. (2002) described breccias from the central parts of the dome that display poorly-defined margins against wallrocks and clasts, and a granofelsic matrix texture indicative of high-temperature crystallization and/or annealing (Fig. 7c). They compared these breccias to the so-called lunar granulites, which have been previously interpreted as contact-metamorphosed fragmental breccias, and proposed that both the high temperatures and indistinct boundaries signify breccia formation at shock pressures that were sufficiently high to induce localized shock melting of feldspars and hydrous ferromagnesian minerals. They speculated that frictional heating may have played a role in creating the breccias; however, given the background shock heating, it does not seem necessary to invoke large amounts of frictional heat. More recent work has indicated that thin breccias are commonly compositionally highly heterogeneous along their length, and may even display virtually monomineralic segments corresponding to particular coarse-grained minerals in the wallrocks (also Reimold 1991). This suggests extremely limited melt mobility, something that is not likely to be favoured by a mechanism that requires sliding to generate the melt in the first place.

Only in rare cases have displacements of 10 to 50 cm along veins of pseudotachylitic breccia in the Vredefort Dome been recorded. Other indicators are the lack of evidence of cataclasis in the wallrocks flanking the breccias. This is in marked contrast to tectonic friction melts that typically occur in fault or mylonite zones up to several orders of magnitude wider than the melts themselves. Finally, Gibson and Reimold (2001b, 2004) documented increasing shock effects on a mm-scale adjacent to several thin breccias, which suggests that the breccias are sites of localized shock pressure "spikes". Recent modeling of shock wave propagation in heterogeneous media has shown similar P spikes up to 2 to 3 times the background shock P. Given the heterogeneous nature of almost any geological target, enhancement of the shock P by reflection and/or refraction of the shock wave is not only plausible but more likely the norm. It remains to be seen whether increasing background shock P increases or decreases the amount of enhancement – our feeling is that at low background shock P, reflection and refraction will enhance the shock wave and lead to progressively more melting as the P increases, until a threshold value above which the amount of work being done at the grain scale is sufficient to prevent extreme P fluctuations (as reflectors are destroyed or weakened), and the amount of localized melting decreases but is replaced by more wholesale shock metamorphic damage. Finally, almost all veins show some level of differential slip along their margins – typically at mm, but sometimes up to a few cm, scales. This could be part of the original mechanism of breccia formation: differential acceleration of the shock wave can trigger syn-shock slip on favourably oriented surfaces; however, it is equally possible that the movement may be triggered by stresses during the subsequent crater modification phase acting on the lubricated surfaces.

VWB breccias known from fault zones in the wider Witwatersrand Basin do also include several breccia types, including cataclasites, mylonites, and friction melt. Sometimes a melt phase only constitutes a small portion within mostly cataclastic breccia (Fletcher and Reimold 1989; Killick and Reimold 1990; Reimold and Colliston 1994; Reimold et al. 1999b; Dressler and Reimold 2004).

In tectonic fault- and shear zone related settings (TFS), a wide variety of fault breccia can be observed, including cataclasite, mylonite and ultramylonite, and pseudotachylite (e.g., Passchier and Trouw 1998).

B. In both the Vredefort Dome and Witwatersrand Basin, several generations of breccias have been observed – admittedly quite rarely. Reimold and Colliston (1994) provided evidence for pre-impact brecciation, in the form of narrow breccia veinlets (resembling altered pseudotachylitic breccia) that are intimately deformed together with their host gneisses of the Archean basement complex that experienced tectonic deformation prior to 2.02 Ga (Lana et al. 2003a,b). Multiple generations of breccias are also known from the large-scale bedding-plane parallel fault zones of the Witwatersrand basin (Killick and Reimold 1990 – a sample from Elandsrand gold mine); other samples with crosscutting relationships between various generations of breccia were recently obtained from mine geologists on Elandsrand Gold Mine, and some examples are shown in Figure 8. In addition, Berlenbach and Roering (1992) discussed the possibility that pseudotachylitic breccias as old as the Ventersdorp Supergroup (ca. 2.7 Ga) occurred in the basin. Recently, Friese et al. (2003) reviewed the chronological data on Witwatersrand fault rocks.

Repeated brecciation along TFS should certainly be possible, as such deformation zones represent zones of weakness in the crust that may be prone to reactivation. Still in modern times, mining-induced rockbursts frequently have produced fault rock in the deep-level gold mines of the Witwatersrand (e.g., Stewart et al. 2000, and references therein). However, to our knowledge no friction melt has been described from such locations yet.

C. In the Vredefort Dome, orientations of breccia veinlets can be controlled by pre-existing zones of weakness, such as lithological boundaries between gneiss bands of different rheologies, mylonite zones, and other structural features. However, comprehensive mapping of vein and dike orientations with regard to Archean fabrics and pre-impact deformation features (Dressler and Reimold 2004) has shown that such relationships are by no means the rule. Also, where an apparent relationship between a (sub)planar feature and a breccia vein occurs, it does not necessarily persist in the third dimension. However, it still cannot be ruled out that, at least in some instances, large breccia developments could be controlled by faulting (for example, where conjugate pairs of straight veins with obvious displacements of host rock banding are observed – e.g., some images shown by Reimold and Colliston 1994). In the Witwatersrand basin, large breccia developments are preferentially related to fault zones.

D. In tectonic settings, pseudotachylite is generally faulting/shearing-related. As stated above, pre-existing zones of weakness, such as lithological boundaries or mylonitic shear zones, could be the preferred loci for pseudotachylite formation. Sibson (1975) and Grocott (1981) provided a detailed account of pseudotachylite geometries observed in fault zones. Similar accounts are abundant in the structural geological literature, for example in papers in Magloughlin and Spray (1992). Reimold and Colliston (1994) and the subsequent work of our group have shown that the entire range of these geometries can be observed in abundance in the breccia developments of the core of the Vredefort Dome. However, comparing tectonic and Vredefort impact-induced network breccias, there is a different scale of magnitude involved. Tectonic network breccias rarely exceed several decimeters in width, whereas some network breccias at Vredefort have been mapped for hundreds and, in select cases, even over thousands of meters extent and hundred meter width. Within such network breccia occurrences at Vredefort, there may well occur apparent faults, linear features along which the host gneissosity can be displaced by centimeters or decimeters – and in exceptional cases by as much as 50 centimeters. And yet, even the detailed mapping of recent years of such sites by B.O. Dressler in conjunction with our group (Dressler and Reimold 2004) has not shown any evidence that would lead to an interpretation of such sites as major generation planes. Rather, during compressive “shattering” of a rock mass, faulting and slip along individual planes of weakness or along conjugate orientations takes place, according to strain distribution through the rock mass, possibly locally facilitated by pre-existing planes of weakness (fractures, pre-impact mylonites or cataclasites, lithological banding). In contrast to many tectonic occurrences of pseudotachylite, the long established relationship between slip rate and thickness of a friction melt vein (Sibson 1975) is not valid for impact-generated pseudotachylitic breccias.

It is generally accepted that the process of friction melt formation along a fault surface will initially result in strengthening of a fault due to increase of the friction coefficient. However, once a fault is completely separated, resistance decreases. This hypothesis is, however, based on significant displacement occurring along a fault, with a magnitude of at least meters, which is generally not observed along pseudotachylitic breccia veins in impact structures. Melosh (this volume) investigates how thick masses of pseudotachylitic breccia (although he refers to them as pseudotachylite), as observed for example at Vredefort and Sudbury, could form as a result of friction melting, in the face of the melt-strengthening of rock observed. Melosh proposes that melt is produced on narrow faults but is extruded into adjacent country rock where it could pool. This interesting suggestion must be examined further in the field by continuing efforts of mapping 3D exposures including likely generation planes (narrow shear/fault planes) and apparent sites of melt pooling. Could it be that the massive, straight melt dikes observed locally in the Vredefort Structure could be dilational sites formed during decompression and then infilled with melt generated elsewhere?

E. Distinct chemical relationships between host rock and pseudotachylite have been established for tectonic breccias, as shown by many publications such as those reviewed by Reimold (1991). These relationships are the result of differential fracture toughness of certain minerals and their specific susceptibility to melting (e.g., Spray 1995). The same relationships between chemical compositions of host rock and breccia pairs have been observed for the Vredefort breccias and tectonic pseudotachylite-host rock pairs (Reimold 1991). Dressler (1984, and literature reviewed in there) has already shown that the same applies to the Sudbury Breccias and their host rocks. However, it has also been shown for the Vredefort case that it is possible to determine compositions for spots in Vredefort breccia veinlets (admittedly very thin ones!) that are identical to the mineral compositions in the immediately adjacent host rock. This suggests that at least some breccias were melted *in situ* and were not formed through injection from a larger generation plane. Furthermore, it indicates that they were not subject to melt fractionation or incomplete mineral melting, or to mixing between minerals or even precursor rock-types.

F. Petrographic characteristics of Vredefort breccias and of many other so-called "pseudotachylites" in impact structures are also similar or even identical to those observed in tectonic pseudotachylite = friction melt. Fine-grained igneous crystallization textures and aphanitic melt matrices are commonly observed, but partial or main presence of cataclastic material also attests to the likelihood that, at least, in some cases, Spray's (1995) model for the generative process of pseudotachylite applies to impact-induced melt breccia as well.

8 Problematics

On the basis of what we now know about pseudotachylitic breccias in impact structures, in comparison to tectonic pseudotachylite, a number of important unresolved problematics, **A – H,** can be listed:

A. The need to distinguish between *bona fide* pseudotachylite = friction melt and other breccia types in general, and in impact structures in particular;

B. How to distinguish in impact structures between *bona fide* pseudotachylite = friction melt and injections of impact melt breccia into the crater floor;

C. How to distinguish between impact and tectonic pseudotachylite = friction melt, especially in the case of very old and large, deeply eroded and multi-deformed impact structures.

Those impact workers that accept Spray's (1998) distinction between "S (shock) and E (endogenic) pseudotachylites". However, we feel that it is (1) important to distinguish between impact and non-impact breccias that may well occur within the region/volume of an impact structure (the Vredefort-Witwatersrand structure is a case in point, as demonstrated in the above), and (2) that, in order to learn about the nature of different pseudotachylite-like (pseudotachylitic) breccias formed by the impact process, it is necessary to investigate in detail and identify the nature of each breccia type first, before the genetic process can be hypothesized – and that this is a prerequisite for learning more about the impact process itself. For this reason alone, it is important to abandon the generalization that is inherent in the term "pseudotachylite" in its current usage in the impact community. Thus, we advocate again (compare Reimold 1995, 1998) that the term "pseudotachylite" only be used where *bona fide* friction melt is refered, and that all other cases be properly investigated to allow correct and detailed categorization; and that a broad, non-genetic term such as *"pseudotachylitic breccia"* (or, where appropriate, *pseudotachylite-like vein* - for example, where a single, glassy veinlet is concerned) be used in all these instances.

D. How do *"impact pseudotachylitic breccias"* form? The evidence reviewed above indicates that there are pure cataclasites, *bona fide* pseudotachylite, impact melt veinlets, and shock melts (note that we differentiate between these last two types of breccia, as shock veins are *in situ* generated shock compression melt (rock), whereas impact melt [rock] veinlets could have been generated elsewhere and then intruded into the now exposed setting). In addition, Kenkmann et al. (2000a) and Langenhorst et al. (2002) have shown that combinations of shearing/faulting and shock melting could be possible mechanisms for the generation of breccias of pseudotachylitic appearance as well.

The problem now is "How to identify the nature and genesis of specific breccia types?" Ultracataclastic or cataclastic breccia in impact structures may represent pre-impact tectonic breccia or impact-produced monomict or polymict lithic (in the older literature the term "fragmental impact breccia" has been used) impact breccia. How to resolve this issue can not be generalized – it depends on the local, case by case, geological setting. A mylonite or ultramylonite is recognized by the pronounced fabric and grain size reduction associated with such material (e.g., Bates and Jackson 1987; Wise et al. 1984: "Matrix produced by syntectonic crystal-plastic processes, ..."). According to A. Deutsch (pers. commun.; also Sazonova and Korotaeva 1995), impact-generated mylonitic rocks ("coptomylonite", according to V. Masaitis) have been described in the Russian literature about the Puchezh-Katunki structure, but apparently only referring to an impact-generated equivalent to protomylonite that, in contrast to ortho- and ultramylonite, is not characterized by extensive recovery (recrystallization) (see, e.g., Wise et al. 1984). As the formation of breccias during the shock compression stage inevitably takes place under high strain rates, it is hardly possible that recovery could take place under impact conditions that involve strain rates of the order of $>10^6$ s^{-1}. (e.g., Spray 1998). It is, however, not entirely inconceivable that impact-generated mylonite (intensely sheared rock, for example produced due to shear movement along blocks of target material) could undergo at least some recovery when formed at depth in the crater basement that, itself, may be covered by a thick, hot pile of impact debris.

At Vredefort, mylonites occur in the granitoid rocks of the core as well as in the collar strata. If one assumes that the bulk of the pseudotachylitic breccia in the Vredefort Dome is the result of the impact event, the granitic mylonites would have to pre-date the impact event, as many such occurrences have been exploited as convenient zones of rock weakness for the generation of pseudotachylitic breccia. And it has been demonstrated by Reimold and Colliston (1994) that at least some pseudotachylitic breccia occuring in the core of the dome must pre-date the Vredefort impact event. Should it be correct to assume that shatter cones form during or immediately after shock compression (e.g., Baratoux and Melosh 2003 and refs. therein), the occurrence of shatter cones overprinting a pseudotachylitic breccia vein shown by Reimold and Colliston (1994) would be further evidence for pre-impact breccia.

A second mode of mylonite presence in impact structures could be the result of post-impact tectonic overprint of an impact structure/formation, whereby the temporal relationships will also have to be understood in order to draw the correct genetic conclusion. The identification of tectonic pseudotachylite = friction melt, in any setting (tectonic or impact), clearly depends on its association with a tectonic fault or shear zone or impact-induced tectonic fault or shear zone. However, a fault association and the presence of melt in the groundmass of such material must be proven, and the temporal status (pre-, syn- or post-impact) of a breccia occurrence needs to be established *before* "tectonic" and "impact pseudotachylite" can be differentiated.

Perchuk et al. (2003) recently embarked on a discussion as to whether some of the dark-matrix breccias, for which the Vredefort impact structure is so famous, should be classified as "ultramylonites" or "pseudotachylite". Their argument in favor of "ultramylonite" is that the "veinlets from Vredefort... do not contain glass. In accordance with the modern classification [Wise et al. 1984], these thin ... viscoelastic deformation features must be regarded as ultramylonites formed through the relaxation of residual stresses...". This observation is, of course, opposed to extensive observation of evidence for melting, crystallization of vein fillings from melt phases, and – as demonstrated by White (1993) for some veinlets – of presence of glass.

An additional complication is presented by the fact that hydrothermal processes are a likely consequence of large impact events, especially if thick sequences of porous sedimentary rocks are part of the target geology. Hot aqueous solutions released as a direct consequence of the impact event are bound to cause alteration of breccia occurrences especially in the upper parts of and around the central uplift of complex impact structures. This aspect is further discussed in a companion paper in this volume (Reimold et al. this volume) that also deals with hydrothermal deposition of metals in impact structures.

Distinction of pseudotachylite = friction melt and other pseudotachylitic breccia from impact melt breccia injections may not be easy. Detailed chemical studies (analysis of bulk vein filling and/or defocused beam electron microprobe analysis, in comparison to host rock chemical analysis) may be required to ascertain whether a vein filling was generated *in situ* or injected. Where comparison of vein filling compositions with the composition of massive and homogeneous impact melt rock is possible (as in the case of Morokweng – Reimold et al. 1999a; Koeberl and Reimold 2003), this distinction may be made easier. In addition, presence or absence of exotic clasts – in comparison with host rock mineralogy – may provide clues, and the presence or absence of shock metamorphic deformation can also provide essential evidence to pinpoint the nature of a breccia as possibly of impact origin or as a possible shock melt alternative to a pseudotachylite or impact melt injection. Although in this regard, one has to bear in mind that impact (shock) -deformed materials may be tectonically overprinted and, vice versa, that pre-impact tectonic breccias may be shock-metamorphosed. Be that as it may – without such detailed detective work we will never fully understand the formation and distribution of different breccia types in impact structures. Outside of the zone of diagnostic shock deformation, all pseudotachylitic breccia (with the exception of impact melt injections) may be slip-related. In this case, detailed kinematic analysis and geochronological evidence may allow one to determine the mode of origin of such breccia or a possible impact relationship.

E. Tectonic and impact workers have to find consensus how the term "pseudotachylite" is to be applied in future. If it is to maintain its genetic meaning (friction melt), impact workers have to take note and abandon their confusing and indiscriminate usage of this term. However, should the term be relegated to purely *descriptive* status, the current use by the impact community would become legitimate. However, we can not see the entire structural geological community reverting to a different application. Thus, the first option must be favored.

F. An interesting aspect raised in the mid-1990s in several papers (e.g., Fiske et al. 1995a) – that "pseudotachylite" formation could contribute significantly to planetary heat budgets – needs to be further evaluated. The idea at the time was that massive melt formation as observed at Sudbury (Sudbury Breccias) and Vredefort could have major consequences for the thermal budgets of planetary bodies, as a result of very large impact events. Clearly, this consideration requires careful distinction of different melt processes, as friction melting on a planetary scale may have vastly different (presumably less?) thermal consequence, compared to large-scale impact melting.

G. The lack of field evidence regarding the nature and genesis of ring features of large, multi-ring impact structures makes evaluation of the current working hypotheses (superfaults involving large-scale friction melting - Spray 1997, and collapse fractures/scarps - Melosh 1989) difficult. Lunar ring features are currently inaccessible for detailed geological analysis, and terrestrial observations are limited to the environs of the very large Sudbury and Vredefort structures. Whereby the latter is more deeply eroded than Sudbury. Whilst evidence at Sudbury seems to favor zonal enrichment of pseudotachylitic breccia, the tectonic (impact-tectonic) significance has only been speculated upon, as the 3D structure of these zones is not known. As already mentioned, the environs of the Vredefort Dome are not extensively exposed for geological analysis. And in any case, this structure is very deeply eroded so that much of the large-scale impact-tectonic phenomena are not preserved. On a positive side, the deep exposures in the gold mines of the Witwatersrand Basin still provide excellent 3D "exposures" for detailed future breccia analysis.

H. Another suggestion made in recent years involves the idea that ubiquitous formation of microscopic "pseudotachylite" could provide an option for the still elusive mechanism by which very large rock volumes in central uplift structures in large impact craters are enabled to be compressed (downwards) and then move upwards by significant distances, before uplift collapse, while maintaining their pre-impact rock fabrics – as demonstrated in the excellent exposures of the core of the Vredefort Dome (Lana et al. 2003a,b). Again, testing of this idea would necessitate understanding of the "pseudotachylite" forming process – and a prerequisite for that is separation of friction and shock melts – or confirmation that both processes are inseparable in the impact cratering process. Our long-time petrographic analysis of Vredefort basement rocks has not shown that a penetrative network of micro-"pseudotachylites" exists at the current exposure level. Breccia veinlets at the micro-scale may occur at millimeter to sub-millimeter spacings, but not in a ubiquitous fashion. To the contrary, numerous specimens at the several centimeter scale have been studied that do not contain any optical discontinuities that one would classify as "pseudotachylitic breccia". Instead, irregular and (sub)planar microfracturing do occur in every sample studied from the Archean gneiss basement, but fracture densities are strongly variable from place to place (compare Reimold 1990) and, possibly, in relation to the spatial distribution of macro-deformation such as macroscopic development of pseudotachylitic breccia.

The notion of some workers that terracing, that is collapse scarp formation on the inner crater rim, may be associated with extensive formation of pseudotachylitic breccia appears a reasonable one, though again, field evidence is lacking - unless the alleged zones of increased breccia development around Sudbury could be linked to this process. The only known massive breccia occurrence in the Rand Anticline at the northern margin of the Witwatersrand Basin can perhaps be related to crater collapse, but it appears from its approximately north-south trend, i.e. radial with respect to the center of the impact structure, this occurrence could also be the result of faulting along a transpressional ridge (Kenkmann and von Dalwijk 2000) rather than along an annular failure zone. The large bedding-parallel fault zones along the southwest-northeast trend in the northern Witwatersrand basin around the

Vredefort Dome seem to dip shallowly (15-20 degrees) towards the Vredefort dome. It has been argued by previous workers (Engelbrecht et al. 1986; Fletcher and Reimold 1989; Killick 1993) that some of these fault zones could have been in existence prior to the impact event and, by implication would have been reactivated in order to carry ca. 2020 Ma massive breccias. It can, however, not be completely excluded that these fault zones could have been – prior to deep erosion – connected to listric, annular fault zones caused by crater rim collapse during the crater modification phase. In both the Sudbury and Vredefort cases, the upper strata and, in essence, the entire crater rim have been eroded and, thus, do not allow complete testing of the 'superfault' hypothesis. And we have to be cognizant of the deep level of erosion at Vredefort. In smaller impact structures, such as the 10 km diameter Bosumtwi crater, faulting has been related to gravity sliding of blocks inwards off the crater rim (Reimold et al. 1998), but to our knowledge, no pseudotachylitic breccias have, to date, been observed in such settings of relatively small crater structures. In conclusion, the data base on which to build a "superfault" hypothesis seems extremely limited and, in fact, insufficient.

9 Conclusions

This work summarizes some pertinent facts in the debate about "what is pseudotachylite?" especially in impact structures, and "what needs to be pursued to further solve the remaining problematics?"; "How are impact pseudotachylitic breccias formed?" Clearly, there are a number of observations and data bases that impact pseudotachylitic breccias and tectonic pseudotachylite have in common. And yet, different processes will, at least in part, be involved in the formation of such breccias under the enormous shock pressures and strain rates that are, during the initial part of the cratering process, much higher than during tectonic deformation. Thus, further work needs to be done on all fronts – with regard to detailed field investigation, petrographic, chemical, structural (i.e., geometries, relationship to precursor rock fabric and deformations), and temporal features and controls, under most careful and critical distinction of individual breccia occurrences. Without such diligent categorization and separation of breccias, we will not be able to learn more about why, how, and where specific pseudotachylitic breccia types may form during the catastrophic impact process. Dressler and Reimold (2004) questioned: "Is there order in chaos?" We will never be able to provide the full answer to this question, if we only recognize two types of impact pseudotachylite – be they referred as types A or B, A1 or A2, or S or E. Even if there would only be two types of pseudotachylitic breccia, none of the existing classifications provide satisfactory distinguishing criteria.

Acknowledgements

This discussion represents the gist of an invited lecture of the same title presented by WUR at the ESF Impact Program Workshop on Impact Tectonics, Mora, Sweden, 1-3 June 2002. The authors would like to thank the Organizers of this workshop for the invitation and the support received that made their attendance of this very fruitful event possible. Extensive discussions with Burkhard Dressler and his comments on an early draft of this manuscript are much appreciated. Reviews by Thomas Kenkmann and Victor Masaitis also assisted in improving this manuscript. This is University of the Witwatersrand Impact Cratering Research Group Contribution No. 55.

References

Allaby A, Allaby M (eds) (1991) Oxford Concise Dictionary of Earth Sciences. Oxford University Press, Oxford-New York: 410 pp

Andreoli MAG, Ashwal LD, Hart RJ, Smith CB, Webb SJ, Tredoux M, Gabrielli F, Cox RM, Hambleton-Jones BB (1995) The impact origin of the Morokweng ring structure, southern Kalahari, South Africa. Abstract, Centenary Geocongress, Geological Society of South Africa, Johannesburg: 541-544

Austrheim H, Boundy TM (1994) Pseudotachylytes generated during seismic faulting and eclogitization of the deep crust. Science 265: 82-83

Bates RL, Jackson JA (eds) (1987) Glossary of Geology. American Geological Institute, Alexandria, Virginia: 788 pp

Berlenbach JW, Roering C (1992) Sheath-fold-like structures in pseudotachylites. Journal of Structural Geology 14: 847-856

Bischoff, L, Oskierski W (1987) Fractures, pseudotachylite veins and breccia dikes in the crater floor of the Rochechouart impact structure, In Pohl J (ed), Research in Terrestrial Impact Structures. Vieweg-Verlag, Braunschweig/Wiesbaden, Germany: 5-29

Brandl G, Reimold WU (1990) The structural setting and deformation associated with pseudotachylite occurrences in the Palala Shear Belt and Sand River Gneiss, Northern Transvaal. Tectonophysics 171: 201-220

Camacho A, Vernon RH, Fitz Gerald JD (1995) Large volumes of anhydrous pseudotachylyte in the Woodroffe Thrust, eastern Musgrave Ranges, Australia. Journal of Structural Geology 17: 371-383

Chen M, Sharp TG, El Goresy A, Wopenka B, Xie X (1996) The majorite-pyrope + magnesiowüstite assemblage: Constraints on the history of shock veins in chondrites. Nature 271: 1570-1573

Chen M, Shu JF, Xie X, Mao H-K (2003) Natural $CaTi_2O_4$-structured $FeCr_2O_4$ polymorph in the Suizhou meteorite and its significance in mantle mineralogy. Geochimica et Cosmochimica Acta 67: 3937-3942

Clarke GL (1990) Pyroxene microlites and contact metamorphism in pseudotachylite veinlets from MacRobertson Land, east Antarctica. Australian Journal of Earth Sciences 37: 1-8

Clarke GL, Norman AR (1993) Generation of pseudotachylite under granulite facies conditions, and its preservation during cooling. Journal of Metamorphic Geology 11: 319-335

Degenhardt JJ, Reid AM, Miller RMcG, Reimold WU (1996) Breccias resembling melt bombs from the Roter Kamm Crater. Meteoritics and Planetary Science 31: 413-415

Dietz RS (1960) Vredefort Ring Structure: an astrobleme (meteorite impact structure). Geological Society of America Bulletin 71:2093

Dressler BO (1984) The effects of the Sudbury event and the intrusion of the Sudbury Igneous Complex on the footwall rocks of the Sudbury Structure. In Pye EG, Naldrett AJ, Giblin PE (eds), The Geology and Ore Deposits of the Sudbury Structure. Ontario Geological Survey Special Volume 1: 97-136

Dressler BO, Reimold WU (2004) Order or Chaos? Origin and mode of emplacement of breccias in floors of large impact structures. Earth-Science Reviews 67:1-54

Dressler BO, Sharpton VL (1997) Breccia formation at a complex impact crater: Slate Islands, Lake Superior, Ontario, Canada. Tectonophysics 275: 285-311

Erismann ThH, Abele G (2001) Dynamic of Rockslides and Rockfalls, Springer-Verlag Berlin-Heidelberg-New York, pp. 32-48

Engelbrecht CJ, Baumbach GWS, Matthysen JL, Fletcher P (1986) The West Wits Line. In Anhaeusser CR, Maske S (eds), Mineral Deposits of Southern Africa, v I: pp 599-648, Geological Society of South Africa, Johannesburg

Ernstson K, Fiebag J (1992) The Azuara impact structure (Spain): new insights from geophysical and geological investigations. Geologische Rundschau 81: 403-427

Ernstson K, Claudin F, Schüssler U, Hradil K (2002) The mid-Tertiary Azuara and Rubielos de la Cérida paired impact structures (Spain). Treballs del Museu de Geologia de Barcelona 11:5-65

Fiske PS, Nellis WJ, Lipp M, Lorenzana H, Kikuchi M, Syono Y (1995a) Pseudotachylites generated in shock experiments: Implications for impact cratering products and processes. Science 270: 281-283

Fiske PS, Nellis WJ, Lorenzana H, Lipp M, Kikuchi A, Syono Y (1995b) Pseudotachylite produced in shock recovery experiments: Implications for impact cratering mechanics. Abstract, American Geophysical Union Spring Meeting 1995. EOS Supplement: 197-198

Fletcher P, Reimold WU (1989) Some notes and speculations on the pseudotachylites in the Witwatersrand Basin and the Vredefort Dome. South African Journal of Geology 92: 223-234

French BM, Koeberl C, Gilmour I, Shirey SB, Dons JA, Naterstad J (1997) The Gardnos impact structure, Norway: Petrology and geochemistry of target rocks and impactites. Geochimica et Cosmochimica Acta 61: 873-904

Friese AEW, Reimold WU, Layer PW (2003) ^{40}Ar-^{39}Ar dating of and structural information on tectonite-bearing faults in the Witwatersrand Basin: Evidence for multi-stage, tectono-thermal activity in the Central Kaapvaal Craton. South African Journal of Geology 106: 41-70

Gibson HM, Spray JG (2000) Shock-induced melting and vaporization of shatter cone surfaces: Evidence from the Sudbury impact structure. Meteoritics and Planetary Science 33: 329-336

Gibson RL, Reimold WU (2001a) The Vredefort impact structure, South Africa (the scientific evidence and a two-day excursion guide). Memoir 92, Council for Geoscience, Pretoria: 110 pp

Gibson RL, Reimold WU (2001b) Petrographic evidence for high shock pressures and shock pressure heterogeneity in the Vredefort impact structure, South Africa. Abstract, 64th Annual Meeting of the Meteoritical Society, Rome, Italy, Meteoritics and Planetary Science 36 supplement: A51-A52

Gibson RL, Reimold WU (2004) Shock pressure distribution in the Vredefort impact structure, South Africa. Proceedings 3rd International Conference on Large Meteorite Impacts, Geological Society of America, Special Paper (in press)

Gibson RL, Reimold WU, Ashley AJ, Koeberl C (2002) Metamorphism on the Moon: A terrestrial analogue in the Vredefort Dome, South Africa? Geology 30: 475-478

Gilmour I, French BM, Franchi IA, Abbott JI, Hough RM, Newton J, Koeberl C (2003) Geochemistry of carbonaceous impactites from the Gardnos impact structure, Norway. Geochimica et Cosmochimica Acta 67: 3889-3903

Grieve RAF, Therriault AM (2000) Vredefort, Sudbury, Chicxulub: Three of a kind? Annual Reviews of Earth and Planetary Science 28: 305-338

Grocott J (1981) Fracture geometry of pseudotachylyte generation zones: a study of shear fractures formed during seismic events. Journal of Structural Geology 3: 169-178

Hall, AL, Molengraaff GAF (1925) The Vredefort Mountain Land in the Southern Transvaal and Northern Orange Free State. Verhandelingen der Koninklijke Akademie van Wetenschappen te Amsterdam (Tweede Sectie), part 24, No. 3: 183 pp

Hart RJ, Andreoli MAG, Tredoux M, Moser D, Ashwal LD, Eide EA, Webb SJ, Brandt D (1997) Late Jurassic age for the Morokweng impact structure, southern Africa. Earth and Planetary Science Letters 147: 25-35

Henkel H, Reimold WU (1998) Integrated geophysical modelling of a giant, complex impact structure: anatomy of the Vredefort Structure, South Africa. Tectonophysics 287: 1-20

Hetzel R, Altenberger U, Strecker MR (1996) Structural and chemical evolution of pseudotachylytes during seismic events. Mineralogy and Petrology 58: 33-50

Heuberger H, Masch L, Preusss E, Schröcker A. (1984) Quaternary landslides and rock fusion in Central Nepal and in the Tyrolean Alps. Mountain Research and Development 4: 345-362

Hilke C (1991) Impaktbreccien der Carswell-Struktur, Sasketchewan, Kanada: Petrographie, Geochemie und Genese. PhD Thesis (unpublished), Westfälische Wilhelms-Universität Münster, Germany: 188 pp and appendices

Kamo SL, Reimold WU, Krogh TE, Colliston WP (1996) A 2.023 Ga age for the Vredefort impact event and a first report of shock metamorphosed zircons in pseudotachylitic breccias and granophyre. Earth and Planetary Science Letters 144: 369-388

Kenkmann T, von Dalwigk I (2000) Radial transpression ridges: a new structural feature of complex impact craters. Meteoritics and Planetary Science 35: 1189-1201

Kenkmann T, Hornemann U, Stöffler D (2000a) Experimental generation of shock induced pseudotachylites along lithological interfaces. Meteoritics and Planetary Science 35: 1275-1290

Kenkmann T, Ivanov BA, Stöffler D (2000b)– Identification of ancient impact structures: Low-angle faults and related geological features of crater basements. In I Gilmour, C Koeberl (editors) Impacts and the Early Earth, Springer Verlag Berlin-Heidelberg-New York, pp 279-307

Killick AM (1993) Pseudotachylites of the West Rand Goldfield, Witwatersrand Basin, South Africa. Ph D Thesis, Rand Afrikaans University, Johannesburg, 273 pp

Killick AM, Reimold WU (1990) Review of the pseudotachylites in and around the Vredefort Dome, South Africa. South African Journal of Geology 93: 350-365

Killick AM, Thwaites AM, Germs GJB, Schock AE (1988) Pseudotachylite associated with a bedding-parallel fault zone between the Witwatersrand and Ventersdorp Supergroups, South Africa. Geologische Rundschau 77:329-344

Koch N, Masch L (1991) Formation of Alpine mylonites and pseudotachylytes at the base of the Silvretta nappe, Eastern Alps. Tectonophysics 204: 289-306

Koeberl C, Reimold WU (2003) Geochemistry and petrography of impact breccias and target rocks from the 145 Ma Morokweng impact structure, South Africa. Geochimica et Cosmochimica Acta 67: 1837-1862

Koeberl C, Reimold WU, Shirey SB (1996) Re-Os isotope and geochemical study of the Vredefort Granophyre: clues to the origin of the Vredefort Structure, South Africa. Geology 24: 913-916

Koeberl C, Hough RB, Boamah D, French BM, McDonald I, Reimold WU (2001) Woodleigh impact structure, Australia: Shock petrography and geochemical studies. Abstract, 64th Annual Meeting of the Meteoritical Society, Rome, Italy, Meteoritics and Planetary Science 36 (supplement): p A101

Lambert P (1981) Breccia dikes: Geological constraints on the formation of complex craters. In Schultz PH, Merrill RB (eds), Multi-Ring Basins, Proceedings of Lunar and Planetary Science 12A, Pergamon Press: 59-78

Lana C, Gibson RL, Kisters AFM, Reimold WU (2003a) Archean crustal structure of the Kaapvaal craton, South Africa – evidence from the Vredefort dome. Earth and Planetary Science Letters: 206:133-144

Lana C, Gibson RL, Reimold WU (2003b) Impact tectnics in the core of the Vredefort dome, South Africa: Implications for central uplift formation in very large impact structures. Meteoritics and Planetary Science 38: 1093-1107

Langenhorst F, Poirier J-P (2000) Anatomy of black veins in Zagami: clues to the formation of high-pressure phases. Earth and Planetary Science Letters 184: 37-55

Langenhorst F, Poirier J-P, Deutsch A, Hornemann U (2002) Experimental approach to generate shock veins in single crystal olivine by shear heating. Meteoritics and Planetary Science 37: 1541-1553

Legros F, Cantagrel J-M, Devouard B (2000) Pseudotachylyte (Frictionite) at the base of the Arequipa volcanic landslide deposit (Peru): Implications for emplacement mechanisms. Journal of Geology 108: 601-611

Leroux H, Reimold WU, Doukhan J-C (1994) A TEM investigation of shock metamorphism in quartz from the Vredefort Dome, South Africa. Tectonophysics 230: 223-239

Macdonald FA, Bunting JA, Cina SE (2003) Yarrabubba – a large, deeply eroded impact structure in the Yilgarn Craton, Western Australia. Earth and Planetary Science Letters 213: 235-247

Magloughlin JF, Spray JG (1992) Frictional melting process and products in geological materials: introduction and discussion. Tectonophysics 215: 197-206

Martini JEJ (1978) Coesite and stishovite in the Vredefort Dome, South Africa. Nature 277: 495-496

Martini JEJ (1991) The nature, distribution and genesis of the coesite and stishovite associated with the pseudotachylite of the Vredefort Dome, South Africa. Earth and Planetary Science Letters 103: 285-300

Masch L, Wenk HR, Preuss E (1985) Electron microscopy study of hyalomylonites – evidence for frictional melting in landslides. Tectonophysics 115: 131-160

Melosh HJ (1989) Impact Cratering: A Geologic Process. Oxford University Press, New York, 245 pp

Mory AJ, Iasky RP, Glikson AY, Pirajno F (2000a) Woodleigh, Carnarvon Basin, Western Australia: a new 120 km diameter impact structure. Earth and Planetary Science Letters 177: 119-128

Mory AJ, Iasky RP,Glikson AY, Pirajno F (2000b) Response to 'Critical comment on: AJ Mory et al. 'Woodleigh, Carnarvon Basin, Western Australia: a new 120 km diameter impact structure'. Earth and Planetary Science Letters 184: 359-365

Moser DE (1997) Dating the shock wave and thermal imprint of the giant Vredefort impact, South Africa. Geology 25: 7-10

Müller W, Kelley SP, Villa IM (2002) Dating fault-generated pseudotachylytes: comparison of $^{40}Ar/^{39}Ar$ stepwise-heating, laser ablation and Rb-Sr microsampling analyses. Contributions to Mineralogy and Petrology 144: 57-77

Nicolaysen LO, Reimold WU (1999) Vredefort shatter cones revisited. Journal of Geophysical Research 104: 4911-4930

Park RG (1961) The pseudotachylite of the Gairloch District, Ross-Shire, Scotland. American Journal of Science 259: 542-550

Passchier CW, Trouw RAJ (1998) Microtectonics. Springer Verlag, Berlin-Heidelberg-New York, 289 pp

Passchier CW, Hoek JD, Bekendam RF, de Boorder H (1990) Ductile reactivation of Proterozoic brittle fault rocks; an example from the Vestfold Hills, East Antarctica. Precambrian Research 47: 3-16

Passchier CW, Bekendam RF, Hoek JD, Dirks PGHM, De Boorder H (1991) Proterozoic geological evolution of the northern Vestfold Hills, Antarctica. Geological Magazine 128: 307-318

Perchuk LL, Sazonova LV, van Reenen DD, Gerya TV (2003) Ultramylonites and their significance for the understanding of the history of the Vredefort impact structure, South Africa. Petrology11: 114-129

Philpotts AR (1964) Origin of pseudotachylites. American Journal of Science 262: 1008-1035

Phillips ME, Bussell MA, McDonald I, Hart RJ, Andreoli MAG (1999) A remote sensing and geoloical investigation of the Vredefort impact structure (South Africa) using Landsat Thematic Mapper imagery [abstract]. Meteoritics and Planetary Science 34 (supplement): A92-A93

Preuss E (1971) Über den Bimsstein von Köfels/Tirol. Fortschritte der Mineralogie 49, Beiheft 1: p. 70

Reimold WU (1990) The controversial microdeformations in quartz from the Vredefort Structure, South Africa. South African Journal of Geology 93: 645-663

Reimold WU (1991) Geochemistry of pseudotachylites from the Vredefort Structure, South Africa. Neues Jahrbuch der Mineralogie Abhandlungen 161: 151-184

Reimold WU (1995) Pseudotachylite in impact structures – generation by friction melting and shock brecciation? – A review and discussion. Earth-Science Reviews 39: 247-264

Reimold WU (1998) Exogenic and endogenic breccias: A discussion of major problematics. Earth-Science Reviews 43: 25-47

Reimold WU, Colliston WP (1994) Pseudotachylites of the Vredefort Dome and the surrounding Witwatersrand Basin, South Africa. In Dressler BO, Grieve RAF, Sharpton VL (eds), Large Meteorite Impacts and Planetary Evolution. Geological Society of America Special Paper 293: 177-196

Reimold WU, Koeberl C (2000) Critical comment on AJ Mory et al "Woodleigh, Carnarvon basin, Western Australia: a new 120 km diameter impact structure". Earth and Planetary Science Letters 184: 353-357

Reimold WU, Miller RMcG (1989) The Roter Kamm Crater, SWA/Namiba. Proceedings of the 19th Lunar and Planetary Science Conference, Cambridge University Press/Lunar and Planetary Institute, Houston: p 711-732

Reimold WU, Colliston WP, Wallmach T (1992) Comment on "Nature, provenance and distribution of coesite and stishovite in the Vredefort Structure" by JEJ Martini. Earth and Planetary Science Letters 12: 213-217

Reimold WU, Oskierski W, Huth J (1987) The pseudotachylite from Champagnac in the Rochechouart meteorite crater, France. Proceedings of the 17th Lunar and Planetary Science Conference, Part 2, Journal of Geophysical Research 92, B4: 737-748

Reimold WU, Brandt D, Koeberl C (1998) Detailed structural analysis of the rim of a large, complex impact crater: Bosumtwi Crater, Ghana. Geology 26: 543-546

Reimold WU, Koeberl C, Brandstätter F, Kruger FJ, Armstrong RA, Bootsman K (1999a) The Morokweng impact structure, South Africa: Geological, petrographical, and isotopic results, and implications for the size of the structure. In Dressler BO and Sharpton VL (eds) Large Meteorite Impacts and Planetary Evolution II, Geological Society of America Special Paper 339: 61-90

Reimold WU, Koeberl C, Fletcher P, Killick AM, Wilson JD (1999b) Pseudotachylitic breccias from fault zones in the Witwatersrand Basin, South Africa: Evidence of autometasomatism and post-brecciation alteration processes. Mineralogy and Petrology 66: 25-53

Reimold WU, Armstrong RA, Koeberl C (2002) A deep drillcore from the Morokweng impact structure, South Africa: petrography, geochemistry and constraints on the crater size. Earth and Planetary Science Letters 201: 221-232

Reimold WU, Buchanan PC, Ambrose D, Koeberl C (2003a) The H4/5 The meteorite fall of 21 July 2002, Lesotho: History of the fall, strewn field determination, and mineralogical and geochemical characterization [abstract]. Meteoritics and Planetary Science 38 (supplement): p A15

Reimold WU, Koeberl C, Hough RM, McDonald I, Bevan A, Amare K, French BM (2003b) Woodleigh impact structure, Australia: Shock petrography and geochemical studies. Meteoritics and Planetary Science 38: 1109-1130

Renne PR, Reimold WU, Koeberl C, Hough R, Claeys P (2002) Critical comment on IT Uysal et al 'K-Ar evidence from illitic clays of a Late Devonian age for the 120 km diameter Woodleigh impact structure, Southern Carnarvon Basin, Western Australia'. Earth and Planetary Science Letters 201: 247-252

Sazonova LV, Korotaeva NN (1995) The genesis and formation parameters of ultramylonites and pseudotachylytes in the central uplift of the Puchezh-Katunki astrobleme. Vestn. Mosk. University, Series 4: Geology, No. 1: 74-81

Shand SJ (1916) The pseudotachylyte of Parijs (Orange Free State), and its relation to "trap-schotten gneiss" and "flinty crush-rock". Quarterly Journal of the Geological Society London 72: 198-221

Sibson RH (1975) Generation of pseudotachylyte by ancient seismic faulting. Journal of the Royal Astronomical Society 43: 775-794

Simpson, C (1981) Occurrence and orientation of shatter cones in Pretoria Group quartzites in the collar of the Vredefort "Dome": impact origin precluded. Journal of Geophysical Research 86, B11: 10 701-10 706

Spray JG (1992) A physical basis for the frictional melting of some rock-forming minerals. Tectonophysics 204: 205-221

Spray JG (1995) Pseudotachylite controversy: Fact or fiction? Geology 23: 1119-1122

Spray JG (1997) Superfaults. Geology 25: 579-582

Spray JG (1998) Localized shock- and friction-induced melting in response to hypervelocity impact. In Grady MM, Hutchison R, McCall GJH, Rothery DA (eds), Meteorites: Flux with Time and Impact Effects. Geological Society London Special Publications 140: 195-204

Spray JG, Thompson LM (1995) Friction melt distribution in a multi-ring impact basin. Nature 373: 130-132

Spray JG, Kelley SP, Reimold WU (1995) Laser-probe ^{40}Ar-^{39}Ar dating of pseudotachylytes and the age of the Vredefort impact event. Meteoritics and Planetary Science 30: 335-343

Stepto D (1979) A geological and geophysical study of the central portion of the Vredefort Dome structure. PhD Thesis (unpubl.), University of the Witwatersrand, Johannesburg: 378 pp

Stewart RA, Reimold WU, Charlesworth EG, Ortlepp D (2000) The nature of shear zones and associated fault rock related to a recent rockburst at Western Deep Levels Gold Mine, Witwatersrand Basin, South Africa. Tectonophysics 337: 173-190

Stöffler D, Grieve, RAF (1994) Classification and nomenclature of impact metamorphic rocks: A proposal to the IUGS Subcommission on the Systematics of Metamorphic Rocks. (abstract) Lunar and Planetary Science XXV: 1347-1348

Stöffler D, Keil K, Scott ERD (1991) Shock metamorphism of ordinary chondrites. Geochimica et Cosmochimica Acta 55: 3845-3867

Techmer KS, Ahrendt H, Weber K (1992) The development of pseudotachylyte in the Ivrea-Verbano Zone of the Italian Alps. Tectonophysics 204: 307-322

Therriault AM, Grieve RAF, Reimold WU (1997) The Vredefort Structure: original size and significance for geological evolution of the Witwatersrand Basin. Meteoritics and Planetary Science 32: 71-77

Thompson LM, Spray JG (1994) Pseudotachylitic rock distribution and genesis within the Sudbury impact structure. In Dressler BO, Grieve RAF, Sharpton VL (eds) Large Meteorite Impacts and Planetary Evolution, Geological Society of America Special Paper 293: 275-287

Trieloff M, Reimold WU, Kunz J, Boer RH, Jessberger EK (1994) ^{40}Ar-^{39}Ar thermochronology of pseudotachylites at the Ventersdorp Contact Reef, Witwatersrand Basin. South African Journal of Geology 97: 365-384

Uysal IT, Golding S, Glikson AY, Mory AJ, Glikson M (2001) K-Ar evidence from illitic clays of a Late Devonian age for the 120 km diameter Woodleigh impact structure, southern Carnarvon Basin, Western Australia. Earth and Planetary Science Letters 192: 281-289

Uysal IT, Golding SD, Glikson AY, Mory AJ, Glikson M, Iasky RP, Pirajno F (2002) Reply to "Comment on: 'K-Ar evidence from illitic clays of a Late Devonian age for the 120 km diameter Woodleigh impact structure, Southern Carnarvon Basin, Western Australia'". Earth and Planetary Science Letters 201: 253-260

Van der Bogert CH, Schultz PH (1997) Pseudotachylites in meteorites: Friction melting as an alternative to shock darkening. (abstract) Lunar and Planetary Science XXVII: 1477-1478

Van der Bogert CH, Schultz PH, Spray JG (2000) Similarities between melt-vein microstructures in naturally shocked and experimentally friction-melted ordinary chondrites. Abstract, 63rd Annual Meeting of the Meteoritical Society, Chicago, Meteoritics and Planetary Science 35 (supplement): p A162

Wenk HR (1978) Are pseudotachylites products of fracture or fusion? Geology 6: 507-511

White JC (1993) Shock-induced melting and silica polymorph formation, Vredefort Structure, South Africa. In: Boland JN, Fitzgerald JG (eds), Defects and Processes in the Solid State: Geoscience Applications. The McLaren Volume. Elsevier Science Publishers, Amsterdam: pp 69-84

Wise DU, Dunn DE Engelder JT, Geiser PA, Hatcher RD, Kish SA, Odom AL, Schamel S (1984) Fault-related rocks: suggestions for terminology. Geology 12: 391-394

The Mechanics of Pseudotachylite Formation in Impact Events

H. Jay Melosh

Lunar and Planetary Laboratory, University of Arizona, Tucson, AZ 85721, U.S.A. (jmelosh@lpl.arizona.edu)

Abstract. This paper presents a discussion of the basic constraints controlling the formation of pseudotachylites in the rapidly sheared rocks in the vicinity of a large meteorite impact. The prevailing opinion among many geologists is that pseudotachylites are formed by friction melting of rocks and/or shearing associated with differential shock compression of adjacent rock types. Several physical studies of friction melting have shown that, in theory, small amounts of movement (centimeters or less) are capable of producing very thin veins of melted rock. More realistic models suggest that irregularities on the sliding surface of the order of the grain size may still create primary melt veins up to a few millimeters thick. The principal mystery of pseudotachylite formation is not that friction can cause melting, but that it seems to form thick masses of it, meters to tens of meters wide. However, such thick masses ought to preclude melting by reducing the friction between sliding rock masses. I propose that one possible solution to this conundrum is that the melt produced by sliding on narrow shear zones is extruded into the adjacent country rock, thus keeping the sliding surfaces narrow, while creating thick accumulations of melt in adjacent low pressure zones that open at the end of the shear zones. For this mechanism to operate, the melted rock must be fluid enough to extrude from the shear zone during the time available during crater collapse. This places strong constraints on the viscosity and temperature of the melt. This model may be tested by future careful investigation of the geometry of pseudotachylite occurrences.

1 Introduction

S. J. Shand (1916) first introduced the term pseudotachylite (variant spelling "pseudotachylyte") to describe an enigmatic dark, flinty rock that apparently intruded granites near the Vaal River in what is now known as the Vredefort impact structure. Shand's early chemical analyses showed that the dark intrusive veins and pockets are compositionally similar to the enclosing granite, but he was unable to give a convincing explanation of their origin.

After Shand's discovery, nearly a century of study has clarified some, but not all, of the mysteries surrounding these strange rocks. It is widely supposed that pseudotachylites form as a result of frictional melting. They are found both in impact structures (Reimold 1995, 1998) and in undoubted tectonic settings (Sibson 1975). At Vredefort, it is easy to find thin, pseudotachylite-filled veins across which 1 to 10 cm offsets are clearly visible (Fig. 1). Although experimenters have succeeded in producing melts by direct high-speed frictional sliding (Spray 1995), concern still exists that some melt may be produced by shock compression and release (Reimold 1995), a concern that is bolstered by the formation of pseudotachylite-like melts in shock compression experiments (Fiske et al.

Fig. 1. Pseudotachylite veins cutting granite gneiss rock near Parys, South Africa. Note the clear 1 to 10 cm-scale offsets across the prominent veins. Rock hammer for scale. Courtesy of Elizabeth Turtle.

Fig. 2. Massive pseudotachylite in the same vicinity as Figure 1. Parallel sections of drillholes are approximately 0.5 m apart. Courtesy of Elizabeth Turtle.

1995; Kenkmann et al. 2000). Even in these experiments, however, there is a strong possibility that relative sliding between more coherent regions may have produced friction melt at the interface. Langenhorst et al. (2002) argue that melt veins in shocked single olivine crystals are almost certainly the result of shear melting, not fluctuations in shock pressure.

The problem is, that not all occurrences of pseudotachylite are narrow veins, as one would expect if they were directly produced by friction between adjacent blocks of rock (Figs. 2 and 3). Shand (1916) described large pockets with angular and even rounded inclusions of country rock on scales up to meters wide, a type of occurrence well illustrated by Killick and Reimold (1990) and Reimold and Colliston (1994). Similar occurrences of thick melt zones, in some cases up to 500 m wide and 11 km long, of so-called pseudotachylite, have been described at the Sudbury impact structure (Spray and Thompson 1995). As I argue below, the occurrence of such large melt masses poses a serious problem for any theory of frictional melting.

Lambert (1981) focused on the dike-like character of pseudotachylites. He distinguished two basic types of pseudotachylite occurrences, as did later work by Martini (1991) which was restricted to Vredefort. The first, type A, consist of a network of thin (less than 1 mm to several mm) veins that may have formed during the initial shock compression (Fiske et al.

Fig. 3. Typical outcrop of dense pseudotachylite near Figures 1 and 2. Both thin veins and angular pockets of dark pseudotachylite are visible. 9 cm long pocket knife in photo center for scale. Courtesy of Elizabeth Turtle.

1995). Note, however, that Reimold (1995; 1998) disputes Lambert's identification of Lambert's type A and A1 breccia types as true pseudotachylites, arguing instead that they are impact melts. Type B pseudotachylites are much larger dike or sill-like bodies that range from cm to several hundred m wide and one or more km long. These have been attributed to frictional sliding along very large displacement faults (Spray 1997) that formed well after shock compression ceased.

This interpretation, particularly of the thick, type B, pseudotachylites, raises a serious mechanical problem. Friction between large blocks of rock is certainly capable of raising the temperature of the interface to the melting point (Anderson 1951; Jeffreys 1942; McKenzie and Brune 1972; Spray 1995). However, once melt appears in abundance, it lubricates the interface and decreases the friction coefficient to the point that no further melt is produced (it is a pleasure to note that Harold Jeffreys (1942) was well aware of this problem, but offered no solution). So how are we to account for the reported occurrences of thick zones of pseudotachylites?

The goal of this paper is not to provide a definitive answer to this problem, but instead to define the physical constraints on frictional pseudotachylite formation. There are many aspects of rapid rock deformation that have not been explicitly considered in previous work. In a sense, this paper is a prolegomenon to a theory of pseudotachylite formation in which the main boundary conditions are set. I will also suggest possible directions in which answers to the principal problems may lie.

2 Frictional Heating of Rocks

The observation that friction can raise rocks to incandescence is an ancient one: People have been starting fires with flint and steel for centuries. However, it took the science of the Industrial Age to properly relate mechanical work, heat and velocity. Amontons and Coulomb first formulated the law of solid friction (Palmer 1949; Scholz 1990). They discovered that the resistance to sliding of one block over another is proportional to the normal load and independent of the area of contact. In modern terms, the basic law of friction states that the shear stress τ needed to move one block over another is proportional to the normal stress σ_N multiplied by the coefficient of friction μ:

$$\tau = \mu \sigma_N \tag{1}$$

For most rocks, μ is close to 0.85 at pressures below about 200 MPa. At higher pressure, a slightly more complicated expression applies; τ = 50 MPa + $0.6\sigma_N$. This relation is often called "Byerlee's law" and holds over a wide range of pressures, temperatures and compositions (Byerlee 1978), so long as the rock is actually solid. At high temperatures this coefficient declines as the rock softens (Ohnaka 1995; Stesky et al. 1974).

When slip occurs at a steady rate, the work done is dissipated as heat. Following directly from Newton's deduction that work equals force times distance, the power $\dot{W}$ dissipated per unit area on a sliding fault is given by

$$\dot{W} = \tau\dot{\delta} \tag{2}$$

where $\dot{\delta}$ is the differential rate of slip (= relative velocity) between the two blocks.

The total heat dissipated during sliding, as δ increases from 0 to its final value δ_f, is

$$W = \int_0^{t_f} \tau\dot{\delta}dt \tag{3}$$

For a constant shear stress τ, this is just $W = \delta\ \tau_f$.

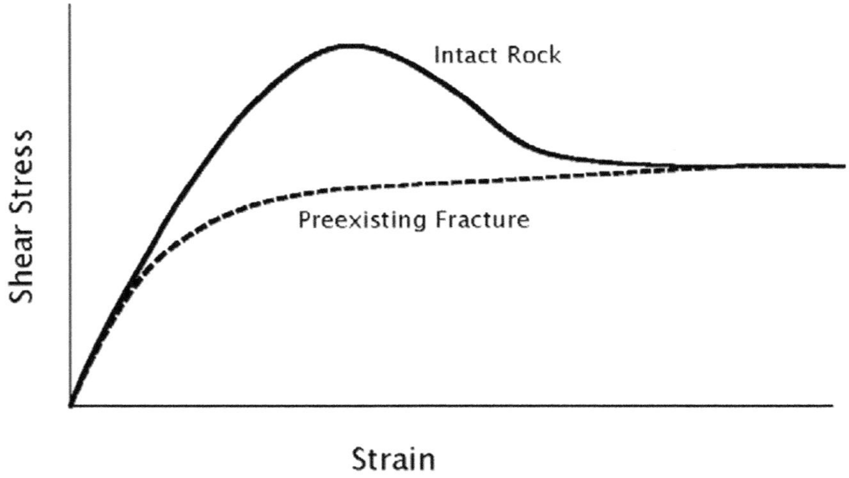

Fig. 4. Typical stress-strain curve for brittle rock. Intact rock (solid curve) shows a well-defined stress maximum, whereas fractured rock (dashed) does not. The peak stress for most rocks occurs at a strain of a few percent, and depends on ambient pressure and temperature, as well as rock composition (Jaeger and Cook 1969).

There has been some confusion in the geological literature about the relative importance of displacement or displacement rate in determining the onset of frictional melting. Certainly, slip at too slow a rate allows time for the heat generated to diffuse away from the shear zone and prevents the attainment of high temperatures. A flint struck too slowly yields nothing but cold flint fragments. However, if the displacement is too small, not enough energy is dissipated to heat the rock. Both displacement and velocity must be within the right range for melting to occur (Jeffreys 1942; Sibson 1975, 1977).

McKenzie and Brune (1972) proposed a simple model of shear heating (a model that I will call the "physicist's model" because of its unrealistic idealizations) to examine the ability of earthquakes to produce melt. Although the original model featured a sophisticated mathematical analysis, such complexity is not necessary to illustrate the basic tradeoffs that lead to melting. The model's chief idealization is that sliding begins on an infinitesimally thin shear zone. As heat is dissipated it flows laterally away from the heated fault plane, obeying Fourier's law (Turcotte and Schubert 1982). It spreads out over a region of width $y = \sqrt{\kappa t}$, where t is the time after heating began and κ is the thermal diffusivity, equal to about 10^{-6} m^2/sec for most rocks. If we assume that slip occurs at a constant rate $\dot{\delta}$ and that the energy dissipated fills the region of width y uniformly (this is not actually true—this is where the sophisticated mathematics comes in, but getting this exactly right is not essential to display the main structure of the solution), then the energy density (energy/unit volume) **E** in the slip zone is:

$$\mathbf{E} = \frac{\tau \dot{\delta} t}{y} = \tau \dot{\delta} \sqrt{\frac{t}{\kappa}} \quad (4)$$

This energy appears as heat. The energy density necessary to achieve melting is given approximately as $\mathbf{E} = \rho C_P (T_S - T_0)$, where ρ is the rock density, C_P is the heat capacity at constant pressure, T_S is the solidus temperature and T_0 is the initial temperature of the ambient rock. Inserting this definition into equation (4) and noting that $\delta = \dot{\delta} t$, the amount of slip δ_M necessary for the onset of melting is:

$$\delta_M = \frac{\kappa}{\dot{\delta}} \left[\frac{\rho C_P (T_S - T_0)}{\tau} \right]^2 \quad (5)$$

Beware that this equation implicitly assumes that the frictional stress τ is independent of temperature! If the frictional stress decreases as the temperature rises, the amount of slip will increase.

It is clear that the amount of slip needed for melting is inversely proportional to the slip velocity, so both factors play important roles. Inserting typical values of the parameters in equation (5) for rock, ρ = 3000 kg/m^3, C_P = 1000 J/kg and (T_S - T_0) = 1000 K, and further suppose that τ is equal to the lithostatic pressure of 3 km of rock, τ = 10^8 Pa, we find that, at a slip rate $\dot{\delta}$ of 1 m/sec, melting begins after a total slip of only 1 mm. This occurs just 1 millisecond after slip begins. However, note that at this time the slip zone is only y = 30 micrometers wide—much smaller than the grain size in almost any rock. This is why I call the model an "unrealistic idealization".

Whatever the shortcomings of this simple model, it does indicate that thin films of melt should form readily during rapid slip of rocks at crustal depths, and it illustrates the basic tradeoffs that must be met for melting to occur at all. Most of the rest of the discussion in this paper is a refinement of the concepts embodied in this model.

3
Brittle Deformation of a Large Rock Mass

Both numerical models and many geologists' depictions of the movement of the rocks beneath a large impact crater often idealize the motion as if it were the flow of a homogeneous fluid. Nothing could be further from the truth, as both modelers and geologists are usually well aware.

Cold, rapidly deformed rock masses respond to applied forces in a brittle manner. That is, deformation is not spread uniformly throughout the rock mass but is, instead, concentrated in narrow zones that separate largely intact regions. The ultimate reason for localization of deformation is simple: A narrow shear zone offers less resistance to sliding than the surrounding intact rocks (actually, the conditions necessary for localization in narrow shear zones are more complex; see Rudnicki and Rice 1975). Whether this lowered resistance is already present in the rock in the form of preexisting fractures (joints or faults), or whether it develops as a result of the large-scale deformation itself, is irrelevant for the final outcome.

At high temperatures and pressures rocks respond to applied stress more homogeneously. Such rocks are described as ductile. The brittle-ductile transition itself is of great interest in understanding the mechanics of earthquakes (Scholz 1990). Brittle faults near the Earth's surface broaden into shear zones at depths of many kilometers. These zones contain sheared rocks such as mylonites. Recognizable shear zones ultimately disappear into broad regions of distributed deformation at the greatest depths. When strain rates are sufficiently high, ductile materials also undergo a kind of strain localization known as adiabatic shear bands (Gruntfest 1963). These shear bands are typically much broader than those that form in response to brittle failure. They arise in any material whose resistance to flow decreases as the temperature increases (Turcotte and Schubert 1982). Adiabatic shear bands have their own large literature, but I explicitly exclude them from consideration here, although they could play an important role in the deformation of the very deepest and hottest rocks surrounding truly gigantic impact craters.

Figure 4 illustrates ultimate cause of localized deformation or strain in brittle rocks. In such rocks the temperature is so low that strain-rate-dependent processes (creep) are unimportant and the stress depends only on strain, not strain rate. This figure plots stress versus applied strain for compressional failure. As strain rises up to the elastic limit, typically less than a few percent, the stress rises monotonically with increasing strain. However, failure begins at somewhat larger strain, and the differential stress reaches a maximum (Jaeger and Cook 1969). In intact rocks, illustrated by the solid curve, failure involves the linking of small fractures that ultimately lower the rock strength to a value determined by the friction of rock-on-rock. The stress thus shows a peak, followed by a more or less rapid decline to a residual stress that regulates further deformation. Pre-fractured rocks, illustrated by the dashed curve, do not show this peak stress: they simply slide when the residual stress is reached. Localization occurs because of the difference in stress between intact and fractured rock.

Strain in brittle rocks is thus very heterogeneous. Nevertheless, it is often useful to describe the deformation in terms of an average strain that applies over a large volume of rock (large compared to the spacing between individual zones of sliding). In this case, average constitutive laws (the relation between stress and strain) can be used and the gross motion of the rock deduced. Nevertheless, as shown in Figure 5, a smooth average strain may very well imply an actual strain concentrated along very narrow zones.

The relation between the average strain, ε, and the displacement along a given fracture zone, δ, is determined by a quantity with the dimension of length. A simple relation links the displacement and strain (Jaeger and Cook 1969):

$$\delta = L\varepsilon \tag{6a}$$

The dimension L may be either the fracture length, or the average spacing between fractures (these two are usually nearly the same, numerically, as parallel fractures relieve differential stresses to distances comparable to their own length (Lachenbruc, 1961).

A corollary of equation (6a) is deduced by taking its time derivative:

$$\begin{aligned} \dot{\delta} &= \dot{L}\varepsilon + L\dot{\varepsilon} \\ \dot{\delta} &\approx L\dot{\varepsilon} \end{aligned} \tag{6b}$$

In compressive failure, where cracks cannot grow in their own plane, the term that includes $\dot{L}$ is negligible and only the second, simpler form will be used here (in brittle failure cracks may extend at the speed of sound and both terms must be retained).

For high strain rates, I will show later that Equation (6b) gives rise to a maximum length for any effective crack because of the finite rate at which elastic disturbances propagate. Nevertheless, in any real mass of rock, multiple length scales are likely to be present and a full analysis must sum over all these scales in some way.

Thus, in a region with an average strain of 3%, this could be the cumulative result of a 3 mm displacement on small fractures spaced 10 cm apart, or of 30 cm offsets on faults spaced 10 m apart. The average strain concept does not specify the size scale *L*, which must thus be determined by other considerations.

Joints are universal in crustal rocks, and it would be tempting to equate *L* with the spacing of pre-existing joint sets. However, due to limitations on the ability of pre-existing joints to accommodate arbitrary strains, it seems inevitable that new fractures must form in any large impact event.

Among metallurgists, it has long been understood that a minimum of five independent slip systems is required for a material to accommodate an arbitrary strain (Taylor, 1938). Precisely which crystallographic directions are used in any given metal varies from case to case and is irrelevant to the need for five independent systems: this constraint is purely geometrical. Most ductile f.c.c. and b.c.c. metals like iron, copper and aluminum possess the necessary five systems, whereas brittle metals such as zinc, which crystallizes into a hexagonal lattice, do not (Groves and Kelly 1963). Most geologic materials have such complicated lattice structures that they rarely possess five independent slip systems, except at very high temperatures (silver chloride is one of the few ionic solids that is ductile at room temperature). Cold rock, of course, generally does not possess a sufficient number of degrees of freedom. Application of these ideas to the deformation of large rock masses (Reches 1978) indicates that three or four sets of faults must be present to accommodate an arbitrary three-dimensional strain. Because it is unlikely that, in a given region, there will be three or four pre-existing joint sets with the right orientation to accommodate the strain imposed by impact crater excavation or collapse, we must anticipate that new fractures will develop during large-scale deformation.

The development of fractures under compression is a complex process. Unlike the much simpler case of tensile failure (Mode I cracks), for which Griffith formulated his famous fracture criterion (Lawn and Wilshaw 1975), shear cracks (Mode II and III cracks) cannot grow in their own plane. Thus, instead of the continued lengthening of a properly oriented tensile flaw, slip on a shear crack produces stress concentrations at its tip that initiate new cracks at a steep angle to the plane of the original crack (Ashby and Sammis 1990; Horii and Nemat-Nasser 1986). The new fractures propagate parallel to the most compressive stress axis and open as tensile fractures (Fig. 6). The formation of the familiar inclined shear zone in rock mechanics experiments seems to be the result of the growth of small flaws (mainly grain-boundary and internal cleavage cracks), linkage of the out-of-plane secondary cracks as they encounter one another, and their ultimate union into a complex, rough-surfaced band of shearing (Bombolakis 1973). This view of shear zone formation makes it clear how unrealistic the previous section's "physicist's model" of shear melt formation is, in which shearing is supposed to occur on an infinitely narrow plane.

4 Stresses and Strains in an Impact Event

The details of how stresses and strains vary as an impact crater opens and then collapses are highly complex functions of many variables, including details of the flow field, time after impact, rock structure and properties (Collins et al. 2004). Nevertheless, it is possible to give a general overview of the general magnitudes of stress and strain through this process. Impact crater formation is broadly divided into three major stages: Contact and compression, excavation, and collapse (Melosh 1989). Contact and compression, the stage in which the kinetic energy of the projectile is transformed into heat and kinetic energy of the target, is irrelevant to shear zone formation so long as the impact velocity exceeds a few km/s. Most rocks undergo at least partial bulk melting at shock pressures larger than about 50 GPa, a pressure that is exceeded in a basalt-on-basalt impact at a velocity of about 5.2 km/s. Since most impacts in the solar system occur at much higher velocities (the average asteroidal impact velocity on Earth is 17 km/s), I ignore this stage.

The excavation stage includes both the expansion and dissipation of the shock wave and the opening of the crater cavity. Although included in the same stage, these two processes occur at very different rates.

Stresses, strains, and strain rates are highest in the strong compressive shock wave that propagates away from the site of the initial impact. Unfortunately, it is not easy to say precisely how high they become. However, volume strains seldom exceed 2: In a very strong shock the heat added by compression increases the resistance to compression and thus limits the total volume strain achievable by shock (Zel'dovich and Raizer 1967). Strain rates are determined by the rate at which rock engulfed by the shock wave is compressed. This rate is related to the rise time, t_r, of the shock wave or, equivalently, to the width of the shock front w divided by the wave speed U. The widths of shock fronts in gases are tiny: w is comparable to the mean free path of an air molecule, about 10^{-6} m (Zel'dovich and Raizer 1967). However, due to inelastic processes, in metals and single-crystal oxides shock widths are broader, about 10^{-3} m (Swegle and Grady 1985). Study study of the rise time of shock waves from underground nuclear tests suggests that w in rocks may reach hundreds of meters, but depends inversely on the particle velocity v_p behind the shock front (Melosh 2003).

Direct gauge observations of the shock wave from the 62 kT PILEDRIVER nuclear detonation in the granitic Climax Stock at the Nevada Test Site showed that the rise time depends on the particle velocity v_p according to the relation:

$$t_r = (0.01 \pm .01)\left(\frac{15\, m/\sec}{v_p}\right) \sec \tag{7}$$

The particle velocity in these measurements ranged from 100 m/s down to 15 m/s. Because the particle velocity itself declines as approximately the –1.87 power of the range r from the explosion center (Melosh 1989), I estimate that in the impact of a projectile of radius a, the particle velocity declines as

$$v_p(r) = \frac{v_i}{2}\left(\frac{a}{r}\right)^{1.87} \tag{8}$$

where v_i is the impact velocity. Combining equations (7) and (8), the strain rate in the shock is given by

$$\dot{\varepsilon} = \frac{v_p}{w} = \frac{v_p}{V_P t_r} \cong \frac{v_i^2}{0.5 V_P}\left(\frac{a}{r}\right)^{3.74} \sec^{-1} \tag{9}$$

where V_P is the compressive wave speed in granite, about 5.1 km/s in this case. Inserting some typical values in this equation, the strain rate $\dot{\varepsilon}$ at a particle velocity of 100 m/s (the highest for which data exists) is about 13 s^{-1}. If this can be extrapolated back to a particle velocity of 3.6 km/s, at which the shock pressure reaches the 50 GPa limit in granite, then $\dot{\varepsilon}$ exceeds 10^4 s^{-1}. At the other end of the process, using equation (9) to estimate the strain rate at the edge of the transient crater at $r \sim 20\, a$ for a 20 km/s impact, $\dot{\varepsilon}$ would be approximately 2 s^{-1}. Thus, strain rates in the compressive shock wave itself range from more than 10,000 s^{-1} down to a few s^{-1}.

The corresponding strains can be computed by multiplying the strain rate times the rise time. Strain magnitudes range from 0.7 at the highest velocity to 0.01 near the crater rim.

Stress may be estimated from the second Hugoniot equation, which states that the pressure jump across the shock $(P-P_0) = \rho_0\, v_p\, V_P$, where ρ_0 is the uncompressed density of the rock. These stresses range from 50 GPa, just below bulk melting of the rock, down to about 1 GPa at the crater rim.

Estimation of the strain rates in the excavation flow is more straightforward. It is well established that, during most of the excavation

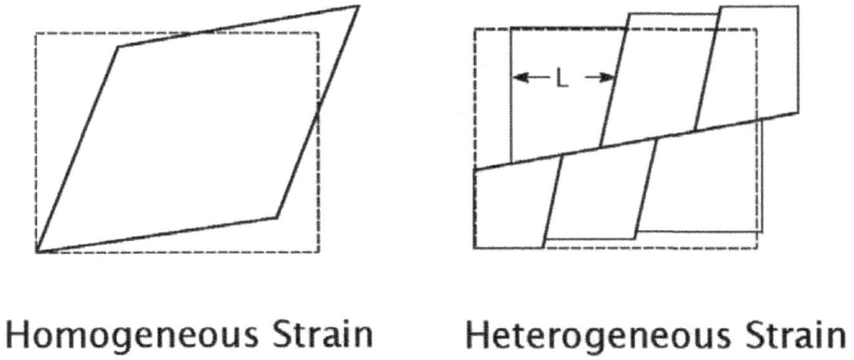

Fig. 5. Homogeneous versus heterogeneous strain. Although the total strain is the same in both cases, the strain is localized across a number of narrow shear zones in the heterogeneous strain case. The characteristic spacing of fractures is given by length parameter *L*. Note that a minimum of two independent slip systems is necessary to accommodate arbitrary strain in the two dimensions shown here.

stage, the radius *R* grows as a power of the time, $R(t) \sim t^{2/5}$ (Melosh 1989). The entire duration of the growth process is given by the formation time of a gravity-dominated crater, $T_f = \sqrt{D/g}$, where *D* is the diameter of the transient crater and *g* is the acceleration of gravity.

The highest strain rate is near the rim of the growing crater. The overall strain is of the order of *R/a* and the maximum strain rate at any given time *t* is determined from the ratio of the derivative of the radius to the radius itself:

$$\dot{\varepsilon} = \frac{\dot{R}}{R} = \frac{21}{5t} \tag{10}$$

This strain rate applies within a distance R of the growing crater and falls as some power with greater distances (approximately given by the famous Z-model; Maxwell (1977)). Assuming that the power law growth model holds from the edge of the projectile, $R = a$, to the final transient crater diameter at $R \sim 20\ a$, maximum and minimum strain rates can be derived:

$$\dot{\varepsilon}_{max} = \frac{21800}{5\ T_f}$$
$$\dot{\varepsilon}_{min} = \frac{2\ 1}{5T_f} \tag{11}$$

The formation time for a crater the size of Vredefort, in which the transient crater diameter $D \sim 100$ km (Turtle and Pierazzo 1998), is of the order of 100 sec and so the strain rate ranges from about 7 s^{-1} down to a low of 4 x 10^{-3} s^{-1}. Strain rates in the excavation flow are thus generally lower than in the shock wave itself, but there is no large gap between the two regimes, except to note that the highest strain rates in both cases occur at the smallest radii. The innermost rocks, close to the impact site, always experience higher strain rates and larger strains than the more distant rocks.

Finally, during the collapse portion of the cratering flow, the overall time scale is the same as the formation time T_f. Because the transient crater starts out with a depth/diameter ratio of about 0.3 and collapses nearly flat (for large craters), the strain is of order 0.3 and so the strain rate is nearly the same as $\dot{\varepsilon}_{min}$ in equation (11). Although the magnitude of the strain rate during collapse is thus similar to that during excavation, the *sign* of the strain rate is reversed. Cratering motions are grossly outward during excavation, but inward during collapse (O'Keefe and Ahrens, 1999). As a result, strains that develop during excavation may be partly, or wholly, reversed during collapse. It is thus important to distinguish the maximum strain that occurs during the overall cratering process from the final strain after all motion ceases.

An estimate of the stresses that develop during the final stages of crater excavation and collapse is not straightforward. On the surface, I could simply accept the results of rock mechanics tests on real rocks (e.g., Fig. 4) and relate the stress directly to the strain. However, it has been known for some time that, if this is done, the observed collapse of large impact craters cannot be reproduced (Melosh and Ivanov 1999). It is clear that some kind of strength degradation must occur in the near vicinity of the crater (less than one crater radius) that drastically reduces the strength of the surrounding rock during collapse. The apparent strength must drop to only a few MPa (Melosh 1977) for most of the duration of crater collapse. I have proposed the idea that strong vibrations created by scattering of the shock wave may briefly reduce the strength of the rock by the process of "acoustic fluidization" (Melosh, 1979). Whether this is the detailed mechanism or not, it is clear that most of the strain during late excavation and collapse must take place at stress levels nearly 100 times lower than those expected from normal rock friction. For this reason, many of the results below must be taken with caution. However, in the final stages of flow when the vibrational energy has dissipated but while the rocks are still in motion, the normal friction coefficient may again assert itself and some high-stress shearing may take place, albeit at strains only equal to a fraction of those computed from the general motion of the rock debris.

Well outside the transient crater rim, no strength degradation occurs and large stresses may develop in this region in response to the mass deficiency in the crater interior. The rings that develop around large impact basins probably form in this regime. This is also the region where massive pseudotachylites have been reported (Spray and Thompson 1995).

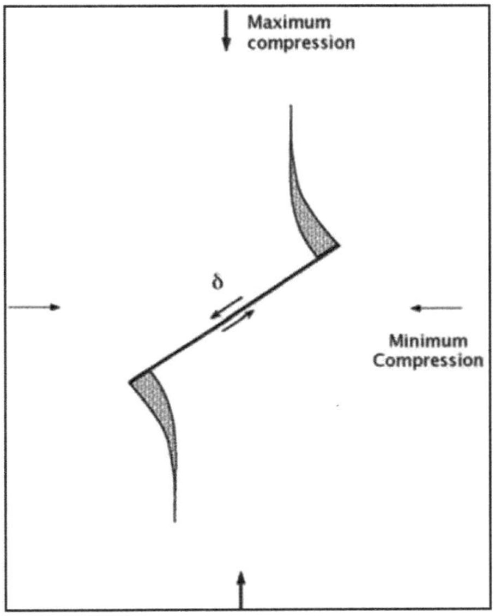

Fig. 6. Unlike tension cracks, shear cracks cannot grow in their own plane and so have difficulty extending. Instead, stresses at their tips open tension cracks at a steep angle to the original shear fracture and parallel to the most compressive stress direction in this idealized picture of shear fault growth.

5
Fault Slip in a Deforming Rock Mass

A large, brittle, mass of rock under compression deforms internally along a large number of variously oriented, discrete slip zones that, together, contribute to the average strain of the mass. Each of these slip zones is a potential site for frictional melting. Evaluation of this potential requires information about the slip zones themselves and both the maximum displacement and velocity of slip across them. Tectonically generated pseudotachylites often form in well-defined orientations with respect to the regional stress field (Grocott 1981; Sibson 1975). No such preferred orientations have been reported for the Vredefort pseudotachylites (Killick and Reimold 1990). This may be the result of the rapid, reversing motions (and therefore stresses) that characterize impact crater excavation and collapse.

There is a well-defined upper limit to the speed at which a fault can slide. This limit is set by the rate at which the elastically strained material containing the fault can feed energy to the slipping surface. For a fault of length L, the minimum time over which slip can occur is simply L/V_S, where V_S is the shear wave velocity in the material (because most internal strain is by shear, rather than by volume change, this is the appropriate speed; the corresponding shear modulus is $G = \rho V_S^2$). Noting, from Equation (6b), that the slip velocity $\dot{\delta} = L\dot{\varepsilon}$, the strain rate is simply the maximum strain at failure, $\varepsilon_f = \tau / G$, divided by the slip time L/V_S. The resulting maximum slip velocity is independent of L:

$$\dot{\delta}_{max} = \frac{\tau}{G} V_S = \frac{\tau}{\rho V_S} \tag{12}$$

The total amount of slip occurring at this rate is simply $\dot{\delta}_{max}$ multiplied by the slip time L/V_S, or $\delta_{fast} = \tau L / G$.

Subsequent to this fast elastic failure, the displacement δ may continue to increase at a slower rate determined by the overall strain rate given by Equation (6b); see Figure 7. This occurs if, after the initial rapid failure (the "bang" often noted in rock mechanics experiments at the moment that fracture occurs), the rock surface slides stably as strain increases. On the other hand, in many experiments it is observed that sliding of rock-on-rock occurs in a "stick-slip" regime (Scholz 1990), in which the average strain rate is accommodated by a large number of rapid elastic jumps, each occurring at a rate determined by Equation (12). In this case, all of the strain occurs at the slip rate given by (12), although in intermittent fashion. The number of jumps is approximately given by $n = \varepsilon G / \tau = \varepsilon/\varepsilon_f$.

Equation (12) is in good agreement with many observations of earthquakes (Scholz, 1990). In earthquakes, the tectonic strain rate is very low, typically 10^{-13} sec^{-1} or less, so that all of the high velocity motion occurs during the rapid elastic unloading during the earthquake itself. Because of the low average strain rates, the interval between elastic slip events may be decades or even millennia. Stress drops τ are typically about 10^7 Pa in earthquakes (Hanks 1977), from which Equation (12) predicts a slip velocity of order 1 m/sec, in good agreement with many seismic observations. The reason that earthquake stress drops are so low is an enduring mystery (Zoback et al. 1987), but if the more rapid loading of rock masses near an impact results in a stress change approaching the high-pressure rock strength of 10^8 to 10^9 Pa, then slip rates $\dot{\delta}_{max}$ may approach 100 m/s.

The existence of a speed limit for fault slip puts a constraint on the maximum length possible for a coherently slipping fault. Thus, combining (12) and the derivative of (6), results in

$$L_{max} = \frac{\tau \, V_S}{G \, \dot{\varepsilon}} = \frac{\varepsilon_f}{\dot{\varepsilon}} V_S \tag{13}$$

Equation (13) essentially states that faults cannot be longer than the distance that sound travels in the time it takes to exceed the yield stress.

This inverse dependence of fault length on strain rate means that, early in the cratering process when strain rates are very high, faults are short and closely spaced (assuming that the average fault spacing is also of order L). In the shock, strain rates of 10^4 sec^{-1} imply that fault length and spacing may be only centimeters. During excavation, lengths of hundreds of meters become possible, while during the final phases of collapse, fault spacing can be as large as tens of kilometers. It is thus possible to have early, high strain rate fault networks overprinted by later, coarser networks. This estimate supports the common observation that large fault networks often arise from the linkage of much smaller faults (Segall and Pollard 1983), although it is not clear that fault network formation at slow tectonic rates is the same as rock fragmentation at the much higher strain rates near an impact.

6
Friction Melting in Shear Zones

The amount of friction melt around an impact crater can be computed, now that the displacement and slip velocity on internal shear surfaces are known as functions of the gross deformation of the rock mass. Instead of estimating the shear zone width from thermal diffusion, as in the "physicists' model" described above, I will take the width of the shear zone r to be some fraction of the grain size of the rock itself. I thus suppose that r is of the order of millimeters to a centimeter. This assumption is supported by theoretical models of compressional failure (Ashby and Sammis 1990), and by direct observation in experiments that produced frictional melt (Spray 1995; Tsutsumi and Shimamoto 1997).

For present purposes, I assume that r is so large that thermal diffusion can be completely neglected on the time scale of crater collapse. This may not be quite correct: The thermal time constant of a 1-millimeter-wide shear zone is 1 second, and that of a 1 centimeter shear zone is 100 seconds (Sibson, 1975). Nevertheless, I presume that thermal conduction does not play any truly essential role in the friction heating process. Some of my results may thus be off by a factor of 2 or so. Other factors that may be important in controlling the shear zone width r are abrasion and thermal spallation in the very high thermal gradients adjacent to the zone (Moore and Sibson 1978). Both of these factors only serve to increase r and therefore make melting more difficult and thermal diffusion less relevant.

Repeating the arguments that led to the estimate of melting in the "physicists' model", but now with a fixed shear zone width r, one equates the energy density at the onset of melting, $\mathbf{E} = \rho C_P (T_S - T_0)$, to the energy density generated by frictional slip at the end of the elastic rebound, $\tau \delta_{fast} / \mathrm{r} = \tau^2 L / G\mathrm{r}$. The resulting equality yields a minimum fault length, L_{min}, which is just capable of generating friction melt:

$$L_{min} = \left[\frac{\rho C_P (T_S - T_0)}{\tau^2 / G} \right] \mathrm{r} \tag{14}$$

The dimensionless factor in brackets is effectively the ratio between the energy density necessary to cause melting and the elastic strain energy in the rock at failure. It is thus always much greater than 1. If τ is as large as 10^9 Pa, then this factor is about 30, whereas for more typical stresses of 3 x 10^8 Pa, this factor rises to about 300. The minimum fault length required to form frictional melt is thus between 3 and 30 cm, for r of 1 mm, to 3 and 30 m for r of 1 cm. The corresponding displacements are from 10% to 3% of the fault lengths.

Equation (14) also permits a crude estimate of the volume fraction of friction melt in the rock. Because the melt is confined to a zone of length L_{min} and width r, and faults are spaced at intervals of approximately L_{min}, the ratio between the melt volume r L_{min}^2 and the rock volume L_{min}^3 is simply r $/ L_{min}$. This ratio thus ranges between 1/30 and 1/300, so that friction melt volume percentages range from about 3% down to about 0.3%. Although faults longer than L_{min} may also contribute melt, the fractional abundance of melt is always largest for the shortest and most abundant faults, so consideration of much longer faults will not greatly alter this estimate.

This low volume of melted rock gives little hope that pseudotachylite formation will alter the gross rheologic properties of the rocks. The small volume of melt is trapped in a much larger volume of cold country rock and so will cool rapidly and the rock will remain strong.

7
The Pseudotachylite Conundrum

We are now ready to face the central mystery of pseudotachylite formation. We have seen that, at strain rates and under stresses likely to occur in the vicinity of large impact craters, it is plausible that shear zones with widths of mm to cm may begin to develop melt by friction (narrower veins have been observed in meteorites, but these are likely produced in the very high strain rates occurring in shock waves). I have deduced a minimum fault length, L_{min}, on which friction melts just form. But what about the larger faults that can easily attain the melt threshold and then keep sliding? Do these continue to produce larger and larger quantities of melt? To answer these questions we must address the difficult subjects of the rheology of partially molten granitic rocks (Petford, 2003) and the dynamics of rapidly sheared rock surfaces, both of which pose more questions than answers at the moment.

It seems obvious that, once the solidus temperature T_S of the rock is reached and melt starts to appear, the resistance to sliding should decrease (Jeffreys 1942; Sibson 1975). But because the dynamics of fault slip ensures a nearly constant slip velocity $\dot{\delta}$, a lowered resistance means less heat dissipation and hence a slower heating rate. Friction melting should be self-limiting.

The only relevant experiment on the friction coefficient during melting (Tsutsumi and Shimamoto 1997) showed complicated results: Just before melt appeared on the sliding surfaces, the friction coefficient declined. However, as melt appeared the friction coefficient increased dramatically. The experimenters noted that during the experiments melt actually sprayed out of the shear zone, suggesting that the expected self-limiting feature could not develop because the melt flowed away as quickly as it formed. Unfortunately, it is not clear just what was melting in these experiments. The rock contained a mixture of plagioclase, pyroxene, hornblende and biotite, all of which probably melt at different temperatures: Under the short time scales accompanying frictional melting there is little time for chemical equilibration.

What about deeply buried shearing surfaces? In the experiments of Tsutsumi and Shimamoto the sliding surface was only about 1 cm away from the edge of the sample and melt could thus escape readily. Is a similar escape possible for sliding surfaces many km below the surface?

It is easy to estimate the volume discharge $\dot{Q}$ (per unit width of the sheet) of fluid with Newtonian viscosity η from a narrow sheet of thickness r (Batchelor 1970):

$$\dot{Q} = -\frac{1}{12}\frac{\mathrm{r}^3}{\eta}\frac{\partial P}{\partial x} \tag{15}$$

where P is the pressure and x is in the plane of the sheet.

Applying this idea to the crack illustrated in Figure 6 suggests that perhaps the friction melt appearing on a fault of length L (larger than the minimum length) might flow along the surface of the fault and accumulate in the wedge-shaped pockets at the fault tip. Since the crack-tip pockets open in a tensional (mode I) configuration, the pressure in them during sliding is almost zero, thus providing a strong pressure gradient out of the fault as well as a volume into which friction melt might flow. Indeed, the thicker pseudotachylite occurrences along fault planes often have the appearance of wedge-shaped pockets (Killick and Reimold 1990; Shand 1916). Sibson (1975) also noted a distinction between narrow sliding zones and adjacent "injection dikes" that suggested tensional failure. The question then becomes, can the melt flow out of the shear zone fast enough to keep the friction coefficient high? The answer is clearly a function of the fault length: If the fault is too long, the discharge will be too small to remove a substantial volume of melt during the time available for sliding. On the other hand, the duration of rapid sliding increases with fault length.

To estimate the importance of this effect, compare the volume of melt squeezed out of the shear zone, $\dot{Q}L/V_S$, to the volume of melt remaining in the zone, $\mathrm{r}\,L$, per unit length of the zone in the case of fault slip at the maximum rate. The pressure gradient is of order $\tau/(L/2)$. The ratio, **R**, between these two volumes is:

$$\mathbf{R}\,(fast) = \frac{volume\ of\ melt\ squeezed\ out}{volume\ of\ shear\ zone} = \frac{1}{6}\frac{\mathrm{r}^2\tau}{\eta V_S L} \tag{16}$$

Turning this relation around, the minimum viscosity η necessary for **R** to exceed 1 in this fast slip regime is given by.

$$\eta_{min}(fast) = \frac{1}{6}\frac{\mathrm{r}^2\tau}{V_S L} \tag{17}$$

Inserting rather generous estimates of $\mathrm{r} = 1$ cm, $\tau = 3 \times 10^8$ Pa and $V_S = 3000$ m/s, reveals that the viscosity must be less than $1/L$ (m) Pa-s. Thus, even 1 m long faults require melts with viscosity of only about 1 Pa-s, about equivalent to motor oil. A 100 m long fault would require viscosities similar to that of water. This appears to pose a serious problem for this mechanism. Because of the high viscosities of silica-rich melts, it seems to rule out this mechanism for the granitic rock that underlies the Vredefort and Sudbury craters, although the high initial temperatures of the rocks deep beneath these craters (500-900 °C) may help make the lavas more fluid.

The only obvious way to redeem this process is to note that the time interval assigned to the outflow was that associated with the fast elastic

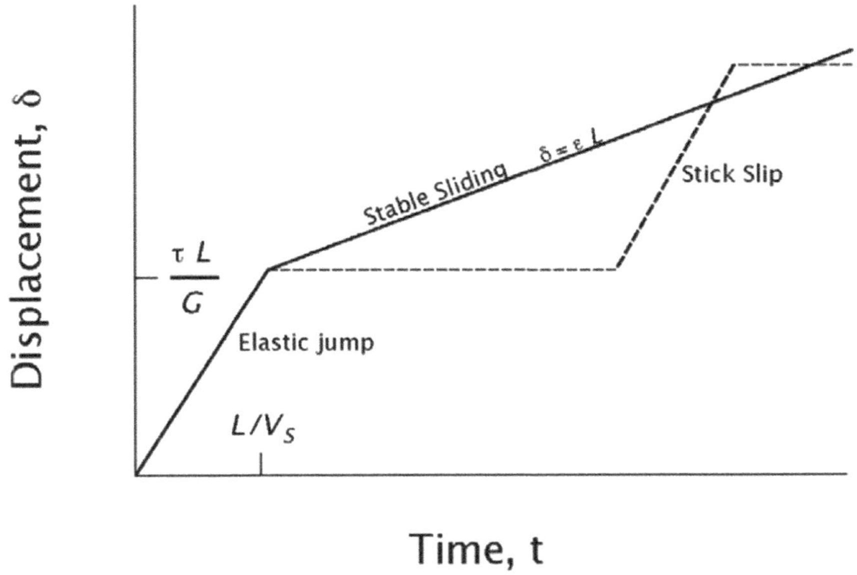

Fig. 7. When a shear zone fails, the initial displacement increases at a rate determined by the speed at which the adjacent rock can feed elastic energy into the developing shear zone. Later displacement may be governed either by stable sliding at a rate determined by the deformation of the enclosing rock mass, or by stick-slip jumps that produce the same average displacement rate.

slip of the fault. If, as I suggest in Fig. 7, slip continues for a much longer time (either in steady or stick-slip mode), then perhaps viscous squeeze-out of the melt might become important. Another possibility, suggested by observations of tectonic pseudotachylites (Sibson 1975) is that, instead of melt flowing to the end of the slipping fault, it escapes at numerous locations along the face of the fault. In this case the appropriate value for the distance of squeeze-out is much less than L, while the timescale for flow is still determined by L. This does improve the situation, but field observations support this proposal only for tectonic fault zones (Sibson 1975). If, instead of the elastic slip time, L/V_S, I substitute the crater formation time T_f (which is the longest time scale available during crater formation and collapse), the ratio of the volume squeezed out to that in the shear zone becomes

$$\mathbf{R}\,(slow) = \frac{1}{6}\frac{\mathrm{r}^2\tau}{\eta L^2}T_f \tag{18}$$

From which the estimate of the minimum viscosity to reach R ~ 1 rises to

$$\eta_{min}(slow) = \frac{1}{6}\frac{\mathrm{r}^2\tau}{L^2}T_f \tag{19}$$

For a 100 km diameter transient crater T_f is about 100 seconds and, for the previous estimates of r and τ, the minimum viscosity is 5 x $10^5/L^2$(m) Pa-s. For fault lengths of a few meters this suggests that squeeze-out is possible even in granitic rocks. The proviso is that it must occur on the long time scale of crater collapse, not on the scale of elastic fault slip. It also suggests that this process is important only for the very largest craters, for which both τ and T are large enough to provide both the stress and time for relatively viscous granitic melts to squeeze out of frictionally slipping cracks. This long time scale also precludes slip zones r as narrow as 1 mm, as they would completely solidify on the time scale for melt expulsion.

Applying Equation (19) to the experiments of Tsutsumi and Shimamoto (1997), in which L was about 1 cm and r was about 1 mm, the minimum viscosity for the extraction of melt over the ca. 10 s duration of melting is 3 x 10^4 Pa-s, well within the range of possibility for a gabbroic melt. Thus, the simple estimate for the occurrence of melt expulsion from a sliding surface does seem to agree with the limited experimental data in hand. The only other controlled experiment on friction melt formation (Spray 1995) used a much more silica-rich rock (Westerly Granite) in a configuration with dimensions similar to those of Tsutsumi and Shimamoto. In that experiment Spray did not report any movement of the melt, consistent with its probably much greater viscosity. A poorly controlled "experiment" by Killick (1990) reported melt production by an unlubricated drill bit, but offered too little information to estimate melt mobility.

8 Conclusions

The detailed mechanics of how pseudotachylites arise is not obvious. Under crustal conditions frictional melts arise rather easily during rapid deformation of brittle rocks. Friction between large blocks of rock is certainly capable of raising the temperature of the interface to the melting point. However, these melts are most likely to be very thin (mm to cm) and in the very act of thickening they tend to shut off the very process that created them.

The maximum shear stress τ_{max} that can be sustained across a viscous layer of viscosity η and thickness r shearing at a velocity $\dot{\delta}$ is

$$\tau_{max} = \eta\dot{\delta} / \mathrm{r} \tag{20}$$

If the viscosity η is sufficiently large, it is possible that even fully molten material could sustain a shear stress comparable to that implied by frictional sliding. This is, in fact, the basis for the construction of the popular "strength envelope" curves that define different stress regimes in the lithosphere (Scholz 1990). For example, for a frictional stress τ_{ma} of 3 x 10^8 Pa, melt in a 1 cm thick shear zone slipping at 10 m/sec would exert a resistance to shear equal to the frictional stress if the viscosity of the material in the shear zone exceeds 3 x 10^5 Pa-sec. Not only is this within the range of possibility for silica-rich melts not far above the solidus, but the presence of abundant crystal fragments may also raise the viscosity of the mixture to a much higher value than that of the liquid alone (the enhancement factor is given approximately by the Einstein-Roscoe relation; McBirney and Murase 1984; Petford 2003). We can define a critical viscosity for the validity of this friction-viscous shear equality:

$$\eta_* = \tau_{max} \mathrm{r} / \dot{\delta} \tag{21}$$

The problem with this idea is that that viscosity of silicate melts depends very strongly on temperature: A temperature increase of 100° C is usually enough to drop the viscosity by an order of magnitude. Similarly, a substantial increase in r or decrease in sliding rate $\dot{\delta}$ is likely to virtually stop energy dissipation in the shear zone.

This is the principal mystery of pseudotachylite formation: the problem is not producing melt by friction, but producing enough of it. It is possible that presently unknown rheologic properties may arise during sliding in hot, crystal-laden partially molten rock that solve this difficulty. Stick-slip motion may be one way to bridge this gap. However, such mechanisms have yet to be spelled out in detail.

In this paper I have suggested that one way to keep τ_{max} high is to keep r low. This can occur if the melt's viscosity is sufficiently low that much of the melt that forms is extruded from the plane of sliding into adjacent open fissures and pockets. The proposed appearance of both thin veins and larger, intrusive masses of pseudotachylite is at least in qualitative agreement with observations.

The mechanics of pseudotachylite formation deserves a great deal more attention than it has received to date. I hope that this short prolegomena will motivate such further work.

Acknowledgement

The author has benefited greatly from discussions with Boris Ivanov. This work was supported by NASA grant NAG5-11493. Uwe Reimold, Thomas Blenkinsop, and Christian Koeberl provided helpful reviews.

References

Anderson EM (1951) The Dynamics of Faulting. Oliver and Boyd, Edinburgh, 206 pp
Ashby MF, Sammis CG (1990) The damage mechanics of brittle solids in compression. Pure and Applied Geophysics 133: 489-521
Batchelor GK (1970) An Introduction to Fluid Dynamics. Cambridge University Press, Cambridge, 615 pp
Bombolakis EG (1973) Study of the brittle fracture process under uniaxial compression. Tectonophysics 18: 231-248
Byerlee J (1978) Friction of Rocks. Pure and Applied Geophysics 116: 615-626
Collins GC, Melosh HJ, Ivanov BA (2004) Damage and deformation in numerical impact simulations. Meteoritics and Planetary Science 39: 217-231
Fiske PS, Nellis WJ, Lipp M, Lorenzana H, Kikuchi M, Syono Y (1995) Pseudotachylites generated in shock experiments: Implications for impact cratering products and processes. Science 270: 281-283
Grocott J (1981) Fracture geometry of pseudotachylite generation zones: A study of shear fractures formed during seismic events. Journal of Structural Geology 3: 169-178
Groves GW, Kelly A (1963) Independent slip systems in crystals. Philosophical Magazine 8: 877-887
Gruntfest IJ (1963) Thermal feedback in liquid flow: Plane shear at constant stress. Transactions of the Society of Rheology 7: 195-207
Hanks, TC (1977) Earthquake stress drops, ambient tectonic stresses and stresses that drive plate motion. Pure and Applied Geophysics 115: 441-458
Horii H, Nemat-Nasser S (1986) Brittle failure in compression: Splitting, faulting and brittle-ductile transition. Philosophical Transactions of the Royal Society of London A 319: 337-374
Jaeger JC, Cook NGW (1969) Fundamentals of Rock Mechanics. Chapman and Hall, London, 515 pp
Jeffreys H (1942) On the mechanics of faulting. Geological Magazine 79: 291-295
Kenkmann T, Hornemann U, Stöffler D (2000) Experimental generation of shock-induced pseudotachylites along lithological interfaces. Meteoritics and Planetary Science 35: 1275-1290
Killick AM (1990) Pseudotachylite generated as a result of a drilling "burn-in". Tectonophysics 171: 221-227

Killick AM, Reimold WU (1990) Review of the pseudotachylites in and around the Vredefort 'Dome', South Africa. South African Journal of Geology 9: 350-365
Lachenbruch, AH (1961) Depth and spacing of tension cracks. Journal of Geophysical Research 66; 4273-4292
Lambert P (1981) Breccia dikes: Geological constraints on the formation of complex craters. In: Schultz PH, Merrill RB (eds) Multi-ring Basins, Proceedings of Lunar and Planetary Science 12A. Pergamon Press, New York, pp 59-78
Langenhorst F, Poirier J-P, Deutsch A, Hornemann U (2002) Experimental approach to generate shock veins in a single crystal olivine by shear melting. Meteoritics and Planetary Science 37: 1541-1553
Lawn BR, Wilshaw TR (1975) Fracture of Brittle Solids. Cambridge University Press, New York, 204 pp
Martini JEJ (1991) The nature, distribution and genesis of coesite and stishovite associated with the pseudotachylite of the Vredefort Dome, South Africa. Earth and Planetary Science Letters 103: 285-300
Maxwell, DE (1977) A simple model of cratering, ejection, and the overturned flap. In: Roddy DJ, Pepin RO, Merrill RB (eds) Impact and Explosion Cratering, pp 1003-1008. Pergamon, New York
McBirney AR, Murase T (1984) Rheological properties of magmas. Annual Reviews of Earth and Planetary Science 12: 337-357
McKenzie D, Brune JN (1972) Melting on fault planes during large earthquakes. Geophysical Journal of the Royal Astronomical Society 29: 65-78
Melosh HJ (1977) Crater modification by gravity: A mechanical analysis of slumping. In: Roddy DJ, Pepin RO, Merrill RB (eds) Impact and Explosion Cratering, pp. pp. 1245-1260. Pergamon Press, New York
Melosh HJ (1979) Acoustic fluidization: A new geologic process? Journal of Geophysical Research 84: 7513-7520
Melosh HJ (1989) Impact Cratering: A Geologic Process. Oxford University Press, New York. 245 pp
Melosh HJ (2003) Shock viscosity and rise time of explosion waves in geologic media. Journal of Applied Physics 94: 4320-4325
Melosh HJ, Ivanov BA (1999) Impact crater collapse. Annual Reviews of Earth and Planetary Science 27: 385-415
Moore HE, Sibson RH (1978) Experimental thermal fragmentation in relation to seismic faulting. Tectonophysics 49: T9-T17
O'Keefe JD, Ahrens TJ (1999) Complex craters: Relationship of stratigraphy and rings to the impact conditions. Journal of Geophysical Research 104: 27,091-27,104
Ohnaka M (1995) A shear failure strength law of rock in the brittle-plastic transition regime. Geophysical Research Letters 22: 25-28
Palmer F (1949) What about friction? American Journal of Physics 17: 181-187
Petford N (2003) Rheology of granitic magmas during ascent and emplacement. Annual Reviews of Earth and Planetary Science 31: 399-427
Reches Z (1978) Analysis of faulting in three-dimensional strain field. Tectonophysics 47: 109-129
Reimold WU (1995) Pseudotachylite in impact structures--generation by friction melting and shock brecciation?: A review and discussion. Earth Science Reviews 39: 247-265
Reimold WU (1998) Exogenic and endogenic breccias: A discussion of major problematics. Earth Science Reviews 43: 25-47

Reimold WU, Colliston WP (1994) Pseudotachylites of the Vredefort Dome and the surrounding Witwatersrand Basin, South Africa. In: Dressler BO, Grieve RAF, Sharpton VL (eds) Large Meteorite Impacts and Planetary Evolution, Geological Society of America Special Paper 293, Boulder, CO, pp. 177-196

Rudnicki JW, Rice JR (1975) Conditions for the localization of deformation in pressure-sensitive dilatent materials. Journal of the Mechanics and Physics of Solids 23: 371-394

Scholz CH (1990) The Mechanics of Earthquakes and Faulting. Cambridge University Press, Cambridge. 439 pp

Segall P, Pollard DD (1983) Nucleation and growth of strike slip faults in granite. Journal of Geophysical Research 88: 555-568

Shand SJ (1916) The pseudotachylyte of Parijs (Orange Free State) and its relation to 'trap-shotten gneiss' and 'flinty crush-rock'. Quarterly Journal of the Geological Society of London 72: 198-221

Sibson RH (1975) Generation of pseudotachylyte by ancient seismic faulting. Geophysical Journal of the Royal Astronomical Society 43: 775-794

Sibson RH (1977) Fault rocks and fault mechanisms. Journal of the Geological Society of London 133: 191-213

Spray JG (1995) Pseudotachylyte controversy: Fact or friction? Geology 23: 1119-1122

Spray JG (1997) Superfaults. Geology 25: 579-582

Spray JG, Thompson LM (1995) Friction melt distribution in a multi-ring impact basin. Nature 373: 130-132

Stesky RM, Brace WF, Riley DK, Bobin PY (1974) Friction in faulted rock at high temperature and pressure. Tectonophysics 23: 177-203

Swegle JW, Grady DE (1985) Shock viscosity and the prediction of shock wave arrival times. Journal of Applied Physics 58: 603-701

Taylor GI (1938) Plastic strain in metals. Journal of the Institute of Metals 62: 307-324

Tsutsumi A, Shimamoto T (1997) High-velocity frictional properties of gabbro. Geophysical Research Letters 24: 699-702

Turcotte DL, Schubert G (1982) Geodynamics: Applications of Continuum Physics to Geological Problems. John Wiley and Sons, New York. 450 pp

Turtle EP, Pierazzo E (1998) Constraints on the size of the Vredefort impact crater from numerical modeling. Meteoritics and Planetary Science 33: 483-490

Zel'dovich YB, Raizer YP (1967) The Physics of Shock Waves and High Temperature Hydrodynamic Phenomena. Academic Press, New York

Zoback MD, Zoback ML, Mount VS, Suppe J, Eaton JP, Healy JH, Oppenheimer D, Reasenberg P, Jones L, Raleigh CB, Wong IG, Scotti O, Wentworth W (1987) New evidence for the state of stress on the San Andreas fault system. Science 238: 1105-1111

Silicified Cone-in-Cone Structures from Erfoud (Morocco): A Comparison with Impact-Generated Shatter Cones

Stefano Lugli [1], Wolf U. Reimold [2] and Christian Koeberl [3]

[1]Dipartimento di Scienze della Terra, Università degli Studi di Modena e Reggio Emilia, Largo S. Eufemia 19, 41100 Modena, Italy (lugli.stefano@unimore.it)

[2]Impact Cratering Research Group, School of Geosciences, University of the Witwatersrand, Private Bag 3, P.O. Wits 2050, Johannesburg, South Africa (reimoldw@geosciences.wits.ac.za)

[3]Department of Geological Sciences, University of Vienna, Althanstrasse 14, A-1090 Vienna, Austria (christian.koeberl@univie.ac.at)

Abstract. Several geological features, including sedimentary cone-in-cone structures and percussion marks, may resemble impact-generated shatter cones. Especially inexperienced workers may mistake such features for impact deformation. In 1997, our group investigated an alleged occurrence of shatter cones in the Hamada area of southeastern Morocco and found that these are actually cone-in-cone structures, probably from the Lower Visean Merdani Formation. Here, a detailed discussion of cone-in-cone structures, as well as a short review of shatter cone characteristics, is presented in an effort to clarify some distinguishing criteria. Important differences include: (1) Shatter cone striations are of distinctly roundish shape, whereas cone-in-cone striae are step-like; (2) shatter cones never show scaled surfaces; (3) broken cone-in-cone structures invariably produce one surface with striated cone features, but its opposite side would display scaled cone cups; (4) shatter cones do not telescope out of the bedding-plane, as cone-in-cone structures may do; (5) at the thin section scale, the internal structure of cone-in-cone features is well preserved, even after complete silicification of the primary carbonate. Thus, careful observations should allow unambiguous decision whether certain rocks contain shatter cones or cone-in-cone structures.

1 Introduction

Shatter cones are a mesoscopic fracture phenomenon that is widely considered a macroscopic shock metamorphic feature and, as such, a diagnostic indicator of impact structures (e.g., French 1998). Indeed, many impact structures have first been noted because of observations of shatter cones. However, there are several geological phenomena that either resemble shatter cones closely, including percussion marks/cones, such as those described by Reimold and Minnitt (1996) from the eastern environs of the Bushveld Complex in South Africa, or cone-in-cone structures that have also, in the past, been occasionally mistaken for shatter cones (Amstutz 1965).

Cone-in-cone structures are among the most spectacular and enigmatic geological features. Their origin and mechanism of formation have been debated for more than a century and are still essentially unresolved. Cone-in-cone structures may undergo silicification, but siliceous cones are only sporadically mentioned in the literature. During a visit to Morocco, the first author was alerted to the possible existence of shatter cone-like features in the Erfoud area, southeastern Morocco (Fig. 1). This resulted in 1997 in a short expedition by our group into these parts, when the alleged shatter cones were revealed as a large occurrence of cone-in-cone structures.

The aim of this paper is to provide a detailed description of these silicified cone-in-cone structures from Morocco and to discuss similarities with and differences to shatter cones.

2 Cone-in-Cone Structures

Calcareous cone-in-cone structures (Fig. 2) are spectacular features that have intrigued scientists since their discovery in the 19th century. Bates and Jackson (1987) provided the following definition: "*A minor sedimentary structure in thin, generally calcareous layers of some shales and in the outer parts of large concretions, esp. septaria; it resembles a set of concentric, right circular cones fitting one into another in inverted positions (base upward, apex downward), commonly separated by clay films and consisting usually of fibrous calcite and rarely of siderite or gypsum...*". Rare cone-in-cone-like structures consist also of pyrite (Carstens 1984), gypsum or siderite (Mozley 2003), but in these cases

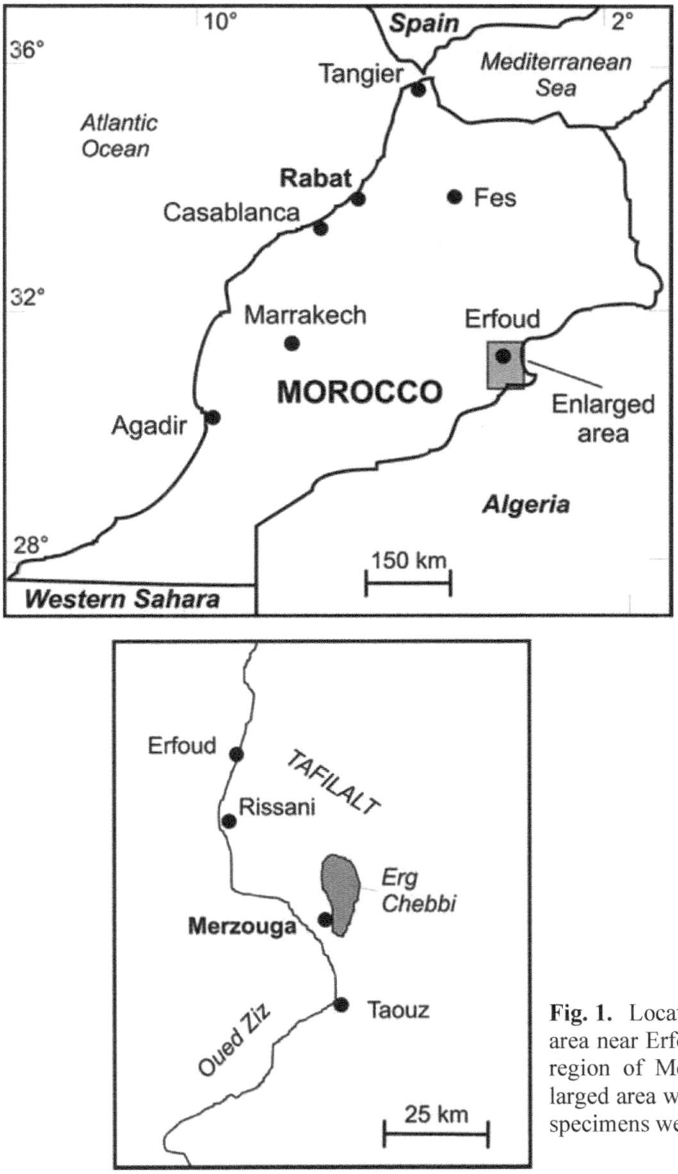

Fig. 1. Location map of the area near Erfoud in the Tafilalt region of Morocco (top). Enlarged area where cone-in-cone specimens were sampled (left).

(Pettijohn 1957) their characteristics and origin appear to differ significantly from those of the calcareous variety. The following description of cone-in-cone structures is based mostly on the works of Gresley (1894), Tarr (1922, 1932), Gilman and Metzger (1967), Franks (1969), and Selles-Martinez (1994).

2.1 Macroscopic Characteristics

Cone-in-cone structures consist of cones fitting one into another (nested cones) arranged in lenticular beds of several decimeter thickness and width, in calcite-cemented sandstones, or at the edges of disc-like to ellipsoidal concretions several decimeters in thickness and up to about 10 m long, in shales (Fig. 2). The orientation of the cones is mostly vertical with apices directed downwards in singular cone layers (sometimes called "beef"), but subradial arrangements with apices pointing toward the core of a concretion have also been described. Individual cones range from less than 1 to about 12 cm in height. Cone apical angles vary from 12° to 100°, but most reported angles fall between 20° and 60°. Shorter cones tend to

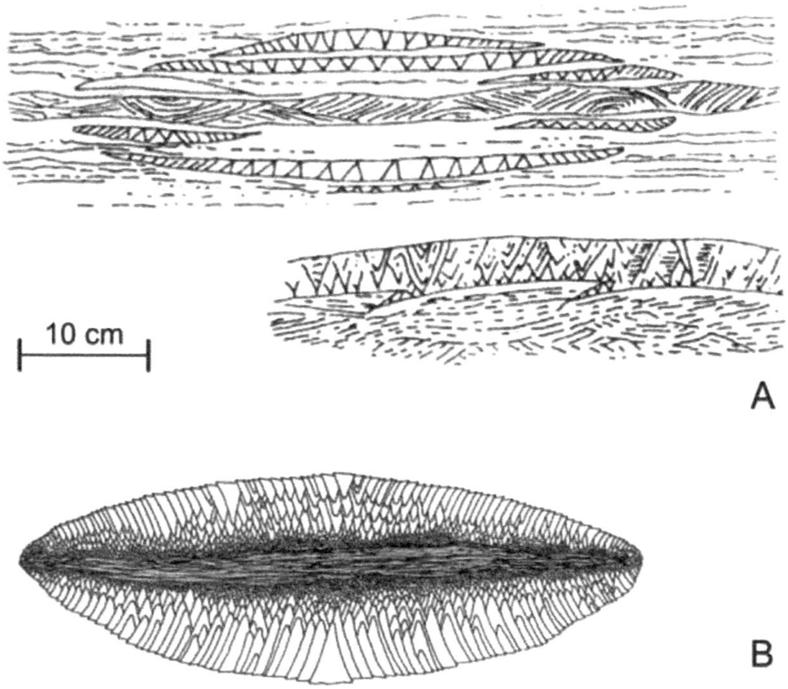

Fig. 2. Schematic cross section of cone-in-cone structures found in shale: A) Individual layers in the Devonian of Pennsylvania (redrawn by Selles-Martinez 1994, after Gresley 1894), and B) ellipsoidal concretion, Canadaway Fm., Upper Devonian, Western New York; cone axes are perpendicular to bedding in surrounding shale and to the central layer, which consists of silt (simplified after Gilman and Metzger 1967).

have relatively larger apical angles.

The cones are separated by striated argillaceous films forming annular or anastomosing rings that are more common in shorter cones. These annular depressions filled by argillaceous material produce a ribbed or scaly appearance (corrugation) on the surface left by the removal of the cones (cone cups). Cone terminations are visible on the top of the layers, where cones commonly project out (telescoping of cones). Important characteristics of cone-in-cone structures are summarized in Table 1.

2.2 Cone-in-Cone Petrography

Most cone-in-cone structures are composed of fibrous calcite and films of argillaceous material. The cone structure consists of nested and interfering cone-shaped, plumose aggregates of calcite fibers. The individual fibers are generally less than 6 μm wide but may be up to 3 cm long; plumose aggregates of fibers may attain 7 cm in length. The long axes of the plumose aggregates are oriented parallel to the long axes of the cones. Individual cones and cone segments are separated by argillaceous films that are up to 0.2 mm thick. Thinner argillaceous films may separate individual plumose aggregates of fibrous calcite and define the conical layering of the cone-in-cone structure. Small amounts of carbonaceous matter that can also be associated with pyrite and/or marcasite are commonly concentrated along the argillaceous films. Locally, the argillaceous material has been concentrated in angular patches corresponding to the filling of the annular rings around the cone cups. These fillings mark the termination of the fibers of a plumose aggregate against those of an adjacent aggregate or cone.

Fig. 3. Silicified cone-in-cone structures from the Hamada region southwest of Erfoud (Morocco). The sample is a loose fragment from a thick individual concretionary layer. Note that striae on cone surfaces are either straight (marked S) or fan-like (F). Scale bar length: 1 cm.

The cones have a tendency to break apart along the striated argillaceous films, resulting in elongated, curved and tapered fragments called "leaves". These characteristics have led some workers to interpret the spaces filled by argillaceous material as fractures (Tarr 1932; Gilman and Metzger 1967; Gilman 1968; Selles-Martinez 1994). Supporting evidence for this interpretation would be the apparent offset of cones and conic scales on the counterpart of the striated surfaces, as well as the telescoping of cones outward with respect to adjacent external cones.

Rare siliceous cone-in-cone structures have been reported in the literature and interpreted as the result of silica replacement of calcite structures (Woodland 1964). However, Becq-Giraudon (1990) proposed a primary origin for the siliceous cone structures of the Lower Ordovician in the Montagne Noir (Massif Central, France).

Table 1. Important characteristics of cone-in-cone structures.

	Cone-in-cone
Host rock	shales, sandstones and pedogenic rocks
Associated feaures	septaria, carbonate concretions
Size of cones	mm to dm
Composition of cones	calcite, quartz (replacement)
Cones defined by	calcite fibers, concentrations of argillaceous material, fractures?
Orientation of cone axes	perpendicular to bedding
Orientation of cone apices	toward substrate layer: upward and downward
Apical angles	14° to 100°
Striae on cones	straight, radial from apices
Formation time	unknown, up to million years?
Origin	early diagenesis? pressure solution? crack-seal mechanism?

2.3 Origin of Cone-in-Cone Structures

The origin of cone-in-cone structures has been strongly debated, but many aspects of their formation have remained unclear. A common definition (Bates and Jackson 1987) states: *"The structure appears to be due to pressure aided by crystallization and weathering (solution) along intersecting conical shear zones"*. The important hypotheses for the formation of calcareous cone-in-cone structures can be divided into two groups: those favoring early displacive formation of concretions in soft sediment and those that emphasize late fracturing of concretions with or without excess pore fluid. Most workers reported the association of cone-in-cone structures with other concretions such as septaria and also with organic-rich sediments. Some authors relate their growth to degradation of organic matter (Brown, 1954) or possible microbial activity (MacKenzie 1972; Aso et al. 1992; McBride et al. 2003), in particular for pedogenic rocks (Aassoumi et al. 1992). In summary, the following proposals have been made (see also Mozley 2003):

Fig. 4. Back of the sample illustrated in **Fig. 3** showing annular rings (corrugations) and cone cups left by the removal of cones (marked C and by arrows). Erfoud (Morocco) sample. Scale bar length: 1 cm.

Recrystallization of fibrous aragonite to calcite forming open cone fractures intruded by argillaceous matter (Tarr 1922; Gilman and Metzger 1967).

Differential pressure solution on calcite fibers along conical shear planes induced by the weight of overlying strata (Tarr 1932; Pettijohn 1957).

Settling and volume shrinkage during the slow dewatering of highly saturated and loosely packed subaerially-exposed sediments (Shaub 1937).

Precipitation of fibrous calcite along fractures or in concretionary masses. The fibers would then be deformed by "tractional forces" during deformation of the host strata or during concretionary growth of the calcite

Fig. 5. Another portion on the back of the sample illustrated in **Fig. 3** showing relatively wide annular rings. Erfoud (Morocco) sample. Scale bar length: 1 cm.

component (Bonte et al. 1947).

Crystallization of calcite fibers under gravity-induced differential stress, during gradual compaction of the sediment. The angles of cone apices would be determined by the plasticity of the enclosing sediment: large apical angles would indicate relatively low plasticity (Woodland 1964).

Early displacive growth of cone-shaped plumose aggregates of fibrous calcite (Franks 1969).

Diagenetic water-escape structures promoting the "reordering of phyllitic elements within a sediment undergoing diagenetic compaction and which exhibits differences in porosity and competency in its lithological composition; cone in cone can be compared to a schistosity acquired in the realm of hydroplastic deformation" (Becq-Giraudon 1990).

Brittle fracturing of crystalline calcite aggregates, grown by a crack-seal mechanism in overpressured environments, and induced by a decrease in pore pressure of the plastic host sediment (Selles-Martinez 1994).

Shallow burial (<1500 m) precipitation from modified marine pore fluids (Hendry 2002).

Early diagenetic growth (depth <40 m), probably microbially mediated, in sandstone units beneath flooding surfaces and sequence boundaries (McBride et al. 2003).

Fig. 6. Top view of the sample shown in **Fig. 3-5**. Note telescoping cones (white arrows) and conical surfaces forming cone cups (black arrows). Scale bar length: 1 cm.

Fig. 7. Silicified cone-in-cone structure consisting of nested and interfering cones. Note section of intersecting and overlapping circular arcs of former fibrous calcite forming cone "branches" originally separated by argillaceous material (arrow). The interference of these conical surfaces produces the striated appearance of the cone surfaces. Erfoud (Morocco) sample. Scale bar length: 1 cm.

Fig. 8. Silicified cone-in-cone structures from the Hamada region southwest of Erfoud (Morocco). Note wide cone bases. The sample is a loose fragment from a thin individual concretionary layer. Scale bar length: 1 cm.

Fig. 9. Top view of the sample shown in **Fig. 8**. Note telescoping cones (white arrows) and a hollow cone cup (black arrow). Scale bar length: 1 cm.

3 Silicified Cone-in-Cone Structures from Erfoud, Morocco

3.1 Location of Samples

The alleged "shatter cone" site investigated by us in 1997 is located in the Hamada area southwest of the town of Erfoud in the Tafilalt region of Morocco, close to the border with Algeria. Many loose samples of silicified cone-in-cone structures were collected on the Hamada surface at a site located a few kilometers west of Merzouga village, close to the sand dune field known as the Erg Chebby (Fig. 1). Unfortunately, layers containing the cone-in-cone structures are not exposed in the area and no observations on the original stratigraphic and three-dimensional relationships of the presumed concretions could be made. According to the Geological Map of Morocco (Carte Géologique du Maroc, 1986), the basement below the Hamada area is formed by the Carboniferous (Lower Visean) Merdani Formation. These strata include dark gray shales with fine-grained sandstone intercalations, as well as extensive red iron-rich shale layers, in which cone-in-cone structures were reported.

3.2 Macroscopic Description

The collected samples are fragments (loose debris) of what are probably individual concretionary layers/lenses. The specimens vary in length from 2 to 10 cm and in thickness from 2 to 7 cm (Fig. 3-9). The cone-in-cone structures are dark brown to black in color and often display a strong luster. Individual cones range in size from a few millimeters to 6 cm. Apical angles are variable and range from 30° to 60°. Scales on cone cups are well developed and may be as wide as 0.5 cm (Fig. 4 and 5). In thin layers, the cones have a strong tendency to telescope out of the sample surface (Fig. 9), whereas relatively thicker layers have an uneven relief formed by interference of large cone tops that project out from the bedding plane by 1 to 2 mm (Fig. 6). The bedding plane relief, thus, has a step structure formed by intersecting and overlapping circular arcs separated by argillaceous films (Fig. 7 and 9). These circular arcs are segments of the nested conical surfaces forming the cone-in-cone structure, and the intersection and interference of these conical portions produce the striated appearance of the cone surfaces (Fig. 3 and 7). Striae are generally straight

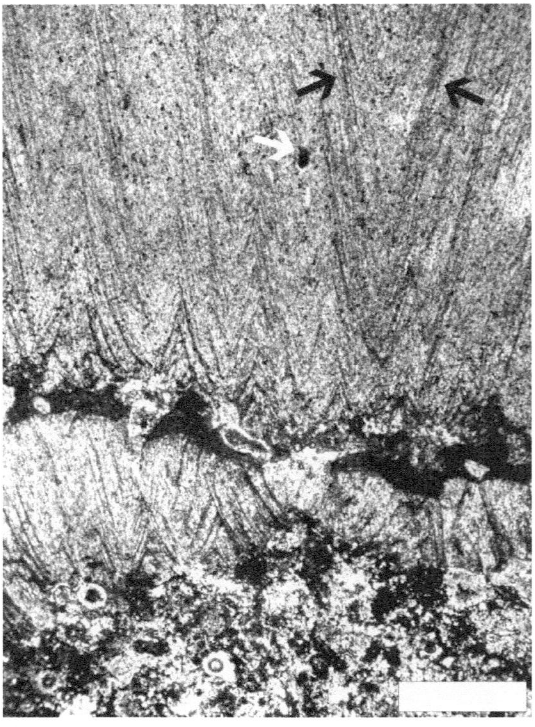

Fig. 10. Photomicrograph of two separate silicified layers in cone-in-cone structures (top) grown on layers of dark ferruginous material including shell fragments and detrital quartz (bottom). Apices of cone structures point downward in photograph, but original orientation is unknown. Dark gray to black areas contain argillaceous material that defines the conical layering (black arrows) and outlines plumose aggregates of former fibrous calcite. Minute, disseminated, dark spots are pyrite framboids (white arrow). Erfoud (Morocco) sample. Round spots are thin sectioning artifacts (bubbles; in lower part of photograph). Section cut parallel to cone axes. Plane-polarized light; scale bar length: 0.5 mm.

Fig. 11. Photomicrograph of two separate silicified layers of minute cone-in-cone structures grown in opposite directions from two different substrate beds and meeting in the middle of the photograph. The substrates are layers of dark ferruginous material including shell fragments and detrital quartz (at top and bottom). Erfoud (Morocco) sample. Round spots are thin sectioning artifacts (bubbles; in bottom right corner). Section cut parallel to cone axes. Plane-polarized light; scale bar length: 0.5 mm.

but, especially in the case of prominently telescoped cones, show a fan-like arrangement and rapidly widen in major parts of cones (Fig. 8).

The cones are silicified and do not have the tendency to break apart along curved surfaces (so-called “leaves“ – Franks 1969), as is common for cone-in-cone structures in carbonate.

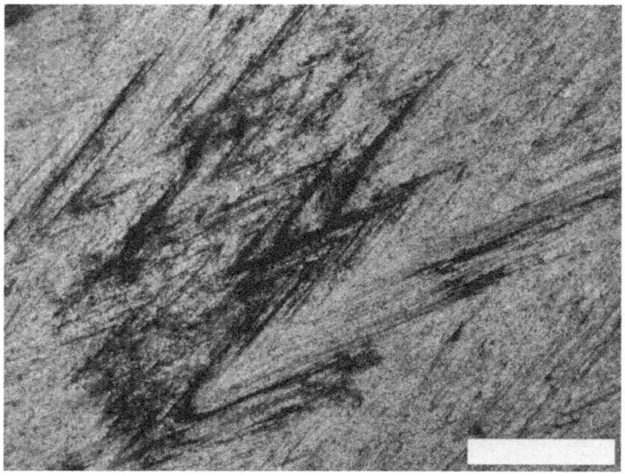

Fig. 12. Photomicrograph of silicified cone-in-cone structures grown on an unknown substrate. Apices of cone structures point to the left in photograph, but original orientation is unknown. Black areas contain argillaceous material and abundant pyrite and Fe-oxide/hydroxide that outline interference of nested conical surfaces. Erfoud (Morocco) sample. Section cut parallel to cone axes. Plane-polarized light; scale bar length: 0.5 mm.

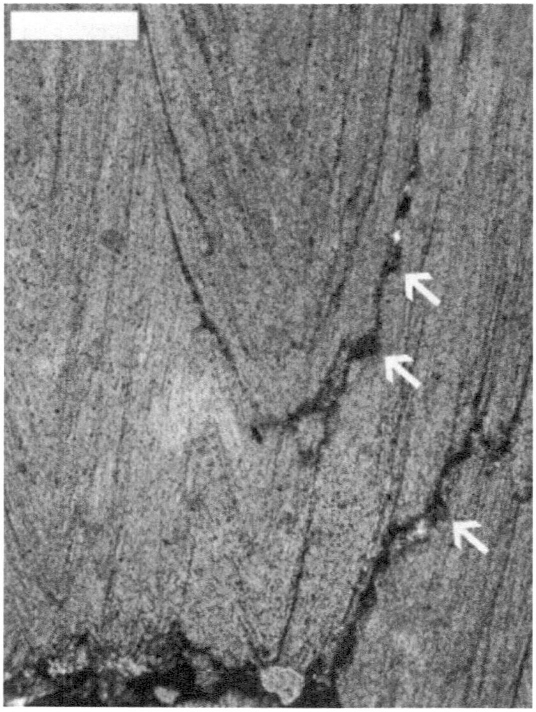

Fig. 13. Photomicrograph of cone-in-cone structures grown on layer of dark ferruginous material (bottom). Apices of cone structures point downward in photograph, but original orientation is unknown. Dark gray to black areas consist of argillaceous material that is particularly concentrated in angular patches corresponding to fillings of annular rings (arrows). Erfoud (Morocco) sample. Section cut parallel to cone axes. Plane-polarized light; scale bar length: 0.5 mm.

3.3 Petrography

Under the optical microscope cones can be seen to originate from layers of dark ferruginous material including unidentified shell fragments and thin layers of detrital quartz (Figs. 10 and 11). In most studied fragments, cones show a uniform growth direction, but in a few layers several millimeter-sized cones have grown in opposite directions from two different substrate laminae (Fig. 11). The conical layering is marked by argillaceous and carbonaceous material including minute, disseminated pyrite framboids, which more commonly occur around cone apices (Fig. 10-13). Alteration of pyrite to form iron oxide and/or hydroxides appears to be responsible for the dark color of the rock. Annular argillite-filled depressions of a scaly appearance on the cone cups are well developed and form angular patches as wide as 1.25 mm (Fig. 13 and 14).

Sections of cone-in-cone structures cut perpendicular to cone axes show nested circular shapes formed by interference and intersection of smaller arc segments separated by argillaceous films (Fig. 15). The bases of cones

Fig. 14. Photomicrograph of cone-in-cone structures showing composite angular argillaceous fillings of annular rings (dark). The shapes of the former calcite fibers now replaced by quartz are still visible. Note that the "stepped" profile is only on one side of the cone surface (right), whereas its counterpart is smooth (left). Cone-in-cone structures commonly break apart along these surfaces producing cone cups with annular rings (right) and striated cones (left). Section cut parallel to cone axes. Plane-polarized light; scale bar length: 0.5 mm.

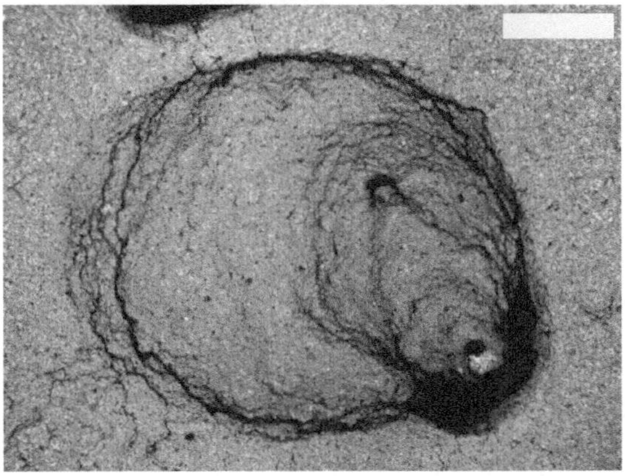

Fig. 15. Photomicrograph of cone-in-cone structures cut perpendicular to cone axes. Note that nested circular shapes are formed by interference of smaller arc segments. Plane-polarized light; scale bar length: 0.5 mm.

are commonly composed of microcrystalline quartz, whereas zones around cone apices contain a mosaic of larger quartz crystals up to 0.1 mm in size. A few samples contain minute relicts of granular calcite crystals partly replaced by quartz, which occur commonly away from cone apices. Although the carbonate relicts are not composed of fibrous calcite, some degree of preferred orientation can be observed in thin section. The c-axes of the calcite crystals tend to be oriented parallel to the cone axes or parallel to the conical layering. These relic fibrous structures indicate that silicification of the cone-in-cone structure postdates the recrystallization of original fibrous calcite to form a microgranular carbonate fabric. Unfortunately, no indications on the timing of these processes are available, as no observations on the general stratigraphy and mesoscopic cross-cutting relationships of the cone-in-cone structure-bearing layers could be made in the field.

Although silicification affected the studied samples to various degrees, the internal structure typical of cone-in-cone structures is well preserved. This suggests that the only example of siliceous cone-in-cone structures described in detail in the literature (from the Lower Ordovician of the Montagne Noir, France) is not necessarily a primary structure, as suggested by Becq-Giraudon (1990). The observation that no carbonate relicts are present in the siliceous structure can not be considered evidence that cone-in-cone structures formed by precipitation of silica, but rather that the original carbonate may have been completely silicified.

4
Shatter Cones

4.1
Background and General Characteristics

The conical fracturing phenomenon known as *shatter cones* (Figs 16 and 17, 20 and 21) is widely considered shock (impact) diagnostic. Since the early pioneering work by Dietz (1947, 1959, 1961, 1963, 1968), Hargraves (1961), Manton (1962, 1965), Milton (1977), Milton et al. (1972, also 1996), and Roddy and Davis (1977), shatter cones have been described from many impact structures and have been, in many cases, the first indication for the presence of an impact structure.

French (1998) defined shatter cones as *"distinctive curved, striated fractures that typically form partial to complete cones"*. It must, however, be emphasized that we ourselves have not come across many complete cones – though several allegedly complete specimens from Vredefort have been described in the literature (Albat and Mayer 1990; see also discussion by Nicolaysen and Reimold 1999). Sizes of shatter cones reportedly range from just a few millimeters in length to more than 1 m. Our own observations from Vredefort indicate that most features are a few to 15 cm

Fig. 16. Photograph of the front of a well-developed shatter cone specimen from the Steinheim Basin meteorite crater, southern Germany. The coin allows to compare the orientation of this sample with that of the same, reversed specimen in **Fig. 17**, illustrating that the apex orientations of cone fractures visible on both sides of this sample do not consistently point into the same direction. Also note several incomplete cone fractures that nevertheless still indicate highly diverse directions in which striae seemingly converge. Sample length ca. 15 cm.

Fig. 17. Photograph of the back of the same specimen from Steinheim shown in **Fig. 16**.

in length, and the largest ones observed have lengths of up to 50 cm and diameters of up to 50-60 cm. Partial cones may represent segments of cones not larger than 10°-20°, or may subscribe up to >270° segments. Workers are urged to investigate whether such significant proportions of cones really represent single geometries, or whether they – as favored by Nicolaysen and Reimold (1999) – represent combinations of smaller cone segments on intersecting curviplanar joints that have combined to apparent larger cone segments.

Shatter cones have generally been described from rocks of central uplifts of complex impact structures, but isolated fragments have been observed in impact breccia units as well (e.g., in clasts in Sudbury Breccia of South Range exposures). Because of this, it is generally thought that shatter cones form during the early shock compression stage, but some evidence of shatter cones post-dating other impact deformation (such as fault gouge or pseudotachylitic breccia - Simpson 1981 and Reimold and Colliston 1994, respectively) has, for example, been presented from Vredefort.

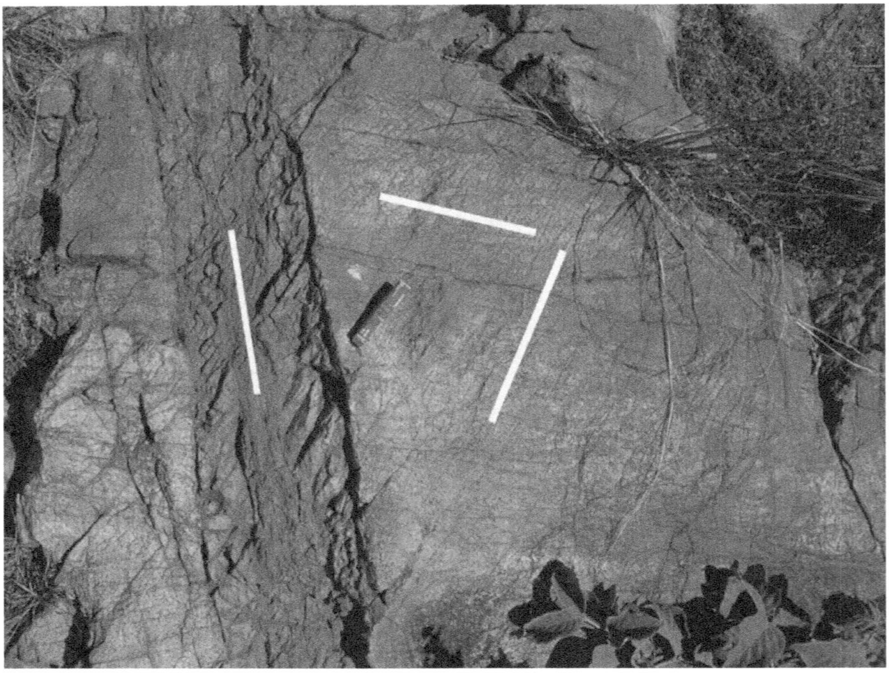

Fig. 18. Several sets of closely-spaced joints – termed multipli-striated joint sets by Nicolaysen and Reimold (1997) – in Hospital Hill Quartzite from the inner collar of the Vredefort Dome. Locality: Smilin Thru resort, north of Parys. Short white lines illustrate the orientations of important joint sets. Gum packet, for scale, 7 cm long.

Shatter cones may occur in the form of individual cone features or as groups of many, then often partially overlapping, features. Striations on a specific cone segment may come together at a well defined apex or may emanate from a several millimeter to >1 cm wide apex area. They diverge distinctly, so that the base of a cone segment is generally at least 2-3 times as wide as the apex area. Striations are generally not straight but are curved and often meandering, and the microtextural analysis by Nicolaysen and Reimold (1999) has shown that striae may indeed meander around grain boundaries. It should be noted, however, that this is not the rule, as striae cross-cutting and possibly displacing mineral grains have also been observed. Near the base of a cone segment, striae may diverge even more into a pattern that has been described as "horse-tailing". French (1998) remarked that *"parasitic cones commonly occur on the surfaces of both complete and partial shatter cones, forming a "nested" texture"*. Apices of many cones may, at a given locality, point into a uniform direction, which has led many workers to adhere to a theory that the apex

Fig. 19. Another example of MSJS, from the same locality as the example shown in **Fig. 18**.

direction can be related to the shock wave propagation direction. In this "master cone" hypothesis (see Nicolaysen and Reimold 1999 – for a detailed but critical discussion and references of earlier work), it is proposed that cone apices, after rotation of the host strata into the presumed pre-impact orientation, should point into the direction from which the shock wave originated. Nicolaysen and Reimold (1999), however, indicated that while apices at Vredefort may be preferentially aligned in a direction that was predetermined by a strong anisotropy (namely the bedding-plane) in the supracrustals of the Vredefort dome, other cone fractures may have opposite or strongly divergent orientations. Specimens from Steinheim Basin (southern Germany) demonstrate how varied the directions of cone apices may be on the multi-facets of a single specimen (Fig. 16, 17, and 21).

Sagy et al. (2002) presented some results on Vredefort shatter cones and proposed that the angle between outer striae on parasitic cones followed a distinct relationship with distance of sampling site from the center of an impact structure. They claimed that this angle decreased according to a definite relation with increasing distance of sampling sites from the center of the Vredefort dome. This hypothesis was further investigated by Wieland and Reimold (2003) and Wieland et al. (2003), who concluded that (a) no specific "striation angle" could be measured on any given shatter cone specimen and (b) both the hypothesis and trend promoted by Sagy et al. (2002) could not be reproduced.

Nicolaysen and Reimold (1999) drew further attention to the fact that shatter cones at Vredefort – and Sudbury – are intimately related to a pervasive system of curviplanar fractures occurring - in the case of Vredefort - in up to 12 orientations at a given site. These authors termed these fracture sets "MSJS" – multipli-striated joint sets (Figs 18 and 19). These curviplanar joints/microjoints are spaced at several mm to 1 cm, thus forming an intense and pervasive fracture system affecting large rock volumes. Cone segments and their striae were directly related to intersections and orientations of such joints (MSJ) by Nicolaysen and Reimold (1999). This observation was also already reported by Manton (1962). Nicolaysen and Reimold (1999) also illustrated the presence of MSJS in the Sudbury Structure.

Shatter cones can be observed in many different rock types, whereby finer-grained lithologies (e.g., limestone, siltstone) generally exhibit better examples of this deformation phenomenon than relatively coarser-grained materials. However, even at 0.5 cm grain size, as in some granitoids of the Vredefort basement, admittedly crude cone fractures have been observed. Nevertheless, the beautiful examples in limestone from Houghton Dome,

Fig. 20. A circa 40 cm long shatter cone from the Booysens Shale locality (Gibson and Reimold 2001) in the northwestern collar of the Vredefort Dome. Note the pronounced horsetailing of striae on one side of the sample. Scale bar in centimeters.

Steinheim, or Ries, and from the fine-grained metapelites and sandstones of the Vredefort and Sudbury structures are legendary.

Some workers maintain that it is not necessary to rely on shock pressures to generate shatter cone-like fractures in rock. However, from shock deformation scaling in impact structures and some shock experimentation, it has been established that low shock pressures, typically 2-10 GPa (perhaps as high as 30 GPa), can lead to shatter cone formation. It is also important to note that shatter cones from a number of structures have been shown to have surficial melt products (e.g., Gay 1976; Gay et al. 1978; Gibson and Spray 1998; Nicolaysen and Reimold 1999). Various workers (e.g., Martini 1991) have related this to friction melting and, thus, occurrences of pseudotachylitic breccia in impact structures. Nicolaysen and Reimold (1999) showed that distinct displacements can be observed in thin sections on MSJS, and that thin melt films may occur on such microjoints, as well as in intersections of microjoints of different orientations (see also below – section 4.4 on Microscopic Textures Related to Shatter Cones).

Fig. 21. Another shatter cone specimen from the Booysens Shale locality. Scale bar in centimeters. Note the different orientations of apices of shatter cones in this example.

4.2 Shatter Cone Formation

Different mechanisms have been proposed for the formation of shatter cones. This includes the idea by Johnson and Talbot (1969) that the elastic precursor of a shock front is scattered by a heterogeneity in a rock (see also Sharpton et al. 1996). Gash (1971) proposed that shatter cones were produced from the interaction of an incident shock wave with a tensile wave reflected from a reflective source such as a local heterogeneity in a rock (for example a relatively large grain, pore space, or fracture). Milton (1977) suggested that shatter cones formed during relaxation after peak compression. And Baratoux and Melosh (2003) have just recently advocated that while cone initiation takes place at heterogeneities in a rock, this would occur after *"the passage of the main plastic compression pulse"* (ibid). It is obvious that more work, combining geological (i.e., field work) and mineralogical investigations with theoretical advance, is required before a comprehensive model for the formation of this impact phenomenon can be established. Nicolaysen and Reimold (1999) concluded from their macroscopic to microscopic analysis of shatter cones and joint phenomena at Vredefort that *"formation of shatter cones appears to be invariably linked with the occurrence of the MSJS micro-jointing phenomenon"*. One serious impediment related to the investigation of the pervasive joint structures in the Vredefort Dome is the lack of temporal constraint on the number of generations of joints and the timing of joint formation with respect to the short-term impact process.

4.3 Shatter-cone like Phenomena

Abundance of shatter cones certainly seems to be diagnostic of impact. However, cone fractures of amazing similarity to impact produced shatter cones have been described from fossil water-falls (Reimold and Minnitt 1999) in the eastern Transvaal (Mpumalanga Province) of South Africa and from a location in the southeastern collar of the Vredefort Dome, and from southern Botswana. These percussion marks may comprise complete or major segments of circular fractures (up to 360 degree features were recorded). These authors also showed that joint structures at the sites in Klapperkop Quartzite of Mpumalanga and in the southeastern sector of the collar of the Vredefort Dome and the relation of these structures to the percussion marks closely resemble shatter cone and MSJS relationships of impact origin.

Well-developed shatter cone specimens are rare in impact structures. In the past, rather poorly developed, more or less well striated, fractures have also been mistaken for shatter cones. Where only curviplanar to linear fractures with not well developed or only subparallel striations are observed, one should be cautious in drawing an immediate conclusion that this could be evidence for impact. A case in point is Rohleder's (1936) report of shatter coning from Bosumtwi, Ghana. This worker concluded from one rather ambiguous occurrence that the crater could be of impact origin, in analogy to the Steinheim structure (Germany), which had been studied by him as well. In the meantime, the impact origin of Bosumtwi has long been confirmed, but our own efforts in 1996 to detect shatter cones in the rocks of the well exposed crater rim of the Bosumtwi crater were unsuccessful. Only a handful of fractures with possible slickensides were observed.

An early report of shatter cone-like features from a volcanic vent in New Mexico was made by Elston and Lambert (1965).

The detailed work carried out at the Vredefort Dome (references as listed above) has established that a whole variety of fracture phenomena exhibits more or less well-defined partial cone or more planar geometries, all of which are characterized by striations. Besides cone fractures, more linearly trending fractures with striations of variable divergence are abundantly observed in the collar rocks of the dome. In such occurrences, it is only sometimes possible to observe somewhat better defined garben of striae that may resemble cones. Generally, striae may have more or less parallel arrangements and, consequently, may resemble slickensides. Care must then be taken to look for the typical step features one would expect to find on slickenside surfaces and that can serve as movement indicators and could provide a criterion distinguishing slickensides from poorly striated surfaces related to shatter coning. On linearly trending fractures, garben of striae may have apices in one direction only, or they may alternate in a tight pattern. Whether all such features observed at Vredefort are true impact-derived deformation, or have been formed as the result of 3 Ga (since deposition of the Witwatersrand Supergroup host rock) tectonic deformation, is not clear. Thus, caution must be observed when identifying shatter cones and when deciding which other fracturing phenomena could be of impact or unrelated origin.

There remains enormous scope for further detailed field and laboratory study of shatter cones in impact structures, also in comparison with similar deformation phenomena. While in theory scattering of shock waves on material heterogeneities seems to explain the shatter cone formation process, a link between macro- and microscopic characteristics and formation is still missing.

4.4 Microscopic Features Related to Shatter Cones

A comprehensive review of petrographic observations on shatter cones was presented by Nicolaysen and Reimold (1999). Lilly (1981) concluded that *"shock effects due to shatter coning are only present to a depth of one or two grain diameters away from the shatter cone surface"*. Carter (1965) investigated planar microdeformation features in shatter cone specimens, work that resulted in one of the first detailed studies of PDFs. A related recent study is that by Hargraves and White (1996) on shatter cones from the Beaverhead impact structure (Montana). Gay (1976) and Gay et al. (1978) investigated manifestations of glass on the surfaces of shatter cones from Vredefort and compared them with similar manifestations on slickenside surfaces. More detail on glass manifestations on shatter cone surfaces was provided by Gibson and Spray (1998).

Nicolaysen and Reimold (1999) summarized their micropetrographic findings as follows:

- Presence of glass or melt on shatter cone surfaces was confirmed. Glass was also observed at intersections of their shatter cone-related MSJS.
- MSJS were described to be characterized by annealing in the immediate contact zone in host rock and cataclasis in a somewhat wider zone. Friction on these microjoints was supported by frequent observations of displacements and glass films.
- Their scanning electron microscopic comparison of micro-features of shatter cone and slickenside surfaces resulted in the conclusion that "both have many common features but that shatter cones are distinct by their divergent striae patterns".
- A step-like pattern on shatter cone surfaces, perpendicular to the direction of striae, was interpreted as a result of MSJ planes cross-cutting a shatter cone surface.

These authors also studied thin sections cut perpendicular to shatter cone surfaces and MSJ planes and parallel as well as perpendicular to shatter cone striae. They noted that microstructural damage could be related directly to narrow zones, at maximum a few millimeters wide, just below a shatter cone surface. In these narrow zones, intergranular cracks occurred that generally trended subparallel to the cone surface. Other fractures were oriented at high angles to the surface. Nicolaysen and Reimold (1999) studied both sericitic matrix-supported quartzite and mature quartzite specimens and concluded that the brittle deformation style in these narrow zones was the same irrespective of host material. They found that annealing along shear fractures is prominent. MSJs frequently display melt along their extensions or melt pools at intersections between MSJ of different (cross-cutting) sets. Most displacements along MSJ were measured at 50-100 μm, but displacements by as much as 1 mm were also noted. Where individual MSJ were studied over a considerable distance (about 1 cm), displacements were found to be locally variable (between no obvious displacement and several hundred μm). They reported that many MSJ intergranular fractures meandered around grains, but that this was not the rule.

It is obvious from the literature about shatter cone characteristics reviewed here that shatter cone surfaces and related multipli-striated joint sets (MSJS) give ample evidence of relative movement between two rock volumes on either side of the shatter cone surface. Melting phenomena, as well as displacements and annealing along MSJS, are evidence for friction along these fractures (as already proposed by Martini 1991).

5
Comparing Cone-in-Cone and Shatter Cone Structures

The smoothness, black color and hardness of silicified cone-in-cone structures may appear like the result of melting of a rock, as is observed on surfaces of shatter cones. In particular, silicified cone-in-cone structures do not break apart along curved surfaces ("leaves"), as their carbonate precursors might. Although iron impregnated, silicified cone-in-cone structures may resemble shatter cones, but the two features can be easily distinguished.

Some difficulties could arise in those cases where only small rock fragments may have become available and where no detailed information on outcrops or geological background can be provided. Where only insufficient stratigraphic and structural information is available, the origin of cone-in-cone structures may not be recognizable (as was the case for our visit of the Hamada region of Morocco) or the relationship of striated cone fractures to fossil water-falls may not be obvious, as it was the case in the initial stages of the Reimold and Minnitt (1996) study of percussion marks in Mpumalanga and in the southeastern Vredefort dome.

The key criteria to distinguish between silicified cone-in-cone and shatter cone structures are: (1) Shatter cone striations are roundish; those of cone-in-cone structures are distinctly step-like; (2) shatter cones never show scaled surfaces (cone cups – Figs 4 and 5); (3) cone-in-cone structures that have broken along a main plane of weakness (represented by the clay films bordering the cones) invariably produce two different surfaces, but one is not the imprint of the other; one surface always shows striated cone features, whereas on the opposite side only scaled cone cups are present; (4) shatter cones do not telescope out of the bedding-plane, as cone-in-cone segments often do; (5) cone-in-cone apices may display opposite orientations above and below the concretionary layers/lenses from which they grew; and, finally, (6) at the thin section scale, the internal structure of cone-in-cone features is well preserved, even after complete silicification of the primary carbonate; in particular scaled cone traces are well-preserved and do not resemble any phenomenon associated with shatter cones.

Acknowledgments

WUR's research and his participation in the expedition to Morocco have been supported by the National Research Foundation of South Africa and by a grant from the Research office of the University of the Witwatersrand. This is University of the Witwatersrand Impact Cratering Research Group Contribution No. 67. CK's research and field work in Morocco were supported by the Austrian Science Foundation, project Y58-GEO. We appreciate field support by D. Jalufka. The shatter cone specimen from the Steinheim structure, which is shown in Fig. 16 and 17, was generously provided by W. Reiff (Stuttgart, Germany). Critical reviews by H. Dypvik and C. Talbot are gratefully acknowledged.

References

Aassoumi H, Broutin J, El Wartiti FP, Koeniguer J, Quesada C, Simancas F, Toutin-Morin N (1992) Pedological nodules with cone in cone structure in the Permian of Sierra Morena (Spain) and central Morocco. Carbonates and Evaporites 7: 140-149

Albat HM, Mayer JJ (1990) Shatter cones in Vredefort rocks – imagination or reality? South African Journal of Geology 93: 547-548

Amstutz GC (1965) A morphological comparison of diagenetic cone-in-cone structures and shatter cones. Annals of the New York Academy of Sciences 123: 1050-1056

Aso E, Gisbert TJ, Valero Garces B (1992) Type septaria-cone in cone nodules in the Stephano-Permian of the Catalan Pyrenees. Carbonates and Evaporites 7: 132-139

Baratoux D, Melosh HJ (2003) A new model for the formation of shatter cones: Consequences for the interpretation of shatter cone data in terrestrial impact structures. Earth and Planetary Science Letters 216: 43-54

Bates R.L. and Jackson J.A. (1987) Glossary of Geology. American Geological Institute, Alexandria, Va, USA,788 pp

Becq-Giraudon JF (1990) Cone in cone structure in an environment of siliciclastic deposition. New observation on Lower Ordovician cone-in-cone structures of the Montagne Noire, south French Massif central. Géologie de la France 2: 11-19

Bonte A, Denaeyer ME, Goguel J (1947) Les facteurs mécaniques dans la genèse de la structure "cone-in-cone". Comptes Rendus Sommaire des Sciences de la Société Géologique de France 9: 182-184

Brown RW (1954) How does cone-in-cone material become emplaced? American Journal of Science 252: 372-376

Carter NL (1965) Basal quartz deformation lamellae – A criterion for recognition of impactites. American Journal of Science 263: 786-806

Carstens H (1984) Early diagenetic cone-in-cone structures in pyrite concretions. Journal of Sedimentary Petrology 55: 105-108

Carte Géologique du Maroc, Tafilalt-Taouz (1986) Sheet No. 244, Ministère de l'Energie et des Mines, Direction de la Géologie, 1 : 200 000.
Dietz RS (1947) Meteorite impact suggested by the orientation of shatter-cones at the Kentland, Indiana, disturbance. Science 105: 42-43
Dietz RS (1959) Shatter cones in cryptoexplosion structures (meteorite impact?). Journal of Geology 67: 496-505
Dietz RS (1961) Vredefort Ring structure: Meteorite impact scar? Journal of Geology 69: 499-516
Dietz RS (1963) Cryptoexplosion structures: A discussion. American Journal of Science 261: 650-664
Dietz RS (1968) Shatter cones in cryptoexplosion structures. In: French BM, Short NM (eds) Shock Metamorphism of Natural Materials, Mono Book Corporation, Baltimore, Maryland, pp 267-285
Elston WE, Lambert PW (1965) Possible shatter cones in a volcanic vent near Albuquerque, New Mexico. Annals of the New York Academy of Sciences 123: 1003-1016
Franks PC (1969) Nature, origin, and significance of cone-in-cone structures in the Kiowa Formation (early Cretaceous), north-central Kansas. Journal of Sedimentary Petrology 39: 1438-1454
French BM (1998) Traces of Catastrophe. Lunar and Planetary Institute, Houston, Contribution No. 954, 120 pp
Gash PJS (1971) Dynamic mechanism of the formation of shatter cones. Nature Physical Sciences 230: 32-35
Gay NC (1976) Spherules on shatter cone surfaces from the Vredefort structure, South Africa. Science 194: 724
Gay NC, Comins NR, Simpson C (1978) The composition of spherules and other features on shatter cone surfaces from the Vredefort structure, South Africa. Earth and Planetary Science Letters 41: 372-380
Gibson HM, Spray JG (1998) Shock-induced melting and vaporization of shatter cone surfaces: Evidence from the Sudbury impact structure. Meteoritics and Planetary Science 33: 329-336
Gibson RL, Reimold WU (2001) The Vredefort Impact Structure, South Africa (The Scientific Evidence and a Two-Day Excursion Guide). Council for Geoscience, Pretoria, Memoir 92: 110 pp
Gilman RA (1968) Kink bands of non-tectonic origin. Journal of Geology 76: 702-709
Gilman RA, Metzger WJ (1967) Cone-in-cone concretions from western New York. Journal of Sedimentary Petrology 37: 87-95
Gresley WS (1894) Cone-in-cone; how it occurs in the Devonian series in Pennsylvania, U.S.A., with further details of its structure, varieties, etc. Quarterly Journal of the Geological Society of London 50: 731-739
Hargraves RB (1961) Shatter cones in the rocks of the Vredefort Ring. Transactions of the Geological Society of South Africa 64: 147-154
Hargraves RB, White JC (1996) Micro-shock deformation adjacent to the surface of shatter cones from the Beaverhead impact structure, Montana. Journal of Geology 104: 233-238
Hendry JP (2002) Geochemical trends and palaeohydrological significance of shallow burial calcite and ankerite cements in Middle Jurassic strata on the East Midlands Shelf (onshore UK). Sedimentary Geology 151: 149–176

Johnson GP, Talbot R (1964) A theoretical study of the shock wave origin of shatter cones. Masters Thesis, Air Force Institute of Technology, Wright-Patterson Air Force Base, Dayton, Ohio, USA, 92 pp

Lilly PA (1981) Shock metamorphism in the Vredefort collar: Evidence for internal shock sources. Journal of Geophysical Research 86: 10,689-10,700

MacKenzie WS (1972) Fibrous calcite, a middle Devonian geologic marker, with stratigraphic significance, district of MacKenzie, Northwest Territories. Canadian Journal of Earth Sciences 9: 1431-1440

Manton WI (1962) The Orientation and Implication of Shatter Cones in the Vredefort Ring Structure. MSc Thesis (unpublished), University of the Witwatersrand, Johannesburg, South Africa, 167 pp

Manton WI (1965) The orientation and origin of shatter cones in the Vredefort Ring. In: Whipple HE, Spitzer MI (eds) Geological Problems in Lunar Research. Annals of the New York Academy of Science 123: 1017-1049

Martini JEJ (1991) The nature, distribution and genesis of the coesite and stishovite associated with the pseudotachylite of the Vredefort Dome, South Africa. Earth and Planetary Science Letters 100: 285-300

McBride EF, Picard MD, Milliken KL (2003) Calcite-cemented concretions in Cretaceous sandstone, Wyoming and Utah. Journal of Sedimentary Research 3: 462-475

Milton DJ (1977) Shatter cones – An outstanding problem in shock mechanics. In: Roddy DH Pepin RO, Merrill RB (Editors) Impact and Explosion Cratering, Pergamon, New York, pp 703-714

Milton DJ, Barlow BC, Brett R, Brown AR, Glikson AY, Manwaring EA, Moss FJ, Sedmik ECE, Van Son J, Young GA (1972) Gosses Bluff impact structure, Australia. Science 175: 1199-1207

Milton DJ, Glikson AY, Brett R (1996) Gosses Bluff, a latest Jurassic impact structure, central Australia. Part 1: Geological structure, stratigraphy, and origin. AGSO Journal of Australian Geology and Geophysics 16: 453-486

Mozley P (2003) Diagenetic structures. In: Middleton GV (ed), Encyclopedia of Sediments and Sedimentary Rocks. Kluwer Academic Publishers, Dordrecht: 219-225

Nicolaysen LO, Reimold WU (1999) Vredefort shatter cones revisited. Journal of Geophysical Research 104: 4911-4930

Pettijohn FJ (1957) Sedimentary Rocks. Second edition. Harper Brothers. New York, 718 pp

Reimold WU, Colliston WP (1994) Pseudotachylites of the Vredefort Dome and the surrounding Witwatersrand Basin, South Africa. In: Dressler BO, Grieve RAF, Sharpton VL (eds) Large Meteorite Impacts and Planetary Evolution. Geological Society of America Special Paper 293: 177-196

Reimold WU, Minnitt RCA (1996) Impact-induced shatter cones or sedimentary percussion features in the Klapperkop Formation of the Eastern Transvaal? South African Journal of Geology 99: 299-308

Roddy DJ, Davis LK (1977) Shatter cones formed in large-scale experimental explosion craters. In Roddy DH, Pepin RO, Merrill RB (eds) Impact and Explosion Cratering, Pergamon, New York, pp 715-750

Rohleder HPT (1936) Lake Bosumtwi, Ashanti. Geographical Journal 87: 51-65

Sagy A, Reches Z, Finegerg J (2002) Shatter cones: Dynamic fractures generated by meteorite impacts. Nature 418 : 310-313

Selles-Martinez J (1994) New insights in the origin of cone-in-cone structures. Carbonates and Evaporites 9: 172-186

Sharpton VL, Dressler BO, Herrick, RR, Schnieders B, Scott J (1996) New constraints on the Slate Island impact structure, Ontario, Canada. Geology 24: 851-854

Shaub B.M. (1937) The origin of cone-in-cone and its bearing on the origin of concretions and septaria. American Journal of Science 34: 331-344

Simpson C (1981) Occurrence and orientation of shatter cones in Pretoria Group quartzites in the collar of the Vredefort 'Dome': impact origin precluded. Journal of Geophysical Research 86: 10,701-10,706

Tarr WA (1922) Cone-in-cone. American Journal of Science 220: 199-213

Tarr WA (1932) Cone-in-cone. In: Twenhofel WH (ed) Treatise on Sedimentation. Williams and Wilkins, Baltimore: 716-733

Wieland F, Reimold WU (2003) Field and laboratory studies on shatter cones in the Vredefort Dome, South Africa, and their genesis [abs.]. Meteoritics and Planetary Science 38: A16

Wieland F, Reimold WU, Gibson RL (2003) New evidence related to the formation of shatter cones: with special emphasis on structural observations in the collar of the Vredefort Dome, South Africa [abs.]. 3rd International Conference on Large Meteorite Impacts, Nördlingen, Germany, Lunar and Planetary Institute, Houston, CD-ROM, abstract # 4008

Woodland BG (1964) The nature and origin of cone-in-cone structure: Fieldiana Geology 13: 187-305

Redistribution of Lithologies in Impact-induced Dikes of Impact Structures

Victor L. Masaitis

Karpinsky Geological Institute, Sredny prospekt 74, St.-Petersburg, 199106, Russia (vicmas@vsegei.sp.ru)

Abstract. Lithic breccias and impactites, which occur as dikes and other intersecting bodies in the disturbed crater basement and in the crater fill, bear rock clasts derived from all layers of the target including uppermost and lowermost ones. Three groups of dikes bearing such redistributed clasts can be distinguished: injected dikes (composed of impact melt rock and polymict lithic breccia), squeezed dikes (monomict lithic breccias), and dikes in filled by gravitation (sandstone, claystone). These groups differ by the formation during different consecutive stages of cratering, by their respective spatial occurrence inside impact structures and in their vicinity, by the type of strain that induced fissures, by the directions of clast displacement, by the composition and aggregate state of the transporting systems that carried these clasts, etc.

The study of redistributed clasts, and dikes that bear them, allows reconstructing various features of an impact structure, and some conditions of its formation (a character of brittle deformation of host rocks, a depth of crater excavation and penetration of fissures into its true and apparent floor, a mode of clast transportation by fluidized systems). All these data are important for the investigation of "instant tectonics" taking place during impact cratering.

1 Introduction

Rock fragments (or so called "xenoliths") are usually found within dikes inside and outside of impact craters. They are different in lithology from the country rocks of the dike walls. These fragments may be named "wandering clasts", and are derived from different horizons of the target, and transported for a long distance during cratering, sometimes in opposite directions. The dimensions of clasts may vary from mm to tens of cm, they can be angular or rounded. Wandering clasts, although rare, are discernable by their peculiar lithologies, which allow to reconstruct their initial position in the vertical section of layered target rocks, and thus to estimate the trajectory and amplitude of their displacements. Various clast-bearing dikes have been distinguished in many impact craters (Wilshire et al. 1972, Halls and Grieve 1976, Grieve and Robertson 1976, Stöffler 1977, Lambert 1981, Mashchak and Ezersky 1982, Bischoff and Oskierski 1987, Stöffler et al. 1987, Müller-Mohr 1992, Rondot 1994, 1995, Hunton and Shoemaker 1995, Therriault et al. 1997, Sturkell and Ormö 1997, Warme and Kuehner 1998, Masaitis 1999, and others). Three main groups of dikes, that differ by their respective mode of origin and that bear wandering clasts, can be distinguished: dikes formed by injection, dikes formed by squeezing, and dikes formed by infilling (Fig. 1, Table 1). In impact craters and in their vicinity other groups of dikes exists (e.g., pseudotachylites and cataclasites, apophyses of thick impact melt sheets etc.), which formed variably during excavation and modification stages. As they show only small-scale displacements of clasts, they are not considered here.

The first group comprises clast-bearing dikes that formed during the excavation stage by downward injection of brecciated and melted material. Three varieties that differ in the composition of matrix, occurrence in the crater, and mode of transportation are distinguished. The second group is formed mainly during the compression and excavation stages, due to

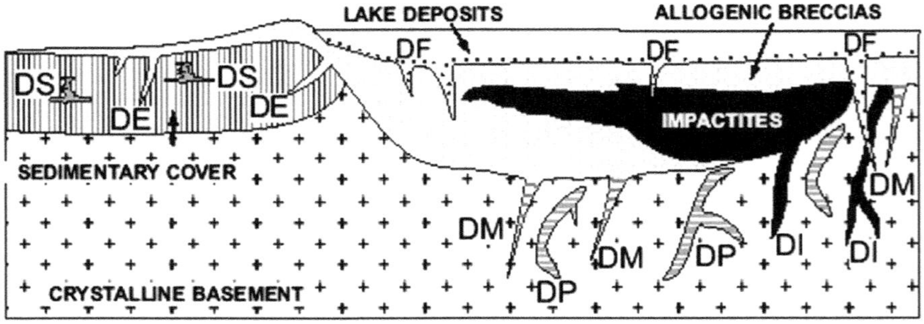

Fig. 1. Schematic occurrence of different groups of dikes within impact craters and in their vicinity. A complex crater with a central uplift in the two-layered target is shown. Dikes formed by injection: DI: impactites, DM: mylolisthenites and polymict lithic breccias, DE: polymict lithic breccias. Dikes formed by squeezing: DS: monomict lithic breccias. Dikes formed by infilling: DF: sandstones, clays. Pseudotachylite dikes: DP.

increased pressure in the water-saturated target layers in the vicinity of a growing crater, and by squeezing out of these unconsolidated deposits. The third group of dikes forms at the late modification stage by infilling of open fissures in the apparent crater floor. The dikes are indicated with letters, according to the type of dike-forming rocks (impactite or impact melt rock - DI, mylolisthenite (see below) and polymict lithic breccia - DM), or mode of origin (ejection - DE, squeezing - DS, infilling - DF).

2
Clast-bearing Dikes Formed by Injection

The most widespread group of injected dikes forms at the excavation stage during the transient cavity growth and radial tension of its floor. In the true crater floor two varieties of such dikes may be distinguished: a) DI - impactite dikes (impact melt matrix), and, 2) DM - mylolisthenite (from Greek myle – mill, and olistainein – to slide, Rondot 1994) and polymict lithic breccia dikes (matrix composed of clastic material, which is fluidal in the case of mylolisthenite), but some transitional types may exist. Some dikes formed involving multiple injections. Spatial distribution of these varieties in the crater floors is often radial and circumferential, but also depends on the initial irregularities within the target rocks - contacts, faults

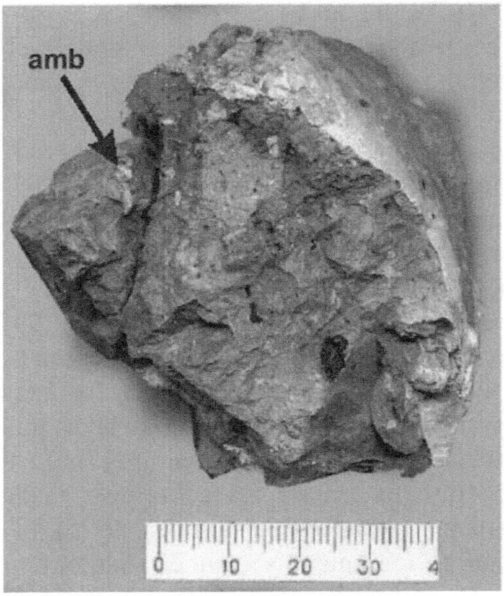

Fig. 2. Zhamanshin impact structure, Kazakhstan. Core sample of the polymict lithic breccia dike from the drill hole in the central uplift from 177 m depth. The breccia contains some particles of amber (amb) and organic debris derived from Paleocene deposits. Scale in mm.

etc. The thickness of these dikes is usually from cm to several meters, but in large impact craters they can reach up tens of meters. The larger fragments are usually concentrated in the central part of such dikes. The rock fragments are often shock metamorphosed, and sometimes show thermal and hydrothermal alterations. These dikes usually are partially disturbed at the early modification stage during the collapse of the transient cavity, and are displaced back upward together with the host rocks.

Injected dikes of tagamite (or another impact melt rock), mylolisthenite and polymict lithic breccia comprising wandering clasts are widespread in many impact structures. They are found in large structures whose diameters are from hundreds of km to many tens of km: Vredefort (French et al. 1989, Therriault et al. 1997, Henkel and Reimold 1998, Gibson and Reimold 2001), Sudbury (Dressler 1984, Müller-Mohr 1992), Popigai (Masaitis et al. 1998), Manicouagan (Currie 1972, Grieve and Floran 1978, Dressler 1990), Puchezh-Katunki (Masaitis and Pevzner 1999), Charlevoix (Robertson 1968, 1973,), Siljan (Hjelmquist 1966, Rondot 1976) etc., as well as in the middle-sized craters with the diameters from some tens to ten km: Sierra Madera (Wilshire et al. 1971, 1972), Slate Island (Halls and Grieve 1976, Sharpton et al. 1996, Dressler and Sharpton, 1997), Rochechouart (Lambert 1981, Bischoff and Oskierski 1987, Rondot 1995), Carswell (Pagel et al. 1985), Ries (Hüttner 1977, Stöffler 1977, Chao et al. 1978, Stöffler et al. 1987), Zhamanshin (Masaitis et al. 1993), Wells Creek (Roddy 1977), etc., and small ones, with diameters of only several km, e.g. Flynn Creek, Steinheim (Roddy 1977), and others.

The lithology of wandering clasts allows to estimate their stratigraphic displacements or depth of penetration in the brecciated crater basement, including its central uplift.

In the Sierra Madera impact structure the wandering clasts, found in polymict lithic breccia dikes and lenses, were displaced by about 1200 m (Wilshire et al. 1972). Numerous clast-bearing dikes were mapped in the Rochechouart impact structure (Lambert 1981, Bischoff and Oskierski 1987). Dikes in the floor of this crater are represented mostly by mylolisthenites, composed of crushed crystalline rocks (Rondot 1995). Polymict lithic breccia and other dikes occur abundantly in the central part of the Slate Island impact structure. They include specific rock fragments, showing that the amplitude of stratigraphic displacement may be as much as 5 km (Halls and Grieve 1976, Dressler and Sharpton 1997). In the Charlevoix impact structure mylolisthenite dikes rarely include sedimentary fragments (Robertson 1968, 1973, Rondot 1976). The estimated penetration of clasts into the basement is about of 1 km or more. The most frequent dikes (their arrangement is radial and concentric) occur close to the center of the impact structure and in the ring trough, but they are absent directly in the center (Rondot 1976).

In the largest and oldest impact structure known in the world, the Vredefort structure (e.g. Henkel and Reimold 1998, Gibson and Reimold 2001), wandering clasts (as such carbonates, quartzites) from the sedimentary cover of the target region were trapped by impact melt (granophyre), and transported downwards at the distance of some tens of km from their initial position. These relatively thick granophyre dikes resulted from melt injection into the central part of crater bottom, and have radial and circumferential arrangement (French and Nielsen 1990, Therriault et al. 1997, Gibson and Reimold 2001). Specific lithology of wandering clasts have been determined for the relatively small and fresh Zhamanchin crater (Boiko et al. 1991). Clasts of amber and other Paleocene siliciclastic rocks were found in an injected polymict lithic breccia dike, that cuts the brecciated rocks of the central uplift (Fig. 2). The particles of amber are now located about 100 m below the original position of this organic mineral in the undisturbed Paleocene beds, which intitially overlain folded Paleozoic rocks of the target (Masaitis et al. 1993).

Table 1. Dikes bearing wandering lithic clasts (Compiled after Wilshire et al. 1972, Halls and Grieve 1976, Stöffler 1977, Mashchak and Ezersky 1982, Bischoff and Oskierski 1987, Lambert 1987, Stöffler et al. 1988, Müller-Mohr 1992, Rondot 1994, 1995, Hunton and Shoemaker 1995, Sturkel and Ormö 1997, Warme and Kuehner 1998, Masaitis and Pevzner 1999).

	Matrix lithology	Location	Source of clasts	Direction of clasts displacements	Stage of cratering	Mode of origin
DI	Impactite (impact melt rock)	Beneath the true crater floor, in the ejected blocks	Uppermost target layers	Downwards (+ upwards)	Excavation	Injection into fissures
DM	Mylolisthenite, polymict lithic breccia					
DE	Polymict lithic breccia	In the crater wall	Uppermost and Lowermost target layers	Downwards and upwards		
DS	Monomict lithic breccia	In the water saturated target rock	Local target layers	Upwards	Compression, excavation, early modification	Squeezing out
DF	Sandstone, sand, clay	Beneath the apparent crater floor (in breccias and impactites)	Debris from crater lake bottom and dike walls	Downwards	Late modification	Infilling of open fissures

The deepest registered penetration of melt injection into crater basement rocks occurs in the Puchezh-Katunki crater (Fig. 3). Brecciated crystalline rocks of the central uplift are intersected by numerous mylolisthenite (Fig. 3) and tagamite (impact melt rock) dikes. The latter were traced on cores from deep drill hole (5374 m) to a depth of about 4.3 km (Fig.4). Sedimentary clasts in the breccia dike were found at the depth of about 0.5 km from the surface of central uplift (Masaitis and Pevzner 1999). According to the analysis of this core, the depth of impact melt penetration exceeds several times the depth of penetration of polymict breccia injections comprising wandering clasts.

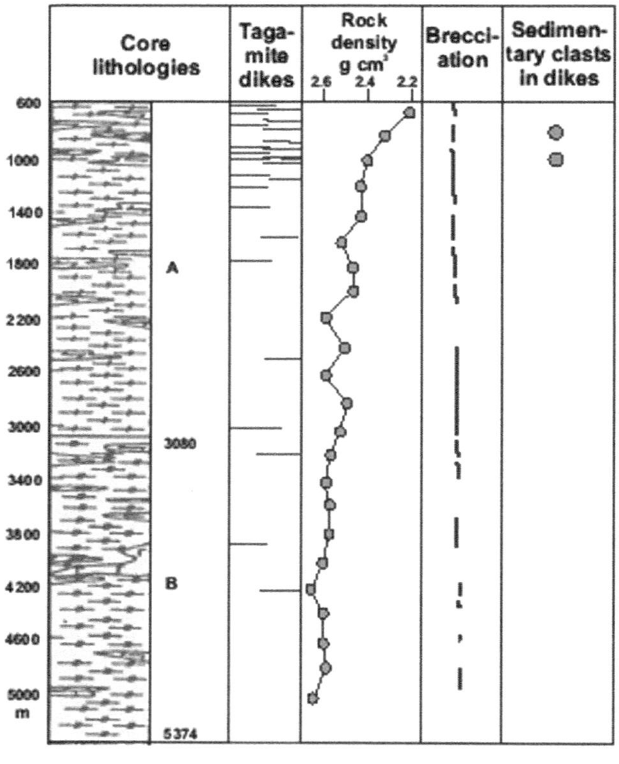

Fig. 3. Puchezh-Katunki impact structure, Russia. Schematic section of the lower part of the deep drill hole in its central uplift. Core lithologies: A = gneisses, amphibolites; B = gneisses, amphibolites, schists, quartzites, calciphyres. The amount and thickness of tagamite dikes decrease downwards, as well as the intensity of brecciation. In contrast, rock density increases downwards. Sedimentary clasts are found in mylolisthenite dikes to a depth of 1050 m. The upper part of the section (0-600m), which consists of crater lake deposits, suevites and allogenic breccia, is not shown (from Masaitis and Pevzner 1999, modified).

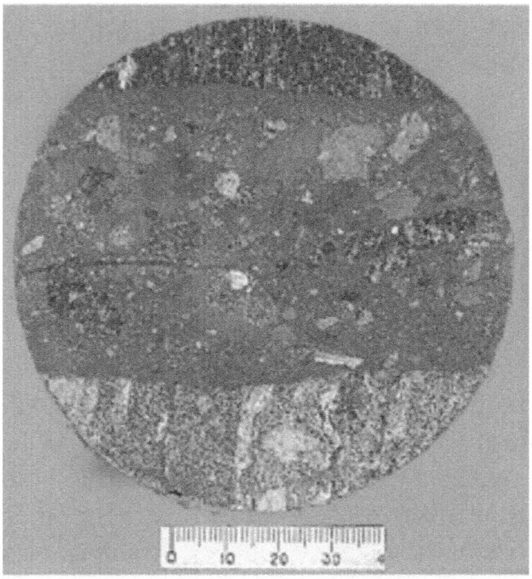

Fig. 4. Puchezh-Katunki impact structure, Russia. Core sample of a mylolisthenite dike in the gneiss of the central uplift. The drill hole is located 2.3 km south of the center of the impact structure, depth 356 m. Scale in mm.

The DI and DM injection dikes are formed during the short interval when downward motion of the transient cavity floor reverses, and fractures are caused by tensile stresses (Broberg 1988, Stöffler et al. 1987, Melosh 1989). The dike formation results from downward injection of gas- or melt-saturated,). crushed and compressed material following the Z-flow model (Melosh 1989. The low viscosity of this material is evident from its propagation over long distances, its ability to inject into very thin cracks and from the zoned inner structure of such dikes caused by hydrodynamic forces.

Blocks of target rock injected with these clast-bearing dikes and derived from the transient cavity floor are sometimes ejected, and can be incorporated within the allochthonous breccia lens. Such blocks cut by DM and DI dikes are found in the Ries and Popigai craters (Stöffler et al. 1987, Masaitis et al. 1998). In the Popigai crater these dikes cut brecciated

Fig. 5. Popigai impact structure, Russia. A core sample of a fluidal mylolisthenite dike composing of crushed leucocratic crystalline rock and including clasts of Upper Proterozoic quartzite (qt). The dike cuts brecciated pyroxene gneiss (dark gray on the photograph) of the annular uplift in the northwestern sector of the structure. Drill hole 0928, depth 313 m. Scale in mm.

blocks and contain wandering clasts from all principal lithologies known in the sedimentary target rocks: from uppermost Cretaceous to lowermost

Upper Proterozoic layers (Fig. 5, 6). The clast material derived from overlying beds (that initially occurred in normal stratigraphic sequence) is injected in all subjacent rocks, which compose now the ejected blocks. Some specific structural features of DM type dikes allow to reconstruct their mode of origin. For example, the clast dimensions sometimes exceed the thickness of a host dike, showing that open fractures collapsed (or closed) after infilling (Fig. 7). Very thin branching veins of microbreccia in the gneiss and limestone blocks (Fig. 6) probably formed by absorption of debris into opening fissures at the time of underpressurization just behind the propagating compression wave. These microbreccia veins are mostly composed of loose sandy material derived from Cretaceous beds, originally located in the upper portion of the target section. At the moment of injection this material must have been in a state of suspension under low viscosity.

The DI and DM groups of dikes mark the extent of fracturing within the crater floor, and also the extent of downward displacement of clasts from higher target levels. In the case of strongly eroded impact structure, these dikes may serve as the last indicators of the former crater.

Fig. 6. Popigai impact structure, Russia. A Cambrian limestone block of allogenic breccia that is cut by numerous branching microbreccia dikes, composed of sandy material, that contains organic debris derived from Cretaceous beds of the target. Western sector of the structure, Variegated Rocks cliff.

Fig. 7. Popigai impact structure, Russia. A polymict microbreccia dike cutting the cataclased gneiss of a large block in allogenic breccia. The dike consists of sandy matrix saturated with organic debris derived from Cretaceous beds. Embedded are slightly rounded fragments of Cretaceous (?) carbonificated wood (w), that shows evidence of desiccation, Permian black slate (s), Cambrian limestone (l), angular gneiss clasts and small clasts of impact glass are also present. Western sector of the structure, Variegated Rocks cliff. Scale in mm.

The third variety of injected dikes, e.g. DE type dikes, formed by injection of ejected polymict lithic breccia, and can be observed in crater walls or in fractured target rock outside of craters. Probably these dikes originated from ground surge transportation of ejected clastic material, represented by a mixture of various rock fragments. These polymict breccias are similar in lithology to allogenic lithic breccias forming lensoid bodies inside craters and ejecta blankets outside of them. Downward injection of these dikes causes downward displacement of some clasts. In some cases dikes may contain clastic material ejected from lowermost levels of the target as well. The dikes composing of such breccia were found in the vicinity of Lockne impact structure too (Sturkel and Ormö 1997), probably such clast-bearing dikes also occur in some other impact craters.

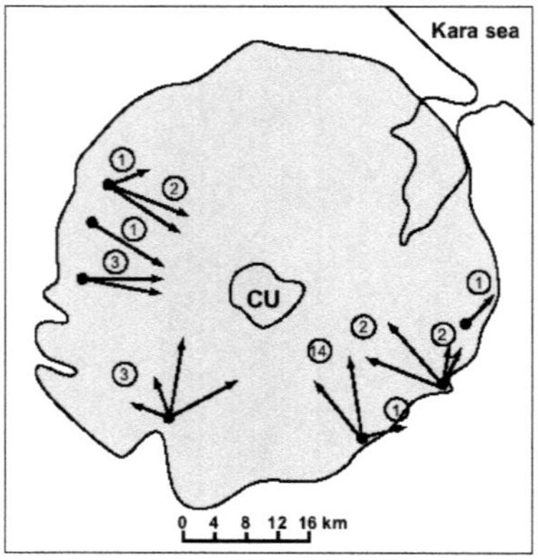

Fig. 8. Kara impact structure, Russia. Radial and circumferential orientations of dikes filled in by sandy material are shown by arrows; all dikes are located close to the crater edge. Numbers of dikes are indicated in circles. CU= central uplift of the structure (from Mashchak 1990).

3 Clast-bearing Dikes Formed by Squeezing

This special group of dikes and other bodies (sills, pipes) form in sedimentary target rocks mostly outside of an impact crater, but in its vicinity. They form in the water-saturated layers of the target due to pressure increase during the compression and excavation stages, or due to fracturing created by rarefaction wave and induced underpressure. The fractures propagate upwards, and clasts from lowermost layers are transported upwards in flow suspension. The dikes mostly made up of monomict breccias of local sedimentary rock fragments or loose deposits. Shocked fragments are rare.

The dikes may be zoned and may contain larger clasts in their interior. Dike thickness may reach tens of meters, and amplitude of upward displacement of wandering clasts may be as much as hundred meters or more.

There are a few examples of such dikes in the Sierra Madera srucrure (Wilshire et al. 1971, 1972), in the vicinity of Lockne (Sturkell and Ormö 1997), and in the Upheaval Dome (Hunton and Shoemaker 1995, Kriens et al. 1997) impact structures, also close to Alamo crater (Sandberg et al. 1997, Warme and Kuehner 1998). All of those dikes, sills and other bodies may be considered as the result of squeezing out.

Only in the Suffield explosion craters (Prairie Flat and Snowball), small craters formed in alluvium by high energy explosion tests in alluvium, dikes of such origin were mapped in detail (Jones 1977). The water borne sand was squeezed out from below along radial and circumferential fissures at the compression stage or not long after, and formed a network of thin dikes around the crater pit.

4 Clast-bearing Dikes Formed by Infilling

Another group of dikes forms during the relatively long-term late modification stage of cratering. These dikes caused by doming of the brittle crater fill, its adjustment to slow upward movements and the development of the tension fractures in the crater floor. Open fissures and cracks in the apparent crater floor are later infilled with debris due to gravitation. This debris consists of poorly cemented clast-bearing sandy and clayey material forming the fillings of such openings, and was derived from redeposited impactites and allochthonous lithic breccia. During redeposition at the bottom of the crater lake, or fissure walls destroying the clasts are transported downward into these open fractures together with water-saturated silt and sand, and are partly rounded. The lithology of fragments usually mirrors the whole target rock diversity; representatives of all target horizons may be found. Some fragments are shocked, and admixture of glass particles and melted material may take place. Formerly

Fig. 9. Kara impact structure, Russia. The repeatedly filled sandy dike contains fragments of various lithologies. The dike, that strikes parallel to the hammer handle, dissects suevite (upper right and lower left). Southeastern sector of the structure, Kara River. Hammer for scale.

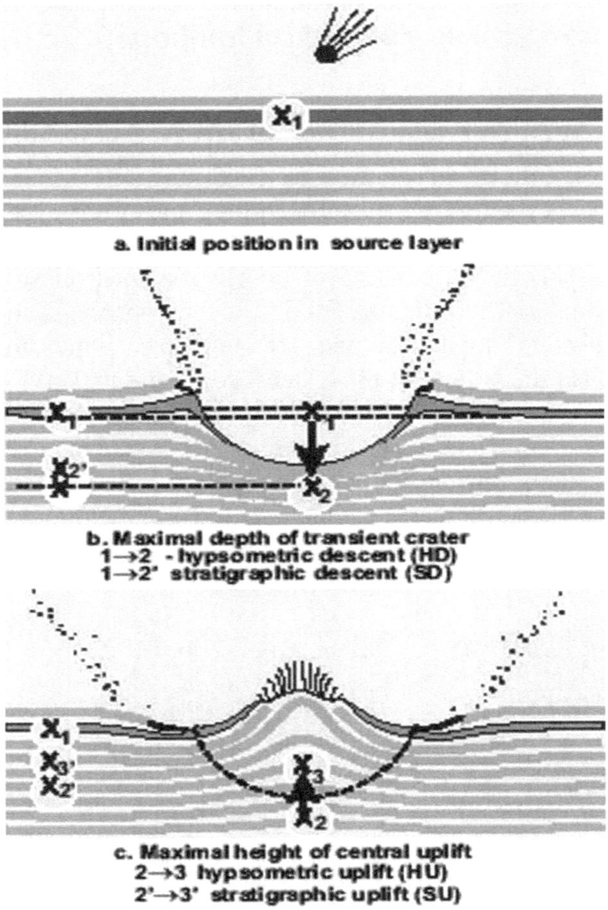

Fig. 10. Wandering clasts: amplitudes of displacement in an impact crater. X –positions of clast, derived from the uppermost layer (dark gray): **a**. initial position, **b**. downwards displacement, **c**. upwards displacement. The displacements measured in normal stratigraphic section are shown on the left sides of the schemes (2', 3').

some of these dikes were described as "clastic dikes" (Mashchak and Ezersky 1980, 1982, Mashchak and Fedorova 1987). They are similar to neptunic dikes, known in some geological regions.

Dikes formed by infilling of debris are found in the Kara and Ust-Kara impact craters, where they dissect the suevite sequence, occur all around and close to the crater edge (Mashchak and Ezersky 1980, 1982). In these craters such dikes (up to several meters thick), have radial and circumferential arrangements (Mashchak 1990, Fig. 8). The dikes cut the

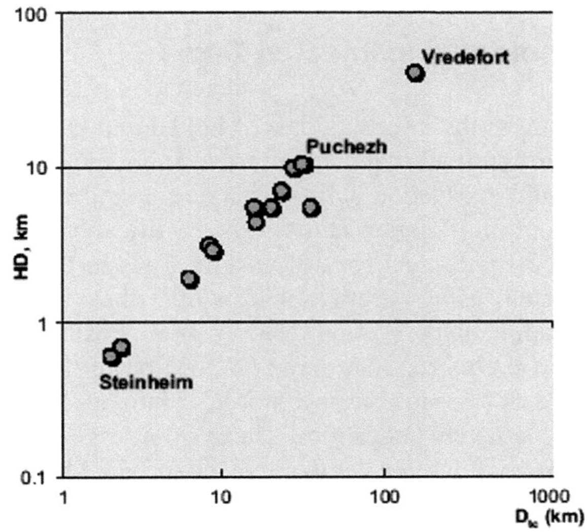

Fig. 11. The diagram shows estimated amplitudes of downward clasts (+ melt) displacement vs. diameter of transient crater. HD = hypsometric descent of wandering clasts, D_{tc} = diameter of transient crater.

host suevites and split and thinned out downward, and sometimes they show repeated infilling (Fig. 9). Individual dikes have been traced by drilling to a depth of about 400 m. This observation indicates, that secondary descent of clasts during the late modification stage may be significant. Such dikes also occur in the Popigai (Mashchak and Fedorova 1987, Masaitis et al. 1998), and Ries (Chao et al. 1978) craters. In the Puchezh-Katunki impact structure these dikes cross-cut suevites and allogenic breccias, and sometimes penetrate the underlying brecciated parautochthonous rocks of the true crater floor (Masaitis and Pevzner 1999).

This dike type formed by late infilling is possibly related to radial and concentric patterns on the floor of some lunar craters, so called "floor-fractured" craters, e.g. Humboldt (Wilhelms et al. 1987). These patterns caused by uplift of underlying layer due to viscous relaxation after the crater formation (Melosh 1989).

5 Displaced Lithologies in Injected Dikes

The clasts occurring in the injected dikes of the DI and DM types undergo complex downward and subsequent upward displacement, that begins at the compression and excavation stages of cratering and comes to the end when the crater floor uplift ceases at the early modification stage.

The downward displacement (or descent) of clasts in the injected dikes is usually estimated using stratigraphic width, that is the measured thickness of a sedimentary section below initial position of the clast lithology source to the present stratigraphic position of the clast. But this estimation is not exact because it does not take into account the thinning and stretching of the layers beneath the crater floor, especially close to its center. These deformations are temporary and occur when the transient crater reaches its maximum depth. More precise would be estimation of hypsometric or an absolute amplitude of descent. Additional contribution from the thinning and deflection of layers may be significant, thus, true hypsometric displacement always exceeds the stratigraphic ones (Fig. 10). Similar consideration concern the estimation of the uplift of clasts and clast-bearing host rocks in the crater floor. In some cases stretching may lead to increase of twice the layer thickness in a central uplift. cavity floor. The HD of wandering clasts may be estimated from the initial depth of clast source in the target, the depth of final transient cavity and depth penetration of clast-bearing dike into crater floor. In the HD estimation the ratios between final crater diameter (D), diameter and depth of final transient crater (D_{tc} and H_{tc}) may be used, including ratios D_{tc}=0.5-0.65D (Grieve et al. 1981), and H_{tc}/D_{tc}=0.24-0.32 (Melosh 1989).

The most significant depth of hypsometric descent (and penetration) of wandering clasts occurs in the central part of the craters. Clast-bearing dikes (DI and DM types) in the true crater floor are more abundant and thicker around the central uplift, but directly in the center they seem to be rare, thinner and may be entirely absent (Rondot 1994). Such character of dikes allocation may be caused by closure of fissures in the crater center due to stress caused by inward and upward motion of the rock mass. Still plastic clast-loaded material that fills in fissures may be partially squeezed upward, and this displacement should be taken into account too.

The principal dimensional parameters of the impact craters together with the estimate of hypsometric descent (including the penetration depth of clast-bearing dikes into the crater floor) can be used for compiling a plot D_{tc} vs. HD (Fig. 11), that is based on data for more than a dozen complex craters with diameters from 3.8 to 250 km, where injection dikes are present. This plot may be considered as a first approximation and requires additional and more precise data.

The diagram (Fig. 11) shows that HD is about half a kilometer for small craters, e.g. Steinheim (Roddy 1977). Study of the deep drill-hole in the central uplift of the Puchezh-Katunki structure showed, that thin injections of impact melt with small downward displaced clasts can be traced to a depth of about 4.8 km, but all fissures, that are related to decrease of rock densitydisappear at a depth of about 3.5 km (Masaitis and Pevzner 1999). The hypsometric descent may reach about of 40 km during formation of giant complex Vredefort impact structure (Gibson and Reimold 2001). The strength of the rocks of the crater floor and beneath to the significant depth at the moment of fissuringrequires special viewing however it leaves for framework of this paper.

According to to data presented on Fig.11 the regression line is HD = 0.504+0.282 D_{tc}. The ratio HD/D_{tc} is about 0.3-0.4 and is independent of crater size. In general, the curve HD vs. D_{tc} reflects the upper limit of hypsometric descent of clasts drawn by propagating dike material. Some of the dikes observed on the surface are in many cases relatively thick, and obviously are continuing to significant depths.

The upward motion and its hypsometric amplitude (HU) of the clast-bearing injected dikes together with the transient cavity floor may be evaluated as well. In every case absolute hypsometric uplifts are about 5-20% lower then the HD values. The total distance of downward and upward displacements (T_{dis}) of the wandering clasts embedded in injected dikes in large craters can reach tens of kilometers. Taking into account the time span of the crater formation (about of 1-5 minutes in the case of large craters) it is possible to calculate the average velocity of the motion of clast-bearing material of 200-500 m/s during the excavation and early modification stages. Additional downward displacements of some clasts may occur later during the late modification stage, and these clasts formerly embedded in injected dikes may be redeposited into newly opening fissures.

6 Conclusion

Long-distance downward and upward displacements of rock fragments induced by impact cratering may be distinguished using their specific lithology compared with source layers of target rocks in their original position. These fragments are used as markers of transportation of fragmented material, and are called wandering clasts.

1. Wandering clasts are found in dikes of monomict and polymict lithic breccias and impact melt rocks, that penetrate crater floors, crater fills and in the surroundings of impact crater. These dikes are produced at different stages of cratering by multiple injection, squeezing, and open fracture gravitational infilling.
2. An amplitude of hypsometric descent of wandering clasts (HD) that penetrate into the fractured true crater floor at the excavation stage directly depends on the diameter of the transient crater (D_{tc}). The total distance of downward and upward displacement of wandering clasts can reach tens of kilometers in the large impact structures. While in motion clasts velocity may exceed some hundreds m/s.
3. Wandering clasts and various dikes comprising them are considered as important structural feature mirroring "instant tectonics" of impact cratering, and subsequent relaxation events.

Acknowledgements

Dr. Herbert Henkel prompted the key point of this paper to the author, who is grateful to him for it and for the fruitful discussion on the subject. Author thanks for the technical assistance of A.T. Maslov with the preparation of figures and processing. Reviews by Dr. Ph. Claeys and Dr. W.U. Reimold and editorial handling by Dr. C. Koeberl led to significant improvement in the manuscript. The author appreciates their efforts.

References

Bischoff L, Oskierski W (1987) Fractures, pseudotachylite veins, and breccia dikes in the crater floor of the Rochechouart impact structure, South West France, as indicators of crater-forming processes. In: Pohl J (ed) Research in terrestrial impact structures, Vieweg & Son, Braunschweig, pp 5-29

Boiko YaI, Mashchak MS, Raikhlin AI (1991) Impact crater Zhamanshin. Guide of geological excursion. All-Russia Geological Research Institute (VSEGEI) Press, St.-Petersburg, 29 pp

Broberg KB (1987) Physical aspects on cratering and slumping. In: Boden A, Erickson KG (eds) Deep drilling in crystalline bedrock. I. Springer, Berlin, pp 261-276

Chao ECT, Hüttner R, Schmidt-Kaler H (1978) Principal exposures of the Ries meteorite crater in Southern Germany. Bayerisches Geologishes Landesamt, München 84 pp

Currie KL (1972) Geology and petrology of the Manicouagan resurgent caldera, Quebec. Geological Survey Canada Bulletin, 198, 153 pp

Dressler BO (1984) The effects of the Sudbury event and the intrusion of the Sudbury Igneous Complex on the Footwall rocks of the Sudbury structure. In: Pye EG, Naldrett

AJ, Giblin PE (eds) The geology and ore deposits of the Sudbury structure. Ontario Geological Survey, Special Volume 1, pp 97-138
Dressler BO (1990) Shock metamorphic features and their zoning and orientation in the Precambrian rocks of the Manicouagan structure, Quebec. Tectonophysics 171: 229-245
Dressler BO, Sharpton VL (1997) Breccia formation at a complex impact crater: Slate Islands, Lake Superior, Ontario, Canada. Tectonophysics 275: 285-311
French BM, Nielsen RL (1990) Vredefort bronzite granophyre: chemical evidence for origin as meteoritic impact melt. Tectonophysics 171: 119-138
Gibson RL, Reimold WU (2001) The Vredefort impact structure, South Africa. The scientific evidence and a two-day excursion guide. Council for Geoscience, South Africa. Memoir 92, 111 pp
Grieve RAF, Floran RJ (1978) Manicouagan impact melt, Quebec. 2. Chemical interrelations with basement and formational processes. Journal of Geophysical Research 83: 2761-2771
Grieve RAF, Robertson PB (1976) Variations in shock deformation at the Slate Islands impact structure, Lake Superior, Canada. Contributions to Mineralogy and Petrology 58: 37-49
Grieve RAF, Robertson PB, Dence MR (1981) Constraints on the formation of ring impact structures, based on terrestrial data. In: Schultz PH and Merrill RB (eds) Multi-ring basins: Formation and Evolution. Proceedings, Lunar Planetary Science 12A, pp 37-57
Halls HC, Grieve RAF (1976) The Slate Islands, a probable complex impact structure in Lake Superior. Canadian Journal of Earth Sciences 13:1301-1309
Henkel H, Reimold WU (1998) Integrated geophysical modeling of a giant, complex impact structure: anatomy of the Vredefort structure, South Africa. Tectonophysics, 287: 1-20
Hjelmqvist S. (1966) Beskrivning till berggrundskarta over Kopparbergs lä. Sverige Geologiska Undersöning Ser. Ca, 40: pp 217
Hunton PW, Shoemaker EM(1995) Roberts Rift, Canyonlands, Utah, a natural hydraulic fracture caused by comet or asteroid impact. Ground Water 33: 561-569
Jones GH (1977) Complex crater in alluvium. In: Roddy DJ, Pepin RO, Merrill RB (eds) Impact and explosion cratering. Pergamon Press, New York, Oxford, Toronto, Sydney, Frankfurt, pp 163-184
Kriens BJ, Shoemaker EM, Herkenhoff KE (1997) Structure and kinematics of a complex impact crater Upheaval Dome, southeast Utah. Brigham Young University, Geology Studies, 42, p.1. pp 19-81
Lambert P (1981) Breccia dikes: geological constraints on the formation of complex crater. In: Schultz PH, Merrill RB (eds) Multi-Ring Basins: Formation and Evolution. Proceedings, Lunar and Planetary Science 12A. Pergamon, NewYork, pp 59-78
Masaitis VL (1999) Impact structures of northeastern Eurasia: the territories of Russia and adjacent countries. Meteoritics and Planetary Science 34: 691-711
Masaitis VL, Pevzner LA, eds (1999) Deep drilling in the Puchezh-Katunki impact structure (in Russian). All-Russia Geological Research Institute (VSEGEI) Press, St-Petersburg, 392 pp
Masaitis VL, Raikhlin AI, Utkina YuD (1993) Coptogenic complex of the Zhamanshin crater (results of the deep drilling) (in Russian). Meteoritika 50: 137-141

Masaitis VL, Mashchak MS, Raikhlin AI, Selivanovskaya TV, Shafranovsky GI (1998) Diamond-bearing impactites of the Popigai astrobleme (in Russian). All-Russia Geological Research Institute (VSEGEI) Press, St- Petersburg, 178 pp
Mashchak MS (1990) Morphology and inner structure of the Kara and Ust- Kara astroblemes (in Russian). In: Masaitis VL (ed). Impact craters on the Mesozoic - Cenozoic boundary. Nauka Press, Leningrad, pp 37-55
Mashchak MS, Ezersky VA (1980) Clastic dykes of the Kara crater (Pai- Khoi). [abs] Lunar and Planetary Science XI: 680-682
Mashchak MS, Ezersky VA (1982) Clastic dikes in the impactites and allogenic breccias of the Kara astrobleme (northeastern slope of the Pai- Khoi Range) (in Russian). Lithology and Economic Minerals 1: 130-136
Mashchak MS, Fedorova IG (1987) The composition and mode of origin of clastic dikes in the tagamites of the Popigai astrobleme (in Russian). Meteoritika 46: 124-127
Melosh HJ (1989) Impact cratering. A geological process. Oxford University Press, New York, Clarendon Press, Oxford. 245 pp
Müller-Mohr V (1992) Breccias in the basement of a deeply eroded impact structure Sudbury, Canada. Tectonophysics 216: 219-226
Pagel M, Weathley K, Ey F (1985) The origin of the Carswell circular structure. In: Laine R, Alonso D, Svab M (eds) The Carswell structure uranium deposits, Saskatchevan. Geoogical Association of Canada, Special Paper, 29, pp 214-223
Robertson PB (1968) La Malbaie structure, Quebec – a Palaeozoic meteorite impact site. Meteoritics 4: 1-23
Robertson PB (1973) Zones of shock metamorphism at the Charlevoix impact structure, Quebec. Geological Society of America Bulletin 86: 1630-1638
Roddy D (1977) Flynn Creek impact crater, United States, Steinheim impact crater, Germany, and Snowball explosion crater, Canada. In: Roddy DJ, Pepin RO, Merrill RB (eds) Impact and explosion cratering, Pergamon Press, New York, Oxford, Toronto, Sydney, Frankfurt. pp 125-162
Rondot J (1976) Comparison entre les astroblèmes de Siljan, Suède, et de Charlevoix, Quebec. Bulletin Geological Institute University Uppsala 6: 85-92
Rondot J (1994) Recognition of eroded astroblemes. Earth-Science Reviews, 35: 331-365
Rondot J (1995) Les impacts météoritiques à l'exemple de ceux du Quebec. MNH Incorporated, Beauport, 157 pp
Sandberg CA, Morrow JR, Warme JE (1997) Late Devonian Alamo impact event, global kellwasser events, and major eustatic events, Eastern Great Basin, Nevada and Utah. Brigham Young University, Geology Studies, 42, pp 129-160
Sharpton VL, Dressler BO, Herrick RR, Schnieders B, Scott J (1996) New constraints on the Slate Islands impact structure, Ontario, Canada. Geology, 24: 851-854
Stöffler D (1977) Research drilling Nördlingen 1973: polymict breccias, crater basement and cratering model of the Ries impact structure. Geologica Bavarica 75, pp 443-458
Stöffler D, Bischoff L, Oskierski W, Wietz B (1988) Structural deformation, breccia formation, and shock metamorphism in the basement of complex impact craters: implications for the cratering process. In: Boden A, Erickson KG (eds) Deep drilling in crystalline bedrock. I. Springer, Berlin. pp 277-297
Sturkell EFF, Ormö J (1997) Impact related clastic injections in the marine Ordovician Lockne impact structure, central Sweden. Sedimentology 44: 793-804
Therriault AM, Reimold WU, Reid AM (1997) Geochemistry and impact origin of the Vredefort granophyre. South African Journal of Geology 100: 115-122

Warme JE, Kuehner H-C (1998) Anatomy of an anomaly: the Devonian catastrophic Alamo impact breccia of Southern Nevada. International Geological Review 40: 89-216

Wilhelms DE, McCanley JF, Trask NJ (1987) The geologic history of the Moon. United States Geological Survey, Professional Paper 1348, 302 pp

Wilshire HG, Howard KA, Offield TW (1971) Impact breccias in carbonate rocks, Sierra Madera, Texas. Geological Society of America Bulletin 82: 1009-1018

Wilshire HG, Offield TW, Howard K, Cummings D (1972) Geology of the Sierra Madera cryptoexplosion structure, Pecos County, Texas. United States Geological Survey, Professional Paper 599-H, 42 pp

The Preliminary Analysis of Polygonal Impact Craters within Greater Hellas Region, Mars

Teemu Öhman[1], Marko Aittola[2], Veli-Petri Kostama[2] and Jouko Raitala[2]

[1]Institute of Geosciences, Department of Geology, P.O. Box 3000, FIN-90014 University of Oulu, Finland. (teemu.ohman@oulu.fi)

[2]Planetology Group, Division of Astronomy, Department of Physical Sciences, P.O. Box 3000, FIN-90014 University of Oulu, Finland

Abstract. The polygonal planimetric shape of impact craters has been known for a long time, but has not been discussed much in the past. Polygonal craters exist on all kinds of celestial bodies that have fractured rigid crusts. They are also found on the Earth (e.g., Meteor Crater, Söderfjärden). Polygonal craters are thought to have been formed by two possible mechanisms. Simple polygonal craters, such as the square-shaped Meteor Crater, result when the excavation flow opens the crater, tearing the target more easily along pre-existing fractures or other planes of weakness. Complex polygonal craters are supposed to have formed during the modification stage, when rocks of the crater rim slump along the fractures in the target. Both mechanisms lead to straight segments of the rim. However, in simple craters the straight rim is typically at some angle, usually about 45°, to the direction of the fractures, whereas in complex craters the rims are thought to be parallel to them. Thus, the regional fracture trends are easier to deduce from complex polygonal craters.

We have studied the distribution and rim orientations of polygonal craters within the geologically versatile greater Hellas region in the southern hemisphere of Mars. Our results (rose diagrams of straight rim segments) indicate the existence and dimensions of radial and concentric fracture patterns around the impact basins of Hellas and Isidis. Such deformation patterns have been observed previously using other indicators of tectonism (grabens, etc.). Other major causes of fracture directions visible in our rose diagrams are the volcanoes in or near the study area.

Especially Elysium Mons has created a prominent radial fracture system. Our results correlate well with those obtained by other methods, thus indicating that the use of polygonal craters in finding, studying, and mapping structural properties of a planet's crust is justified.

1 Purpose of This Study

The existence of polygonal craters, i.e., craters that are more or less angular instead of circular or ellipsoidal, has been known at least for about a century (Fielder 1961; Kopal 1966). Given their frequency on the surface of the Moon, they probably have been known as long as detailed observations of the lunar craters have been available. We have noted that polygonal craters are common, or at least present, not only on the Moon, but also on the Earth, Venus, Mercury, Mars and several asteroids (e.g., 433 Eros) and icy moons (e.g., Europa) – i.e., on all kinds of bodies with rigid crusts and impact craters. Despite this, in the more recent impact crater literature polygonal craters are conspicuously absent.

The aim of our present work is 1) to give a brief review of some of the essential earlier studies concerning the observations and the formation of polygonal craters, 2) to study the distribution of polygonal craters in the greater Hellas region, Mars, and 3) to check whether or not polygonal craters can be used as a tool in the investigation of the structural and tectonic features of Martian crust. The goal of this work is also to revive the concept of polygonal craters in the impact cratering community.

2 Earlier Studies and the Formation of Polygonal Craters

Theories of the origin of hexagonal lunar craters were discussed already in 1908 by Puiseux (Fielder 1961; Kopal 1966). The 1960's, however, seems to have been "the golden age" of polygonal crater studies. Gilbert Fielder, a strong advocate of the lunar craters' volcanic origin, fundamentally stated that polygonally shaped craters are common (Fielder 1961). Later Fielder also noted the connection between square-shaped and hexagonal craters with orthogonal and three-directional fracturing, respectively (Fielder 1965). From the structural and tectonic point of view, Kopal's (1966) remark that "...many of the hexagonal craters...are similarly orientated and their sides more or less parallel – whether they are adjacent or separated by some distance." is of major significance.

Eugene Shoemaker's detailed work (Shoemaker 1963) on Meteor (Barringer) Crater, Arizona, was essential for the eventual understanding of the formation mechanism of small (simple) polygonal craters. He noted that two orthogonal sets of regional joints form approximate diagonals across the almost square-shaped Meteor Crater. The rim of the crater is bisected by nearly vertical tear faults – parallel to regional joint directions – and that largest tears occur in the corners of the crater. The regional lines of weakness thus controlled the shape of the crater. Shoemaker, however, did not explain the actual mechanism.

Another prominent figure in the study of impact craters, Ralph Baldwin (1963), wrote about terrestrial meteoritic craters that they are often polygonal rather than circular. About lunar craters he noted that craters with four, five, or six straight sides are common, or that at least they have a tendency toward such shapes. Based on Eugene Shoemaker's work on the Meteor Crater, Baldwin emphasized that earlier structural lines of weakness, not later distortional forces, were the essential factor in the formation of polygonal craters. (Baldwin 1963)

Ronca and Salisbury (1966) also thought about the possibility of non-circular lunar craters being a result of an impact into a densely jointed target, deeming it, however, to be an unlikely scenario. Adler and Salisbury (1969), continuing the work of Ronca and Salisbury (1966), found that their "subcircular" (no reference to the polygonality of the craters is made in either of the aforementioned papers) craters were concentrated close to the lunar maria, which were known to be the centers of extensive radial fracture systems.

In an attempt to study the alleged lunar tectonic grid, Elston et al. (1971) measured a large number of polygonal crater walls from the highlands and maria of the lunar near side. They concluded that the polygonal craters (and the lunar grid) are controlled by tensional and conjugate shear fractures. They didn't completely rule out the possibility of the observed fracture pattern emerging from early major impacts, but certainly preferred an endogenic origin for it.

On Earth not only the Meteor Crater has been observed to be of polygonal outline. For example, Grieve et al. (1988) mentioned that in satellite images the Bigach structure in eastern Kazakhstan appears as a polygonal rimmed area. To our knowledge, the best sample of a hexagonal impact structure on Earth is the Söderfjärden structure in western Finland (e.g., Laurén et al. 1978, Abels 2003). It displays a very distinct hexagonal plan view discernible from aerial photography, satellite images and topographic maps. Highly eroded impact structures, like Saarijärvi in northern Finland (Pesonen et al. 1997, 1998), can reveal their true polygonal shape only in geophysical properties (Öhman 2002; Öhman et al. 2003).

Polygonal craters on Mars were identified already from the Mariner images in the late 1960's – early 1970's, but didn't receive more than a brief mention from Pike (1971, 1980). R. A. Schultz (1985), in his study of ancient Mars tectonics, used a small number of "polygonal crater wall-scarps" along with other tectonic features to investigate Tharsis-centered tectonism (see Fig. 8 in Schultz 1985), but did not give any details or descriptions of the craters he studied.

Of utmost importance for the study of cratering mechanics were the early cratering experiments conducted at the NASA Ames Vertical Gun Ballistic Range, the results of which are commented on by, e.g., Gault et al. (1968). Polygonal craters were also studied, but they were described only very briefly. The most interesting finding by Gault et al. (1968) concerning polygonal craters was that two orthogonal sets of fractures can give rise not only to square-shaped craters, but also to hexagonal craters. In the case of a square-shaped crater the fractures formed diagonals across the crater, just as at Meteor Crater. The relationship between the fractures and the rim directions of the hexagonal crater were, unfortunately, not described.

The unfortunate failure to study the pre-existing fracture pattern's relationship to polygonal craters holds also for high-explosive cratering studies conducted on Buckboard mesa at the Nevada Test Site. These explosions in jointed basalt resulted in both partly and completely hexagonal and pentagonal craters (Johnson 1962, cit. Fulmer and Roberts 1963). Much smaller explosions (Boeing Oso Cratering Experiment) in jointed argillite produced roughly square-shaped craters. Fulmer and Roberts' (1963) summary of these ~2–3 m diameter craters shows, however, that a clear connection between crater shape and the strike of joints cannot be postulated, since there are joints parallel both to craters' rims and to their diagonals.

In addition to experimental craters, Fulmer and Roberts (1963) studied lunar craters, noting that different types of polygons, from square to at least an octagonal shape, are present on the Moon's surface. An important conclusion of their experimental and photogeological work is that where craters display a distinctly polygonal shape, the target material is rigid with a well-developed simple fracture pattern. Dominantly circular craters are formed either in relatively soft materials, in rigid material with a complex set of very closely spaced fractures, or in rigid material with very widely spaced or non-existent fractures (Fulmer and Roberts 1963).

Schultz (1976) studied the morphology of lunar craters and observed that the polygonal shape of craters with diameters between 1–15 km is more easily identified than that of larger craters. In craters less than 1 km in diameter the polygonal outline becomes increasingly difficult to recognize because of rubbly nature of the rims. He also noticed that the plan views of lunar craters larger than 15 km in diameter are usually either polygonal or scalloped. In Schultz's study (1976) the connection between straight segments of the crater rim and regional structural trends was clearly manifested. The time of the formation of the polygonal plan view was attributed both to the excavation (craters smaller than 15 km in diameter, ~simple craters) and modification (larger than 15 km, ~complex craters) stages of cratering process. In the case of smallest, weakly developed polygonal craters, Schultz (1976) attributed the formation of the polygonal plan view partly to later erosion and encroachment by lava.

Pike (1977) gave a brief review of the studies concerning crater circularity or polygonality. Although non-circularity and polygonality of a crater is not the same thing, one can conclude from Pike's (1977) review and the aforementioned earlier studies of lunar craters, that larger craters tend to be polygonal more often than smaller ones. This somewhat contradicts the results of Eppler et al. (1977), who carried out the only truly mathematical and statistical studies of the planimetric shapes of lunar craters that we are aware of. Eppler et al. (1977, 1983) used Fourier analysis to evaluate the relationships between variation in crater shapes and variation in, e.g., crater age, size, and crustal properties. One of their main results was that for craters larger than 18 km in diameter, the crater size does *not* affect the planimetric shape, although they rather ambiguously also state that larger craters tend more toward circularity than smaller ones. In addition, Eppler et al. (1977) noted that highland craters are less circular than mare craters. The reasons behind the variation in the planimetric shapes of lunar craters were speculated to be, e.g., target lithology and regional joint sets.

Eppler and co-workers continued and expanded their original study (Eppler et al. 1977) and summarized their work in an important but largely unknown paper (Eppler et al. 1983). They presented two models for the formation of polygonal craters. The crucial factor in both of them is of course the existence of fractures, joints, faults or other zones of weakness (henceforth "fracture" should not be understood literally, but rather as a collective term comprising all the above-mentioned and similar features in rocks or other rigid materials, regardless of their origin or exact definition) in the target rocks (Eppler et al. 1983). Thus, a fractured crust is a pre-requisite for the formation of polygonal craters. The first model, based largely on field evidence from Meteor Crater (e.g., Shoemaker 1963), explains the formation of polygonal simple craters. They are thought to form when the excavation flow opens the crater tearing the target more easily along pre-existing fractures. In other words, the excavation of the crater progresses more easily along some plane of weakness than in other directions, leading to a square-shaped outline (Eppler et al. 1983). Thus, at least in the case of the Meteor Crater the fracture directions form diagonals across the crater, i.e., the crater rims form approximately an angle of 45° with the fracture directions (Shoemaker 1963). This model is supported also by the relationship of small polygonal crater rims and linear structural elements on the Moon. However, the experiments by Gault et al. (1968) showed that two perpendicular pre-existing fracture directions can also lead to three rim orientations, i.e., a hexagonal crater. Therefore, the information gained from the study of small polygonal crater rims cannot be unambiguously transferred to regional fracture directions (see also Fulmer and Roberts 1963).

The second model of Eppler et al. (1983) is an attempt to explain polygonal craters in the size range of complex craters. They present perhaps a much more straightforward case than simple craters. Complex polygonal craters seem to be the result of slumping in the modification stage of the cratering process: the collapse of the over-shooting rim takes place along some plane of weakness in the target rocks (Eppler et al. 1983). Thus, the dominant fracture directions in the area can be directly measured from the orientations of the straight rim segments in complex polygonal craters.

From a theoretical point of view, however, slumping along pre-existing fractures represents a problem, since the collapse of a crater should take place in a medium that has gone through an extreme degradation of strength (e.g., Melosh and Ivanov 1999 and references therein). Thus, the effect of pre-existing zones of weakness is supposed to be small. Yet, as an exact mechanical model of impact crater collapse is still lacking (Melosh and Ivanov 1999) and no better formation mechanism for complex polygonal craters has to our knowledge been proposed, we feel free to use the idea of fracture-dominated slumping of the crater rim as a working hypothesis. In short, we assume that in simple polygonal craters the straight segments of the crater rim bisect the regional fractures in some angle, whereas in complex craters the straight segments are parallel to them.

Eppler et al. (1983) also discussed the possibility of, e.g., erosion or flooding being a cause for the polygonal outline of craters. The result of their study is straightforward: degradation or aging of the crater increases the small-scale *irregularity* of craters, but does not affect large-scale polygonality. According to Eppler et al. (1983), "fundamental shape characteristics... are likely to persist relatively unchanged by degradational processes". Flooding of craters by lava near lunar maria could change the shape of some craters, but actually it slightly increases the craters' circularity. Other variables like topography or layered target have only a very minor effect on the planimetric shape of any larger population of impact craters. As the polygonal outline is a permanent feature of craters, it can reveal old structural patterns of the crust that can no longer be perceived otherwise (Eppler et al. 1983).

One more indication in favour of the hypothesis that polygonal shape is not an erosional feature, but originates from the cratering process itself, comes from the study of Venus. Venus' small population of impact craters is geologically young and in a very pristine state (e.g., Phillips et al. 1992), yet we have observed polygonal craters also there. Of course it is possible and even probable that erosion can create polygonal features for some impact craters, but we do not think it plays a major role in any study dealing with a large area, or a large number of craters.

The observations and conclusions presented above show that although some extensive studies concerning polygonal craters have been carried out in the past, there are still major gaps in our understanding. The cratering experiments (e.g., Fulmer and Roberts 1963; Gault et al. 1968) clearly indicate that the relationship between pre-existing fractures in the target and the crater's straight rims is not as straightforward as often presented (e.g., Eppler et al. 1983; Melosh 1989) even in the case of simple craters. The theoretical basis for the formation of complex polygonal craters is somewhat flimsy, although it serves well as a working hypothesis for the studies of areal tectonism. In such investigations it would be interesting to study the ratio of polygonal and circular craters, rather than just the number of polygonal craters. Rigorous mathematical and statistical analysis (Eppler et al. 1977, 1983) should be incorporated to reduce the effect of subjectivity in the results and conclusions based on them.

3 Impact Structures and Fracturing

Fractures in a planet's crust can, of course, be created by several different means, but one of major significance in cratered landscapes is impact cratering. Cratering creates mainly two systems of fractures, one radial and one concentric to the crater. This relatively shallow directly impact-induced fracturing may extend up to one crater diameter beyond the crater rim. This has been confirmed by geological and geophysical methods from several terrestrial craters and also by cratering experiments. (e.g., Fulmer and Roberts 1963; Curran et al. 1977; Gurov and Gurova 1982; MacKinnon and Tanaka 1989; Henkel 1992; Pilkington and Grieve 1992)

The largest craters, i.e., impact basins, can produce radial and concentric deformation patterns not only by direct fracturing during the early stages of the cratering process, but also, and mainly, by subsequent infilling of the basin by denser material and isostatic equilibration. This long-lasting deformation of the basin and its surroundings can lead to the formation of ridges, grabens and both radial and concentric fractures extending deep to the crust. These indications of the deformation can extend thousands of kilometers away from the center of the basin. (e.g., Schultz et al. 1982; Wichman and Schultz 1989)

4
The Schematic Geology of the Greater Hellas Region

Our study area (latitude 2°N–66°S, longitude 8°W–208°W) covers a large portion of the southern hemisphere of Mars (see Fig. 1 and 2). The region is predominated by the Hellas impact basin some 2000 km wide and 9 km deep, possibly caused by an oblique impact with a trajectory from northwest (Leonard and Tanaka 1993, 2001; Tanaka and Leonard 1995). Another major impact basin, Isidis, is immediately to the north of our study area.

The region around Hellas impact basin is a geologically diverse area. The highland units and especially the western rim of the Hellas basin, Hellespontus Montes, represent the oldest material in the region (Tanaka and Leonard 1995). They include the wide old cratered highland west of Hellas (Noachis Terra) and the eastern highland region. The very shallowly sloped large volcanic edifices of Tyrrhena and Hadriaca Paterae east of Hellas, and Peneus and Amphitrites Paterae south of Hellas represent some of the most ancient recognizable post-impact Hellas volcanism on Mars (Plescia and Saunders 1979). These paterae with highly degraded appearance of the slopes portray evidence of pyroclastic eruptions (Tanaka and Scott 1987; Crown and Greeley 1993; Tanaka and Leonard 1995). They are surrounded by the low-lying volcanic plains of Hesperia and Malea Plana (Tanaka et al. 2002). In the north our study area is on the verge of the volcanic plains of Syrtis Major Planitia. Also Elysium Planitia, a major center of Martian volcanism, reaches the northeastern corner of our study area. The border of Elysium Planitia also marks the global dichotomy boundary.

The Hellas internal rim boundary displays interesting delta formations of several large outflow channels as well as shorelines resulted in standing water body or in an ancient ice-covered lake (Moore and Wilhelms 2001). Although channels implying fluvial activity can be observed on both sides of Hellas, particularly the low eastern plains region bears softened features adjacent to large channel formations. It has been suggested that the origin for the large outflow channels in the northeastern Hellas area was due to the late-stage effusive volcanism of Tyrrhena Patera (Leonard and Tanaka 2001). It may have triggered collapse and outflow erosion producing the valleys of Dao, Niger, and Harmakhis (Squyres et al. 1987), which are roughly radial to Hellas suggesting a possible connection to Hellas-centered tectonism, i.e., mainly fracturing (Crown et al. 1992). The eastern Amphitrites area has also a wide, channel-like formation connecting to the Hellas basin.

The youngest geological unit consists of the material transported by eolian activity and deposited mainly as dune fields within craters. This unit has distinctly concentrated to the western side of the basin. The absence of large dune fields in the eastern region is surprising because the whole Hellas region is proposed to provide dust storm materials (Martin and Zurek 1993).

The volcanic plains of Hesperia and Malea Plana bear a rich collection of wrinkle ridges similar to lunar mare ridges (e.g., Raitala 1988; Crown et al. 1992). Many of these are induced by the paterae, but the numerous ridges radial and concentric to Hellas imply tectonic control by the basin. Some of the ridges in Hesperia Planum continue to the adjacent highland areas (Raitala 1988). The wrinkle ridges have their origin mainly in the sinking of the Planum, which lead to compression and faulting. The trends of the ridges are probably reflecting older crustal fractures and other similar structures (Raitala 1988 and references therein).

The investigation of the various craters of the region and combining the crater distribution of morphologically classified craters (rampart, collapsed, craters modified by fluvial activity) (Aittola et al. 2002) with a geological analysis of the region has increased our knowledge of the geology and evolution of the region (Kostama et al. 2002). The already obtained crater distributions indicate that the Hellas area displays a versatile sample of Martian impact craters.

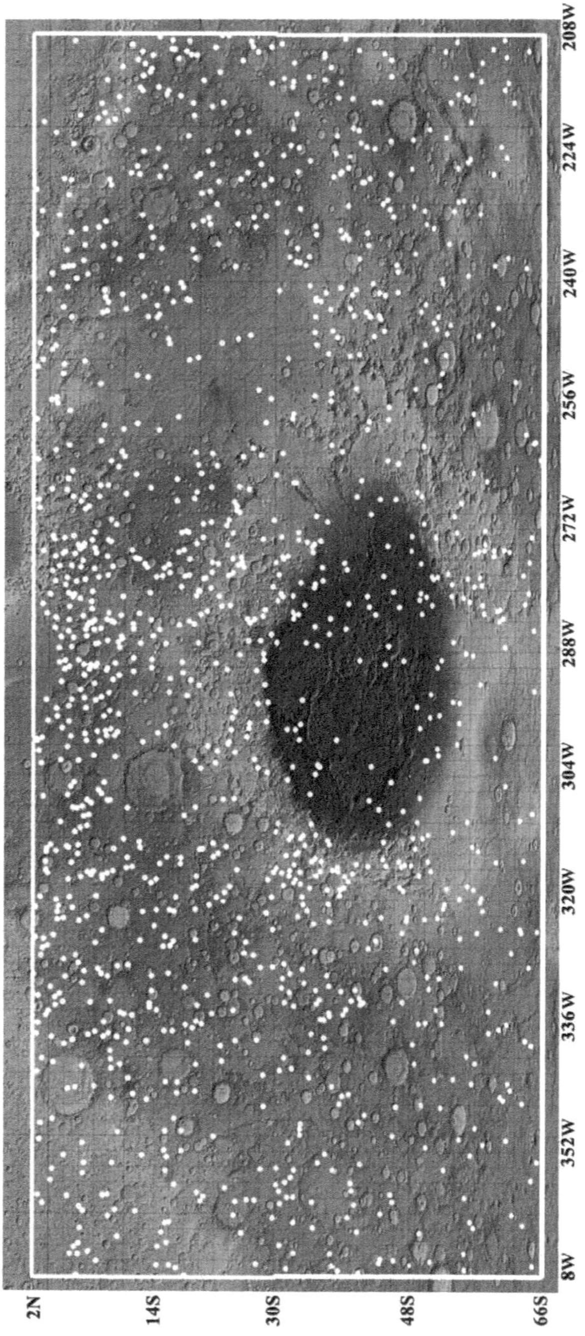

Fig. 1. The distribution of polygonal impact craters (white dots) in greater Hellas region plotted on MGS-MOLA topographic data. The white rectangle indicates our study area (2°N–66°S, 8°W–208°W).

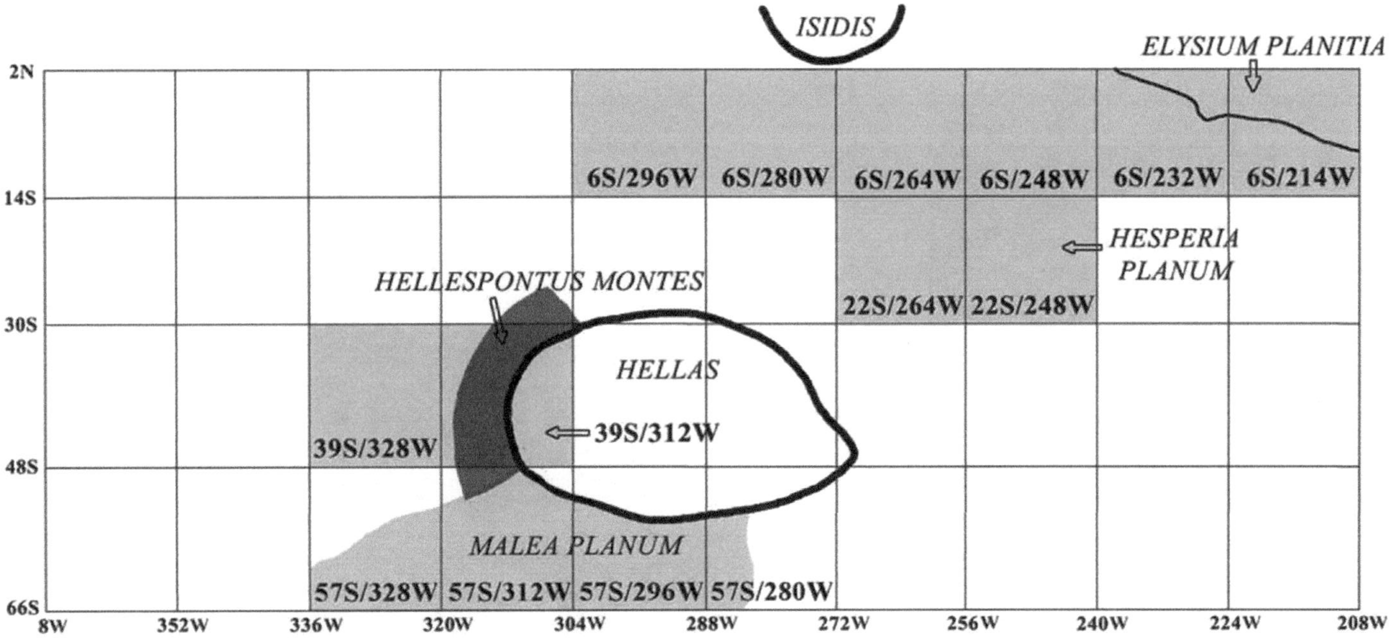

Fig. 2. The block division of the study area with outlines of some major regions discussed in the text. The shaded areas indicate the blocks in which rim strike measurements were made. Central coordinates of each block are used as their identification number.

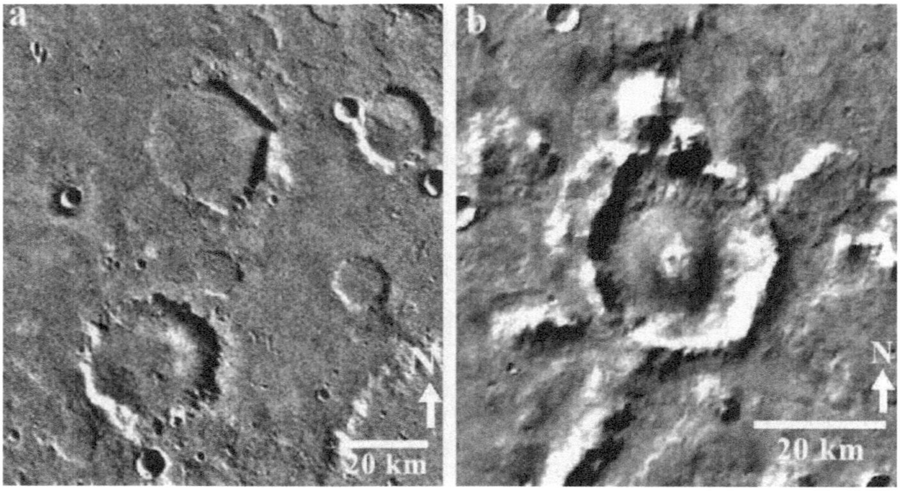

Fig. 3. (**a**) (left) Hexagonal craters north from Hellas (25°S/277°W). Note that many of the crater rims are parallel to each other. (**b**) (right) A very symmetric hexagonal crater north from Hellas (28°S/283°W). Images from Viking Orbiter MDIMs.

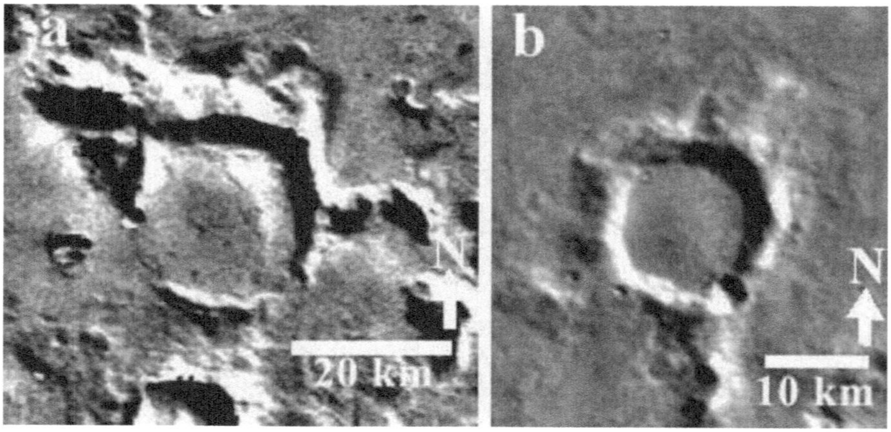

Fig. 4. (**a**) (left) An unusually large square-shaped crater north from Hellas (24°S/298°W). (**b**) (right) Polygonal crater northwest from Hellas (29°S/317°W) with some tendency towards a pentagonal outline. Images from Viking Orbiter MDIMs.

5 Methods

We have done a detailed mapping of polygonal impact craters in the greater Hellas region (2°N–66°S, 8°W–208°W) using the Viking Orbiter Mars Digital Image Mosaic (MDIM). The resolution of the images is 0.2314 km per pixel in the equator, but gets worse further south. The new higher resolution images, such as Mars Global Surveyor's Mars Observer Camera (MOC) or Mars Odyssey's Thermal Emission Imaging System (THEMIS) images still have a far too limited coverage to be utilized in a systematic study over such a vast area as presented here.

The study area was divided into 40 blocks (see Fig. 2), each of which was examined independently by two researchers. We defined "polygonal crater" as being a crater with at least two straight rim segments and a clearly discernible angle between them. We disregarded craters where polygonal outline was distinctively the result of post-impact processes like erosion, sedimentation, or younger impact. In addition, all craters that we accepted as being truly polygonal had to be classified as polygonal by both of the researchers studying that particular block of the area.

The strikes of the rim directions were measured from three major parts of the study area: the highland area south from Isidis (reaching the volcanic plains of Elysium Planitia in the northeastern corner of our study area), Hellespontus Montes west from Hellas and the plains of Malea Planum. In the area of Malea Planum only those polygonal craters that were formed in the plains material were included in the measurements. In other words, in each of the four blocks of the study area that are in the region of Malea Planum, we disregarded all polygonal craters in highland material. In other studied regions (i.e., around Hellespontus Montes and the large area south and southeast of Isidis), we measured directions from all polygonal crater rims regardless of their target material. The strikes were then plotted in rose diagrams using 10° azimuth intervals. The use of 15° azimuth intervals wouldn't change the general trends in the diagrams. The precision of the strike measurements is estimated to be about ±5°. Note that in the rose diagrams the frequencies of the directions are given as percentages, not as absolute counts. The frequencies were rounded to the nearest percent.

The transition diameter between simple and complex craters on Mars depends to some extent on the properties of the target material. However, usually the transition from simple to complex craters takes place in the diameter range of 5–8 km (Pike 1980). Practically all of the craters we were able to designate as being polygonal craters were at least about 10 km in diameter and, thus, were the size of complex craters. Therefore, the measured rim strikes are assumed to be parallel to the directions of fractures in the target, as described above.

Figures 3 and 4 show some typical polygonal craters in the greater Hellas region, as seen in MDIM. To rule out the possibility that polygonal shape of some smaller craters included in this work is only an artifact of poor resolution, we studied higher resolution Mars Odyssey THEMIS visual (19 m/pixel) and infrared (100 m/pixel) images. Fig.s 5 and 6 show the same polygonal craters northwest from Hellas displayed in Viking Orbiter MDIM (Fig. 5a and 6a) and Mars Odyssey THEMIS nighttime infrared images (Fig. 5b and 6b). This comparison clearly indicates that the resolution of MDIM is sufficient for this study, but also emphasizes that different degrees of polygonality are seen in different wavelengths. Very small polygonal craters are not abundant, but can be seen in some THEMIS visual images.

An inherent problem of photogeology is that illumination can enhance certain features while others – even prominent ones – can be significantly subdued (e.g., Schultz 1976). Of course, this applies also to polygonal craters and this study to some extent. We cannot rule out the possibility that some craters now included in our study would not fulfill our relatively strict criteria for polygonality, or display slightly different rim strikes under other lighting conditions. However, the craters we accepted as being truly polygonal had straight rim segments in at least two directions. In our opinion, this combined with the facts that each polygonal crater had to be accepted by two researchers independently, and that the number of craters studied was quite high, diminish the illumination bias in the resulting rose diagrams (Fig. 7–12) to negligible.

Different erosional processes could be speculated to be the cause of the polygonal shape of impact craters. However, given that polygonal craters are very common in many parts of the study area, we feel that it is highly unlikely that some erosional process could cause the polygonal outline for any significant number of craters, especially when distinctively modified craters with some polygonal features were discarded. It seems much more plausible that the cause of the polygonal shape is related to the cratering process itself, i.e., mechanisms outlined above. This is also supported by the observations and conclusions of Eppler et al. (1983) described above.

6 Results

6.1 Morphology and Distribution

The vast majority of polygonal craters in greater Hellas region is, or has a tendency towards being hexagonal. In addition, pentagons and also parts of octagons occur relatively often, but clearly square-shaped craters tend to be rare. Square-shaped craters are usually small complex craters, but larger complex craters exhibit almost exclusively hexagonal or pentagonal shapes.

Polygonal craters are most common in the northern part of the study area close to the Isidis impact basin. Near Isidis there are regions where more craters are polygonal than circular. In the volcanic plains of Hesperia Planum and Malea Planum the number of polygonal craters is low. Also worth noting is that the polygonal craters in Malea Planum are generally not as clearly polygonal as in other parts of the study area, and they are also typically smaller. The concentration of polygonal craters in certain areas becomes evident in Fig. 1, which displays the distribution of polygonal craters plotted on MGS-MOLA (Mars Global Surveyor – Mars Orbiter Laser Altimeter) topographic data.

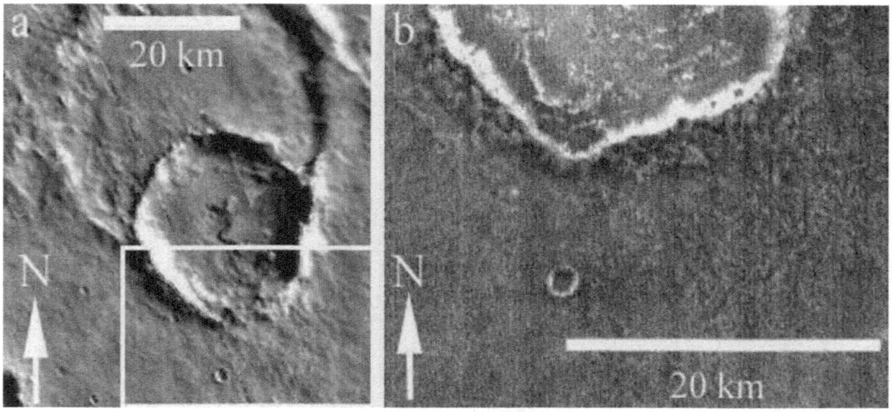

Fig. 5. (**a**) (left) An eroded polygonal crater northwest from Hellas (23°S/324°W) in Viking Orbiter MDIM. The white box indicates the area shown in (**b**). (**b**) (right) Southern part of the crater in (**a**) as seen in Mars Odyssey THEMIS nighttime infrared image. Polygonal appearance of the southern part is markedly enhanced in the infrared image. A subframe of THEMIS image I02140006.

6.2 Rim Strikes

In *Hellespontus Montes* (block centered at 39°S/312°W), on the western rim of the Hellas basin, the rims of polygonal craters display a prominent easterly direction (Fig. 7b, number of measurements n=118). The strikes are reported here by one direction only, i.e., a crater rim striking east–west is usually reported as "east", especially in the case of intercardinal directions, and for instance a strike 040°–220° is reported as 040°. The rose diagrams, however, are plotted as bidirectional because we feel that strikes are easier to grasp from bidirectional rose diagrams compared to unidirectional diagrams. The azimuth interval 080°–100° gathers about 21% of all the rim direction measurements. Another major rim direction in the area is north-northeast. A number of rims are oriented in a wide azimuth range around 280°–340°, whereas in the direction 040°–080° only very few rims exist.

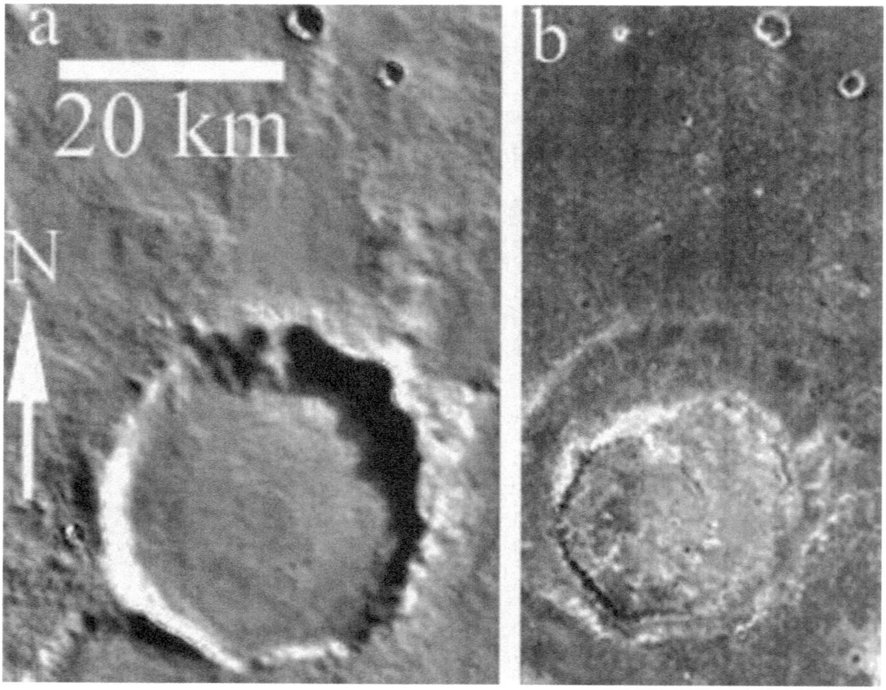

Fig. 6. The same polygonal crater northwest from Hellas (27°S/324°W) in Viking Orbiter MDIM (**a**) (left) and Mars Odyssey THEMIS nighttime infrared image (**b**) (right). Note that the actual rim is rather vaguely discernible in infrared. A subframe of THEMIS image I02140006.

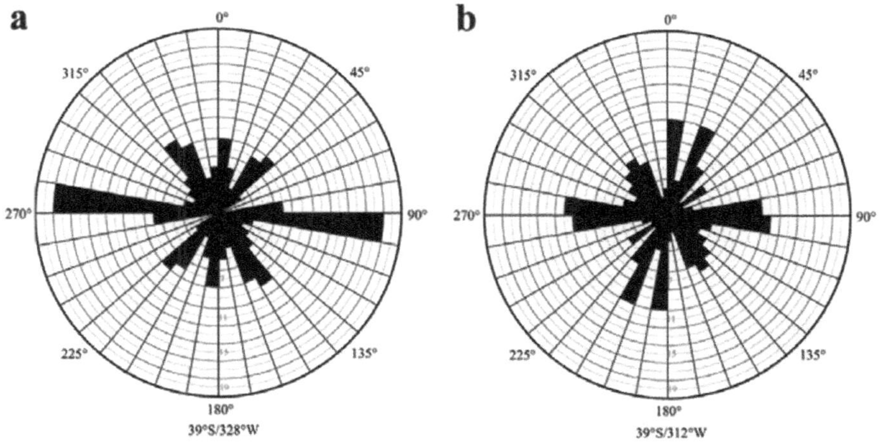

Fig. 7. (**a**) (left, block 39°S/328°W, number of measurements n=76) and (**b**) (right, block 39°S/312°W, n=118). Bidirectional rose diagrams with 10° azimuth intervals, depicting the percentage of polygonal crater rim strikes west from Hellespontus Montes (**7a**, **see Fig. 2** for location) and from Hellespontus Montes itself (**7b**). In the rose diagrams the outermost circle represents 20% and the darker circles inside mark every 2% except where otherwise stated. Note that the strong E–W peak radial to Hellas is present in both diagrams, whereas the NNE–SSW peak concentric to Hellas and parallel to grabens is predominate in Hellespontus Montes (**7b**) but diminishes further west (**7a**).

The block 39°S/328°W around crater *Rabe* immediately west from Hellespontus Montes (Fig. 7a, n=76) displays a pattern very similar to that of Hellespontus Montes. Rims striking east towards Hellas predominate with 18% of all the measurements. The northern trend is significantly weaker compared to the prominent north-northeast peak in Hellespontus Montes. The approximately northwest striking component is much more clearly clustered than in Hellespontus Montes. Also in this block crater rims oriented 050°–080° are rare.

The blocks situated between Hellas and Isidis at latitude 6°S display a very fascinating pattern (Fig. 8–10). Block 6°S/296°W (Fig. 8a, n=129) just on the southern edge of *Syrtis Major Planitia* shows a strong peak (12%) in azimuth interval 020°–030°(–050°). In the azimuth range 270°–360° rims seem to be rather randomly oriented. It is worth noting that there are gaps in the diagram on both sides of the largest peak at azimuths 000°–020° and 050°–070°.

Figure 8b (n=173) depicting the block 6°S/280°W east from crater *Fournier* and immediately southwest from Isidis shows a beautifully symmetric rim orientation distribution. The rose diagram seems to be a "refined" version of Fig. 8a. The dominant rim strike is 020°–030° (radial to Isidis) with 16% of all the measurements. Other peaks are at 280°–300°

and 330°–350° with about the same percentage spread in a twice as wide azimuth interval. The peaks have obvious gaps between them, and the same gaps can also be seen in Fig. 8a. The peaks evidently reflect the hexagonal outlines of the craters in the area.

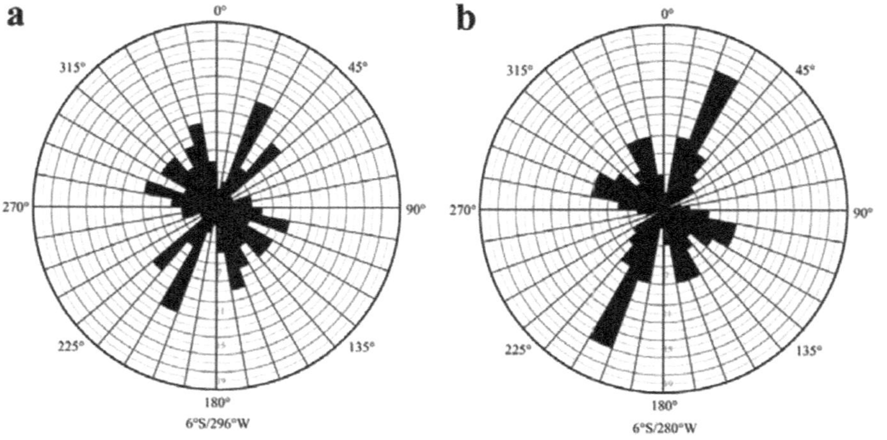

Fig. 8. (**a**) (left, block 6°S/296°W, n=129) and (**b**) (right, block 6°S/280°W, n=173). Rose diagrams of the percentage of polygonal crater rim strikes southwest from Isidis basin (see **Fig. 2** for location). Note the strong 030° trending peak present in both diagrams. This direction is radial to Isidis. Compare to **Fig. 9a** and **9b**.

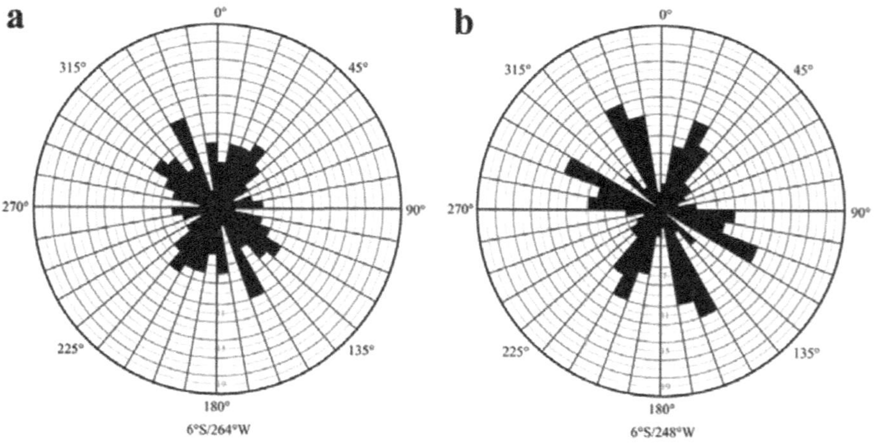

Fig. 9. (**a**) (left, block 6°S/264°W, n=84) and (**b**) (right, block 6°S/248°W, n=74). Rose diagrams of the percentage of polygonal crater rim strikes southeast from Isidis basin (see **Fig. 2** for location). Peaks at about 330° radial to Isidis are noteworthy, but also other major peaks are present. Compare to **Fig. 8a** and **8b**.

Block 6°S/264°W is situated southeast from Isidis. The corresponding rose diagram in Fig. 9a (n=84) has a lot more scatter than its counterpart on the other side of Isidis (Fig. 8b). The strongest peak (only 10%) is in the direction 330°–340° (radial to Isidis). Despite the fact that this appears as the strongest peak in the diagram, the wide petals of the rose in 290°–320° and 010°–050° gather more measurements. Roughly east–west trending polygonal crater rims are almost completely lacking.

The rose diagram in Fig. 9b (n=74) depicts block 6°S/248°W hosting the crater *Escalante* on its northern margin. The diagram is again beautifully symmetric with three prominent peaks separated by roughly 60°. The strongest peaks each with about 22–25% of measurements point towards 330°–350° (radial to Isidis), 270°–300° and 010°–040°. Rims oriented in other directions are scarce. Despite some differences, the similarity between Fig. 9a and 9b is evident.

In Fig. 10a (n=65), depicting the block 6°S/232°W mainly west from crater *Knobel*, the strongest peak (12%) emanates from rims oriented 030°–040°. However, there is plenty of scatter and lots of measurements gather on both sides of 320°–330°. Very few rims have a roughly east–west orientation. An important point to notice is that blocks 6°S/232°W and especially 6°S/214°W differ from previous blocks in that they are partly on the volcanic plains of Elysium Planitia. This also means that they are on the global dichotomy boundary.

In contrast to block 6°S/232°W, block 6°S/214°W in the northeastern corner of our study area hosts plenty (14%) of polygonal craters with rim segments trending 270°–280° (Fig. 10b, n=52). Another prominent peak is at 030°–040°. The major directional gap is at 050°–080°.

In order to see if the directions that are predominate in Fig. 9a and 9b are also present further south from Isidis, we measured rim orientations in blocks 22°S/264°W (Fig. 11a, n=51) and 22°S/248°W (Fig. 11b, n=38). These blocks are actually closer to Hellas than to Isidis, and are partly on Hesperia Planum. Both diagrams are quite symmetrical with three peaks. In block 22°S/264°W, just north from *Hadriaca Patera* and west from *Tyrrhena Patera*, the peaks in 330°–340° and 020°–040° correspond to peaks in the blocks closer to Isidis. However, the wide east–west peak (260°–290°) in 22°S/264°W has no clear match in the northern blocks. Almost the same applies for the block 22°S/248°W, which hosts *Tyrrhena Patera*: the rose diagram has a clearly defined peak (13%) at 340°–350°(–360°) and wider peaks at 020°–050° and 070°–100°.

The rim strike measurements from *Malea Planum* (Fig. 12a–12d) are scarce and thus discussed below in chapter 7.2.

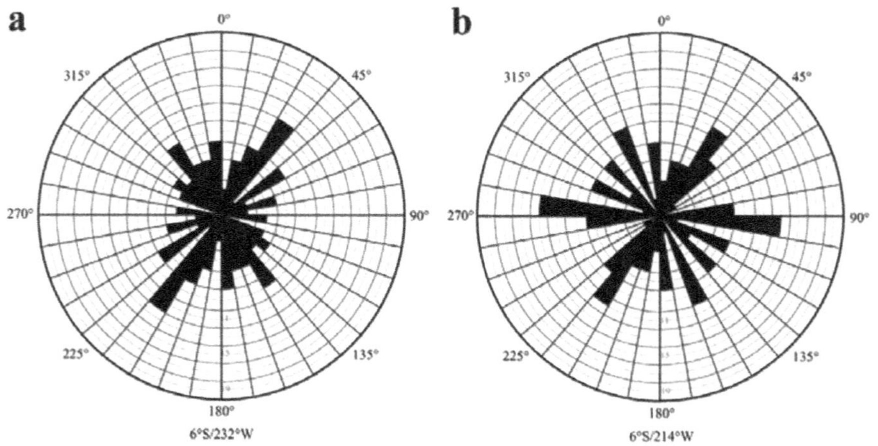

Fig. 10. (**a**) (left, block 6°S/232°W, n=65) and (**b**) (right, block 6°S/214°W, n=52). Rose diagrams of the percentage of polygonal crater rim strikes from the southern margin of Elysium Planitia (see **Fig. 2** for location). Note that the strong E–W peak in **Fig. 10b** is conspicuously missing in **Fig. 10a**. This might be related to the global dichotomy boundary trending roughly E–W being more dominating in the eastern block (**Fig. 10b**). The 040° peaks may be due to fracturing induced by Elysium Mons just north of our study area.

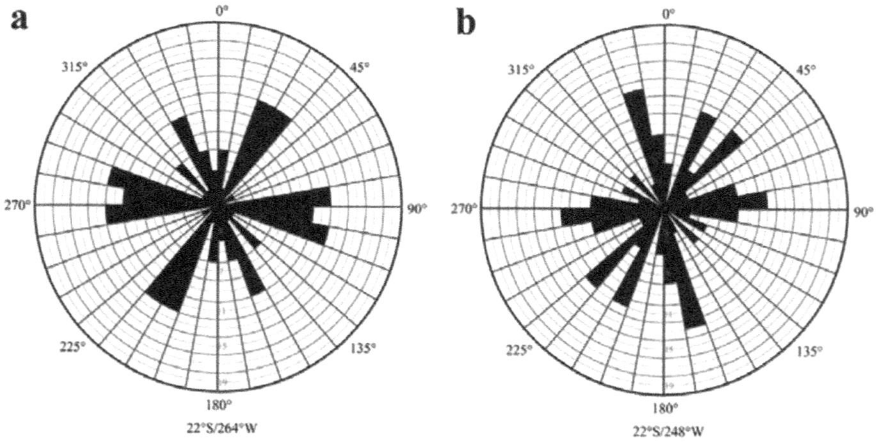

Fig. 11. (**a**) (left, block 22°S/264°W, n=51) and (**b**) (right, block 22°S/248°W, n=38). Rose diagrams of the percentage of polygonal crater rim strikes northeast from Hellas (Hesperia Planum, see **Fig. 2** for location). The roughly E–W and NNW–SSE trending crater rims have their parallel counterparts in the wrinkle ridges of Hesperia Planum (Raitala 1988). The NE–SW peaks may be an indication of volcano-tectonism or Hellas centered radial fracturing.

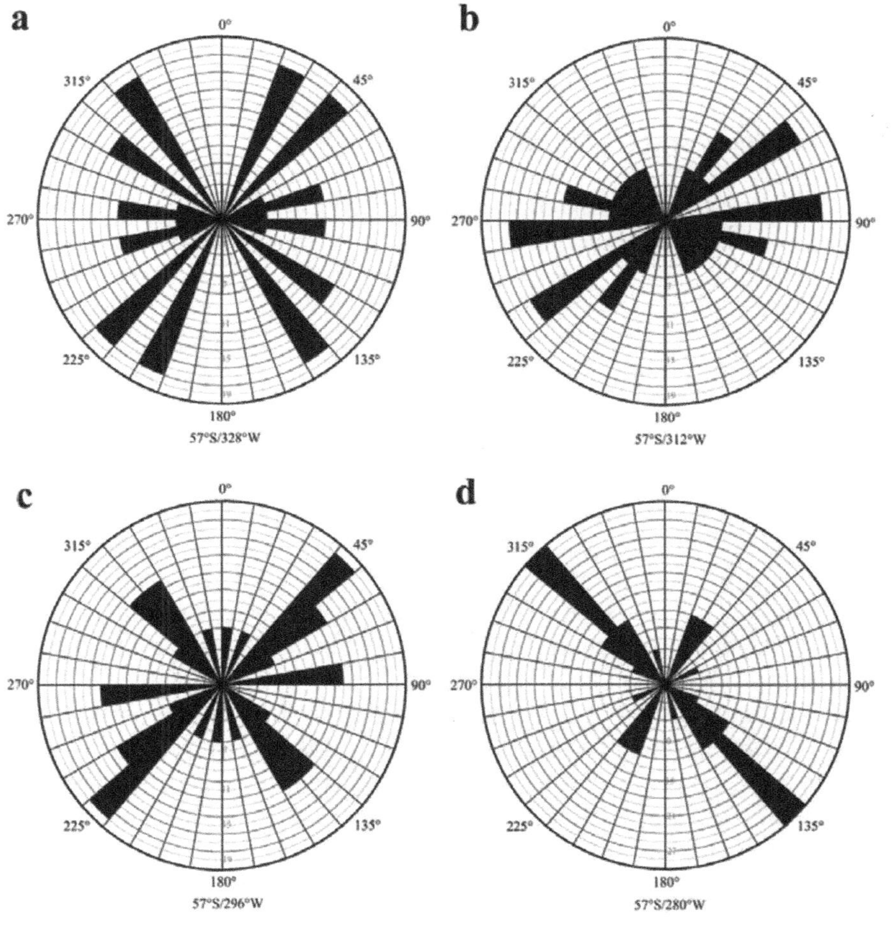

Fig. 12. (**a**) (upper left, block 57°s/328°W, n=22), (**b**) (upper right, block 57°S/312°W, n=18), (**c**) (lower left, block 57°S/296°W, n=16) and (**d**) (lower right, 57°S/280°W, n=16). Rose diagrams of the percentage of polygonal crater rim strikes from Malea Planum (see **Fig. 2** for location). In Fig. 12a–12c the outermost circle represents 20% and darker inner circles mark every 2% as in the other rose diagrams. In Fig. 12d, however, the outermost circle represents 30% and the darker inner circles mark every 3%. Only craters formed in the plains material were included. Note the low number of measurements, which prohibits reliable conclusions. Similar trends (e.g., NW–SE), however, can be seen.

7 Discussion and Conclusions

7.1 Morphology and Distribution

In many of the rose diagrams the dominance of hexagonal craters is manifested by peak intervals of about 60° (see, e.g., Fig. 8b and 9b). Most polygonal craters in greater Hellas region have a tendency towards hexagonal outline, whereas *clearly* square-shaped craters are more or less a rarity, even among small polygonal craters. Large square-shaped craters are absent. "Perfect" pentagons are no anomaly, although they are not nearly as common sight as "perfect" hexagons. Typical are also craters where two rim segments have "overgrown" turning a "would-be-hexagon" into a pentagonal crater. Several craters also exhibit features that justify the use of the name "octagonal" crater.

The distribution of polygonal craters (Fig. 1) shows that in greater Hellas region they are generally more abundant in the northern part of the study area. This results in part from the resolution of the images, which gets worse the further the image is from the equator. The principal reason for the distribution is simply the fact that in the southern part of the study area the total amount of craters is much smaller than in the north, mainly due to volcanic and fluvial deposits covering large areas around Hellas. We believe, however, that the target properties, especially the existence of prominent fracture directions, also play an important role in the distribution of polygonal craters. A good example of this is the concentration of polygonal craters in Hellespontus Montes, a known locality of intense tectonic modification (e.g., Wichman and Schultz 1989). Figure 13 displays the main tectonic features from the study area along with the information from the rose diagrams.

The fact that not all impact craters display polygonal features requires some attention. We are well aware that the distribution pattern of polygonal craters displayed in Fig. 1 has errors due to the subjectivity of recognizing polygonal craters. The true amount of polygonal craters in the study area is larger, since we only accepted craters, which were indicated as such by both researchers dealing with the particular block. Many polygonal craters were left out simply because of this subjectivity. Therefore, Fig. 1 displays only the minimum estimate of polygonal craters in greater Hellas region. In any case, a huge number of craters are more or less circular, and obviously not affected by, e.g., fracturing caused by Hellas and Isidis impact basins. Several factors can explain this. In some places, thick layers of younger sediments or volcanic deposits may have covered the ancient fractured crust so that impacts have not penetrated deep enough to be influenced by old fracturing. Some areas, like Malea Planum, might be composed of mainly pyroclastic material that has been too porous for a dominant fracture pattern to develop. In other localities, impacts and endogenic processes could have created so dense and complex fracture pattern that no dominating directions are present anymore (see Fulmer and Roberts 1963). Given the highly versatile nature of the geology in greater Hellas region, such local variations are not surprising.

7.2 Rim Strikes

The dominant east–west striking rim directions in *Hellespontus Montes* (Fig. 7b) and the area immediately to west from it (Fig. 7a) are easily explained as an indication of radial fracturing caused by the Hellas basin. The approximately north–south component, which is stronger closer to Hellas, most likely originates from concentric fracturing around Hellas. This Hellas concentric fracturing in Hellespontus Montes is a well-known phenomenon and can clearly be seen as roughly north–south trending grabens and scarps (e.g., Wichman and Schultz 1989; Leonard and Tanaka 2001). It is interesting to see that Hellas radial fracturing seems to extend further than concentric, or that at least radial fracturing is easier than concentric to see using polygonal craters.

The concentration of polygonal craters close to the Isidis basin and the fracture directions indicated by the crater rims can be understood as a result of intense radial fracturing related to Isidis basin. This can be seen best southwest from Isidis, where north-easterly rim directions strongly predominate (Fig. 8a and 8b). Southeast from Isidis the picture is not so clear (Fig. 9a and 9b, see also Fig. 11a and 11b), but the radial fracturing seems to exist also there. Weaker concentric fracturing also appears to be present on both sides of the Isidis basin.

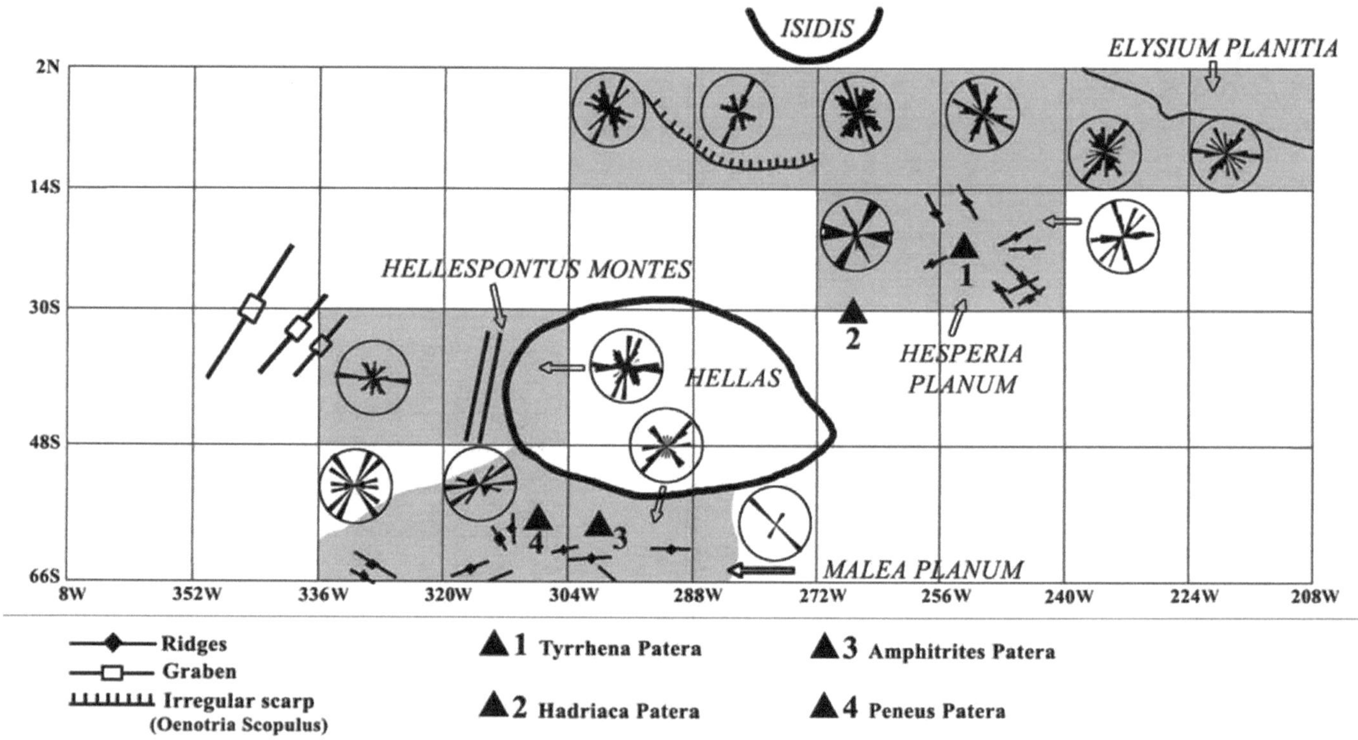

Fig. 13. A sketch map of major geologic and tectonic features (basins, paterae and the volcanic plains, ridges, grabens, lineaments) of the study area with the rose diagrams presenting strikes of polygonal crater rims. Correlation between, e.g., the structural pattern of Hellespontus Montes or Hesperia Planum and the rim orientations is striking. Note the rim strikes radial and concentric to Isidis and Hellas basins, interpreted to indicate basin-centered radial and concentric fracturing, respectively. Shading denotes areas where rim strike measurements were made.

According to Wichman and Schultz (1989), a trough and scarp system radial to Isidis can be traced over 1000 km southeast of Isidis. This distant indication of Isidis–centered deformation seems to appear also on our data as a peak of northwestern rim directions in Fig. 11a and 11b, although there are also other stronger peaks. This is interesting because the blocks 22°S/264°W and especially 22°S/248°W are situated mainly on the volcanic plains of Hesperia Planum, not on older highlands. Nevertheless, our results seem to support those gained by Wichman and Schultz (1989). The same northwesterly trend can also be seen in Raitala's study (1988; see also Crown et al. 1992) of the wrinkle ridges in Hesperia Planum. It is important to note in Fig. 11a and 11b that in these blocks a fracture system radial to Isidis has approximately the same direction as a fracture system concentric to Hellas would have, and vice versa. Therefore in this area it is impossible to differentiate between fractures radial to Isidis and concentric to Hellas using polygonal craters alone. It seems likely that the strong 020°–040° peak in the block 22°S/264°W (Fig. 11a) is caused by a fracture system radial to Hellas. This Hellas radial fracturing can also be present in block 22°S/248°W (Fig. 11b) as the 020°–050° peak. The numerous east–west trending rims in the blocks are tentatively interpreted as resulting from fracturing caused by the volcano-tectonism of Hadriaca and Tyrrhena Paterae. This prominent rim direction peak coincides with one major direction of wrinkle ridges in Hesperia Planum (Raitala 1988), emphasizing its significance.

The two blocks in the northeastern corner of our study area, 6°S/232°W (Fig. 10a) and 6°S/214°W (Fig. 10b), sample not only craters in old highland material, but also in the volcanic plains of Elysium Planitia. The rim strike peaks at 030°–040° are roughly radial to Hellas and could be induced by it. However, we feel that radial fracturing caused by the rise of Elysium Mons is a much easier way to explain this dominant fracture direction. The strong east–west trending peak in 6°S/214°W could also have its origin in the concentric fracturing caused by the volcano-tectonic evolution of Elysium Planitia. Another possibility is that the peak reflects the global dichotomy, since the boundary between the southern highlands and the northern lowlands trends also approximately east–west.

The relatively few polygonal craters on *Malea Planum* display a directional pattern (Fig. 12a–12d), which, on a first look, differs from other studied areas. However, the number of measurements in each four blocks in Malea Planum is very low – only about 10–20% of the number of measurements in other blocks – which may lead to random errors (57°S/328°W n=22 (Fig. 12a), 57°S/312°W n=18 (Fig. 12b), 57°S/296°W n=16 (Fig. 12c), 57°S/280 n=16 (Fig. 12d)). Thus, conclusions drawn from the data of polygonal craters on Malea Planum should be taken as speculative, since the number of measurements does not completely match the statistical criteria. The conclusion that can be drawn is that the distribution of rim directions does not appear to be completely random. One may, with some confidence, hypothesize that the direct influence of the Early Noachian Hellas impact itself is rather weak and that the major cause of the stresses in Malea Planum is related to the rise of the shield of Amphitrites Patera during Late Noachian – Early Hesperian. However, the effect of later Hellas-centered deformation on polygonal craters in Malea Planum cannot be ruled out, because the formation of probably basin-related wrinkle ridges continued into Early Hesperian (see, e.g., Leonard and Tanaka 2001 and references therein).

8
Summary and Outlines for Future Studies

From our review of polygonal craters it becomes evident that polygonal craters are common on different bodies in our solar system and also their possible modes of formation are known – and have been for a long time. However, in the more recent impact literature, awareness of – or perhaps interest towards – polygonal craters has diminished.

Due to their mode of formation, polygonal craters can be used in determining the systematics in the directions and distributions of fractures or other zones of weakness in the crust. This method is applicable to all bodies with a rigid cratered crust, since we have observed that polygonal craters exist on all terrestrial planets and several asteroids and icy satellites. Thus, in the context of planetary geology, polygonal craters can reveal some properties of the target material and be an additional tool in determining the regional stresses.

We have studied the distribution and fracture directions revealed by polygonal impact craters in the region of Hellas basin, Mars. Polygonal craters are most common in the northern part of the study area, especially around Isidis impact basin. There the crater rims reveal beautifully the development of radial fractures caused by the Isidis basin. In Hellespontus Montes the radial fracturing caused by the Hellas impact basin is evident, but concentric fracturing, seen in polygonal crater rims as well as grabens, is present too. In Hesperia Planum polygonal craters reflect distant fracturing radial to Hellas with later overprinting of volcano-tectonic fracturing. This largely coincides with results from the studies of wrinkle ridges. In general it seems that utilizing polygonal craters radial basin-centered fracturing can be traced further than concentric. The polygonal craters in the volcanic plains of Malea Planum display varied rim orientations with no apparent contribution from Hellas. This most likely is the consequence of tectonic overprinting by the stresses created by the rise of Amphitrites Patera. Fracturing induced by the rise of Elysium Mons is the probable cause for the rim strikes in the northeastern corner of our study area. It is possible that the global dichotomy boundary also affects the strikes of polygonal crater rims.

The greater Hellas region is in general relatively poor in indications of tectonism (ridges, grabens, lineations). Thus, the tectonic history is not so easily deciphered. Polygonal craters, however, are abundant. This is especially true for highland areas where otherwise most prominent tectonic indicators – wrinkle ridges – are few or lacking. Polygonal craters can, therefore, be very useful in studies of regional tectonism.

Polygonality is a significant but largely forgotten feature of impact craters. We have shown that polygonal craters can be utilized in the study of regional tectonism on bodies with rigid fractured crusts. The method presented here could easily be developed further. Instead of considering the actual abundance of polygonal impact craters in a particular area, it would probably be more useful to study the ratio of polygonal and circular craters. Statistical analysis should be used to determine the actual significance of the results. Thus, our preliminary work forms the basis of a further study, in which the constantly accumulating Mars Global Surveyor MOC and Mars Odyssey THEMIS datasets will be utilized to characterize in detail the effect of geologic context and crater size (simple vs. complex) to crater rim strikes. Another goal of our future work is to gain a better understanding of the cratering mechanics in fractured targets by carrying out a series of cratering experiments. However, already this study has shown that the direction distribution of the polygonal crater rims can be an effective and easily applicable tool in determining the fracture patterns and possibly some target material properties on cratered surfaces, as has been proposed by earlier workers (e.g., Fulmer and Roberts 1963; Eppler et al. 1983). The method would be best used alongside traditional means of mapping tectonic deformation. The abundance and "universality" of polygonal craters indicate that the effect of structural control should be incorporated to cratering models in order to produce realistic scenarios of the impact cratering process.

Acknowledgements

The manuscript was greatly improved by critical and constructive reviews by Boris Ivanov and Horton Newsom. Financial support from the Wihuri Foundation, the Väisälä Fund of the Finnish Academy of Science and Letters, the Space Institute of Oulu University, the Magnus Ehrnrooth Foundation, the Seppo Säynäjäkangas Science Foundation, the Graduate School of Astronomy and Space Physics and the Finnish Graduate School in Geology is gratefully acknowledged. A substantial portion of data was obtained from NASA through the Nordic Regional Planetary Image Facility (NRPIF) of Oulu. This research has made use of NASA's Astrophysics Data System Bibliographic Services.

References

Abels A (2003) Investigation of impact structures in Finland (Söderfjärden, Lumparn, Laäääjärvi) by digital integration of multidisciplinary geodata. Ph.D. thesis, Westphalian-Wilhelms University Münster, Germany, 292 pp

Acuña MH (2003) Martian crustal magnetism: What have we learned after 6 years of MGS observations? [abs.] 6th International Conference on Mars: abs. #3206 (CD-ROM)

Adler JEM, Salisbury JW (1969) Circularity of lunar craters. Icarus 10: 37–52

Aittola M, Öhman T, Kostama V-P, Raitala J (2002) Impact craters establish geological diversity within Hellas region, Mars [abs.]. Lunar and Planetary Science 33: abs. #1485 (CD-ROM)

Baldwin RB (1963) The Measure of the Moon. The University of Chicago Press, Chicago – London, 488 pp

Connerney JEP, Acuña MH, Wasilewski MH, Kletetschka G., Ness NF, Rème H, Lin RP, Mitchell DL (2001) The global magnetic field of Mars and implications for crustal evolution. Geophysical Research Letters 28: 4015–4018

Crown DA, Greeley R (1993) Volcanic geology of Hadriaca Patera and the eastern Hellas region of Mars. Journal of Geophysical Research 98: 3431–3451

Crown DA, Price KH, Greeley R (1992) Geologic evolution of the east rim of the Hellas basin, Mars. Icarus 100: 1–25

Curran DR, Shockey DA, Seaman L, Austin M (1977) Mechanisms and models of cratering in earth media. In: Roddy DJ, Pepin RO, Merrill RB (eds) Impact and Explosion Cratering, Pergamon Press, New York, pp 1057–1087

Elston WE, Laughlin AW, Brower JA (1971) Lunar near-side tectonic patterns from Orbiter 4 photographs. Journal of Geophysical Research 76: 5670–5674

Eppler DT, Nummedal D, Ehrlich R (1977) Fourier analysis of planimetric lunar crater shape – Possible guide to impact history and lunar geology. In: Roddy DJ, Pepin RO, Merrill RB (eds) Impact and Explosion Cratering, Pergamon Press, New York, pp 511–526

Eppler DT, Ehrlich R, Nummedal D, Schultz PH (1983) Sources of shape variation in lunar impact craters: Fourier shape analysis. Geological Society of America Bulletin 94: 274–291

Fielder G (1961) Structure of the Moon's Surface. Pergamon Press, Oxford – London – New York Paris, 246 pp
Fielder G (1965) Lunar Geology. Lutterworth Press, London, 184 pp
Fulmer CV, Roberts WA (1963) Rock induration and crater shape. Icarus 2: 452–465
Gault DE, Quaide WL, Oberbeck VR (1968) Impact cratering mechanics and structures. In: French BM, Short NM (eds) Shock Metamorphism of Natural Materials, Mono Book Corp., Baltimore, pp 87–99
Grieve RAF, Wood CA, Garvin JB, McLoughlin G, McHone JF (eds) (1988) Astronaut's guide to the terrestrial impact craters. LPI Technical Report 88-03, Lunar and Planetary Institute, Houston, 89 pp
Gurov EP, Gurova EP (1982) Some regularities of the areal spreading of fractures around Elgygytgyn impact crater [abs.]. Lunar and Planetary Science 13: 291–292
Henkel H (1992) Geophysical aspects of meteorite impact craters in eroded shield environment, with special emphasis on electric resistivity. Tectonophysics 216: 63–89
Kopal Z (1966) An Introduction to the Study of the Moon. D. Reidel Publishing Company, Dordrecht, 466 pp
Kostama V-P, Aittola M, Öhman T, Raitala J (2002) Geological units of the Hellas basin region, Mars [abs.]. Lunar and Planetary Science 33, abs. #1486 (CD-ROM)
Laurén L, Lehtovaara J, Boström R, Tynni R (1978) On the geology and the Cambrian sediments of the circular depression at Söderfjärden, western Finland. Geological Survey of Finland, Bulletin 297, 81 pp
Leonard GJ, Tanaka KL (1993) Hellas basin, Mars: Formation by oblique impact [abs.] Lunar and Planetary Science 24: 867–868
Leonard GJ, Tanaka KL (2001) Geologic map of the Hellas region of Mars – Map I-2694. U.S. Department of the Interior, U.S. Geological Survey
MacKinnon DJ, Tanaka KL (1989) The impacted Martian crust: structure, hydrology, and some geologic implications. Journal of Geophysical Research 94: 17359–17370
Martin LJ, Zurek RW (1993) An analysis of the history of dust activity on Mars. Journal of Geophysical Research 98: 3221–3246
Melosh HJ (1989) Impact cratering: a geologic process. Oxford University Press, New York, 245 pp
Melosh HJ, Ivanov BA (1999) Impact crater collapse. Annual Review of Earth and Planetary Sciences 27: 385–415
Moore JM, Wilhelms DE (2001) Hellas as a possible site of ancient ice-covered lakes [abs.]. Lunar and Planetary Science 32: abs. #1446 (CD-ROM)
Öhman T (2002) The indications of cratering process in Saarijärvi impact crater, northern Finland. Unpublished M.Sc. thesis, Institute of Geosciences, University of Oulu, Finland, 180 pp
Öhman T, Pesonen LJ, Elo S, Uutela A, Tuisku P, Raitala J (2003) The origin and evolution of the Saarijärvi impact structure [abs.]. Meteoritics and Planetary Science 38, A52
Pesonen LJ, Lehtinen M, Tuukki P, Abels A (1997) The lake Saarijärvi structure, Taivalkoski – a new meteorite impact crater in Finland [abs.]. In: Kaikkonen P, Komminaho K, Salmirinne H (eds) Sovelletun geofysiikan XI neuvottelupäivät, Oulun yliopisto, Oulu: 8–9
Pesonen LJ, Lehtinen M, Tuukki P, Abels A (1998) The lake Saarijärvi – a new meteorite impact structure in northern Finland [abs.]. Lunar and Planetary Science 29: abs. #1262 (CD-ROM)

Phillips RJ, Raubertas RF, Arvidson RE, Sarkar IC, Herrick RR, Izenberg N, Grimm RE (1992) Impact craters and Venus resurfacing history. Journal of Geophysical Research 97: 15923–15948

Pike RJ (1971) Genetic implications of the shapes of Martian and Lunar craters. Icarus 15: 384–395

Pike RJ (1977) Size-dependence in the shape of fresh impact craters on the moon. In: Roddy DJ, Pepin RO, Merrill RB (eds) Impact and Explosion Cratering, Pergamon Press, New York, pp 489–509

Pike, RJ (1980) Formation of complex impact craters: evidence from Mars and other planets. Icarus 43: 1–19

Pilkington M, Grieve RAF (1992) The geophysical signature of terrestrial impact craters. Reviews of Geophysics 30: 161–181

Plescia JB, Saunders RS (1979) The chronology of the Martian volcanoes. Proceedings of the 10th Lunar and Planetary Science Conference, pp 2841—2859

Raitala J (1988) Superposed ridges of the Hesperia Planum area on Mars. Earth, Moon and Planets 40: 71–99

Ronca LB, Salisbury JW (1966) Lunar history as suggested by the circularity index of lunar craters. Icarus 5: 130–138

Schultz PH (1976) Moon morphology. The University of Texas Press, Austin, 626 pp

Schultz PH, Schultz RA, Rogers J (1982) The structure and evolution of ancient impact basins on Mars. Journal of Geophysical Research 87: 9803–9820

Schultz RA (1985) Assessment of global and regional tectonic models for faulting in the ancient terrains of Mars. Journal of Geophysical Research 90: 7849–7860

Shoemaker EM (1963) Impact mechanics at Meteor Crater, Arizona. In: Middlehurst BM, Kuiper GP (eds) The Moon, Meteorites and Comets. The University of Chicago Press, Chicago - London, pp 301–336

Squyres SW, Wilhelms DE, Moosman AC (1987) Large-scale volcano – ground ice interactions on Mars. Icarus 70: 385–408

Tanaka KL, Leonard GJ (1995) Geology and landscape evolution of the Hellas region of Mars. Journal of Geophysical Research 100: 5407–5432

Tanaka KL, Scott DH (1987) Geologic map of the polar regions of Mars – Map I-1802-C. U.S. Department of the Interior, U.S. Geological Survey

Tanaka KL, Kargel JS, MacKinnon DJ, Hare TM, Hoffman N (2002) Catastrophic erosion of Hellas basin rim on Mars induced by magmatic intrusion into volatile-rich rocks. Geophysical Research Letters 29: 37-1–37-4, doi 10.1029/2001GL013885

Wichman RW, Schultz PH (1989) Sequence and mechanisms of deformation around the Hellas and Isidis impact basins on Mars. Journal of Geophysical Research 94: 17333–17357

BP and Oasis Impact Structures, Libya: Remote Sensing and Field Studies

Christian Koeberl[1], Wolf Uwe Reimold[2], and Jeff Plescia[3]

[1]Department of Geological Sciences, University of Vienna, Althanstrasse 14, A-1090 Vienna, Austria (christian.koeberl@univie.ac.at)

[2]Impact Cratering Research Group, School of Geosciences, University of the Witwatersrand, Johannesburg 2050, South Africa (reimoldw@geosciences.wits.ac.za)

[3]MP3-E163, Applied Physics Laboratory, Johns Hopkins University, 11100 Johns Hopkins Road, Laurel, MD 20723-6099, USA (jeffrey.plescia@jhuapl.edu)

Abstract. Remote sensing images and recent field studies provide important morphological and structural constraints for the BP and Oasis impact structures in Libya. The BP structure (25°19' N, 24°20' E) is highly eroded and appears on satellite images in the form of a set of concentric rings. The rocks are composed of sandstones of the Jurassic to Cretaceous Nubia Formation. The Landsat image allows division of the structure into inner, middle, and outer rings. The ERS1 and Radarsat radar images do not show much detail, but the rings are more pronounced than in the visible wavelengths, probably because the radar penetrates the sand cover. In the field it became obvious that the cited diameter of 2.8 or 3.2 km for the outer ring does not represent the actual crater diameter, as the "middle ring" represents the actual crater rim with a diameter of about 2 km. The middle ring is characterized by a distinct series of hills of up to about 30 m elevation above the surrounding desert, with rocks dipping at 30 to 50° outwards. Some parts of the rim are characterized by intense folding and faulting. The "outer ring" is a discontinuous ring of low hills with 1-2 m average elevation, dipping at 10° inwards. The "inner ring" is interpreted as a small central uplift structure (ca. 500 m in diameter) with steeply upturned strata that are intensely folded, faulted, and – in places – brecciated.

For Oasis (24° 35' N, 24° 24' E) diameters of 11.5 km or even 18 km (from radar images) have been suggested in the past. However, the topographically most prominent feature of this structure is a ring of discontinuous (up to 100 m high) hills 5.1 km in diameter. The rocks in this prominent ring dip outward and locally are intensely folded. It can be debated whether or not this feature represents the actual crater rim.

1 Introduction

The BP and Oasis impact structures in southeast Libya are highly eroded. They appear as concentric ridges of deformed rocks that rise above the surrounding desert plain by only a few tens of meters. Both structures were first mentioned by Kohman et al. (1967), who described their location based on space and aerial photography. Martin (1969) provided additional detail of the BP structure, which is the smaller of the two. BP and Oasis are centered at 25°19' N and 24°20' E, and 24°35'N and 24°24'E, respectively (Figs. 1, 2); both are located close to the Libyan-Egyptian border. The names resulted from the fact that, in the 1960s, the first research visit to the Oasis structure was made by a field party of the Oasis Oil Company of Libya; BP was visited around that time by a team from the B.P. Exploration Company. The impact origin of the structures was supported in the early 1970s by French et al. (1972, 1974) on the basis of observations of planar deformation features (PDFs) in some of the crater rocks. However, the number of observations of these features was very limited. Underwood and Fisk (1980) suggested that both craters formed simultaneously by a double impact, but apart from the relative proximity (ca. 80 km distance) and a similar degree of erosion, no real evidence to support such a suggestion is available. At this time the absolute ages for these structures are not known due to lack of datable material, although an attempt was made to isolate apatite from sandstone samples for fission track dating (R. Giegengack, pers. comm.); so far not enough material has been available. The ages can only be constrained to post-Nubian; i.e., postdating the Jurassic to Lower Cretaceous target rocks; less than ca. 120 Ma.

1.1 Libyan Desert Glass

Based on the proximity of the BP and Oasis structures to the area of the Libyan Desert Glass (LDG), which is found in a limited strewn field about 150 km to the east (Fig. 1), and the lack of disturbed Jurassic to Lower Cretaceous sandstone strata in the area of the occurrence of the glass, several workers (e.g., Martin 1969; Underwood and Fisk 1980; Murali et al. 1988; cf. Koeberl 1997) have considered the possibility that the LDG might be related to one or both of the eastern Libyan impact structures. However, Diemer (1997) speculated that the BP and Oasis structures could

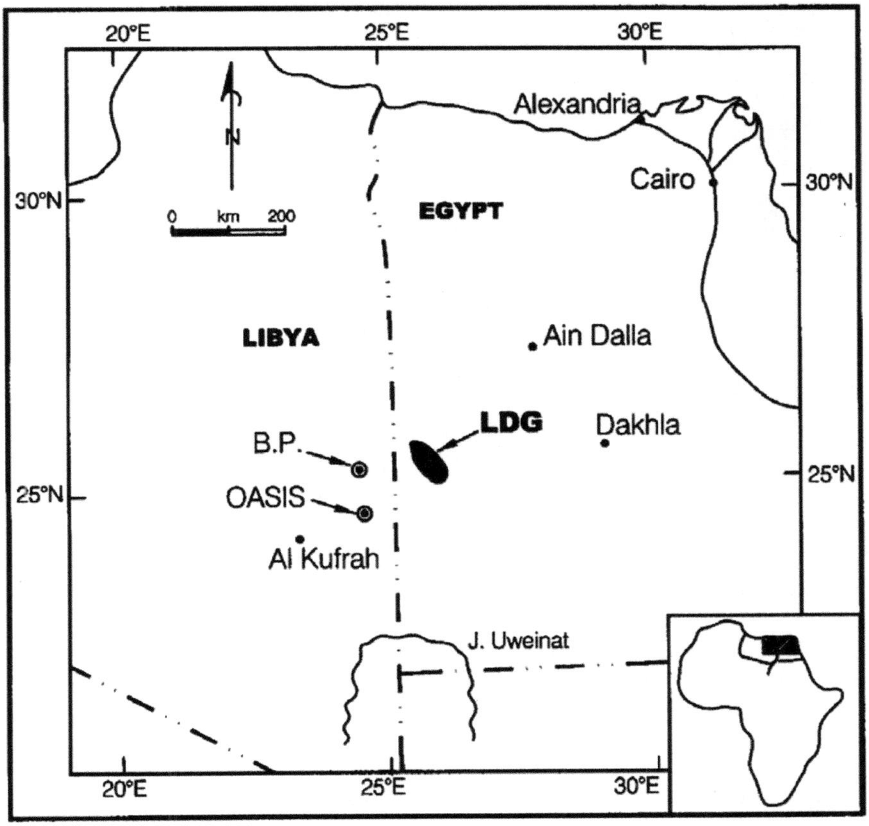

Fig. 1. Location map, showing parts of Egypt and Libya, after Abate et al. (1999). The position of the LDG field (in western Egypt) is shown in relation to the BP and Oasis impact structures in neighboring Libya. No LDG has been found in Libya.

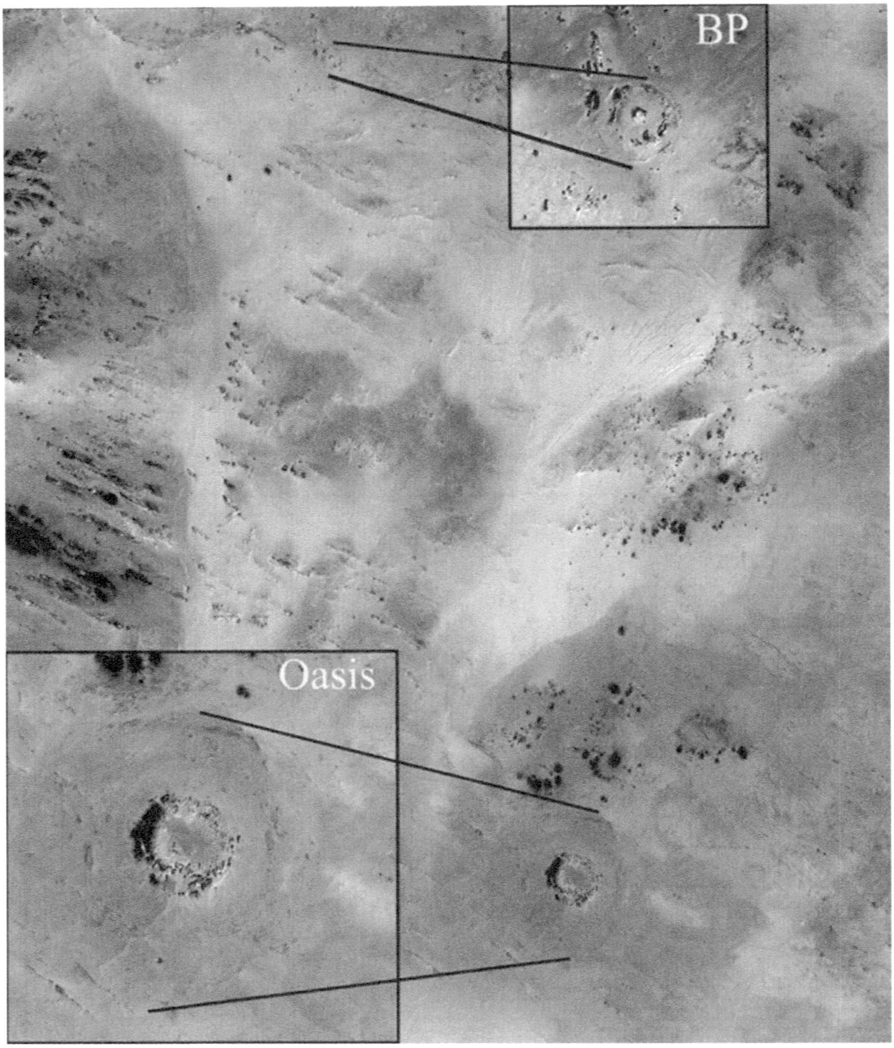

Fig. 2. Regional Landsat image, showing the BP structure at the upper edge of the image, the Oasis structure at the center. Inset shows the structures at slightly higher resolution. North is at the top for all satellite images.

be older than the LDG. Giegengack and Underwood (1997) also discussed that the target rocks currently exposed at the BP and Oasis structures could have been covered at the time of crater formation under about 400 meters of younger sediment.

Libyan Desert Glass is an enigmatic natural glass that is found in an area of about 3000 to 6000 km^2 (the exact numbers vary between different publications), similar to lag deposits, on the floor of corridors between sand dunes of the southwestern corner of the Great Sand Sea in western Egypt, near the Libyan border (Fig. 1). The glass occurs as centimetre- to decimeter-sized, irregularly shaped, and strongly wind-eroded pieces. The total preserved quantity of the glass has been estimated at $1.4 \cdot 10^9$ grams, but it is quite likely that the original mass was much larger (Barnes and Underwood 1976; Diemer 1997). Attempts to determine the age of the LDG were made using the K-Ar and fission-track methods. Due to the low K content of the glass, the errors on these K-Ar ages are too high to be meaningful (Matsubara et al. 1991; Horn et al. 1997). The only precise ages of the LDG come from fission track age determinations, which gave 28.5 ± 2.3 Ma (Storzer and Wagner 1971) to 29.4 ± 0.5 Ma (plateau age; Storzer and Wagner 1977), and 28.5 ± 0.8 Ma (Bigazzi and de Michele 1996; see also Bigazzi and de Michele 1997, and Horn et al. 1997) ages.

The origin of the LDG has been the subject of much debate since its discovery (cf. Clayton and Spencer 1934; Weeks et al. 1984; Diemer 1997), and a variety of exotic processes were suggested, e.g., a hydrothermal sol-gel process (Jux 1983 and Feller 1997); or a lunar volcanic source (Futrell and O'Keefe 1997). However, there is abundant evidence of an impact origin of these glasses, including the presence of schlieren (texture with discontinuous almost parallel bands) and partly or completely digested minerals, such as lechatelierite (a high temperature SiO_2 melt phase); baddeleyite (a high temperature break-down product of zircon; Kleinmann 1969; Storzer and Koeberl 1991; Horn et al. 1997); and the likely existence of a meteoritic component (Murali et al. 1987, 1988, 1989, 1997; Rocchia et al. 1996, 1997; Barrat et al. 1997). This is also in agreement with Os isotopic data of dark bands in the LDG (Koeberl 2000).

The source material of the glass remains a mystery. Storzer and Koeberl (1991) suggested from Zr/U and REE data that none of the sands or sandstones from various sources in the region are good candidates to be the sole precursors of the LDG. Compositional data for surface sands (Koeberl 1997) show significant differences from the average LDG composition. There is some chemical and isotopic similarity to rocks from the BP and Oasis impact structures in Libya (Abate et al. 1999). However, a recent Rb-Sr and Sm-Nd isotopic study of the LDG (Schaaf and Müller-Sohnius 2002) suggests that "Nubian" rocks are not likely precursors of the LDG. To complicate things even further, Barakat (2001) and Kleinmann et al. (2001) found some shocked quartz-bearing breccias in the LDG strewn field, but – so far – no evidence for an actual crater has been found in this area.

The possible connection to the origin of the Libyan Desert Glass led us to start an investigation of BP and Oasis. We quickly realized the lack of detailed geological and structural information, and of representative samples from these structures. The only geological studies were those of some oil exploration geologists and of Underwood and co-workers in the 1960s and 1970s, and all previous workers spent only up to a few hours at each site. Thus, in the late 1990s plans for an expedition were made by two of us (CK, WUR) to study both structures and obtain fresh samples. Unfortunately a number of logistical and bureaucratic problems delayed the actual fieldwork until January 2002. Here we describe first results from our field studies in comparison with remote sensing investigations using Landsat and radar satellite images. Koeberl and Reimold (2002) and Koeberl et al. (2002) reported some of our early results in abstract form.

2
General Geology of BP and Oasis Impact Structures

Both features are located in the Al Kufrah basin of southeastern Libya. At the present erosion level, the rocks that crop out at both structures are sandstones (more or less ferruginous) with minor conglomerates and siltstones. These rocks are strata of probably Early Cretaceous age (Goudarzi 1970; Tawadros 2001; Hallett 2002), which is generally characterized as a ferruginous, fine- to medium-grained sandstone containing abundant cross-beds, mud cracks, ripple marks and silicified wood. There has been some confusion about the name of the rock units. In the literature, the target rocks in the BP-Oasis-Kufra area have been called Nubian Sandstone, or as belonging to the "Nubia Group" or "Nubia Formation" (e.g., Klitzsch 1978). There seems to be quite some discussion regarding the names of these rocks in the general area, as summarized by Tawadros (2001). Regionally, this group/formation is characterized by buttes and mesas of more or less horizontal beds, rising up to some 100 m above the surrounding plains, and some, mostly northwest-trending, ridges. The color of these strata is variably white, yellow, brown, and orange, but in many places it is black, even on a fresh surface. The dark color derives from iron- and manganese-oxide mineralization that is so intense that blocks of Nubian Sandstone can resemble basalt. The thickness of the Nubian Sandstone in the region of these two impact sites is unknown, but it reaches a maximum thickness of 1700 m elsewhere (e.g., Tawadros 2001). The sandstones have experienced low-grade metamorphism and were subject of regional deformation (Underwood 1975, 1976; Underwood and Fisk 1980). Northwest-trending ridges have been suggested to be clastic dikes (Underwood and Fisk 1980), but they probably are differentially weathered zones parallel to northwest-striking fractures. There are regional post-Paleozoic northeast-southwest, as well as a strongly developed post-Lower Cretaceous northwest-southeast, trends. The latter is invariably associated with fault and joint development.

Figures 3a and 3b show aerial photographs of BP and Oasis, respectively. The BP structure was suggested to be of meteorite impact origin by Kohman et al. (1967) and Martin (1969). Earlier, it had occasionally been referred to as the "Jebel Dalma" structure. The only geological description available, until now, is that by Underwood and Fisk (1980), who noted that BP is a relatively small structure defined by two discontinuous rings of hills and a central peak (Fig. 3a). According to these authors, the inner ring has a diameter of 2 km with an average height of 30 m, and the outer ring, which is characterized by strata that dip inwards at fairly low angles, has a diameter of 2.8 km and a maximum relief of only ca. 15 m. The rocks at

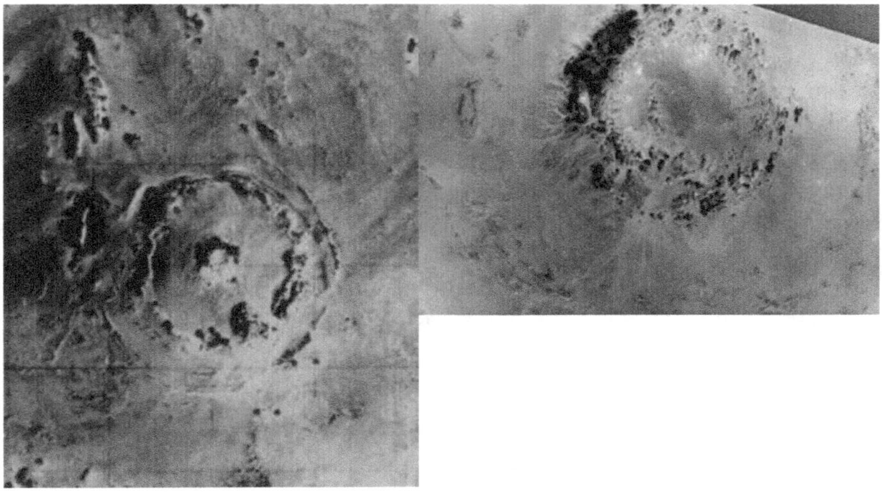

Fig. 3. Aerial photographs of the BP (**a** left) and Oasis (**b** right) impact structures. The image of the BP impact structure shows the two rings of hills and the central peak, whereas for Oasis only the inner ring of hills (diameter: 5.1 km) is clearly visible. The outer set of disturbed strata (11.5 km diameter) is partly visible on the bottom part of the image. North is up in both images.

the center of the structure show intense jointing, and bedding is difficult to discern. Although the outer ring has a diameter of 2.8 km, McHone et al. (1995a,b, 2002) speculated, based on interpretation of space shuttle radar studies, that disturbed beds, which are covered by a thin veneer of sand, extended to a diameter of 3.2 km.

According to Underwood and Fisk (1980), the Oasis structure is a ring-shaped feature with a prominent circular range of hills approximately 100 m high, which forms an inner ring with a diameter of 5.1 km (Fig. 3b). In contrast to BP, the Oasis structure lacks a well defined central peak. Most of the rocks within the central ring of hills are intensely folded, and beds may be vertical or overturned. Previously, a diameter of 11.5 km was assigned to the Oasis structure, based on an outer zone of disturbed strata that is visible in the field (Underwood and Fisk 1980) and from the air (Fig. 3b). However, disturbed strata, covered by a thin veneer of sand, are suggested to extend to a diameter of 18 km, based on Space Shuttle radar imagery (McHone et al. 1995a,b, 2002).

3
Remote Sensing Studies

We have used Landsat and Radarsat images of the BP and Oasis structures in an effort to determine more about the local geology and the morphologic dimensions of the structures. The Landsat data have a resolution of 30 m and were acquired in multiple wavelengths. Radarsat data were acquired at much longer wavelengths attaining a variable resolution of 10 to 100 m, but have the ability to penetrate thin surficial sand and are sensitive to changes in roughness and dielectric constants.

For both the BP and Oasis structures a variety of images were produced in order to discriminate among the lithologies present at the site and to determine various morphometric dimensions. We used: a) false color Landsat images, combining bands 1, 4, 7 and bands 2, 3, 5; b) various stretches (linear, 2%, and Gaussian) of these false color composites that merely exaggerate the contrast of the scene, making features appear brighter or darker than they actually are; and c) spectral classifications of each of the scenes. As there is very little difference in the spectral properties of the materials and there are only a few spectral bands in the Landsat images, the number of spectral classes was limited to five. Several techniques were used, such as Isodata, K-mean, and principal components, to spectrally group the materials. These analyses show that most of the bedrock units have approximately the same spectral character, within the constraints of the limited spectral channels. Finally, (d) radar data were investigated.

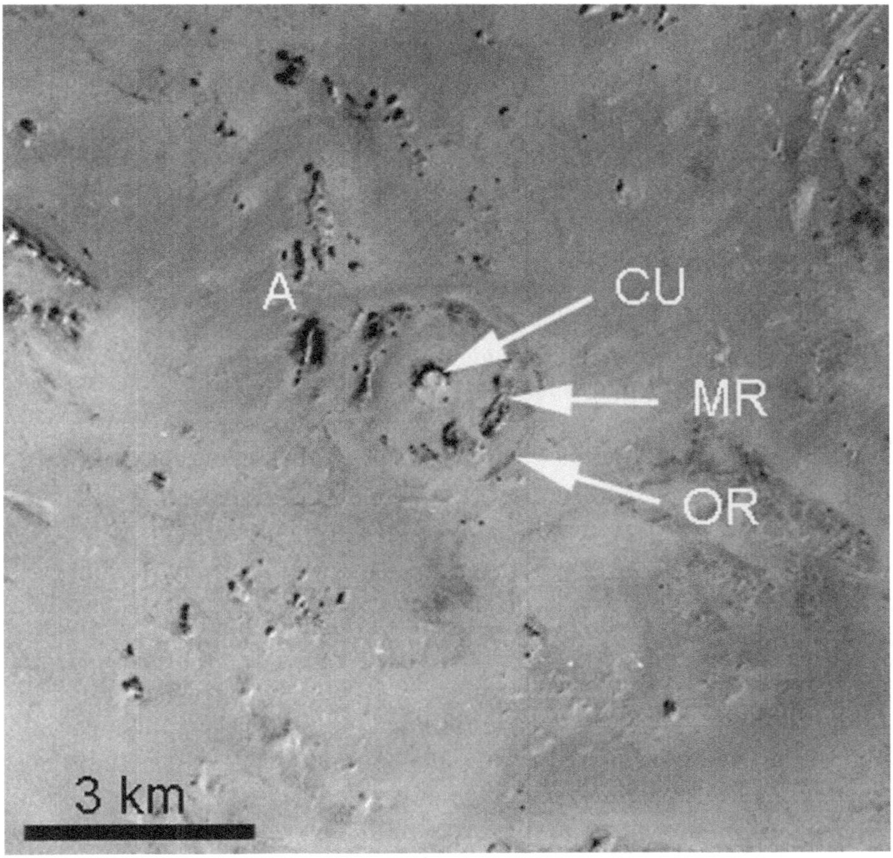

Fig. 4. Landsat image of the BP structure. CU indicates the Central Uplift / Inner Ring; MR indicates Middle Ring; OR indicates the Outer Ring. A denotes the north-trending bedrock ridge discussed in the text that possibly limits the lateral extent of the structure.

4 BP Impact Structure

4.1 General

The BP structure is located in the Gebel Dalma area of Southern Cyrenaica, 165 km northeast of Kufra Oasis, and is centered at 25°19' N and 24° 20' E. It occurs in a broad northeast-trending syncline of Nubian Sandstone. Flanking the syncline are hills composed of Upper Paleozoic shelf clastics, in which neither volcanics nor evaporites have been observed. French et al. (1974) and Underwood and Fisk (1980) described BP as consisting of two discontinuous rings that surrounded a central bloc; the outer ring of hills is 2.8 km in diameter with about 20 m of maximum relief (our observations – see below – indicate only about 10 m maximum relief and, on average, not more than 2 m relief), with sandstone beds dipping inward at 3°-15°. The inner ring of hills is more deformed, has a diameter of about 2 km, and an average relief of 30 m. Most of these beds dip outward at 20°-40°. These two rings form an asymmetric (wider on the eastern side) ring syncline. According to Underwood and Fisk (1980), the inner ring also shows numerous gently plunging folds with axes tangent to the structure and dipping outwards by as much as 70°. The central block is 0.6 km in diameter and has about 38 m of relief. Beds are intensely jointed, and the eroded southern half of the block exposes the oldest rock in the area, a light-colored (purplish to whitish) sandstone that has been complexly folded.

McHone et al. (1995a,b, 2002) noted that much of the surface at BP is sand-covered. As Shuttle radar penetrates the sand, it reveals the underlying bedrock; the data show a nearly complete radar-bright pattern that, according to these authors, was typical of central uplifts. McHone et al. suggested that the radar data indicated a crater diameter of up to 3.2 km. Three radar-bright rings were observed and suggested to be composed of eroded blocks of Nubian sandstone. The structure reportedly is surrounded by a subtle fluvial drainage, not seen in visible imaging.

Within the structure itself, Underwood and Fisk (1980) reported only Nubian Sandstone. These authors noted that the structure consisted of three near-circular concentric rock outcrops. The innermost ring had high-angle and chaotic dips forming a mass of craggy outcrop. The middle ring consisted of uniformly outward-dipping (30°-50°) strata. The third ring dipped inward at 5°-15° and formed a low scarp, for most of its extent barely protruding above the surface. Underwood and Fisk (1980) interpreted these outer two rings to define a ring syncline.

4.2 Landsat Image

On Landsat imagery, the BP structure exhibits three concentric rings with diameters of about 425, 2075, and 3135 meters. Figure 4 illustrates the structure in Landsat images (see also color plate 1). The radar and Landsat images provide various levels of detail. In the visible wavelength images (e.g., Fig. 4 and color plate1), the rings appear to be separated; however, in the radar image (Fig. 5) they appear to be more continuous.

Fig. 5. Radarsat image of BP impact structure. The more continuous nature of the circumferential "rings" when compared with the visible image (**Fig. 4**) is apparent. Note also the interruption of the northwest-trending structure. North is at the top.

Inner Ring: The inner ring (the "central block" of Underwood and Fisk [1980]) is continuous around ~270° of arc, but the southeastern sector is missing. The ring is about 425 m in diameter. In infrared (IR) wavelengths, the ring appears more continuous. The ring is formed by a massif, which is about 150-160 m wide and surrounded by a 145 m wide comparatively flatter region with a few isolated, largely sand-covered hills - especially on the southern side. It is possible that bright sand on the southern side of the ring that obscures the southern half of the ring at visible wavelengths.

Middle Ring: The middle ring (the "inner ring" of Underwood and Fish [1980]) is composed of discontinuous hills of variable width and morphology. It is about 2075 m in diameter. In the visible wavelengths much of the middle ring appears to be composed of two concentric ridges separated by 150-200 m. In the south–southeast and north–northwest quadrants, the ring becomes wider with apparently more complicated structure. The south side has a large (409 x 495 m) massif, and in the northwest a massif is about 500 m across. The bedrock ridges have alluvium extending away for variable distances. In most places the bedrock ridges are less than 100 m across.

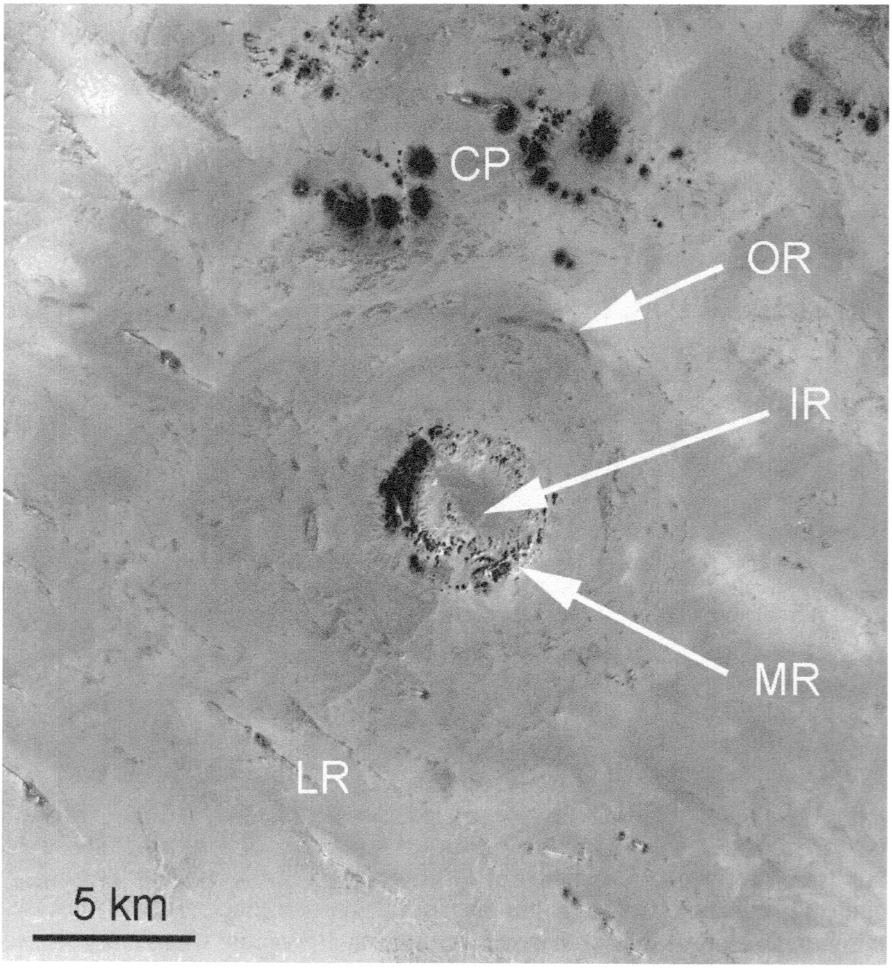

Fig. 6. Landsat images of the Oasis structure: CU indicates the Central Uplift / Inner Ring; MR indicates Middle Ring; OR indicates the Outer Ring. CP are the clastic plugs mentioned by Underwood and Fisk (1980) and discussed in the text.

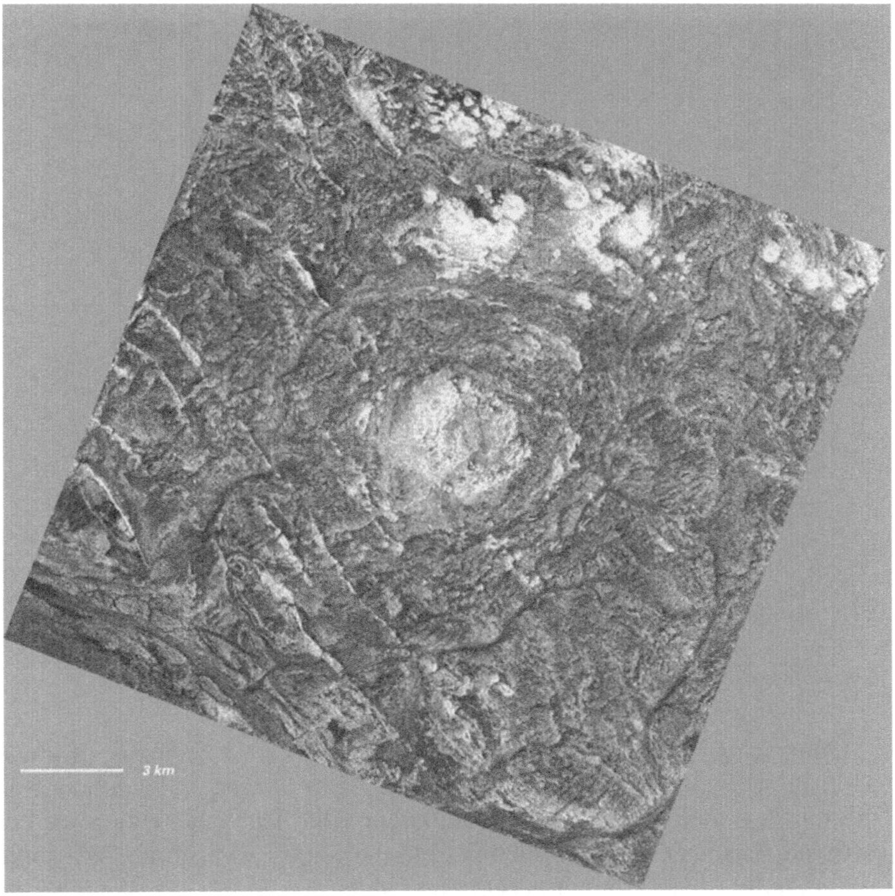

Fig. 7. Radarsat image of Oasis impact structure, projected so that north is up. Compare to the visible image (**Fig. 6**).

On the south-southeast side (at azimuth 155°) there is a major drainage that crosses the middle ring from the interior. At 211°, there is a minor drainage, which seemingly also crosses the middle ring. It certainly drains the hills associated with the middle ring. At azimuth 322°, there is a northwest (N54°W) trending structure that cuts across the middle ring for about 450 m. Farther out are two more linear features that are offset from this feature and trend in the same northwest direction.

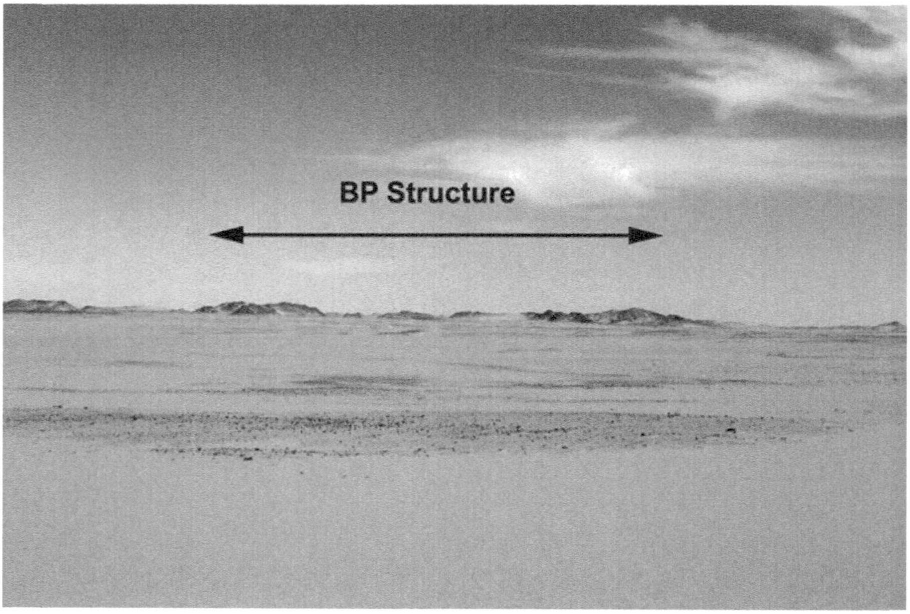

Fig. 8. Approaching the BP structure in the field from the west.

Outer Ring: There is an outer low-relief ring with a diameter of about 3135 m. This outer ring is defined only by low arcuate hills, although in the southeast and northwest there are larger hills. This outer ring is also of variable distance (1350 to 1900 m) from the center. At azimuth 285° there is a large hill that has a northerly trend, which would be on strike with the hills farther north and outside of the structure. However, in the radar scene (Fig. 5) it has a curved appearance as if it were part of the outer ring. In the field (see below) it is clear that this ring has a topography of only a few meters, with shallowly inward dipping strata, i.e., this cannot be the crater rim.

To the northwest of the crater, at a distance of about 2200 m from the center, there are a set of north-northeast (N20°E) trending ridges that do not appear to be affected by the deformation related to the formation of the structure. These features provide a limit on the extent of deformation. The hill to the northwest (N54°W), at 1742 m from the center of BP, does appear to be affected though; it has a slightly more easterly orientation than the other parts of the unit by about 10-20°, particularly at its southern end.

Throughout the rings, the bedrock exposed has very similar spectral character. The largest differences are observed for various surficial units, such as alluvium or sand cover of variable thickness. The material exposed in the middle and inner rings appears similar, and there is some variation in the outer ring, which may be due to aeolian materials.

4.3 Radar Image

The radar image (Fig. 5) has less resolution than the visible images, but because the wavelength is longer, there is penetration of the dry surficial sands. In the radar images, the ring structure appears more continuous than in the visible wavelengths. Even in the radar images, the different rings are separated by areas of low reflectivity. These dark areas could be areas where the sand is thicker (the radar did not penetrate) or where different bedrock materials occur. Presumably the ridges observed in the visible wavelength images represent differentially eroded layers that remain above the surficial sand cover.

Fig. 9. Upturned strata of sandstone on the south side of the inner ring (central uplift) of BP.

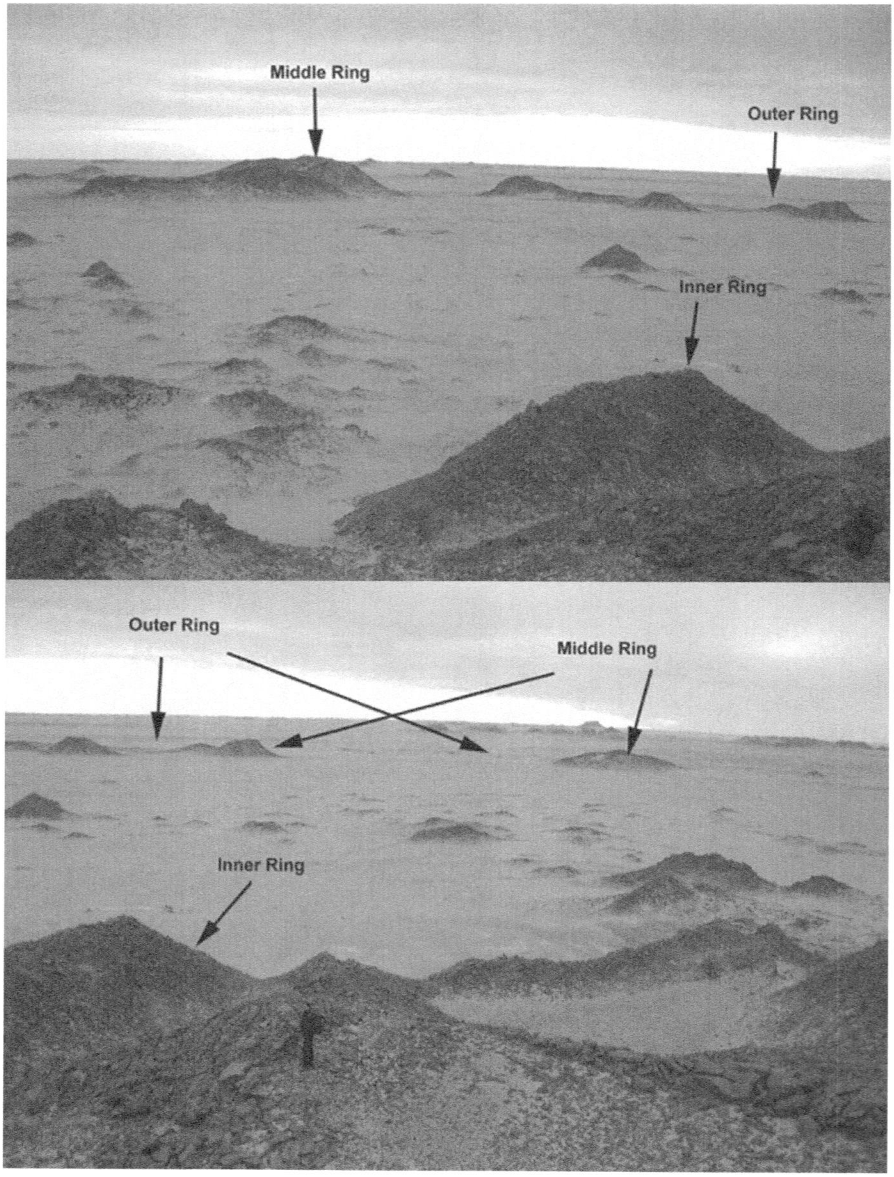

Fig. 10 (a top, b bottom). Overview of the three ring structures at BP, as seen from the top of the inner ring (central uplift), looking southeast and east, respectively.

5 Oasis Impact Structure

5.1 General

The Oasis impact structure, centered at 24° 35' N, 24° 24' E, was reported by Underwood and Fisk (1980) to have a "disturbed zone" of 11.5 km diameter, but the discontinuous hills that compose the topographically prominent part of the structure form a circular array of 5.1 km diameter. Most of the rocks in this prominent ring dip outward, and locally they are intensely folded. Hills reach a maximum height of about 100 m. Outside of this ring, the disturbed rocks have only a meter of relief. No central block, like that at BP, had been observed. McHone et al. (1995a,b, 2002) noted that SIR-C Shuttle radar data indicated several rings at Oasis extending to a diameter of 18 km. According to these authors, the structure is located within a regional zone of radar-prominent linear ridges of Nubian Sandstone; these ridges are about 100 m high and trend northwest. We indeed observed this trend when we continued our route from the Oasis crater structure towards Kufrah Oasis on a southwesterly bearing.

5.2 Landsat Image

As at BP, Oasis is seemingly composed of three rings: a well-defined, prominent ring (referred to as the middle ring), a narrow, partly exposed central ring, and an outer "ring" composed of numerous discontinuous, concentric low hills. The Landsat image (Fig. 6 and color plate 2) indicates an average diameter of about 15.7 km for disturbed strata, and other sources (e.g., radar: Fig. 7) indicate almost 19 km. The limit of deformation on the north side is well constrained in some of the exposures at 8.8-8.9 km from the center (Figs. 6 and color plate 2), and to the northeast at a similar distance where the regional structural trends are undisturbed. Elsewhere, the limit is not well defined.

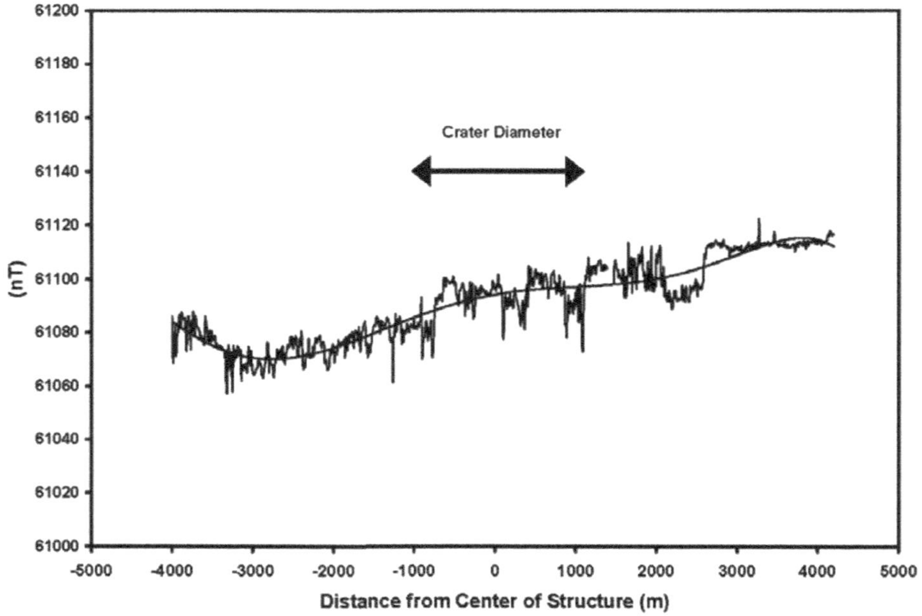

Fig. 11. Ground-based magnetic profile at BP crater, measured with an Envi-Mag proton magnetometer at 10 m spacing. Data were corrected using local base station data, measured at about hourly intervals. The position of the crater rim is indicated.

Inner Ring: The inner ring is ~1100-1300 m in diameter. It is defined by a bedrock arc from the northwest to west and southwest; the northern and northeastern part of the ring are missing or not exposed at the surface. The ring is about 200-250 m wide. It is separated from the middle ring by a region that appears like an alluvial surface or low hilly region. On the southern side, the bedrock exposures of the inner ring seemingly merge with the wide area of bedrock exposures of the middle ring. The distance between the inner edge of the middle ring and the outer edge of the inner ring is about 855 m.

Middle Ring: The bedrock exposures of the middle ring have a 5.5 to 6.2 km diameter. The diameter of the middle ring is variable with azimuth. It is widest and most continuous in the northwest (about 1300 m across). From the north around to the east and south, it is composed of closely spaced hills. Within the northwest part of the ring, the bedrock exposure is composed of concentric topographic highs and lows with the ridges spaced 2.6 to 3.7 km apart. The southwest part of the middle ring contains a large basin 1450 x 1740 m across.

Fig. 12. (**a**, top) Approaching Oasis impact structure from the north; overview of the central ca. 5 km diameter ring of hills. (**b**, bottom) Oasis inner ring as seen from the south.

A major drainage feature occurs in the south-southwest. Here a channel extends from the center of the structure into the surrounding plains. It also has a linear dune associated with it. The drainage on the west-southwest side seems to cuts through the middle ring and extend into the hills between the middle and inner rings.

Outer Ring: The outer ring consists of concentric, discontinuous hills that extend from the middle ring (~ 5700 m in diameter) out to a distance of as much as 9 km from the center. Individual hills extend only short distances and are separated by alluvial/aeolian material. The hills appear to have little topographic expression vis-à-vis the middle ring. They are more obvious on the north and east sides, and more subdued to the southwest.

Fig. 13. Oasis impact structure. Strongly deformed (folded) section of inner ring of hills (on the eastern part of the structure).

In the western quadrant (azimuth 262°) of the outer ring there is an elliptical hill about 370 x 1140 m in size, at 4902 m from the center. It is topographically asymmetric, with a high eastern wall and a low western wall. A series of northwest trending (N61°W - N53°W) linear ridges cut across the region and cut the southern and western part of the Oasis structure.

Various spectral classification techniques (e.g., IsoData, Unsupervised, Principal Component) were used to determine the extent of the spectral variation within the rocks exposed at the surface. The different techniques produce essentially the same result, namely that within the constraints of spatial and spectral resolution all of the bedrock units are the same. The spectral classification of the bedrock associated with the Oasis structure is similar to that of the bedrock exposed to the north. The alluvial units exhibit some spectral variation. This variation probably results from different combinations of material across the alluvial surfaces and may be related to different fluvial systems.

5.3 Radar Image

The Radarsat image of Oasis (Fig. 7) shows a concentric pattern associated with the structure. Similar to the radar image for BP, the radar data for Oasis show that the concentric structure, which appears as discontinuous concentric bands in the visible light data, is a continuous structure both laterally away from the structure as well as circumferentially around it. This pattern suggests uniform deformation around the circumference of the structure. Areas that appear only in the visible wavelengths and which are bright in the radar are presumably differentially weathered material that stands above the sand sheet. The northwest-trending features observed in the visible images are also apparent in the radar as bright linear features, again indicating that most of the feature stands above the sand. These features can be traced along strike where they have lower brightness, suggesting that they are not exposed at the surface but buried by the sand sheet.

The drainages are very well defined in the radar data, particularly along the west and east sides of the structure. They clearly control the distribution of alluvial material seen in the Landsat images. The drainage on the west side of the structure appears to be a single major channel with a few tributaries, compared with the more braided nature of the drainage on the east side of the structure. This difference could be controlled by topography.

Fig. 14. Alleged clastic plugs (Underwood and Fisk, 1980) north of the Oasis structure, apparently outside of limit of deformation visible in the field.

Fig. 15. Deformed ridge near clastic plugs, about 4 km outside the inner ring at Oasis.

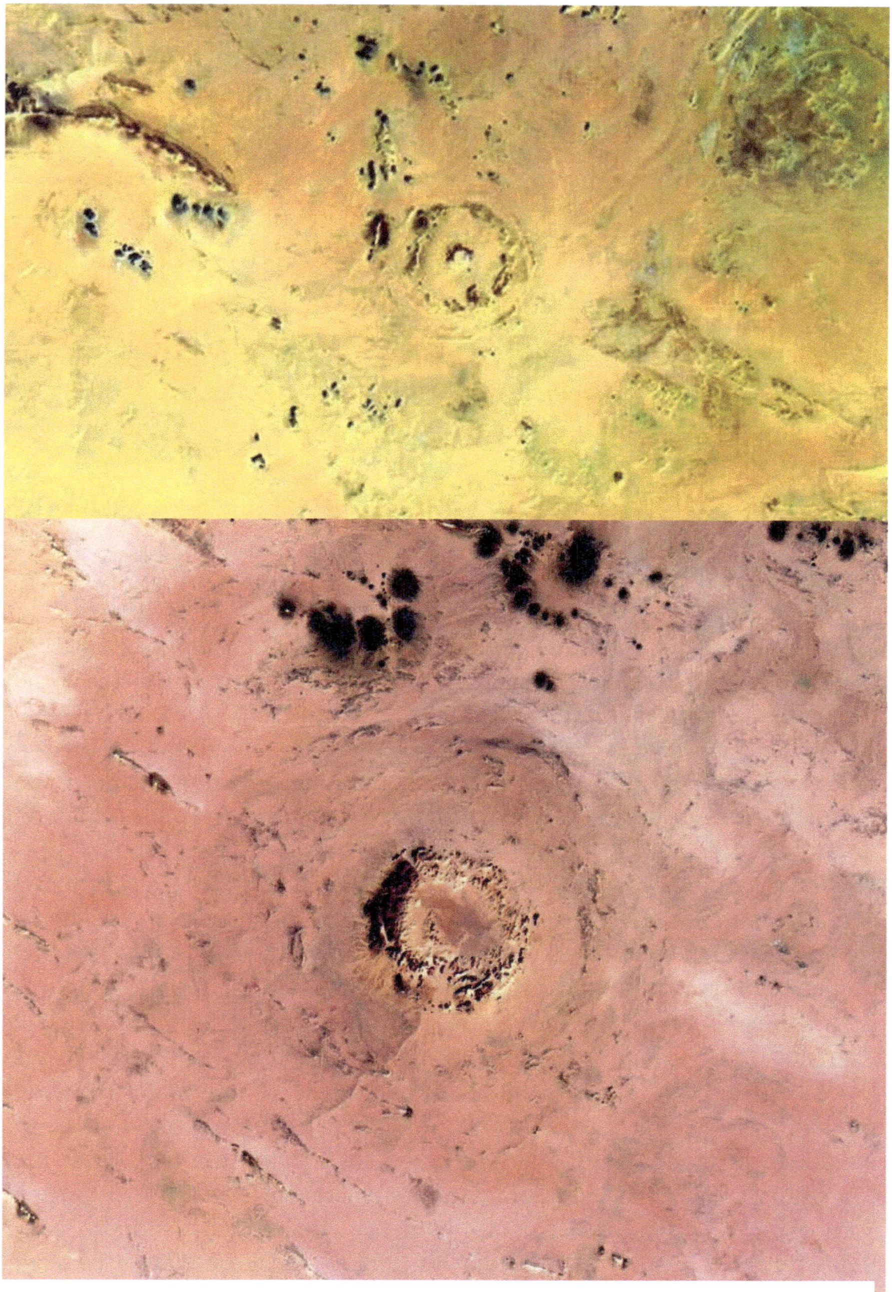

Color Plate: (**1**, top) BP impact structure (visible natural light, Landsat image). (**2**, bottom) Oasis impact structure (visible natural light, Landsat image).

6 Field Work at BP and Oasis – First Observations

In January 2002, two of us (CK and WUR) organized a field trip to the BP and Oasis impact structures in Libya. It was planned to stay about 5 to 7 days at each of the two structures, conduct extensive structural geological and geophysical investigations, and collect samples for petrographic and geochemical work, and to search for potentially dateable material. Due to unforeseen bureaucratic complications, only the BP structure could be studied as planned, whereas the visit to Oasis was limited to less than half a day. Also the sample export posed some problems as discussed in detail by Reimold and Koeberl (2002). Detailed structural evaluation, as well as petrographic and geochemical work, is in progress and will be presented elsewhere. Here we only report some first observations pertinent to the interpretation of the satellite images.

6.1 BP Structure

Upon approach in the field, the structure shows – in comparison to the mesa and butte terrane of the region – limited, but clearly discernible, topography (Fig. 8). The "inner ring" is a central block of ca. 500 m width, with steeply upturned rock units that are intensely folded, faulted, and – in places – brecciated (Fig. 9). A variety of microdeformation has been noted in quartz from these breccias and from other sandstone samples, including a few bona fide planar deformation features (work in progress).

What has previously been referred to as the "middle ring" is in fact the actual crater rim with a diameter of about 2 km. This ring is characterized by a distinct series of hills of up to 30 m elevation above the surrounding desert, with rocks dipping at 30 to 50° outwards (Fig. 10a, b). Some parts of the rim, in particular the southern and eastern part of this zone, are characterized by intense folding and faulting, whereby up to 100 m wide blocks have been rotated and displaced inwards out of the generally circular trend within this ring. The "outer ring" is a discontinuous ring of subdued, generally less than 1-2 m, but locally up to 10 m high, topography. Here, the sandstones dip inwards, at values of ± 10°. No geological or geomorphological evidence that might support a rim diameter of 3.2 km was found in the field. This interpretation of such a large crater diameter is based on the interpretation of radar satellite imagery, whereby it is now clear whether the radar data indicate only structure that occurs underneath sand cover. It is possible that the shallowly inwards dipping structure at 1.6 km from the center represents a fault zone, along which strata could have moved inwards towards the center of the crater, in response to central uplift formation.

"Outer ring structures" have been described from a number of better preserved craters, such as the 11-km-diameter Bosumtwi crater in Ghana, with an outer ring feature (so far of unexplained origin) about 18 km in diameter (Wagner et al. 2002), which clearly is beyond the actual crater rim. Kenkmann (2003) recently discussed the presence of a significant detachment fault at the base of the Upheaval Dome structure in the USA.

A ground-based magnetic profile was measured over a distance of almost 10 km across the BP structure (extending one crater diameter in each direction from the crater rim). Measurements were performed with an Envi-Mag proton magnetometer at 10 m spacing. At regular time intervals (every hour), a base station located on the western crater rim was revisited in order to obtain control on the diurnal variation. Attempts to obtain regional diurnal variation data failed, as the closest magnetic observatories located in Athens and Rome could not provide substantial background information for the BP area. Thus, the data were corrected with only the local base station data. The results show a noisy, but essentially featureless traverse, even through the central uplift area (Fig. 11). Minor fluctuations in the data must be related to the mineralogical variability of the Nubia strata encountered in the structure. In particular, some dark-brown to blackish strata are highly enriched in iron compared to lighter-colored, more silicious sandstones. However, such a profile is consistent with a single rock type exposed at the surface and in the shallow subsurface and the absence of any rock of significantly different magnetic properties (e.g., impact melt occurrence; which, however, is not expected to be extensive in a crater mainly in sandstone).

6.2
Oasis

For Oasis, to which we could only pay a cursory visit, a diameter of 11.5 km has been quoted. However, the structure is characterized by a topographically prominent – in comparison to all other topographic features in the area - and intensely faulted ring of hills (the "middle ring") of about 5-6 km diameter (Fig. 12a, b). The rocks in this prominent ring mainly - but not always - dip outward. Much of the segment of this ring that we traversed is intensely folded (Fig. 13), on a macro- to meso-scale (i.e., hundreds of meters to decimeters, even centimeters).

There are two possibilities to interpret this ring structure: (1) it could represent – in analogy to the structure of Gosses Bluff (e.g., Milton et al.

1996) – the eroded central uplift of a much larger impact structure, whereby the actual crater rim would have been entirely eroded, and (2) alternatively, this prominent ring could be the remnant of the crater rim, whereby one can refer to the Landsat imagery that indicates a central area that is distinct from the ring feature and, itself, could represent a remnant of an eroded central uplift (compare Fig. 6 and color plate).

In the environs of this prominent ring of hills, no further prominent topographic features have been identified, despite the locally strong indications on the Landsat images. However, in a few places to the north, at a distance of about 3 km from the prominent ring feature, a shallowly inwards dipping, low (< 1 m) ridge was observed, which was very reminiscent of the outer ring feature around BP. It is clear that further detailed structural work is required to resolve these open questions.

According to Underwood and Fisk (1980), about 6-8 km north and northeast of Oasis, a field of circular or rounded knobs of sandstone occur. These are 10-100 m in diameter and 10-50 m high. These authors suggested that the knobs might be linked to the crater-forming impact event. They described them as clastic plugs, apparently composed of Nubian sandstone, which had been emplaced by vertical flow of impact-fluidized sand. Our investigation indicated that in some cases (Fig. 14) they have a distinct chimney-like form with sandstone strata in the central areas having near-vertical bedding orientations. However, no evidence for fluidization of sand - as speculated by Underwood and Fisk (1980) - was found. In contrast, several distinct ridges were recognized as erosional remnants of complex fold structures (Fig. 15). The question whether or not this folding is related to the impact event, or reflects regional tectonic deformation, still remains open. It appears reasonable to interpret these plug-like structures as remnants of folds, of which the fold hinges have been completely eroded, leaving the vertically standing strata of the fold cores behind.

7 Summary and Conclusions

The BP and Oasis impact structures in Libya are in geographic proximity to the Libyan Desert Glass (LDG) strewn field in the Libyan Desert, which led to speculation that these craters might be related to the LDG. Here we have discussed remote sensing investigations (satellite images) in comparison with first results from a field trip to both impact structures.

The highly eroded BP structure appears on satellite images in the form of three concentric rings. In radar images the rings are more pronounced than in the visible wavelengths, probably because the radar penetrates the sand cover. The three "rings" are: an "inner ring", which is interpreted here as a small central uplift (ca. 500 m in diameter) with steeply upturned strata that are intensely folded, faulted, and – in places – brecciated; a "middle ring" with a diameter of about 2 km, which ring is characterized by a distinct series of hills of up to about 30 m elevation above the surrounding desert, with rocks dipping at 30 to 50° outwards. Some parts of the rim are characterized by intense folding and faulting. Lastly, there is a discontinuous "outer ring",, ca. 2.8 km in diameter, consisting of low hills with 1-2 m average elevation, dipping at 10° inwards. Earlier suggestions of a crater diameter of 2.8 or 3.2 km (from Shuttle radar data) were not confirmed in our field studies. It became obvious that the "middle ring" represents the actual crater limit, and BP, therefore, has a rim-to-rim diameter of 2 km, making it the smallest known complex impact structure on Earth.

For the Oasis structure, diameters of 11.5 km or even 18 km (from radar images) have been quoted in the past. The topographically most prominent feature of this structure is a ring of discontinuous hills that are up to 100 m high, with an average diameter of about 5.1 km. The rocks in this prominent ring dip outward and locally are intensely folded. It is not clear whether this feature represents the actual crater rim.

We are currently performing detailed petrographic and geochemical analyses of the samples collected during our field work, and are also evaluating the structural data. The ages of the BP and Oasis structures are an open question; it is not known if the two craters are coeval or not. A search in the field has not yielded any obvious material for dating. Thus, the connection to the LDG remains unproven for the time being.

Acknowledgments

We are grateful to Tarek El Mahdy (Dabuka Expeditions, Munich/Cairo) for field trip logistics, to A. Barakat (Geological Survey, Cairo, Egypt) for field assistance, and to H.P. Schönlaub (Geological Survey of Austria) for the loan of a proton magnetometer. The field work was mostly financed by the Austrian Science Foundation, grant Y58-GEO (to C.K.), and a grant from the South African National Research Foundation (to W.U.R.). We appreciate helpful reviews by F. Tsikalas and J. Plado, and editorial comments by H. Henkel. This is University of the Witwatersrand Impact Cratering Research Group Contribution No 79. Plescia's contributions were supported by funding from NASA's Planetary Geology and Geophysics Program.

References

Abate B, Koeberl C, Kruger FJ, Underwood JR Jr (1999) BP and Oasis impact structures, Libya, and their relation to Libyan Desert Glass. In: Dressler BO, Sharpton VL (eds) Large Meteorite Impacts and Planetary Evolution, Geological Society of America, Special Paper 339: 177-192

Barakat AA (2001) Hypervelocity meteorite impact features within the Libyan Glass area. Annals of the Geological Survey of Egypt V: XXIV.

Barnes VE, Underwood JR Jr (1976) New investigations of the strewn field of Libyan Desert Glass and its petrography. Earth and Planetary Science Letters 30: 117-122

Barrat JA, Jahn BM, Amosse J, Rocchia R, Keller F, Poupeau G, Diemer E (1997) Geochemistry and origin of Libyan Desert Glasses. Geochimica et Cosmochimica Acta 61: 1953-1959

Bigazzi G, De Michele V (1996) New fission-track age determinations on impact glasses. Meteoritics and Planetary Science 31: 234-236

Bigazzi G, De Michele V (1997) New fission-track ages of Libyan Desert Glass. In: Proceedings, "Silica '96" Meeting on Libyan Desert Glass and related desert events, Pyramids, Milan, Italy, pp 49-58

Clayton PA, Spencer LJ (1934) Silica glasses from the Libyan Desert. Mineralogical Magazine 23: 501-508

Diemer E (1997) Libyan Desert Glass: an impactite State of the art in July 1996. In: Proceedings, "Silica '96" Meeting on Libyan Desert Glass and related desert events, Pyramids, Milan, Italy, pp 95-110

Feller M (1997) Vitreous silica from the Sahara. In: Proceedings, "Silica '96" Meeting on Libyan Desert Glass and related desert events, Pyramids, Milan, Italy, pp 111-114

French BM, Underwood JR Jr, Fisk EP (1972) Shock metamorphic effects in two new Libyan impact structures [abs.]. Geological Society of America, Abstracts with Programs 4 (7): 510-511

French BM, Underwood JR Jr, Fisk EP (1974) Shock metamorphic features in two meteorite impact structures, Southeastern Libya. Geological Society of America Bulletin 85: 1425-1428

Futrell DS, O'Keefe JA (1997) A brief discussion of the petrogenesis of Libyan Desert Glass. In: Proceedings, "Silica '96" Meeting on Libyan Desert Glass and related desert events, July 18, 1996, Pyramids, Milan, Italy, pp 115-120

Giegengack R, Underwood JR Jr (1997) Origin of Libyan Desert Glass: some stratigraphic considerations. In: Proceedings, "Silica '96" Meeting on Libyan Desert Glass and related desert events, Pyramids, Milan, Italy, pp 37-40

Goudarzi GH (1970) Geology and mineral resources of Libya; a reconnaissance. US Geological Survey Professional Paper 660, 104 pp

Hallett D (2002) Petroleum Geology of Libya. Elsevier, Amsterdam, 508 pp

Horn P, Müller-Sohnius D, Schaaf P, Kleinmann B, Storzer D (1997) Potassium-argon and fission-track dating of Libyan Desert Glass, and strontium and neodymium isotope constraints in its source rocks. In: Proceedings, "Silica '96" Meeting on Libyan Desert Glass and related desert events, Pyramids, Milan, Italy, pp 59-76

Jux U (1983) Zusammensetzung und Ursprung von Wüstengläsern aus der Großen Sandsee Ägyptens. Zeitschrift der deutschen geologischen Gesellschaft 134: 521-553

Kenkmann T (2003) Dike formation, cataclastic flow, and rock fluidization during impact cratering: an example from the Upheaval Dome structure, Utah. Earth and Planetary Science Letters 214: 43-58

Kleinmann B (1969) The breakdown of zircon observed in the Libyan Desert Glass as evidence of its impact origin. Earth and Planetary Science Letters 5: 497-501

Kleinmann B, Horn P, Langenhorst F (2001) Evidence for shock metamorphism in sandstones from the Libyan Desert Glass strewn field. Meteoritics and Planetary Science 36: 1277-1281

Klitzsch E (1978) Geologische Bearbeitung Südwest-Ägyptens. Geologische Rundschau 67: 509–520

Koeberl C (1985) Trace element chemistry of Libyan Desert Glass [abs.]. Meteoritics 20: 686

Koeberl C (1997) Libyan Desert Glass: geochemical composition and origin. In: Proceedings, "Silica '96" Meeting on Libyan Desert Glass and related desert events, Pyramids, Milan, Italy, pp 121-132

Koeberl C (2000) Confirmation of a meteoritic component in Libyan Desert Glass from osmium isotopic data [abs.]. Meteoritics and Planetary Science 35: A89-A90

Koeberl C, Reimold WU (2002) Field studies at the BP and Oasis impact structures, Libya [abs.]. Meteoritics and Planetary Science 37: A79

Koeberl C, Plescia J, Reimold WU (2002) BP and Oasis impact structures, Libya: Remote sensing and field studies [abs.]. In: von Dalwigk I (ed) Abstracts, 8th ESF-IMPACT Workshop "Impact Tectonism", Mora, Sweden, p 36

Kohman TP, Lowman PD Jr, Abdelkhalek ML (1967) Space and aerial photography of the Libyan Desert glass area [abs.]. Paper presented at the 30th Annual Meteoritical Society Meeting, Moffett Field, California

Martin AJ (1969) Possible impact structure in southern Cyrenacia, Libya. Nature 223: 940-941

Matsubara K, Matsuda J, Koeberl C (1991) Noble gases and K-Ar ages in Aouelloul, Zhamanshin, and Libyan Desert impact glasses. Geochimica et Cosmochimica Acta 55: 2951-2955

McHone JF, Blumberg DG, Greeley R, Underwood JR Jr (1995a) Space Shuttle radar images of terrestrial impact structures: SIR-C/X-SAR [abs.]. Meteoritics and Planetary Science 30: 543

McHone JF, Blumberg DG, Greeley R, Underwood JR Jr (1995b) Orbital radar images of Libyan impact structures [abs.]: Geological Society of America, Abstracts with Programs 27 (6): A-209

McHone JF, Greeley R, Williams KK, Blumberg DG, Kuzmin RO (2002) Space shuttle observations of terrestrial impact structures using SIR-C and X-SAR radars. Meteoritics and Planetary Science 37: 407-420

Milton DJ, Glikson AY, Brett R (1996) Gosses Bluff – a latest Jurassic impact structure, central Australia. Part 1: geological structure, stratigraphy, and origin. AGSO Journal of Australian Geology & Geophysics 16: 463-486

Murali AV, Zolensky ME, Carr R, Underwood JR Jr, Giegengack RF (1987) Libyan Desert Glass: Trace elements and gas inclusions [abs.]. Geological Society of America, Abstracts with Programs 19: 782

Murali AV, Zolensky ME, Underwood JR Jr, Giegengack RF (1988) Formation of Libyan Desert Glass [abs.]. Lunar and Planetary Science 19: 817-818

Murali AV, Linstrom EJ, Zolensky ME, Underwood JR Jr, Giegengack RF (1989) Evidence of extraterrestrial component in the Libyan Desert Glass [abs.]. EOS Transactions, American Geophysical Union 70: 1178

Murali AV, Zolensky ME, Underwood JR Jr, Giegengack RF (1997) Chondritic debris in Libyan Desert Glass. In: Proceedings, "Silica '96" Meeting on Libyan Desert Glass and related desert events, Pyramids, Milan, Italy, pp 133-142

Reimold WU, Koeberl C (2002) Traveller's paradise Libya. Geobulletin – Quarterly Newsbulletin of the Geological Society of South Africa 45(1): 24-28

Rocchia R, Robin E, Fröhlich F, Meon H, Froget L, Diemer E (1996) L'origine des verres du désert libyque: un impact météorique. Comptes Rendus Acadademie de Science (Paris) 322 (Ser. IIa): 839-845

Rocchia R, Robin E, Fröhlich F, Ammosse J, Barrat JA, Meon H, Froget L, Diemer E (1997) The impact origin of Libyan Desert Glass. In: Proceedings, "Silica '96" Meeting on Libyan Desert Glass and related desert events, Pyramids, Milan, Italy, pp 143-158

Schaaf P, Müller-Sohnius D (2002) Strontium and neodymium isotopic study of Libyan Desert Glass: Inherited Pan-African age signatures and new evidence for target material. Meteoritics and Planetary Science 37: 565-576

Storzer D, Koeberl C (1991) Uranium and zirconium enrichments in Libyan Desert Glass [abs.]. Lunar and Planetary Science 22: 1345-1346

Storzer D, Wagner GA (1971) Fission-track ages of North American tektites. Earth and Planetary Science Letters 10: 435-440

Storzer D, Wagner GA (1977) Fission track dating of meteorite impacts [abs.]. Meteoritics 12: 368-369

Tawadros EE (2001) Geology of Egypt and Libya. A.A. Balkema Publishers, Rotterdam, 480 pp

Underwood JR Jr (1975) Reconnaissance geology of meteorite impact structures in SE Libya [abs.]. Geological Society of America Abstracts with Programs 7: 242

Underwood JR Jr (1976) Impact structures in the Libyan Sahara: some comparisons with Mars, *in* Proceedings, International Colloquium of Planetary Geology, Geologica Romana (Rome) 15: 337-340

Underwood JR Jr, Fisk EP (1980) Meteorite impact structures, southeast Libya. In: Salem MJ, Busrewil MT (eds) The Geology of Libya, Proceedings, Symposium, 1978, London, pp 893-900

Wagner R, Reimold WU, Brandt D (2002) Bosumtwi impact crater, Ghana: A remote sensing investigation. In: Plado J, Pesonen LJ (eds) Meteorite Impacts in Precambrian Shields, Impact Studies, vol. 2, Springer, Heidelberg, pp 189-210

Weeks RA, Underwood JR Jr, Giegengack R (1984) Libyan Desert Glass: A review. Journal of Non-crystalline Solids 67: 593-619

Late Modification-Stage Tectonic Deformation of the Popigai Impact Structure, Russia

Mikhail S. Mashchak and Mikhail V. Naumov

Karpinsky All-Russia Geological Research Institute (VSEGEI), Sredny prospect 74, 199106 St. Petersburg, Russia (mvn@mail.wplus.net)

Abstract. The 35-Ma-old Popigai impact structure (about 100 km in diameter) is characterized by its good preservation state and exposure, and detailed investigations that were made through deep drilling and mapping. The present contribution shows data on the post-impact faulting in the Popigai crater that were obtained from detailed mapping combined with aerial photo interpretation. The fracturing of impact rock bodies is widespread throughout the crater. Two hierarchic groups of fracture deformations are distinguished: (1) Main radial and concentric impact-related faults, which determine a block structure of the crater; (2) Small-size local faults and fissures from 0.5 to 5.5 km in length and from some centimeters to 10-20 m wide within impact rock sequences. There is no preferred orientation of individual local fractures, but fault systems have near-radial or near-concentric directions. As a typical example, an intensely fractured impact melt body was mapped at the 1:16,000 scale. Local fracture deformations originated during the compaction of strongly inhomogeneous sequences of impact rocks, whereas block faulting has been caused by both regional tectonic movements, i.e., the uplifting of the Anabar Shield, and, to a lesser content, by relaxation processes. Thus, the 35 -Ma-long post-impact modification history of the Popigai crater is determined by the superimposition of the regional tectonics on the long-term relaxation movements. As a whole, the late modification stage tectonics is found to have only an insignificant effect to the Popigai crater, so that both the original structure and the crater topography have been retained in a good state.

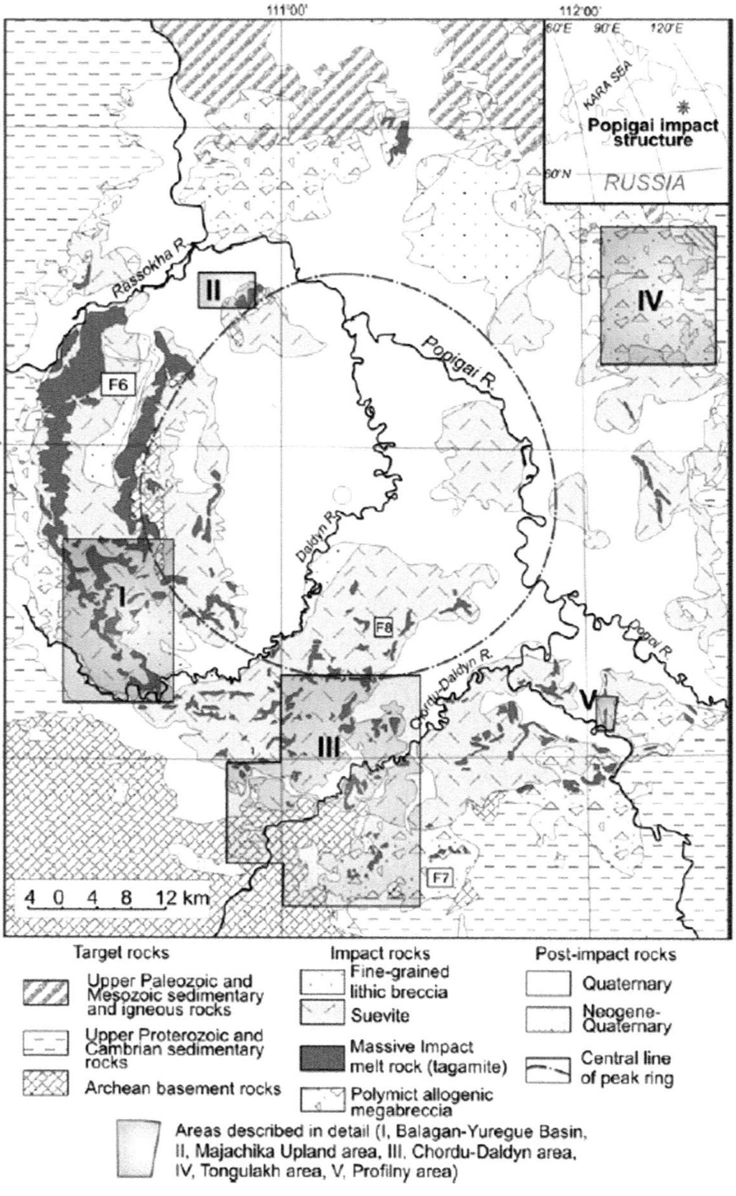

Fig. 1. Schematic geologic map of the Popigai impact structure. Modified after Masaitis et al. (1998). Areas described in detail are shown by shaded quadrangles. The locations of areas that are shown in aerial photographs **(Fig. 5-7)** are shown by white quadrangles.

1 Introduction

The general tectonics of complex impact crater formations, as well as the impact crater mechanics, have been considered in detail by many researchers (e.g., Grieve 1987, Melosh 1989, Melosh and Ivanov 1999, Kenkmann 2002, and others). On the contrary, the post-early modification stage tectonic history of impact structures is neglected, because of the most of them are essentially modified by erosion or buried beneath later sediments. In this paper, we present data on post-impact faulting within a well-studied large impact crater - the about 35-Ma old Popigai structure. The objective is to show real features of post-consolidation brittle deformations within impact rock sequences, obtained

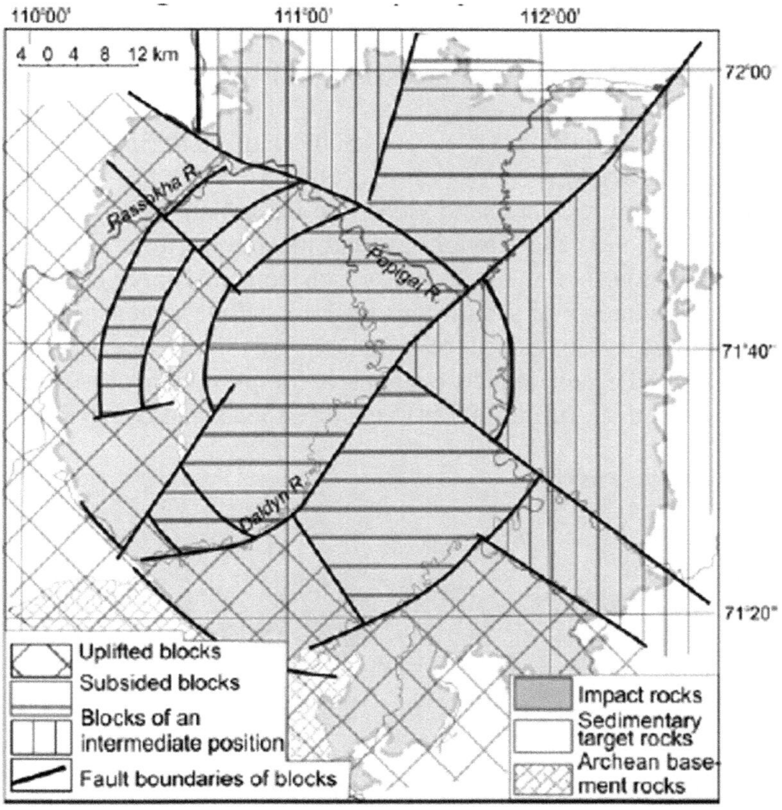

Fig. 2. Post-impact tectonic structure of the Popigai crater. Fault blocks differing in extent and direction of post-impact vertical displacements are distinguished.

from detailed geological mapping combined with aerial photo interpretation, and thereby to draw attention to the problem of the late modification stage evolution of impact structures. In this aspect, Popigai is one of the best objects due to its good preservation and exposure, and detailed investigations that were made through deep drilling and from abundant outcrops of diverse impact lithologies.

2 General Geology of the Popigai Crater

The 100-km-diameter Popigai impact structure (Masaitis et al. 1980, 1998, 1999, and references therein) formed 35.7 Ma ago (Bottomley et al. 1997) at the northeast edge of the Anabar Shield, Northern Siberia. The crater was excavated within Archean crystalline rocks with overlying Proterozoic to Permian sedimentary cover with a thickness of up to 1.5 km. The crater is filled by various kinds of lithic breccias and impact melt rocks - both fragmental (suevites) and massive (tagamites) (Fig. 1). The total volume of impact rocks has been suggested to exceed 5000 km^3, and impact melt rocks contribute up to 2800 km^3.

The impact lithologies are subdivided into three main units with exceptionally variable thickness and strongly gradual transitions, from bottom to top in the vertical section: (1) a lower unit, which contains polymict megabreccia composed of fragments of sedimentary and crystalline target rocks, and crystalline megabreccia composed of fragments of shocked crystalline target rocks; (2) a middle unit, which contains a complex blanket of impact melt rocks, including thick (up to 600 m) tagamite sheets and overlying vitric and

Fig. 3. Panoramic view of a continuous bluff of impact rocks (with a height of 120 m), which forms the right side of the Chordu-Daldyn River valley, 11 km from the mouth. The bluff traces a concentric fault system in the southern part of the crater.

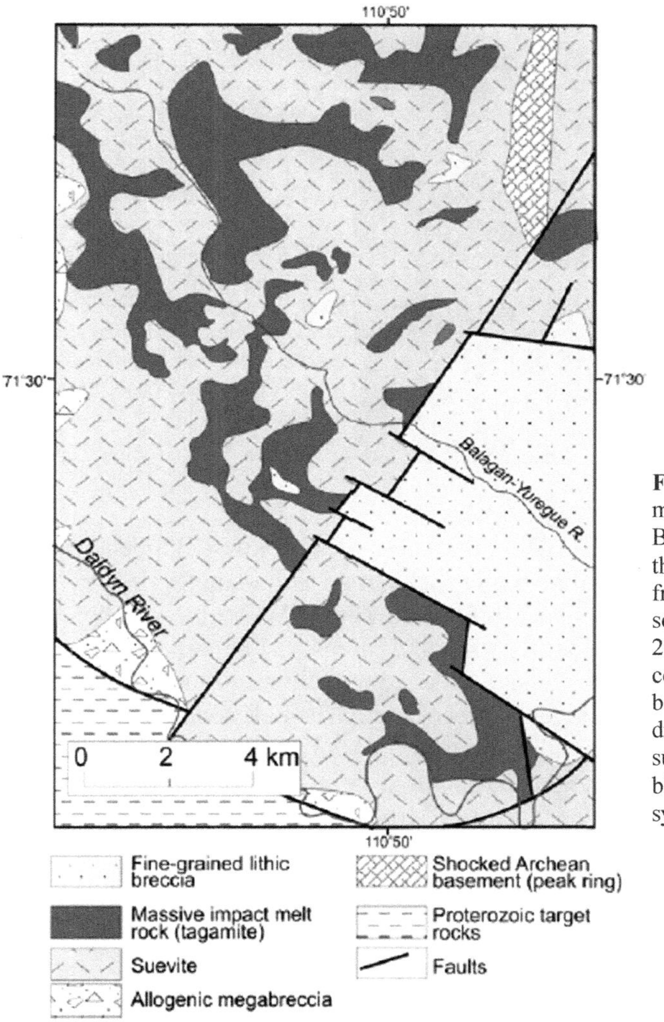

Fig. 4. Schematic geologic map of the Balagan River Basin, southwestern sector of the Popigai crater, compiled from deep drilling data. In the southeastern part of this area a 200-m-thick coptoclastite cover overlying tagamite has been mapped from 34 drillcores, and indicates subsidence of the southeastern block along a radial fault system.

crystal-vitric suevites, which are composed mainly of black glass fragments (vitroclasts) and fragments of shocked crystalline rocks and their minerals; (3) an upper unit containing vitric-lithic suevites, which are composed mainly of greenish-gray vitroclasts and fragments of sedimentary rocks and of fine-grained lithic breccia; thin shallow tagamite bodies also occur.

The main elements of the inner structure of the Popigai crater are: (1) a central depression, 40 km in diameter, filled mainly by vitric-lithic suevites

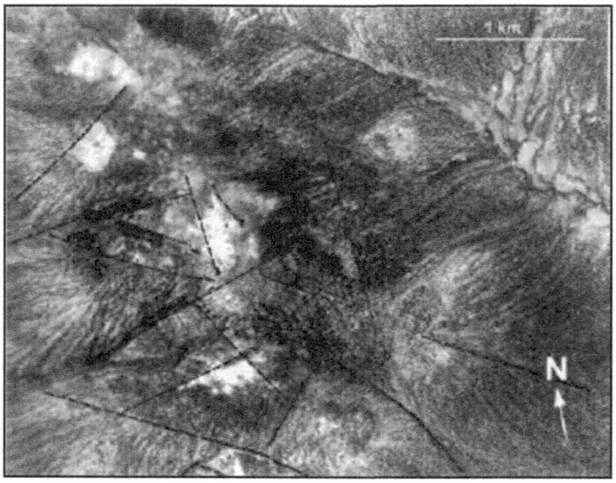

Fig. 5. Image of fault deformations within an allogenic megabreccia field on an aerial photograph (scale 1: 25 000), in the southeastern part of the crater, the Ongtu-Yuregue River valley. The fractures and faults are revealed by linear drainage patterns, zones of abundant vegetation (dark colors), and boundaries between lithologies that differ in their colors. Some photo-interpretation of faults is shown. The location of the area is given in **Fig. 1**.

and fine-grained lithic breccia with a thickness of >1.5 km; (2) a peak ring of disturbed and shocked crystalline target rocks of 45 km diameter; it is exposed below the impactites and impact breccia in the NW sector of the crater (Majachika Upland), as well as in some other areas; (3) an annular trough 15 to 20 km wide, filled with an impact rock sequence with a thickness of up to 1700 m, including coarse allogenic breccia, impact melt rocks (suevites and tagamites), and fine-grained lithic breccia; (4) an outer zone of deformed crystalline and sedimentary target rocks, 10-15 km wide.

In the present topography, the impact structure is expressed as a 100- to 200-m-deep circular depression. The depression is characterized by radial and

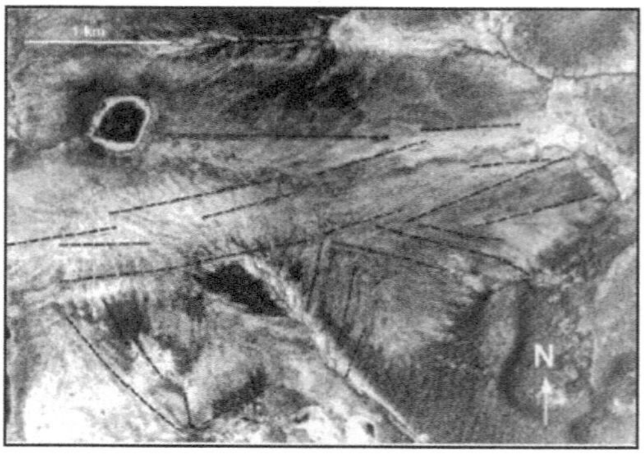

Fig. 6. Image of fault deformations within a suevite blanket on an aerial photograph (scale 1: 25 000) in the western part of the crater, the Namsique-Daldyn River valley. Suevites are indicated by light-colored areas. Some photo-interpretation of faults is shown. The location of the area is given in **Fig. 1**.

concentric features of its inner structure that are emphasized by drainage patterns and near-parallel semi-circular uplands consisting of tagamite in the western part. Impact rocks within the depression are partly covered by Pliocene and Quaternary continental deposits, which are up to 160 m thick, as indicated by drilling data. Remnants of Pliocene-Lower Quaternary sands and gravels up to 20-30 m thick occur mainly along the southern rim of the crater, whereas younger Quaternary deposits are widespread throughout the depression area.

3 Post-Impact Tectonic Structure of the Popigai Crater

During its 35 Myr-long history, the Popigai crater has been slightly modified by denudation and deposition. The maximum depth of erosion is estimated to vary from about 100 m in the inner depression to 250-500 m at the southwestern rim of the crater. According to Plotnikova (1990), four periods of the late modification can be distinguished: (1) Miocene - Early Pliocene: denuda-

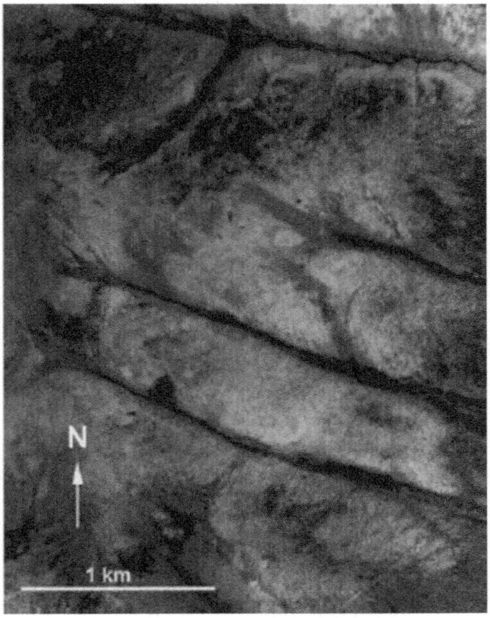

Fig. 7. Aerial photograph of a suevite field, showing well-exposed linear drainage pattern caused by faulting, in the southern part of the crater. The location of the area is shown in **Fig. 1.**

tion of impact rocks, (2) Pliocene-Early Quaternary: deposition of lake sediments (sands with gravel intercalations), (3) Middle and Late Quaternary: Sartan Glaciation and deposition of sands, aleurites, gravels, and clays up to 77 m thick in lake basins; (4) Holocene: formation of river valleys and alluvial deposits (up to 20 m thick). The present distribution of Pliocene and Quaternary deposits is partly controlled by block faulting (Plotnikova 1990).

The post-impact tectonic structure of the Popigai crater involves a series of fault blocks differing in extent and direction of vertical displacements (Fig. 2). The fault boundaries of blocks are traced from linear patterns of river valleys, continuous linear bluffs of impactites (Fig. 3), and linear zones where post-impact sediments are of the increased thickness. Some faults are proposed from sharp changes of the geological structure and sudden changes in thickness of impactites in nearby areas. For example, crystalline rocks of the peak ring form a chain of table-shaped hills, divided by narrow radiate-orientated depressions where the post-impact deposits are rather thick (15-40 m). Long axes of separate ranges deviate from concentric orientations fixing rotational movements of blocks in addition to translational ones.

Two groups of boundary faults can be distinguished: (1) radial and concentric pre-impact (partly revived during the cratering) and syn-impact faults, which have caused the present radial-concentric drainage pattern within the Popigai depression, and (2) northwestward and southwestward orientated post-impact faults that have been caused by regional tectonics (a rise of the Anabar Shield) during the Neogene-Quaternary period. Radial faults are sub-vertical and record predominantly strike slip movements, whereas concentric ones are high-angle faults with both center-down and center-up displacements.

Uneven erosion of different blocks causes a considerable geological difference between separated areas. Inferred from a proposed original impact stratigraphy, subsided and uplifted blocks could be distinguished, and magnitudes of displacements could be evaluated as well. The southern and western parts of the crater, which are adjacent to the Anabar Shield, as well as the western part of peak ring projections, belong to the most elevated blocks; there, impact melt rocks and authigenic breccia are exposed. The central depression and some areas of the annular trough where the thickness of lithic breccias is at a maximum, represent the most subsided blocks. The eastern and northern parts of the crater, which are composed mainly of breccia after carbonate-terrigenous rocks and dolerites, represent blocks of an intermediate position.

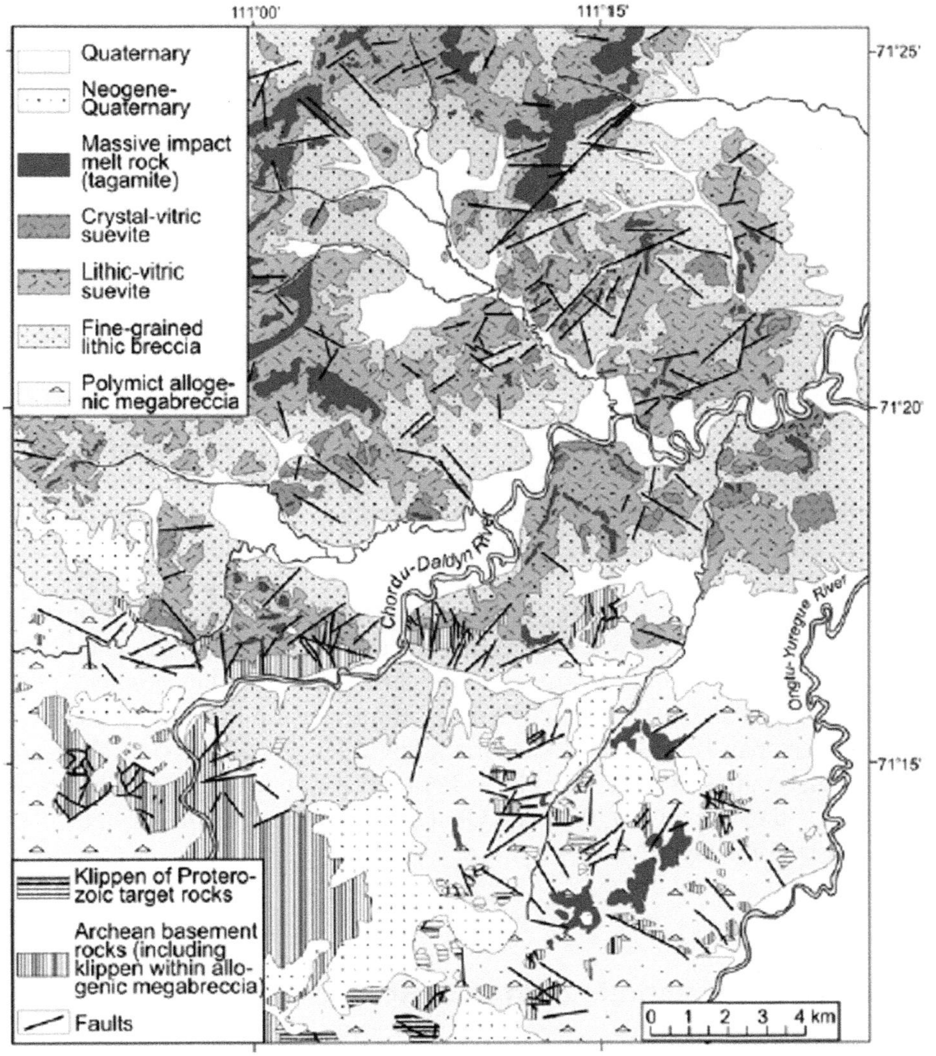

Fig. 8. Geologic map of the Chordu-Daldyn area (the southern part of the Popigai crater). All faults obtained from interpretation of aerial photographs are shown.

A typical radial fault zone is mapped by deep drilling in the southwestern sector of the crater (Balagan-Yuregue area, Fig. 4). There, the roof of the thick impact melt sheet is downfaulted with a vertical shift of above 100 m, so that

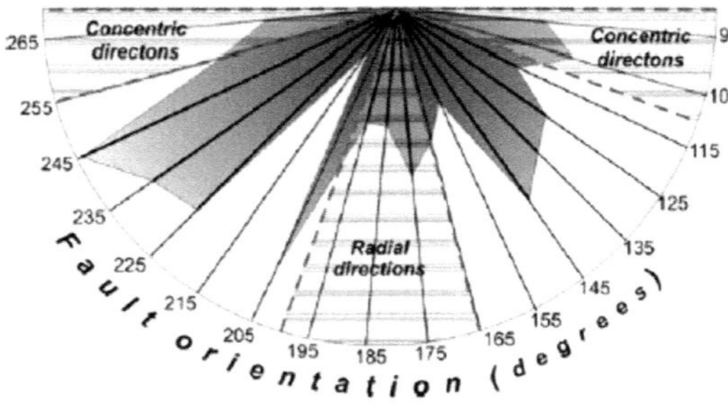

Fig. 9. Orientation distribution data for all 198 fault deformations from the Chordu-Daldyn area. Data are displayed as a rose plot. Areas of radial and concentric orientations are shown.

in the southeastern part of the area a continuous field of fine-grained lithic breccia was mapped. The major fault plane dips southeastward at high angle. The subsided block probably represents the uppermost part of the impact rock sequence that was removed by erosion elsewhere in the Balagan-Yuregue area.

A maximum amplitude of vertical displacements during late modification stage could be estimated from sudden variations in thickness and altitude of post-impact sediments. For example, the thickness of the Pliocene-Quaternary deposits reaches 160 m in the northwestern sector of the annular trough (Majachika Upland area), and the base is at an elevation of -100 m there. In the southern part of the crater this stratigraphic level is 220-300 m above sea level. This indicates that the block faulting continued at least up to the Holocene; the amount of Pliocene-Quaternary vertical displacements, thus, reaches several hundred meters.

4 Fracture Deformations of Impact Rock Bodies

Apart from the block faulting of the Popigai crater, numerous smaller faults and fractures, which cut separate bodies of impact rocks, were discovered during large-scale mapping of selected areas within the crater; these fractures are well-exposed on aerial photographs of areas composed of different impact lithologies, for example, allogenic megabreccia or suevites (Fig. 5, 6). The photogeologic guides of fractures and faults are lineaments of drainage pattern (Fig. 7); zones of abundant vegetation that are indicated by dark colors;

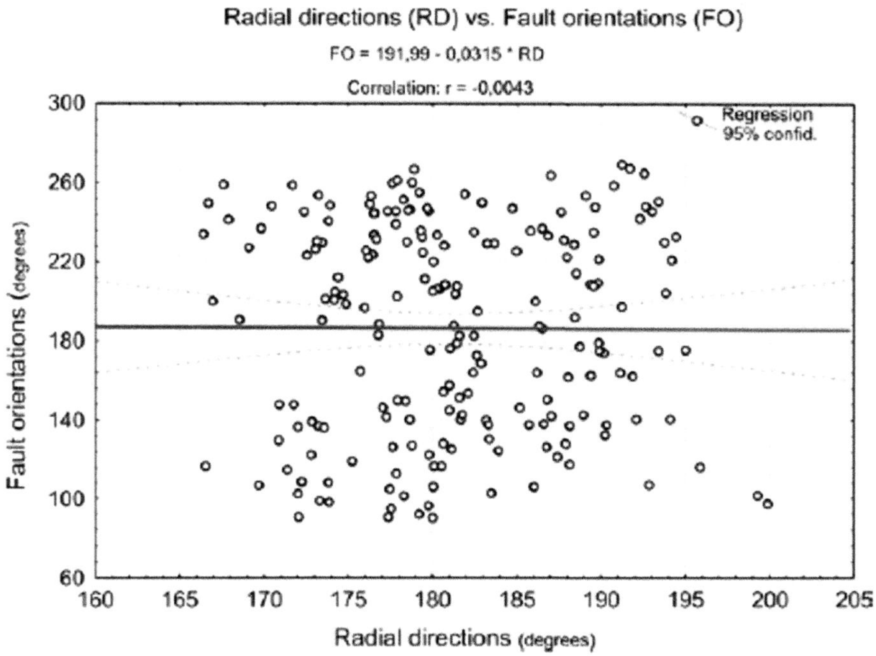

Fig. 10. Correlation plot of fault orientation vs. radial direction for all fault deformations from the Chordu-Daldyn area. The distribution of points shows a chaotic orientation of fractures.

boundaries between lithologies differing in color and specific appearance. As a whole, near-vertical tear faults predominate, usually with minor displacements; however, sometimes heaves more than 10-20 m are observed. The brecciated zones vary in thickness from some centimeters to 20 m, and from several meters to 5 km in length.

The spatial distribution of deformation patterns is uneven; they occur mostly within areas where lithic megabreccias are of maximum thickness, such as the annular trough. As a typical area where the fracturing of impactites is widespread, the Chordu-Daldyn River area in the southern part of the crater (Fig. 1) is considered. This area is located in the midstream of the Chordu-Daldyn River; its general geology is given by Mashchak and Selivanovskaya (1988). It embraces both the annular trough and the outer deformation zone and is composed of diverse impact lithologies (Fig. 8). In the northern part of the area, a thick blanket (more than 400 m) of suevites and fine-grained lithic breccia, with numerous lens-like tagamite bodies with up to 50 m thickness,

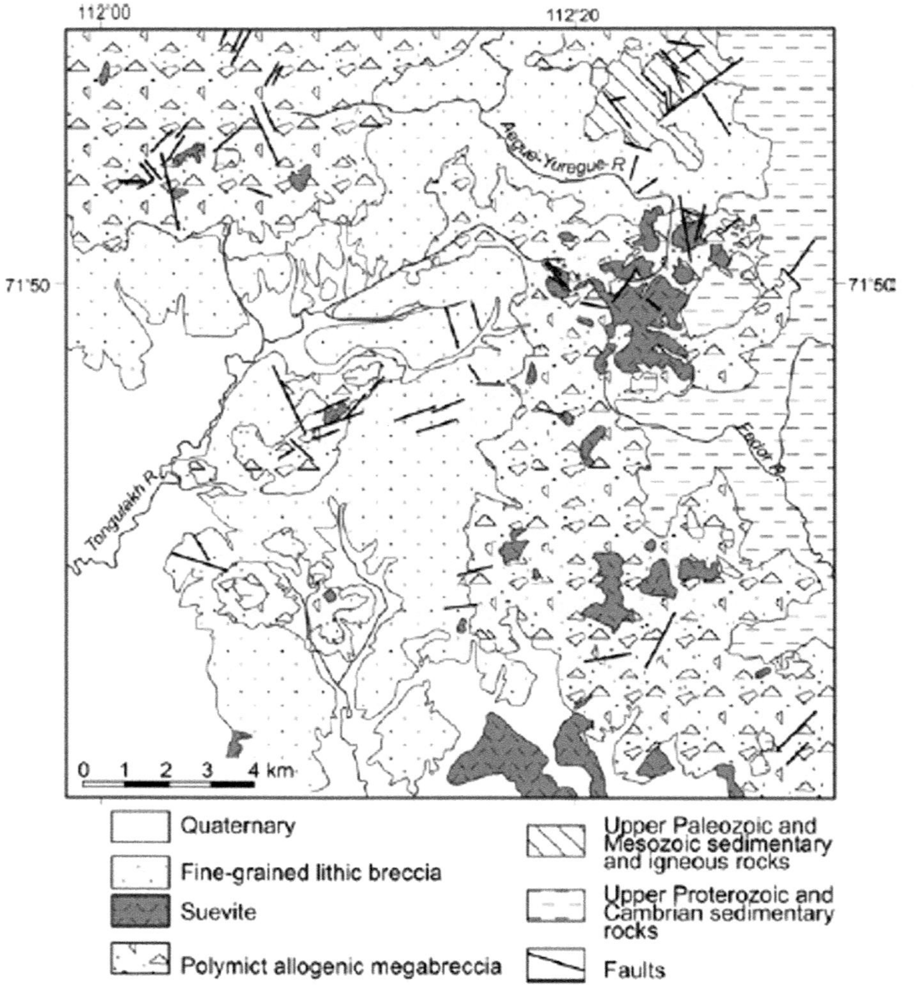

Fig. 11. Geologic map of the Tongulakh area. All faults obtained from interpretation of aerial photographs are shown.

exists. The southern part comprises a radial trench filled by lithic allogenic megabreccia (100-200 m thick) with some tagamite bodies. The northern and southern parts are divided by the narrow zone where a chain of klippen of Archean crystalline rocks emerges through the impact lithologies, possibly fixing the deformed structural rim of the crater. Fractures range from 0.5 to

5.5 km in length. They cut through all impact lithologies; thus, they have formed after the emplacement of these rocks. Some fractures are also widespread within klippen and target rocks of the rim, and they may extend into host allogenic breccias.

The faults exhibit a wide variation in orientation and length throughout the Chordu-Daldyn area (Fig. 9). At this distance from the impact center (22-45 km) those fractures dominate that are oriented obliquely to the radial and concentric directions. A regular orientation pattern of fractures and faults have not been established, and no correlation exists between fracture orientation and radial direction from the crater center (Fig. 10). At the same time, some near-concentric conjugate fracture systems could be determined (Fig. 8).

For comparison with the Chordu-Daldyn area, the fracture pattern of impact rocks in the Tongulakh area is presented. This area is located in the north-eastern part of the crater (Fig. 1). The crater rim is mainly composed of deformed Cambrian carbonate rocks, with a cover of polymict allogenic megabreccia (up to 200 thick). In the area, blocks and klippen of carbonate rocks, quartz arenites of Proterozoic age, and shocked Archean crystalline

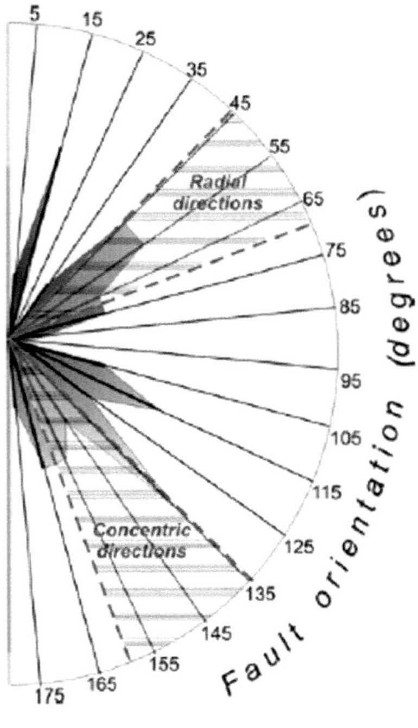

Fig. 12. Orientation distribution data for all 58 fault deformations from the Tongulakh area. Data are displayed as a rose plot. Areas of radial and concentric orientations are shown.

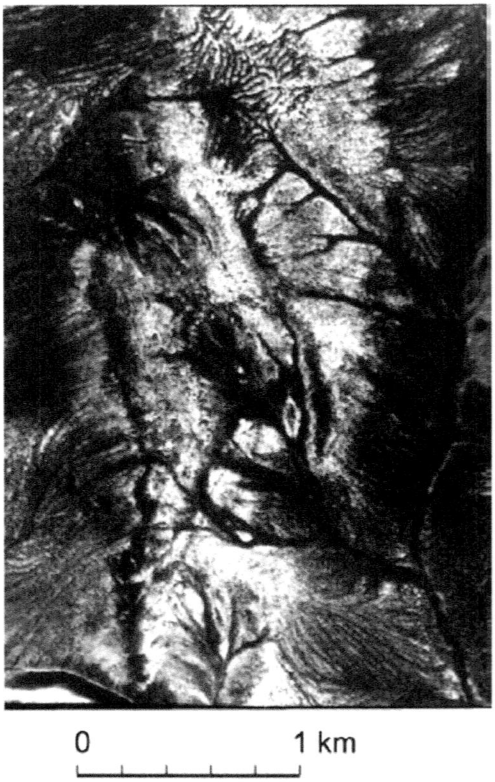

Fig. 13. Aerial photograph of the Profilny area, showing the abundance of topographic and drainage lineaments, which trace fracture deformations. Massive impact melt rocks (tagamites) and suevites are indicated by a homogenous light-gray shade, whereas fine-grained lithic breccia is revealed by the specific patterns formed by alluvial fans.

rocks are common. On top of the breccia, suevites form separate blankets (up to 5 km^2 in area and up to 70 m thick). In contrast with the Chordu-Daldyn basin, fractures within the Tongulakh area are rare and concentrated along some discontinuous radial bands up to 3 km wide (Fig. 11), which possibly correspond to axes of radial trenches. Fractures are from 0.4 to 2.3 km in length (0.8 km on the average). Although fracture deformations vary widely in orientation, they do have preferred radial or concentric orientations (Fig. 12).

For a detailed consideration of the fracture pattern of individual impact melt bodies, mapping at a 1:16,000 scale of a local area (the so-called "Profilny" area) with an extent of 10 km^2 in the southeastern sector of the crater, where fractures are prominent and widespread, has been carried out. In this area, a complex sheet-like impact melt body extends in a meridional direction within a fine-grained lithic breccia field. From both aerial photograph (Fig. 13) and geological map (Fig. 14), the impact melt body is broken into several blocks

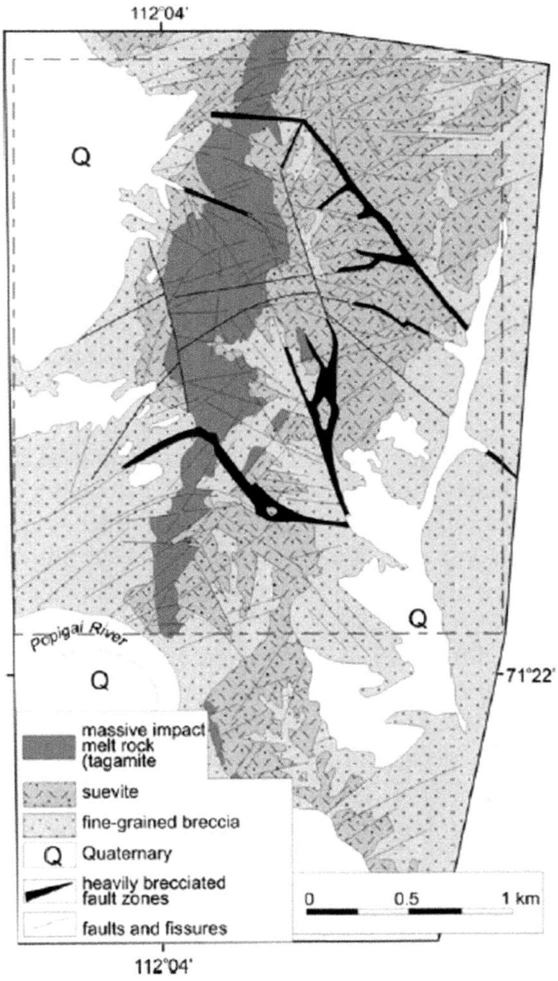

Fig. 14. Geologic map of the Profilny area (mapped by M.S. Mashchak and V.A. Maslov). Shaded line outlines the area covered by the aerial photograph in **Fig. 13**.

with both vertical and lateral displacements; it is apparent that many fractures extend into the enclosing fine-grained lithic breccia. Most geological boundaries are controlled by faults. Fractures are expressed by negative topographic lineaments and are usually traced by narrow lines of larches among the treeless tundra. Locally, fault zones in tagamites and suevites up to 3 m thick with throws up to 6 m are observed in outcrops.

In a large and continuous outcrop of impact rocks on the Chordu-Daldyn River (see Fig. 3), tagamites are cut by several clastic dikes, which are from

0.6 to 3 m thick and up to 30 m in length (Fig. 15). They are composed of rewashed psammitic-aleuritic material, which represents the product of disintegration and redeposition of fine-grained lithic breccias (Mashchak and Fedorova 1987).

5 Summary and Concluding Remarks

Detailed mapping of the Popigai impact structure indicates that faulting within impact rock sequences is widespread throughout the crater. Two hierarchical groups of fractures and faults are distinguished: (1) local faults and fissures within impact rock sequences and (2) main radial and concentric faults sets, which determine the block structure of the crater. The distinguishing of the proper post-impact deformation features from the patterns that belong to the regional tectonic framework on the one side and to the impact on the other side, is frequently ambiguous, because of the post-impact movements exploit syn-impact fault systems, whereas subsequent regional tectonism accentuate impact-related faults as well.

Local fractures are mainly detected from aerial photo interpretation. In some areas with thick sequences of lithic breccia, the density of faults and fractures is as much as 5-10 per 10 km^2. Individual faults do not exceed 6 km in length, and some show a vertical or horizontal separation of up to 10 m; however, tear fractures without separation are predominant. There is no discernible preferred orientation of individual fractures, but fracture systems have near-radial or near-concentric directions, reflecting the subsurface structure of the crater.

These small-scale fractures are specific to the Popigai crater in their density and their geometry and are clearly post-consolidation of the impact melts and breccias. The intensity of impact-induced brittle deformations in the target decreases regularly outwards from the crater and attenuates at the distance of 70-75 km from the crater center (Masaitis et al. 1998). From the 1:200,000 geological mapping data, the Proterozoic and Paleozoic sedimentary rocks outside of this deformed zone are much lesser fractured, compared to impact rocks within the Popigai. Thus, the late modification-stage is most likely dating the formation of local fractures, which probably originated during the compaction of strongly inhomogeneous and unconsolidated thick impact sequences; these include lithic breccia and suevites with tagamite bodies of variable morphology. The thickness of these sequences reaches several hundred meters in the annular trough. Growth of stress along lithological boundaries during compaction caused a sudden ground subsidence as an ultimate strength of matter was exceeded (Rice 1980). The presence of long tear fractures in tagamites indicates that the fracturing occurred during the diagenesis of impact rocks when impact melt became solid to produce brittle deformations.

Post-impact block displacements were controlled predominantly by impact-related radial and concentric faults. The distribution of subsided and elevated

Fig. 15. Clastic dike, 0.8-1.2 m thick, within massive impact melt rock (tagamite). The dike forms a shallow vertical channel in the outcrop wall due to its lesser resistance to weathering compared to tagamite. The location is on the east of the Chordu-Daldyn River valley, 11 km from the estuary.

Fig. 16. Displacement of the Chordu-Daldyn riverbed during 17 years (from 1984 to 2001). The deflection of stream is caused by the neotectonic rise of the southern block of the Popigai structure (at left on the photograph).

blocks, as well as the restriction of an uplift of Pliocene-Early Quaternary sediments to the southern part of the crater and its southern surroundings, and repeated accumulation of sediments in linear areas during a continuous period, suggest that the principal contribution is a regional tectonic overprint (namely, the rise of the Anabar Shield during Neogene-Quaternary period; Novikov 1997), rather than relaxation movements of the post-impact block tectonics of the Popigai crater.

It is important to note that the post-impact tectonic activity in the Popigai has not ceased up until today. For instance, the thickness of recent alluvial deposits in the Popigai River valley is determined by neotectonics (Plotnikova 1990). The Popigai depression differs markedly from surrounding areas by the intense development of thermokarst; during the detailed prospecting of the crater in the last 20 to 30 years, some displacements of riverbeds, changes of shape of lakes, etc., were observed. For example, a displacement of the Chordu-Daldyn riverbed has occurred (Fig. 16). Because the Chordu-Daldyn valley corresponds to a large concentric fault zone within the Popigai crater, this phenomenon indicates a repeated rise of the southern block, which is adjacent to the Anabar Shield.

Thus, the 35-Ma-long post-impact modification history of the Popigai crater is determined by the superimposition of the regional tectonics on the long-term relaxation movements. As a whole, the late modification stage tectonics is found to have only an insignificant effect on the Popigai crater, so that both the original structure and the crater topography have been retained in good condition.

Acknowledgements

Our field works at the Popigai crater were carried out together with our colleagues at VSEGEI, St. Petersburg. We especially thank V.A. Maslov, M.I. Plotnikova, V.L. Masaitis, T.V. Selivanovskaya, I.G. Fedorova, and V.A. Yezersky. The geologists from the Polar Geological Prospecting Enterprise (Khatanga) are also acknowledged for the possibility to study the cores from the Popigai impact structure. We are grateful to Prof. Roger L. Gibson, Dr. Thomas Kenkmann, and Prof. Dr. Christian Koeberl, whose thorough and patient reviews of the original manuscript, criticism and comments led to very considerable improvements of this paper.

References

Bottomley D, Grieve RAF, York D, Masaitis V (1997) The age of the Popigai impact event and its relation to events at the Eocene-Oligocene boundary. Nature 335: 365-368.

Grieve RAF (1987) Terrestrial impact structures. Annual Reviews of Earth and Planetary Science 15: 245-270

Kenkmann T (2002) Folding within seconds. Geology 30: 231–234

Masaitis VL, Danilin AN, Mashchak MS, Raikhlin AI, Selivanovskaya TV, Shadenkov EM (1980) The geology of astroblemes (in Russian). Nedra Press, Leningrad, Russia, 231 pp

Masaitis VL, Mashchak MS, Raikhlin AI, Selivanovskaya TV, Shafranovsky GI (1998) Diamond-bearing impactites of the Popigai astrobleme (in Russian). VSEGEI Press, St.Petersburg, Russia, 182 pp

Mashchak MS, Fedorova IG (1987) Composition and conditions of formation of clastic dikes in tagamites of the Popigai astrobleme (in Russian). Meteoritika 46: 124-127

Mashchak MS, Selivanovskaya TV (1988) Breccia and impactites on the southeastern edge of the Popigai astrobleme (in Russian). Meteoritika 47: 178-188

Melosh HJ (1989) Impact cratering – a geological process. Oxford University Press, New York, 245 pp

Melosh HJ, Ivanov BA (1999) Impact crater collapse. Annual Reviews of Earth and Planetary Science 27: 385-415

Novikov IS (1997) Denudation periods and Cenozoic evolution of relief in the northwestern part of the Anabar anteclise (in Russian). Russian Geology and Geophysics 38: 1465-1474.

Plotnikova MI (1990) Essay on the post-Oligocene history of the Popigai impact structure (in Russian). Meteoritika 49: 154-164

Rice J (1980) The mechanics of earthquake rupture. In: Dziewonsi AM, Boschi E (eds) Proceedings of the International School of Physics "Enrico Fermi", Course 78. Italian Physical Society, pp 555-649

Settings of Meteorite Impact Structures in the Svecofennian Crustal Domain

Väino Puura and Jüri Plado

Institute of Geology, University of Tartu, Vanemuise 46, 51014 Tartu, Estonia. (puura@ut.ee)

Abstract. The Svecofennian Domain, located on the western part of the East European Craton, is a uniform crustal segment with an age of the basement of ~1.9 Ga. Together with younger (1.5 - 1.8 Ga) transition zones it occupies an area of $\sim 1.56 \times 10^6$ km^2. The domain unites shield and platform areas including the submerged Baltic Sea. It includes 22 confirmed, 6 possible, and tens of supposed impact structures that survived recurrent erosional epochs during the long geological history. Signatures of impact-induced tectonics depended much on the syn-impact structure of the target, i.e., presence and thickness of sedimentary deposits covering the basement. Due to differentiated tectonic movements, post-impact scenarios of structures of even the same age differ from place to another within the domain. The present paper analyses the up to 1.2 Ga long post-impact histories of impact structures in accord with the geological development of the domain, i.e. past vertical movements of the crust, which predicted formation and destruction of the sedimentary cover, and variable distribution of shield and platform areas. A great number of at least small- and medium-size craters, which have been located within the past sedimentary cover on top of the crystalline rocks of the Fennoscandian Shield and sediments of the Russian Platform, have been eroded completely or partially. However, between the time intervals with low cratering rates, the Early Palaeozoic is extremely rich in survived structures. It gives hints for unequal erosion history within the Domain and the existence of many hidden (buried) structures. Therefore, we can expect that there is still a large number of partially eroded craters, presently hidden under Quaternary, and also under or within Phanerozoic deposits, to be discovered in the future.

1 Introduction

On Earth, meteorite impact structures are most frequently recognised in cratonic areas, mostly in Precambrian crystalline shields and surrounding platforms with sedimentary cover. The ~1.9 Ga Svecofennian Crustal Domain (SCD; Fig. 1), together with transitional zones to neighbouring domains, occupies part of the Fennoscandian Shield and the neighbouring NW part of the Russian Platform, where the sedimentary cover is usually <2 km thick (Gorbatschev and Bogdanova 1993). Data on 23 (~15 % of known terrestrial impact sites) confirmed and 5 possible impact structures (Table 1, Fig. 1) are used as basis for this analysis. Several authors (Henkel and Pesonen 1992; Puura et al. 1994; Pesonen et al. 2000; Abels et al. 2002) have previously collected and verified the Fennoscandian and Baltic cratering database in detail.

Several impact structures in the SCD are studied by geological, geophysical, and geochemical methods. However, many impact structures require further studies to establish their structure and exact age. Geological evolution of the region has been thoroughly studied (see, e.g., Gorbatschev and Bogdanova 1993; Lahtinen and Huhma 1997), as well as paleogeographic maps have been compiled for late Proterozoic to Mesozoic (Nikishin et al. 1996, Buchan et al. 2000, Torsvik et al. 2001). These data show that the present shield-platform configuration formed after the Tertiary erosion and Pleistocene glaciations whereas the sediments covered much larger areas all over the present shield before the Cenozoic. The thickness of sediments has unevenly grown and reduced through alternating sedimentation and erosion epochs, respectively.

In Europe, the meteorite impact origin was first confirmed for Kaalijärv crater field (Reinwald 1938), ten years after a hypothesis of its impact origin was published (Reinwald 1928, Kraus et al. 1928). In 1938, A. Luha suggested an impact origin for Ilumetsa craters in SE Estonia (Aaloe 1960). Since the 1960ies, the recognition and search for new impact structures in the whole Baltic Sea region became consistent (Abels et al. 2002).

The age of known meteorite impact structures in the region (see Table 1) ranges between 1.2 Ga (Iso-Naakkima; Tynni and Uutela 1984) and <4000 years (Kaalijärv; Veski et al. 2001). The present-day diameter ranges between 0.08 km (Ilumetsa main crater; Raukas et al. 2001) and 65 km (Siljan; Kenkmann and von Dalwigk 2000). However, the eight satellite craters of Kaalijärv crater are only 13 - 39 m in diameter (Aaloe 1960).

Fig. 1. Location of impact structures in the Svecofennian and Karelian Crustal Provinces. Light gray areas mark the marginal zones of the Svecofennian Province, where TTZ - Tornquist-Teysseire Zone. Dark gray indicates anorogenic massifs of rapakivi granites and anorthosites. Paleoproterozoic Transscandinavian Igneous Belt (TIB) is marked by vertical and Karelian-Svecofennian Transition Zone (KSTZ) by horizontal stripes. Boundary between the Precambrian basement exposure of the Fennoscandian Shield and sedimentary areas of the Russian Platform and Bothnian Sea depression is shown by the combination of solid and dotted lines. Contours of local remnants of Proterozoic sedimentary basins (Gulf of Bothnia and Lake Ladoga) are given by stipped lines. Black/white dots are meteorite impact structures penetrating / not penetrating into the crystalline basement.

Table 1. General characteristics of impact structures at the Svecofennian Crustal Domain.

Diameter (km)	Present setting	Age: Stratigraphic	Age: Radiometric	Age: Palaeomagnetic	Target	Primary fill	Recent cover
Name, Country, Coordinates (Selected references)							
Kaalijärv; Estonia; 57°24′ N, 22°44′E (Reinwald 1928, Veski et al. 2001)							
0.11	Platform, exposed	Q_2			Sedimentary (S_3+Q)	B	Lake
Ilumetsa; Estonia; 57°57′ N, 27°24′ E (Aaloe 1969, Raukas et al. 2001)							
0.08	Platform, exposed	Q_2			Sedimentary (D_2+Q)	B	Bog
Logoisk; Belarus; 54°12′ N, 27°48′E (Glazovskaya et al. 1993, Veretennikov et al. 1978)							
17	Platform, buried	Pa_2	42.3±1.1 (K/Ar)		Composite (??+V+D)	B+M+S	Q sed.
Lappajärvi; Finland; 63°10′ N, 23°40′ E (Jessberg and Reimold 1980, Mänttäri and Koivisto 2001, Pipping and Lehtinen 1992)							
23	Shield, exposed	K_2	77.3±0.4 (Ar/Ar) 73.3±5.3 (U/Pb)		Composite (PP+Cbn+O)	B+M+S	Lake+Q sed.
Dellen; Sweden; 61°50′ N, 16°45′ E (Deutsch et al. 1992, Henkel 1992, Müller et al. 1990)							
19	Shield, exposed	K_2	89.0±2.7 (Rb/Sr)		Crystalline (PP)	B+M+S	Lake+Q sed.
Mien; Sweden; 56°25′ N, 14°52′ E (Bottomley et al. 1987, Henkel 1983)							
6.5	Shield, exposed	K_1	121±2.3 (Ar/Ar)		Crystalline (MP)	B+M+S	Lake+Q sed.
Dobele; Latvia; 56°35′ N, 23°15′ E (Puura et al. 1994)							
4.5	Platform, exposed	C_2-P_1			Sedimentary (S+D+C)	B	Q sed.
Vepriai; Lithuania; 55°05′ N, 24°34′ E (Motuza and Gailius 1978, Motuza 1994)							
7.5	Platform, buried	C-P_1			Sedimentary (Cbn+O+S+D?)	B	Q sed.
Mishina Gora; Russia; 58°43′ N, 28°03′ E (Masaitis 1999)							
2.5×4.0	Platform, exposed	C-P			Composite (PP+V+Cbn+O+D)	B	Q sed.

Table 1. (cont.)

Siljan; Sweden; 61°05′ N, 15°00′E (Åberg et al. 1989, Bottomley et al. 1987, Collini 1988, Grieve 1987, Kenkmann and von Dalwigk 2000)							
65	Shield, exposed	D-C_1	368±1 (Ar/Ar) 343±4 (K/Ar)	Devonian	Composite (PP+Cbn+O+S+D?)	B+M	Lake+Q sed.
Karikkoselkä; Finland; 62°15′ N, 25°15′ E (Lehtinen et al. 1996, Pesonen et al. 1999a)							
1.3	Shield, exposed	O?		260-230	Composite (PP+V+Cbn+O)	B	Lake+Q sed.
Lockne; Sweden; 63°00′ N, 14°49′ E (Lindström et al. 1996, Ormö and Miyamoto 2002, Ormö et al. 2002, Therriault and Lindström 1995)							
7.5	Shield, exposed	O_3			Composite (PP+Cbn+O+wat)	B+M	Lake+Q sed.
Kärdla; Estonia; 58°59′ N, 22°40′E (Ainsaar et al. 2002, Plado et al. 1996, Puura and Suuroja 1992)							
4.0	Platform, buried	O_3			Composite (PP+V+Cbn+O+wat)	B	Q marine sed.
Tvären; Sweden; 58°46′ N, 17°25′ E (Flodén et al. 1986, Fredriksson and Wickman 1963, Ormö and Blomqvist 1996)							
2.0	Shield, exposed	O_3			Composite (PP+Cbn+O)	B	Bay, Q sed.
Hummeln*; Sweden; 57°22′ N, 16°19′ E (Flodén et al. 1988, Lindström et al. 1999, Ormö et al. 1999)							
1.2	Shield, exposed	O_2			Composite (PP+Cbn+O)	B	Lake+Q sed.
Granby*; Sweden; 58°26′ N, 14°56′ E (Grahn et al. 1996, Henkel and Pesonen 1992)							
3.2	Shield, exposed	O_2			Composite (MP?+Cbn+O)	B	Q sed.
Sääksjärvi; Finland; 61°25′ N, 22°23′ E (Elo et al. 1992, Kinnunen and Lindqvist 1998, Müller et al. 1990, Papunen 1992)							
4.5	Shield, exposed	Cbn_3	514±12 (Ar/Ar)		Composite (PP+Cbn)	B+M+S	Lake + Q marine sed.
Mizarai; Lithuania; 54°01′ N, 23°54′ E (Motuza and Gailius 1978)							
5.0	Platform, buried	Cbn_3?			Composite (PP?+V+Cbn)	B+M	Sediments
Söderfjärden; Finland; 62°41′ N, 21°35′ E (Abels et al. 2000, Lehtovaara 1982)							
6.6	Shield, exposed	Cbn			Crystalline (PP+wat)	B	Lake + Q marine sed.
Neugrund; Estonia; 59°20′ N, 23°31′ E (Suuroja and Suuroja 2000)							
7.0	Platform, exposed	Cbn_1			Composite (PP+V+Cbn_1+wat)	B	Sea + Q marine sed.

Table 1. (cont.)

Diameter (km)	Present setting	Age: Stratigraphic	Age: Radiometric	Age: Palaeomagnetic	Target	Primary fill	Recent cover
Name, Country, Coordinates (Selected references)							
Suvasvesi N; Finland; 62°41′ N, 28°11′E (Pesonen et al. 1996, Werner et al. 2002)							
4.0	Shield, exposed	Cbn_1?		~250, ~565 or ~775 Ma	Crystalline (PP)	B+M+S	Lake + Q sed.
Suvasvesi S*; Finland; 62°35′ N, 28°14′ E (Lehtinen et al. 2002, Öhman et al. 2003)							
4.0	Shield, exposed	Cbn_1?			Crystalline (PP)	B	Lake + Q sed.
Lumparn; Finland; 60°09′ N, 20°08′ E (Abels et al. 1998, 2000, Winterhalter 1982)							
9.0	Shield, exposed	NP		~580 Ma	Composite (MP+Jn)	B+M	Sea + Q marine sed.
Åvike-bukten*; Sweden; 62°30′ N, 17°41′E (Henkel 1991, Henkel et al. 2002)							
10.0	Shield, exposed	NP			Composite (MP+PP)		Sea + Q marine sed.
Jänisjärvi; Russia; 61°58′ N, 30°55′ E (Masaitis 1975, Müller et al. 1990, Raitala and Halkoaho 1992)							
16.0	Shield, exposed	NP	698±22 (Ar/Ar) 700±5 (K/Ar)		Crystalline (NP)	B+M+S	Lake + Q sed.
Paasselkä; Finland; 62°09′ N, 29°24′ E (Öhman et al. 2003, Pesonen et al. 1999b)							
10.5	Shield, exposed	NP?			Composite? (PP+sed?)	B	Lake + Q sed.
Björkö*; Sweden; 59°24′ N, 17°35′ E (Flodén et al. 1993, 2002)							
8.0	Shield, exposed	MP			Crystalline (NP)	B	Lake + sed.
Iso-Naakkima; Finland; 62°11′ N, 27°09′ E (Elo et al. 1993, Tynni and Uutela 1984)							
2.0	Shield, exposed	MP		1200±100	Composite? (PP+Sed?)	B	Bay + Q sed.

Index: the same is used in Fig. 1. **Name:** *, not proved, but in this paper accepted as impact structure
Stratigraphy: *Q*, Quaternary; *Pa*, Paleogene; *K*, Cretaceous; *PZ*, Palaeozoic; *P*, Permian; *C*, Carboniferous; *D*, Devonian; *S*, Silurian; *O*, Ordovician; *Cbn*, Cambrian; *NP*, Neoproterozoic; *V*, Vendian; *MP*, Mesoproterozoic; *Jn*, Jotnian; *PP*, Paleoproterozoic; *AR*, Archaean
Primary fill: *B*, impact breccia; *M*, impact melt; *S*, suevite. **Recent cover:** *sed.*, sediments

Table 2. Preservation level of meteorite impact structures at the Svecofennian Crustal Domain.

PL	Characteristics	Impact structure and age	Additional signatures
1	Ejecta largely preserved	-	-
2	Ejecta partly preserved	Kaalijärv Q_2	Apparent crater
		Ilumetsa Q_2	Apparent crater
		Kärdla O_3	Buried, rim-related anticline in cover
		Lockne O_3	Incisions, partly hollowed
		Neugrund Cbn_1	Re-exposed, incisions, post-impact fill
3	Ejecta removed, rim partly preserved		
		Dobele C_2-P_1	Buried
		Vepriai C-P_1	Buried
4	Rim largely eroded, crater-fill preserved		
		Logoisk Pa_2	Planated, buried
		Dellen K_2	Hollowed, allochthonous breccia and melt partlysurvived
		Mien K_1	Hollowed, allochthonous breccia and melt partly survived
		Mizarai Cbn_3	Buried
5	Crater-fill products partly preserved		
		Lappajärvi K_2	Hollowed, breccia and melt partly survived
		Mishina Gora C-P	Planated, allochthonous breccia partly survived
		Karikkoselkä O?	Hollowed by glaciers
		Tvären O_3	Hollowed by glaciers
		Hummeln* O_2	Hollowed by glaciers
		Granby* O_2	Hollowed by glaciers
		Sääksjärvi Cbn_3	Hollowed by glaciers
		Söderfjärden Cbn	Hollowed by glaciers
		Suvasvesi N & S* Cbn_1	Hollowed by glaciers
		Lumparn NP	Incisions, partly hollowed by glaciers
		Jänisjärvi NP	Hollowed, allochthonous breccia and melt partly survived
6	Remnants of crater-fill preserved, crater floor exposed		
		Åvike-bukten* NP	Hollowed by glaciers
		Björkö* NP	Hollowed by glaciers
		Paasselkä NP?	Hollowed by glaciers
		Iso-Naakkima MP	Hollowed by glaciers
7	Crater floor exposed, substructure exposed		
		Siljan D-C_1	Planated, ring graben hollowed by glaciers

* not proven

Only a few structures in the SCD have retained easily observable features, whereas the majority are deeply eroded (Table 2) and lost their superstructure. To quantify the rate of survival and erosion of impact structures, Dence (1972) proposed a scale of preservation levels (PL, Table 2). Recognition of roots of impact structures (PL = 6) and even deeply eroded (PL = 4 - 5) craters occurs often to be unsurpassed obstacle. Identification of the only PL = 7 (Siljan, Sweden) is more an exception than a usual practice as was recognised due to its large size and preservation of the outer moat with slumped target sedimentary rocks (Fredriksson and Wickman 1963). In the shield area, Pleistocene glaciations have played an extraordinary important role in destruction of superstructures of almost every crater, as well as buried structures. Therefore, several impact structures in the shield are at the PL = 5, but being completely or partially buried under a Quaternary blanket.

Conditions of post-impact survival or removal, and the structural reshaping, have changed through time, i.e., processes, causing crater preservation or destruction have recurrently replaced each other. This, and its influence to the post-impact development of impact structures in the Baltic Sea region is the main topic of the present paper. Scenarios of formation and post-impact evolution of crater are traced out. *First*, data on geological and physical environments of impact sites (shield or platform, land or sea), *second*, post-impact sequence of possible burial and erosion caused by epeirogenic and syn-orogenic tectonic processes or sea-level changes, and, *third*, Quaternary glacial erosion, accumulation, and flooding of the Baltic erosional depression by post-glacial water-bodies were taken into consideration.

2 Svecofennian Crustal Domain as a Target: Crystalline Basement and Sedimentary Cover

The Svecofennian Crustal Domain is divided into 2 sub-regions (see Fig. 1): (a) SCD within the Fennoscandian Shield and (b) SCD within the Russian Platform in the Baltic States and adjacent areas of Belarus and north-western Russia. The Baltic Sea together with the Gulf of Finland determines the NE-SW-trending boundary between the two sub-regions. Restricted areas with remnants of Meso- to Neoproterozoic (Gulf of Bothnia seabottom with grabens extending into the mainland, Lake Ladoga district) and Lower Palaeozoic (Bothnian submarine district, Lake Siljan area and others) sequences are considered to belong to the shield area (Fig. 1).

Inactive crustal discordances, i.e., transform movements or deformations that cut the primary continuation of the blocks, border the region. The Karelian-Svecofennian Transition Zone (KSTZ, Fig. 1), which was created during the accretion of the Svecofennian orogen to the Karelian palaeocontinent (Korja et al. 1993), forms the northeastern border zone of the region. The SCD is separated from the Sarmatia crustal segment to the southeast by the northeast trending Volhyn-mid-Russian aulacogen system. The Tornquist-Teisseyre Zone (TTZ, Fig. 1) forms the fundamental boundary to southwest. Western frontier of the SCD is represented by the late Paleoproterozoic Transscandinavian Igneous Belt (TIB, Fig. 1), which probably has inherited a transform fault (Gorbatschev and Bogdanova 1993). The Caledonide Orogenic Belt borders the area from northwest and north.

The age of the SCD, a uniform crustal segment containing a range of varied metamorphic and igneous lithologies, is ~1.9 Ga (Gorbatschev and Bogdanova 1993). Although presently the region unites geologically different shield and platform areas, with a submerged Baltic sea-bottom in between, the pre-Quaternary history (Table 3) of it was very much of the same kind. During 1.9 - 1.6 Ga, the Svecofennian orogen went through compression and crustal thickening coupled with folding, metamorphism, intrusion, migmatization, faulting, shearing, and uplift and erosion processes. In places, suites of granulite metamorphism, formed at depths of 10 - 15 km, were exposed at the surface (Koistinen et al. 1996). At 1.65 - 1.5 Ga, faulting and block mountain building and formation of rapakivi-anorthosite igneous centres with central and fault swarm- related volcanic apparatus featured the region (Puura and Flodén 2000). The following erosion and peneplanation led to formation of flat and stable environments favourable for long-term survival of impact structures. From these periods, no impacts are known yet. Since about 1 Ga (during 1.5 - 0.6 Ga) lasting prevalence of continental conditions, eroded mountain environments of post-Svecofennia gradually transformed into a large shield-type area. However, the 1.5 - 1.9 Ga crystalline rocks presently exposed on the shield or sub-cropping under late Vendian-Phanerozoic sedimentary cover were temporarily largely covered by Mesoproterozoic molasse-type (Jotnian, ~1.4 Ga, in the Baltic Sea-Gulf of Bothnia area) or Neoproterozoic (Bothnian Bay-Lake Ladoga area and Volhyn-Orsha-Mid-Russian Aulacogen) siliciclastic deposits (Amantov et al. 1996, Nikishin et al. 1996, Puura and Flodén 1997). Remnants of this cover are preserved in tectonic grabens or depressions (on the bottom of the Bothnian Sea, Bothnian Bay, Lake Ladoga, and Volhyn-Orsha-Mid-Russian Aulacogen).

Table 3. Generalized geological history of the Svecofennian Crustal Domain (SCD).

Age (Ga)	Geological history
1.9 - 1.6	The Svecofennian orogen, compression and crustal thickening
1.65 - 1.5	Faulting and block mountain building, formation of rapakivi-anorthosite igneous centres
1.6 - 1.4	Rapid erosion of block and volcanic mountains, peneplanisation of the territory
1.4 - 1.3	Early platform-type sedimentation (Jotnian)
1.3 - 0.6	Prolonged erosion and peneplanisation
0.6 - 0.3	Accumulation, formation of the sedimentary cover all over the SCD
0.3 - 0.0015	Erosion is dominating, destruction of the sedimentary cover in the western part of the SCD, incision of the basin of the present Baltic Sea, alternating accumulation and erosion processes within the SW marginal zone of the SCD
0.0015 - present	Glacial erosion and accumulation

Late Neoproterozoic erosion and peneplanation led to formation of flat and stable environments on top of differently aged Precambrian structures and lithologies. These processes generated the pre-Late Vendian unconformity that, nowadays, spreads all over the East European Craton.

In the pre-Quaternary times, the boundary between exposed crystalline (paleoshield) and sedimentary (paleoplatform) areas differed from the present as well as varied in time. During the Phanerozoic sedimentation, depocenters were located either in marginal parts of the SCD (from Vendian to Early Cambrian in the east, from Middle Cambrian to Silurian in the NW and SW parts, from Mesozoic to Cenozoic in the SW part) or in its centre (from Devonian to Early Carboniferous). Differentiated erosion, caused by localized tectonic uplifts, repeatedly reshaped the distribution of platform sediments. After the dominant sedimentation periods during Cambrian, Ordovician, and Silurian, almost the whole SCD belonged to the platform.

Starting from the Middle Palaeozoic, late-Caledonian compression and uplift destroyed the previously accumulated sediments and the uppermost part of the Precambrian complexes. In Middle Palaeozoic, local fault-related uplifts were eroded in the internal and distal part of the Baltic Early Palaeozoic sedimentary basin (Baltic synecline in Baltic countries and mid-Baltic seabottom). However, in the NE part of the SCD, from NE Estonia to the Lake Ladoga area, pre-Devonian erosion removed layers of 100 - 400 m thick. In the Lake Ladoga area, pre-Devonian erosion cut off Early Palaeozoic deposits and reached Vendian deposits (Puura et al. 1996).

Devonian marine, partly lagoonal, sediments accumulated and survived in the basin covering most of the Baltic area and the eastern part of the

SCD. During the Devonian, probably, the Caledonides-derived continental sedimentation took place all over the north-western SCD including southern and central Sweden as well as southern and central Finland, whereas shallow evaporate marine basins were characteristic to the present territory of Baltic countries. The area of sedimentation decreased already in the Late Devonian as Early Carboniferous sedimentation continued in the remnant basins in the southern Baltic area, and in the eastern SCD. Large-scale Devonian sedimentation buried the Silurian and older impact structures under up to 1 - 2 km thick cover. At the same time, Devonian became a thick sedimentary target for cratering through Late Palaeozoic and, where survived, Meso- and Cenozoic.

In the SW and W areas of the SCD, Late Carboniferous - Early Permian tectonics (rifting along the Tornquist-Teysseire Lineament and in Oslo Graben) caused uplift of a large area in NE Poland (Mazury Massif) and the southern part of Sweden (Puura et al. 2000), where deposits of Cambrian to Devonian age were eroded. At the same time, central Sweden was unroofed (as dated from Siljan impact structure; see Grieve 1987). In Mazury, up to 2 km thick Late Permian, Meso- and Cenozoic sediments covered the unroofed basement again. Southern Sweden was temporarily covered by Cretaceous, but then unroofed again (Lidmar-Bergström 1996).

The Tertiary uplift of Fennoscandia was another epoch of destruction of the sedimentary cover, including the surroundings of Lake Ladoga, the Gulfs of Finland and Bothnia, the central and northern Baltic Sea, and the NW part of the East Baltic mainland. The final reshaping of the shield-platform distribution took place during the Cenozoic in connection with regional tectonic/palaeogeographic events, such as the beginning of ocean floor spreading in the Norwegian Sea, general uplift and erosion of Fennoscandia, and reshaping of the river systems and watersheds within the neighbouring Russian Platform (Puura and Flodén 1997). The area from NW Poland to North Sea was loaded under clastic sediments whose provenance area was Fennoscandia, the present depression of the Baltic Sea and its surroundings with valley-type landforms. Pleistocene glacial erosion and accumulation diversified and smoothed the valley-shaped landscape. Pleistocene active glacial excavation removed the remnants of sedimentary cover from the present shield area, in many areas even the topmost parts of the basement (as evidenced by large number of erratic material in the SE and S surroundings of the Fennoscandian Shield) and the topmost parts of the pre-Quaternary sedimentary cover in present platform area. On top of the Lauhavuori hills in southern Finland, basal layers of probable Early Vendian have survived (Puura et al. 1996) suggesting the pre-glacial existence of sedimentary cover in southern Finland. Therefore, in large areas, especially in surroundings of the present

Baltic Sea (west coast) and Bothnian Gulf, Tertiary and Quaternary erosion exhumed the Vendian unconformity level.

Generally, duration of continental erosion dominated over the marine accumulation periods. The Devonian continental sedimentation period was considerably short and restricted to the hinterland of Caledonian mountains, NW part of the SCD. The Pleistocene glacial and Holocene accumulation was unevenly distributed forming a blanket of variable thickness.

Although the geological settings and palaeogeographical conditions at the SCD were generally uniform, there were large differences in details. As a result, scenarios of crater formation were variable (see Fig. 2). Two types of impact targets have been possible: (a) crystalline shield and (b) sedimentary platform. The last could have been covered by sea or not. Also, depending on the thickness of sedimentary cover and magnitudes of impacts, the structures may have been penetrated through the sediments (and water) into the crystalline basement.

Scenarios of crater survival and reshaping were related to post-impact evolution of areas hosting impact sites. Often, the scenarios included recurrent erosion and burial. Depending on the vertical extent of impact structures, the presence of covering sediments, and/or vertical extent of erosion, the preservation level (Dence 1972) may vary between 1 and 7.

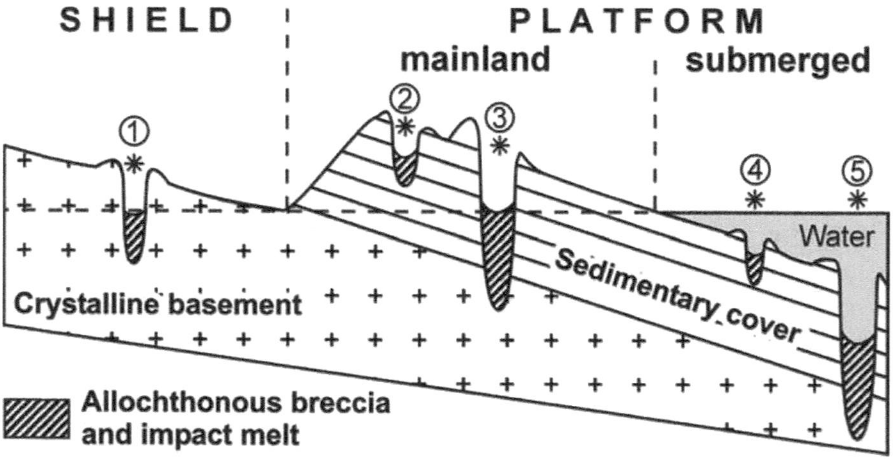

Fig. 2. Typical syn-impact geological settings of meteorite impact structures within the Svecofennian Crustal Domain. 1-3 represent structures on mainland, 4-5 at sea; 1 occurs at the crystalline basement; 2 and 4 have impacted into sedimentary rocks only, 3 and 5 into both, sedimentary and crystalline rocks.

3
Paleogeological and Present-Day Settings of Impact Structures

The area of the SCD, defined in Fig. 1, is ~1.56×10^6 km^2. The Baltic Sea covers an area of ~365 000 km^2 but only one impact structure, Neugrund, is located in the open sea near the mouth of the Gulf of Finland. Three craters, Åvike-bukten, Tvären, and Lumparn, are submerged in sea bays. In the shield area (~880 000 km^2 including northern parts of the Baltic Sea), and platform area (~680 000 km^2 including southern parts of the Baltic Sea), 19 and 9 impact sites are located, consequently.

Within the SCD, the impact structures are preserved irregularly (see also Abels et al. 2002): (a) structures of the same age form groups that are distributed unevenly within the domain, (b) in the platform area, preservation of crater structures is better than in the shield area, and (c) small ($D < 1$ km) craters are rare, indicating a relative loss of smaller structures from the record.

To analyse these irregularities an attempt was made to revise settings of each impact site and their probable transformations during the post-impact geological evolution. Starting from an impact in shield or sedimentary platform environments, on land or at sea, the post-impact sequence of geological processes at crater sites often include single or repeated burial, uplift, exhumation, and/or erosion.

Preservation of crater structures against weathering and erosion depend on the resistance against erosion of target rocks and impact-generated lithologies. In the same circumstances, the impact structures formed into the crystalline rocks have been more resistant against erosion than those formed into the sediments. Direct observations show that relatively resistant rocks, such as impact melt rocks, uplifted crystalline blocks of crater rims, or central uplifts, have been preserved as local positive landforms even in the case of severe epochs of glacial erosion. On the contrary, elevated, soft, or chemically unstable sediments or sedimentary rocks are easily eroded. If (re-)buried after some erosion, the topography of crater sites very much may depend on such lithological circumstances.

However, in more general analysis the partial or complete damage of impact structures in the course of deep erosions has to be taken into the consideration. Up to few tens of kilometers deep erosion of the crystalline basement during the late Palaeoproterozoic and Early Mesoproterozoic naturally damaged the oldest impact structures. Since middle Mesoproterozoic, when the 1.4 Ga old Jotnian siliciclastic deposits formed the first epicratonic basins, two-layered targets, composed of the hard crystalline basement overlain by the soft sediments and/or medium hard sedimentary deposits, were formed. Since Mesoproterozoic through Cenozoic, the selective deep erosion of the sedimentary cover and survival of the monolite crystalline basement became the essential scenario of the development of the land surface in the region (Lidmar-Bergström 1996, 1999; Puura et al. 1983, 1999; Puura and Flodén 1997). In the course of truncation of the sedimentary cover, impact structures nesting in the sedimentary cover only, were destroyed completely. Those structures, penetrating into the basement often survived their lower infrastructure or at least sub-crater roots. Consequently, depending on the existence of sedimentary cover, ant its thickness and erosion processes through the geological history, essential differences in scenarios of crater formation and subsequent survival occurred. In the present morphology of the partially eroded structures, the roles of regional/subregional and local erosional processes as well as lithologies of basement, cover rocks, and impactites are combined. Burial of craters is the best chance for their preservation. However, only very few craters are found to be completely buried.

3.1 Impacts Penetrating into the Precambrian Crystalline Basement

Impacts have hit directly the Precambrian target or penetrated into through the sedimentary cover. Within the SCD and border zones, two main petrographic divisions of the crystalline basement formed the target for impacts: (a) metamorphic and igneous rocks of the Svecofennian orogenic complex, including the KSTZ (Suvasvesi N & S, Paasselkä, and Jänisjärvi structures) and (b) granitoids within the TIB and rapakivi-anorthosite suites (Fig. 1). The Lumparn structure (Winterhalter 1982; Abels et al. 1998) is the only known crater formed in a homogeneous rapakivi suite (Åland Pluton; Puura and Flodén 2000).

The region was subjected to severe erosion during 1.9 - 1.5 Ga. Active tectonic deformation, uplift and deep erosion caused the lack of known impact structures from this age interval. Mesoproterozoic Jotnian cover, presently survived in tectonic depressions only, is an evidence for the deep erosion and planation of the territory at ~1.4 Ga. However, no craters are known from the next 300 Ma time interval (1.5 - 1.2 Ga) either. It is probably due to continuing erosion after 1.5 Ga, and/or, the roots of eroded craters are overlooked yet.

From Precambrian, only the Mesoproterozoic Iso-Naakima structure (Elo et al. 1993) is considered to be formed into the uncovered basement.

Neoproterozoic Jänisjärvi, and possibly Paasselkä (Pesonen et al. 1999b), Lumparn (Abels et al. 2000), Åvike-bukten (Henkel et al. 2002), and Björkö (Flodén et al. 1993, 2002) impacts penetrated into the basement through the Meso- to Neoproterozoic siliciclastic sedimentary cover.

In early Palaeozoic, the Söderfjärden crater (Lehtovaara 1982) formed in an area of exposed basement. Eleven Early Palaeozoic impacts reached the basement through the Vendian to Ordovician sedimentary cover of siliciclastic, clayey and, if present, carbonate (Ordovician) sediments. Mid- to Late Palaeozoic Siljan (Grieve 1987, Åberg et al. 1989) and Mishina Gora (Masaitis 1999), Mesozoic Lappajärvi, and Cenozoic Logoisk (Glazovskaya et al. 1993) structures belong to the family of impacts hitting the basement through sediments as well. In Mesozoic, only two considerable structures, Mien (Henkel 1983) and Dellen (Deutsch et al. 1992), formed into the Precambrian basement unroofed from the previous sedimentary cover.

The on land/sea location and histories of post-impact burial and erosion of basement-reaching structures is presented below while describing the relationships between cratering and sedimentation/erosion. All these data give evidence on large past distribution of sediments all over the SCD.

3.2 Impacts into the Mesoproterozoic and Neoproterozoic Local Sedimentary Areas

Meso- to Neoproterozoic sedimentary basins have survived in downlifted positions within grabens (Satakunta, Gävle, Muhos, Åland Sea), half-grabens (Bothnian Bay, Lake Ladoga district, Landsort Trench district) or gentle syncline-type depressions (central and northern part of the Gulf of Bothnia) (Amantov et al. 1996, Puura et al. 1996, Puura and Flodén 1997). None of craters is located within the survived Meso- to Neoproterozoic sedimentary basins. However, several craters are located near to these and yield Precambrian sedimentary lithologies in allochthonous (Iso-Naakkima, Paasselkä, Jänisjärvi, Åvike-bukten, Lumparn) or even in the (par-) autochthonous (Björkö) impact breccias. These data suggest much wider past distribution of Meso- to Neoproterozoic sediments than presently survived. The northeastern marginal zone of the SCD, along the border of the Karelian massif extending from Lake Ladoga to the Bothnian Bay, was probably loaded under Neoproterozoic clastic sediments (Amantov et al. 1996). In the environments of dominating sedimentation, Paaselkä and Jänisjärvi became likely covered already in late

Neoproterozoic as rifting along the Baltic-Bothnian and Lake Ladoga - Bothnian Bay lines downlifted locally the crustal blocks.

3.3 Impacts into the Phanerozoic Platforms

3.3.1. Late Vendian and Cambrian

In the very beginning of the mature platform stage of development of the SCD, in Late Vendian and earliest Cambrian, the unconformity surface became tilted east, towards the Moscow sedimentary basin. East of the Baltic-Bothnian line, sand, silt, and clay sediments formed a huge uniform blanket that thickened eastwards (Puura et al. 1987, Fig. 3). At the second half of the Early Cambrian, the western part of the SCD was tilted south- and westwards (towards the Iapetus and Tornquist palaeo-oceans), and blanket of sand and silt covered the basement (Fig. 3). Probably, the whole SCD was covered by sediments in the Middle Cambrian (Puura et al. 1987).

A group of Late Vendian and Cambrian craters is found in the area of southeastern Gulf of Bothnia and western Gulf of Finland: Neugrund (Suuroja and Suuroja 2000) in the present marginal platform area within the Lower Cambrian submarine exposure; Söderfjärden and Sääksjärvi (Elo et al. 1992) with shield area (Tables 1 and 2; Fig. 3 and 4). All these structures formed in environments of a thin (<100 m) Late Vendian and Cambrian sedimentary blanket. Two possible Cambrian doublet structures, Suvasvesi N & S (Lehtinen et al. 2002; Werner et al. 2002) are located just northwest of the east-Finland Neoproteozoic crater group (Fig. 1). Lower and Upper Vendian deposits were possibly thickest at the eastern Lake Ladoga and spread at least to Lauhavouri in south-western Finland (Puura et al. 1996). Late Vendian and Cambrian sediments probably covered Iso-Naakkima, Paasselkä, Lumparn, and Åvike-bukten structures. A single Cambrian crater, Mizarai (Motuza and Gailius 1978), formed on-land into the Vendian and Cambrian sediments and buried under subsequent sediments, is found in the southern Baltic area. Most likely, the Neugrund projectile (Suuroja and Suuroja 2000) hit Cambrian sediments at sea, but in the other cases, there is a lack of information to decide on the land-sea distribution.

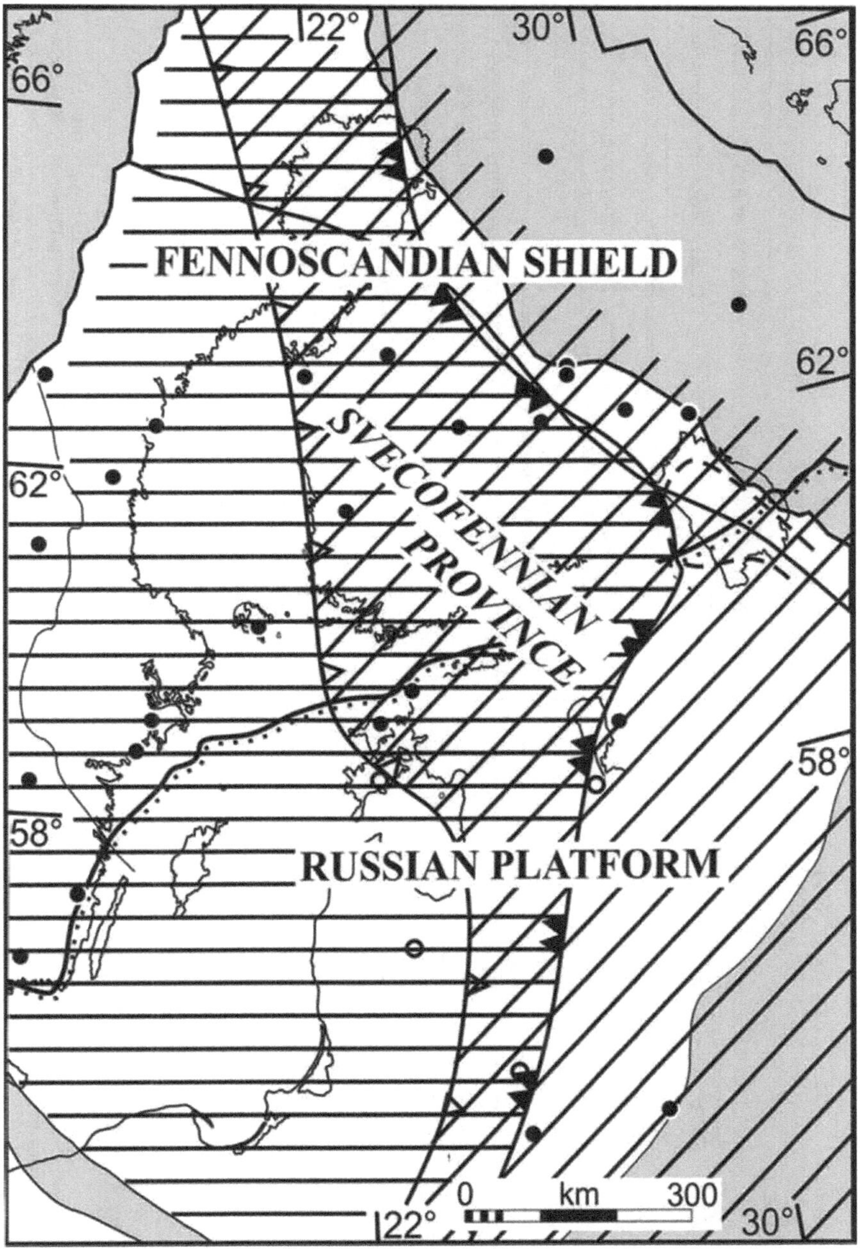

Fig. 3. Past distribution of sedimentary deposits in and around the Svecofennian Crustal Domain (SCD) with its frontiers (white area) after Gorbatschev and Bogdanova (1993). Grey areas mark domains surrounding the SCD. Diagonal lining shows probable distribution of Vendian to earliest Cambrian (Nemakitian-Daldynian) deposits, horizontal, that of late Early Cambrian deposits. Black and white dots are meteorite impact structures penetrating or not penetrating into the crystalline basement, respectively.

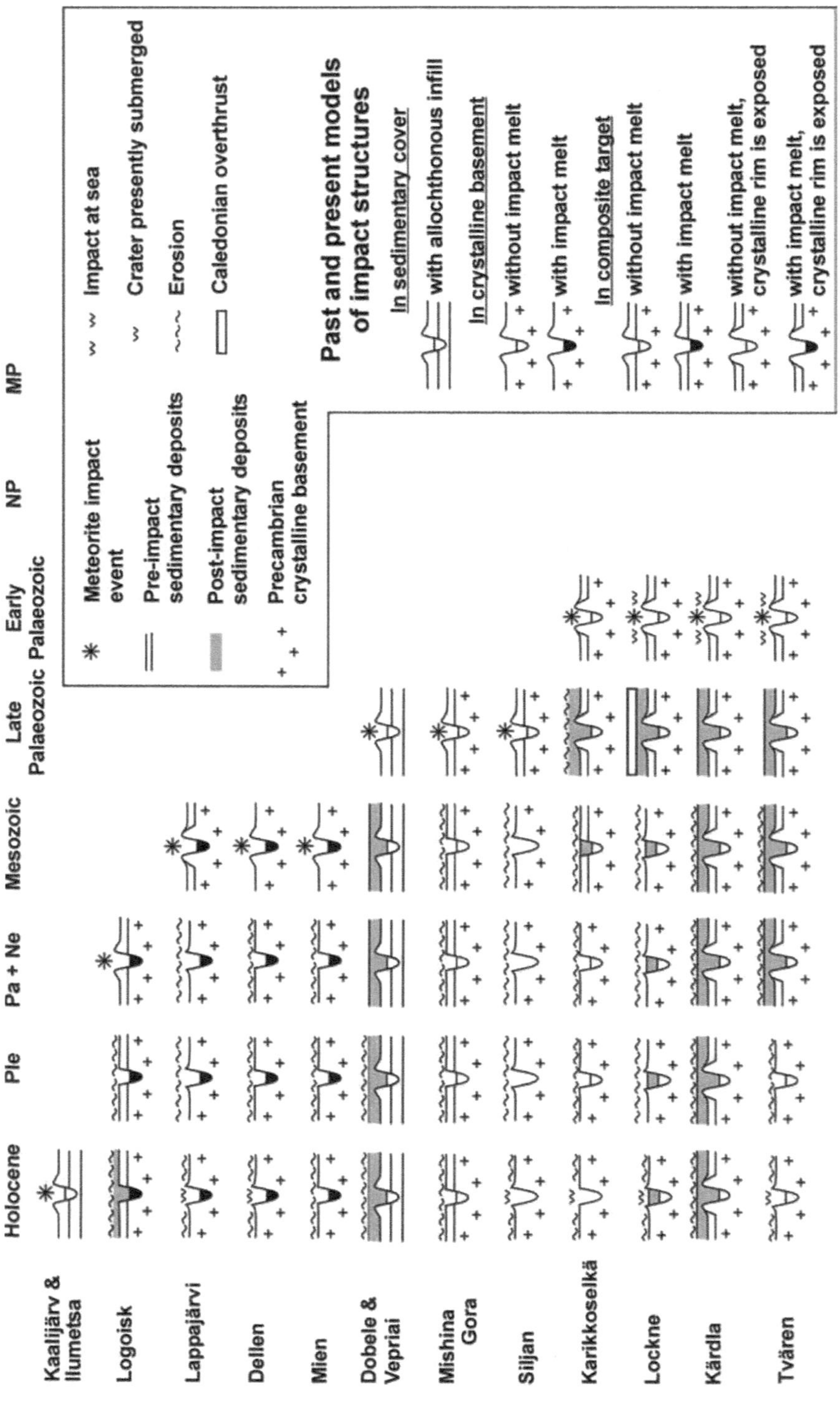

Holocene
Ple
Pa + Ne
Mesozoic
Late Palaeozoic
Early Palaeozoic
NP
MP
Kaalijärv & Ilumetsa
Logoisk
Lappajärvi
Dellen
Mien
Dobele & Vepriai
Mishina Gora
Siljan
Karikkoselkä
Lockne
Kärdla
Tvären
Meteorite impact event
Pre-impact sedimentary deposits
Post-impact sedimentary deposits
Precambrian crystalline basement
Impact at sea
Crater presently submerged
Erosion
Caledonian overthrust
Past and present models of impact structures
In sedimentary cover
with allochthonous infill
In crystalline basement
without impact melt
with impact melt
In composite target
without impact melt
with impact melt
without impact melt, crystalline rim is exposed
with impact melt, crystalline rim is exposed

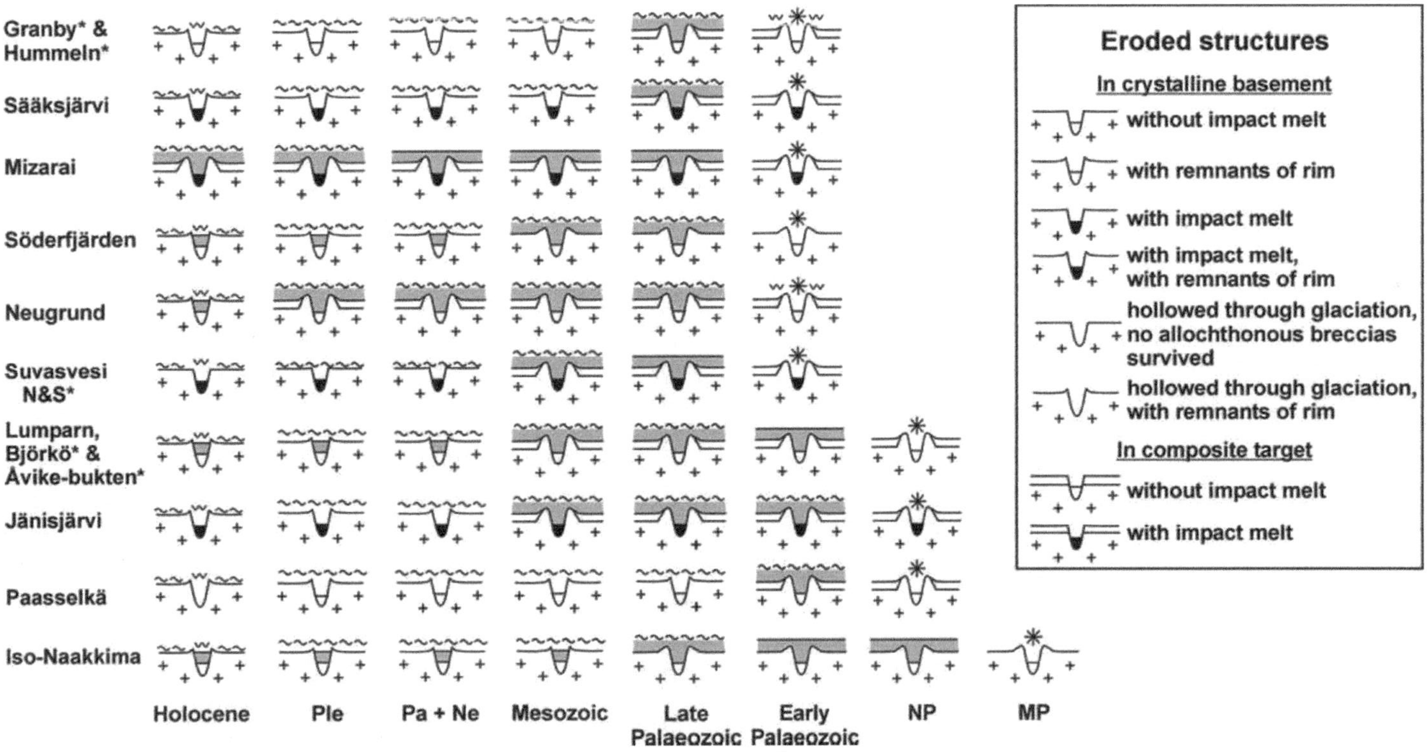

Fig. 4. Cratering record and histories of crater's structural evolution. Fig. 1 indicates the location and Table 1 the properties of structures. See text for details. * = impact origin is not proved.

3.3.2. Ordovician and Silurian

During the Ordovician and Silurian Periods, a gently south-westerly and westerly tilted Baltic sedimentary basin developed within the SCD on top of the thin Cambrian platform. Early and Middle Ordovician sedimentation covered completely the Baltic area. Silurian sedimentation continued in the Baltic area, and in south and central Sweden.

During the Early and Middle Ordovician, numerous impacts hit the target at sea. A disperse group of craters, located at and west of Baltic Sea, formed at the beginning of the transition from Middle to Late Ordovician (according to the International Stratigraphic Chart 2003) Tvären (Ormö 1994, Lindström et al. 1994, Ormö and Blomquist 1996), Kärdla (Puura and Suuroja 1992, Plado et al. 1996, Suuroja et al. 2002), and Lockne (Ormö et al. 2002, Sturkell 1998) structures (Grahn et al. 1996) are presently the best-studied impacts at sea. These impacts penetrated the Middle and Lower Ordovician, and Cambrian sediments into the crystalline basement rocks. Probable impact structures Granby and Hummeln in southern Sweden have also similar histories (Puura et al. 1994, Lindström et al. 1999). A single structure at Karikkoselkä, central Finland (Pesonen et al 1999a), has also a sedimentary ingredient in breccias (L.J. Pesonen 2002, written communication) suggesting that the projectile penetrated Ordovician and Cambrian sediments and hit the basement beneath.

3.3.3. Devonian - Permian

Out of four Late Phanerozoic structures, only the Devonian to Early Carboniferous Siljan structure survived in the shield area. It is the largest in the SCD and most deeply eroded (PL=7) suggesting removal of thick sedimentary cover. Strong regional erosion in late Carboniferous-Permian destroyed its superstructure and infrastructure except deep roots. Therefore, Cambrian and Ordovician deposits are survived only in the down-faulted outer moat of the Siljan impact ring structure, presently partially occupied by Lake Siljan (Collini 1988).

In the less eroded eastern Baltic area, Upper Palaeozoic deposits host Mishina Gora, Vepriai and Dobele structures. The Mishina Gora impact penetrated Upper to Middle Devonian and further down Ordovician, Cambrian and Upper Vendian >400 m thick deposits into the Precambrian basement (Shmayenok and Tikhomirov 1974). The Vepriai and Dobele impacts hit into the sedimentary cover but did not reach the crystalline basement but disturbed the Carboniferous, Devonian, and Silurian deposits (Puura et al. 1994). The considerably deep (>525 m) Vepriai crater penetrated from Devonian through Silurian, Ordovician and Cambrian to Vendian (Motuza and Gailius 1978; Pesonen et al. 2000).

3.3.4. Meso- and Cenozoic

The Early Cretaceous Mien structure was formed in conditions, where the deeply buried basement of southern Sweden was already exhumed during the violent uplift in Late Carboniferous to Permian (Puura et al. 2000), and further on in Mesozoic (Lidmar-Bergström 1996).

The Late Cretaceous Dellen structure (Henkel 1992) formed in an exhumed basement. In the site of the Late Cretaceous Lappajärvi structure the Lower Palaeozoic sedimentary cover was survived at the time of impact (Abels et al. 2002). Therefore, the similar-aged Dellen and Lappajärvi structures are examples of sites with radically different crustal uplift histories west and east of the Gulf of Bothnia, respectively.

The only Tertiary (Middle Eocene) impact crater (Logoisk) penetrated into the basement through the Devonian and Vendian sedimentary deposits. In the flat lowland environments, the structure was smoothed before the accumulation of the Late Quaternary blanket (Veretennikov et al. 1978).

The number of Holocene post-glacial craters in the region is low, whereas craters of >1 km in diameter are unknown. Small post-glacial Kaalijärv and Ilumetsa crater fields are confirmed as meteorite impact structures (see Puura et al. 1994; Abels et al. 2002). These impacts disturbed only the uppermost layers of Silurian and Devonian age, respectively.

4 Post-impact Alteration of Impact Structures

The present boundary of the Fennoscandian Shield with the neighbouring northwestern part of the Russian Platform represents an important frontier between meteorite impact structures with different post-impact histories.

Only a few young or buried (Table 2) impact structures have preserved their primary impact-produced features. Occurring in different succession and intensity, the main factors changing and hiding such features are: (a) erosion, single or recurrent, (b) burial under sediments, single or recurrent,

(c) submergence under water, and (d) deformation (in tectonically mobile areas).

In Fig. 4, individual histories of craters of different age groups from Mesoproterozoic through Holocene are sketched with respect to (a) geological setting of the impact site (sedimentary, crystalline, or at sea), (b) presence of high resistant impact melt rocks in crater interior, (c) burial of the structure, immediate after impact or posterior to primary erosion, (d) erosion histories of cover complex, sedimentary target rocks, and top of crystalline basement, and (f) planation of crater sites or hollowing of craters through the Pleistocene glaciations.

Preservation level according to Dence (1972) of any particular structure is given in Table 2, demonstrating the morphological signatures of crater sites and present setting of craters in an order of increasing PL. Impact structures within the platform area appear to have their PL smaller, i.e., are more preserved compared to structures in the present shield area. Recent (Holocene) craters are best preserved having their ejecta still present (PL = 2). Most of the pre-Pleistocene impact structures of the SCD are preserved at the level 4 - 7, with exceptions of two (Kärdla, Neugrund, and Lockne) at the level 2 and two (Dobele and Vepriai) at the level 3. All these five structures were buried soon after the impact, whereas two, Neugrund and Lockne, were only lately re-exposed. Caledonian orogeny deformed the shape of Lockne structure, but tectonic nappes overriding the impact site protected the crater against erosion (Lindström et al. 1996, Ormö et al. 2002). Craters at the PL 4 (Logoisk and Mizarai) are preserved in platform area, but also the youngest (Cretaceous) structures Mien and Dellen in the shield area. All the rest of craters at PL = 5 - 7, except Mishina Gora, are located in the present shield area, but were, except Iso-Naakkima and Suvasvesi N & S, formed in past platform areas.

The craters formed in past land areas were subjected to erosion immediately after their formation. Depending on the resistance against erosion, their superstructure as well as the surrounding sedimentary blanket were eventually destroyed and removed. In the result, shallowed and smoothed crater sites with partially survived crater interiors are figured out (Dobele, Vepriai, Logoisk, and Mishina Gora).

The smoothing of crater sites before the Pleistocene glaciations was considerable, but not perfect. In many cases in the shield area, after glaciations, the topography of great majority of exposed crater sites became much more complicated than before. Lakes and bays in crater sites are the most typical features. In Fig. 5 and 6 and Tables 1 and 2, it is shown that all craters located on shield at PL = 3 - 7 are hollowed (Fig. 6) or, at least, include glacial incisions (Lumparn, Fig. 5). In platform area, the submarine, re-exposed Neugrund crater is subjected to erosion, which

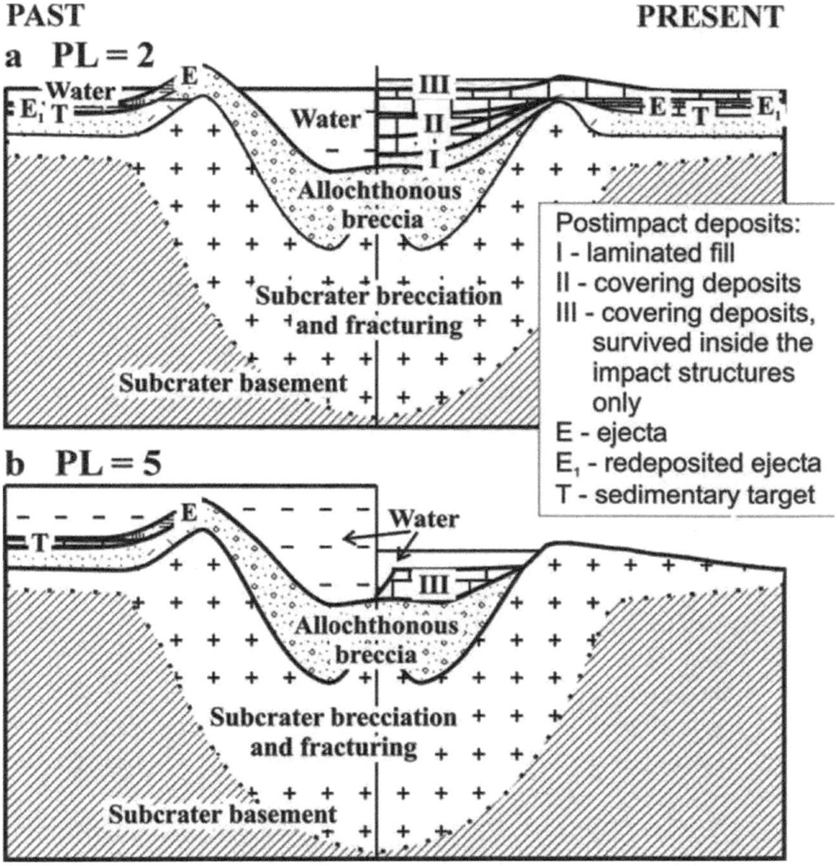

Fig. 5. Some types of past and present morphologies of craters formed at sea (left) and subsequently buried under sedimentary deposits, then exhumed and differently eroded during late Phanerozoic, including Pleistocene glaciations (right). (a) presently buried crater that formed in shallow sea, whereas its rims were slightly eroded immediately after the impact. (b) presently partially submerged incised structures in which the impact and post-impact deposits are survived in the interior only. *PL*, preservation level after Dence (1972).

created deep annular incision channels (Suuroja and Suuroja 2000). Subsurface craters are not subjected to surface lowering, except Kärdla structure, whose post-impact cover of walls, locally composed of hard limestones, is sculptured by glaciers to form drumlins (Fig. 5).

Already the exposed impact structures were eroded during the periods of continental exposition, the selective re-sculpturing by Pleistocene glaciers appears to be mainly responsible for the present shapes of many presently exposed structures. In the region, the glaciers have been formed several valleys along the relatively soft pre-Cambrian fault zones (e.g., Landsort trench in the Baltic Sea) as well as selectively removed the softer impactites and post-impact sediments. The removed clastic impact material, scoured from crater structures by the latest glaciation, is found to form extensive fans in Pleistocene tills south of, e.g., Neugrund (Suuroja and Suuroja 2000), Sääksjärvi (Papunen 1992), and Lappajärvi (Koivisto and Korhonen 1997).

Figure 5 represents the present configuration of a group of craters with similar starting structures, but different post-impact histories. In the sketch (a), an intact structure reminding the Ordovician Kärdla structure (Puura and Suuroja 1992), penetrating into the basement through Middle Ordovician and Cambrian sediments, is given. An impact at sea formed a crater filled with allochthonous breccia and seawater (left side of the figure). Subsequently, laminated carbonate mud (layer I) and limestones (layers II and III) filled and covered the crater. Local specific hard limestones deposited on top and slopes of the elevated crystalline rim wall behaving as shoals at sea, whereas in the crater proper and its surroundings much softer clayey limestones accumulated (Ainsaar et al. 2002). During the Tertiary erosion the structure was almost uncovered. Pleistocene glaciers drumlinized the bedrock relief forming remnant highs composed of hard limestones covering the crystalline rim walls, and lows in-between and in surroundings (see Suuroja et al. 2002). The PL 2 is due to early post-impact partial destruction of the proximal ejecta layer from the elevated parts of the crater rim that formed an atoll-like feature in a shallow sea.

Sketch (b) represents a structure in which allochthonous impact breccias and post-impact deposits are survived in the crater depression only. These structures, pre-Cambrian in age, may be formed into (i) pure crystalline (Iso-Naakkima) or composite (pre-Cambrian crystalline rocks covered by Jotnian sediments; Paasselkä, Jänisjärvi, Lumparn, and Björkö) target. These structures (PL = 5) were eroded either before the burial under Cambrian and Ordovician deposits, and/or afterwards, during the Tertiary unroofing, Quaternary glacial (incisions), and post-glacial erosion.

Analysing the structures formed in the basement before, in the course of, or after the formation of the Phanerozoic sedimentary cover, different, but not historically predicted relationships between the impact and post-impact lithological units may be presented. The most characteristic cases of craters are presented in Fig. 6. For the present platform areas, craters survived (almost) intact (a; PL = 1 - 2) or influenced by erosion (b, PL = 3 - 7) are sketched. The PL of these buried craters should have been reached before the covering platform sedimentation started. In the present shield area (Fig. 6 c-e, g), the crater structures may have been formed either directly in the crystalline basement (e.g. Suvasvesi N & S) or in the

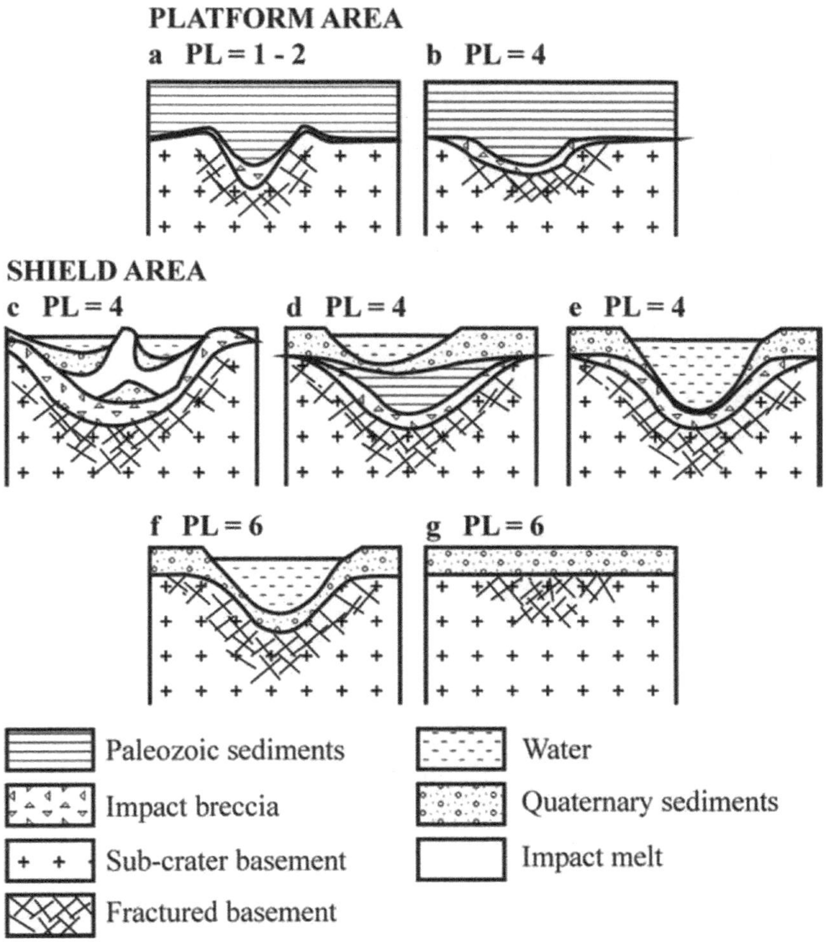

Fig. 6. Most typical types of craters penetrating into basement, subsequently buried under Palaeozoic or other sedimentary rocks in platform areas (a and b), and under Quaternary sediments in shield areas (c - g). a and b indicate craters buried under sedimentary rocks, whereas the structure (a) was not and (b) was previously eroded, (c - f) illustrate craters hollowed and incised due to Quaternary glacial erosion and recently buried under Quaternary sediments and waterbodies (lakes or bays): (c) includes resistant melt-rock bearing interior, (d) includes Phanerozoic post-impact rocks, (e) indicates deeply eroded impact structure with brecciated bottom. Cases (f) and (g) represent deeply eroded impact structures, where allochthonous breccias are totally removed whereas the crater-like structure has survived (f) or not (g).

past composite target. The most common are craters partly hollowed, incised, and filled by impactites with (Fig. 6 c, e; e.g., Jänisjärvi, Sääksjärvi, Mien, Dellen, Lappajärvi) or without melt rocks (Iso-Naakkima, Paasselkä, Åvike-bukten, Granby, Hummeln). All those may be filled with post-impact Phanerozoic sedimentary deposits, but are partially or completely covered with Quaternary deposits and water. Impact structures that are eroded, subsequently filled with Palaeozoic sediments, partially incised during glaciations, and flooded by sea are considerably rare (Fig. 6 d; Lumparn). Siljan is the only proved structure, where the ejecta and within-crater allochthonous breccias are completely destroyed. In Siljan, the down-lifted outer moat is partially incised and filled with water, and the structure is partially covered with Quaternary deposits (Fig. 6 f). Impact structures without distinct morphological signatures, with all the superstructure and crater fill completely destroyed (Fig. 6 g, PL > 7), but with brecciated and fractured sub-crater basement are not yet recognized in the region.

5 Apparent Cratering Rates

From the analyses of < 120 Ma craters, Grieve and Pesonen (1996) postulated that the size-frequency distribution of recognised terrestrial craters principally corresponds to the power law approximation $N_c \propto D^{-2}$, where N_c is cumulative number and D diameter of impact structures. According to their analysis the estimate of the cratering rate is $5.5 \pm 2.7 \times 10^{-15}$ km^{-2} a^{-1} for $D \geq 20$ km, but there occurs a loss of small-size craters. In the SCD the occurrence of only two impact structures (Siljan and Lappajärvi) with $D \geq 20$ km is proved. Therefore, the Phanerozoic cratering rate for $D \geq 20$ km is as small as 1.2×10^{-15} km^{-2} a^{-1}. Grieve and Pesonen (1992) have argued that the geological effects of impacts at simple structures (D < 2 -4 km) are visible to a depth of ≤1/3 the final rim diameter, whereas the depth / diameter ratio is shallower but the absolute depth are often greater. Applying these rules to the SCD show that simple craters may locally have not been penetrating the sedimentary cover because the thickness of sedimentary formations in the region has dominantly been between 0.5 and 3 km (Puura et al. 1996). The larger impacts with final diameter of >10 km have always truncated both the sedimentary cover (if was present at all) and the basement. Generally due (a) to erosion and final destruction of many impact structures and (b) probable existence of many unknown impact structures, hidden under surficial deposits in the platform areas and seabed, we are far from the determination of the real cratering rate in the 1.9 Ga history and frames of the SCD. Therefore, we discuss the apparent cratering rates (ACR) in the SCD and its regions for different epochs of its history.

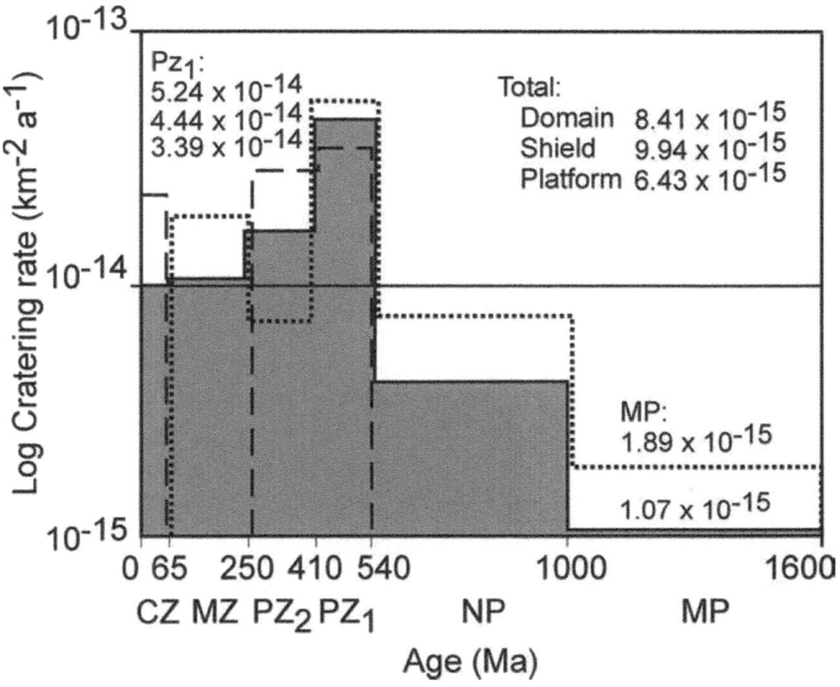

Fig. 7. Apparent cratering rate of meteorite impact structures at the Svecofennian Crustal Domain by age. Solid line represents the total rates; dotted (dashed) line illustrates the rate at shield (platform) area. Abbreviations: *CZ*, Cenozoic; *MZ*, Mesozoic, PZ_2, late Palaeozoic (Devonian - Permian); PZ_1, early Palaeozoic (Cambrian - Silurian); *NP*, Neoproterozoic; *MP*, Mesoproterozoic.

Abels et al. (2002) demonstrated an uneven distribution of Scandinavian crater ages through Mesoproterozoic to Phanerozoic history. Figure 7 presents time- and location-depended rates of cratering in the SCD, whereas proven structures >1 km in diameter are taken into account. The average cratering rates for the 1.56×10^6 km^2 SCD and those for shield and platform (0.88×10^6 and 0.68×10^6 km^2, respectively) areas are calculated for six time intervals from Mesoproterozoic to Cenozoic.

According to the data given in Table 1 and Fig. 7, the observed ACR of structures with D > 1 km (Table 1, Fig. 7) within the SCD ranges within 10^{-15} -10^{-14} for Proterozoic, and 10^{-14} - 10^{-13} for Phanerozoic structures. The Mesoproterozoic ACR (1.07×10^{-15} km^{-2} a^{-1}) is minimum based on one find (Iso-Naakkima) only. The Neoproterozoic ACR is higher (4.18×10^{-15} km^{-2} a^{-1}), but still lower than the average Phanerozoic ACR (2.02×10^{-14} km^{-2} a^{-1}). The maximum ACR (4.44×10^{-14} km^{-2} a^{-1}) comes from Early Palaeozoic (from Cambrian to Silurian). There exists decreasing trend of observed cratering rates from Early Palaeozoic to Cenozoic (Fig. 7).

The average rates in the shield area are much higher than in platform area. To explain it, two factors have to be discussed: a) complete or partial destruction due to erosion and b) hiding under sedimentary cover (or seawater). As described above, erosion of soft sedimentary cover was essential and often recurrent process in many places of the SCD. Due to restricted vertical extent of craters they could not survive erosion of host sedimentary rocks. From another side, in cases of destruction of super- and infrastructure of craters down to the sub-crater levels, still survived impact-induced deformations are still overlooked. Burial is certainly another essential factor of observed cratering rates in sedimentary areas.

The observed Proterozoic cratering record is incomplete. The low Mesoproterozoic rates evidences destruction of impact structures in the shield area. In the platform area, the eroded sub-crater roots are deeply buried, and, thus, search is even more complicated.

Over a half of observed Phanerozoic craters are Cambrian to Ordovician in age indicating favourable conditions for survival of impact structures. Most of these impacts took place in shelf seas or close to the coast, and were rapidly buried under protecting sediments. Also, increased accretion rates of meteorites and cosmic dust for the Lower Ordovician is supposed (see Schmitz et al. 2001). Still, the observed cratering rate is probably less than the real rate during Phanerozoic as the rate of observed craters in the shield area is more than two-fold higher than that in platform area.

The observed Late Palaeozoic (since Silurian), Mesozoic and Cenozoic ACRs are three- to six-fold lower than in Early Palaeozoic. We suppose that beside of latest erosion of the basement in the shield area, the main reason of the low rate is erosion of largely distributed Devonian cover during the Tertiary. The hidden Devonian craters exist only in platform area with survived thick Devonian deposits.

6 Conclusions

The above results suggest that investigations face a lack of impact structures once most probably formed within the SCD. *First*, the number of craters and, respectively, their number per area unit is less than expected. *Second*, the number of observed, expectedly the most frequent small craters (<1 km in diameter) is negligible. *Third*, the aerial distribution of craters from specific age intervals is extremely irregular within the territory. *Fourth*, differently aged craters are distributing unevenly.

Many impact structures are not discovered because a) the considerably thick and widely distributed accumulative Quaternary blanket that hide near-surface structures present in bedrock, b) subsurface craters are hidden within and under the Neoproterozoic and Phanerozoic sedimentary cover, c) suspect structures and evidences observed by geological and/or geophysical studies are not yet checked, d) craters deeply eroded and/or buried under post-impact sequences are complicated to find.

In the present overview, we have not diversified the analysis concerning the complexity of structures (bowl-shaped, complex, ring basins). We neither took into consideration the factors influencing the survival of specific impact signatures (PDFs in minerals, shatter cones, high-pressure polymorphs of minerals, etc.), essential for the recognition of impact sites. In future, respective detailed analysis could support the design of crater search programs. For the recognition of possible impact sites, signatures, which are distinctive not only for impacts, might be important. For example, topographic and geophysical signatures have called attention to many sites afterwards proved as impact structures in Scandinavia (e.g., Pesonen 1996, Henkel and Pesonen 1992) or in its surroundings (Neugrund, Kärdla). Recognition of deeply buried or deeply eroded impact structures by help of complex petrological, remote sensing, geophysical and other techniques is probably the way to new important discoveries (Abels et al. 2002, Gibson and Reimold 2000, Kenkmann et al. 2000, Plado et al. 1999). Geological signatures in sedimentary areas such as uncommon altitudes of geological contacts and bodies (in Kärdla, e.g., uplifted position of crystalline basement), exposures of allogenic breccias (Mishina Gora), sub-surface lithological features as breccia bodies (Vepriai) or breccia layers (Kärdla, Neugrund) and mineralogical (although under discussion, impact spherule layers) raise the interest.

Acknowledgements

We are grateful to A. Abels and J. Kohonen for their careful reviews of the earlier version of the manuscript. We also acknowledge T. Meidla, L. Ainsaar, and T. Flodén for their useful comments, and K. Kübar for designing the figures. The study was supported by Ministry of Education of Estonia (theme # 0551) and Estonian Science Foundation (grants # 4417, # 5500, and # 5817).

References

Aaloe AO (1960) Ilumetsa craters in Estonian SSR (in Russian). Meteoritika 18: 26-31

Abels A, Mannola P, Lehtinen M, Bergman L, Pesonen LJ (1998) New observations of the properties of the Lumparn impact structure, Åland Islands, southwestern Finland [abs.]. Meteoritics and Planetary Science 33: A7-A8

Abels A, Bergman L, Lehtinen M, Pesonen LJ (2000) Structural constraints and interpretations on the formation of the Söderfjärden and Lumparn impact structures, Finland [abs.]. In: Plado J, Pesonen LJ (eds) Meteorite Impacts in Precambrian Shields. Programme and Abstracts, the 4th Workshop of the European Science Foundation Impact Programme, Lappajärvi-Karikkoselkä-Sääksjärvi, Finland, May 24 - 28, 2000. Geological Survey of Finland and University of Helsinki, p 26

Abels A, Plado J, Pesonen LJ, Lehtinen M (2002) The impact cratering record of Fennoscandia - A close look at the database. In: Plado J, Pesonen LJ (eds) Impacts in Precambrian Shields. Impact Studies 2, Springer-Verlag Berlin Heidelberg, pp 1-58

Åberg G, Collini B, Schmitz B (1989) K-Ar isotope analyses from the Siljan Ring meteorite impact structure, Sweden. Geologiska Föreningens i Stockholm Förhandlingar 111: 355-360

Ainsaar L, Suuroja K, Semidor M (2002) Long-term effect of the Kärdla crater (Hiiumaa, Estonia) on Late Ordovician carbonate sedimentation. Deep-Sea Research II 49: 1145-1155

Amantov A, Laitakari I, Poroshin Ye (1996) Jotnian and Postjotnian: Sandstones and diabases in the surroundings of the Gulf of Finland. In: Koistinen T (ed) Explanation to the Map of Precambrian basement of the Gulf of Finland and surrounding area 1 : 1 million. Geological Survey of Finland Special Paper 21: 99-114

Bottomley RJ, York D, Grieve RAF (1978) $^{40}Ar-^{39}Ar$ ages of Scandinavian impact structures: Mien and Siljan. Contributions to Mineralogy and Petrology 68: 79-84

Buchan KL, Mertanen S, Park RG, Pesonen LJ, Elming S-A, Abrahamsen N, Bylund G (2000) The drift of Laurentia and Baltica in the Proterozoic: a comparison based on key paleomagnetic poles. Tectonophysics 319: 167-198

Collini B (1988) Geological setting of the Siljan Ring Structure. In: Juhlin C (ed) Scientific Summary Report of the Deep Gas Drilling Project in the Siljan Ring Impact Structure. Vattenfall Research and Development, Sweden: 11-25

Dence MR (1972) The nature and significance of terrestrial impact structures. 24th International Geological Congress, Section 15: 77-89

Deutsch A, Buhl D, Langenhorst F (1992) On the significance of crater ages: new ages for Dellen (Sweden) and Araguainha (Brazil). Tectonophysics 216: 205-218

Elming S-Å, Bylund G (1991) Palaeomagnetism and the Siljan impact structure, central Sweden. Geophysical Journal International 105: 757-770

Elo S, Kivekäs L, Kujala H, Lahti SI, Pihlaja P (1992) Recent studies of the Lake Sääksjärvi meteorite impact crater, southwestern Finland. Tectonophysics 216: 163-167

Elo S, Kuivasaari T, Lehtinen M, Sarapää O, Uutela A (1993) Iso-Naakkima, a circular structure filled with Neoproterozoic sediments, Pieksämäki, southeastern Finland. Bulletin of the Geological Society of Finland 65: 3-30

Flodén T, Tunander P, Wickman FE (1986) The Tvären Bay struvture, an astrobleme in southeastern Sweden. GFF 108: 225-234

Flodén T, Sturkell E, Söderberg P (1988) Sjön Hummeln – En meteoritkrater I Småland, preliminära resultat av en seismisk undersökning. Department of Geology, Stockholm University Report, 13 pp

Flodén T, Söderberg P, Wickman FE (1993) Björkö, a possible Middle Proterozoic impact structure west of Stockholm, Sweden. GFF 115: 25-38

Flodén T, Bjerkéus M, Henkel H (2002) The Björkö impact structure in Lake Mälaren - results of marine geophysical investigations [abs.]. In: Emelyanov EM (ed) The seventh Marine Geological Conference "The Baltic". April 21-27, 2002, Kaliningrad, Russia. Abstracts. Excursion Guide. Atlantic Branch of PP Shirsov Institute of Oceanology RAS, p 43

Fredriksson K, Wickman FE (1963) Meteorites (in Swedish). In: Lundholm B (ed) Svensk Naturvetenskap 16, Swedish Natural Science Research Council, Stockholm, pp 121-157

Gibson RL, Reimold WU (2000) Deeply exhumed impact structures: A case study of the Vredefort structure, South Africa. In: Gilmour I, Koeberl C (eds) Impacts and the Early Earth. Lecture Notes in Earth Sciences 91, Springer, Heidelberg, pp 249-278

Glazovskaya LI, Parfenova OV, Il'kevich GI (1993) Impactites from the Logoisk astrobleme. Petrology 1: 634-644

Gorbatschev R, Bogdanova S (1993) Frontiers in the Baltic Shield. Precambrian Research 64: 3-22

Grahn Y, Nõlvak J, Paris F (1996) Precise chitinozoan dating of Ordovician impact events in Baltoscandia. Journal of Micropalaeontology 15: 21-35

Grieve RAF (1987) Terrestrial impact structures. Annual Reviews of the Earth and Planetary Sciences 15: 245-270

Grieve RAF, Pesonen LJ (1992) The terrestrial impact cratering record. Tectonophysics 216: 1-30

Grieve RAF, Pesonen LJ (1996) Terrestrial impact craters: Their spatial and temporal distribution and impacting bodies. Earth, Moon, and Planets 72: 357-376

Henkel H (1983) Combined gravity, magnetic and petrophysic interpretation of an impact structure [abs.]. Geoexploration 21: 287

Henkel H (1991) Magnetic crustal structures in northern Fennoscandia. Tectonophysics 192: 57-59

Henkel H (1992) Geophysical aspects of meteorite impact craters in eroded shield environment, with special emphasis on electric resistivity. Tectonophysics 216: 63-89

Henkel H, Pesonen LJ (1992) Impact craters and craterform structures in Fennoscandia. Tectonophysics 216: 31-40

Henkel H, Puura V, Flodén T, Konsa M, Liljequist R (2002) Åvike Bay in east-central Sweden - after all one more impact [abs.]. In: von Dalwigk I (ed) 8th Workshop of the European Science Foundation Program IMPACT, "Impact Tectonism", Mora, Sweden May 31 - June 3, 2002. Stockholm University, p 25

Jessberger EK and Reimold WU (1980) A late Cretaceous ^{40}Ar-^{39}Ar age for the Lappajärvi impact crater, Finland. Journal of Geophysics 48: 57-59

Kenkmann T, von Dalwigk I (2000) Radial transpression ridges: A new structural feature of complex impact craters. Meteoritics and Planetary Science 35: 1189-1202

Kenkmann T, Ivanov BA, Stöffler D (2000) Identification of ancient impact structures: Low-angle faults and related geological features of crater basements. In: Gilmour I, Koeberl C (eds) Impacts and the Early Earth. Lecture Notes in Earth Sciences 91, Springer, Heidelberg, pp 279-308

Kinnunen KA, Lindqvist K (1998) Agate as an indicator of impact structures: An example from Sääksjärvi, Finland. Meteoritics and Plantary Science 33: 7-12.

Koistinen T, Klein V, Koppelmaa H, Korsman K, Lahtinen R, Nironen M, Puura V, Saltykova T, Tikhomirov S, Yanovskiy A (1996) Paleoproterozoic Svecofennian orogenic belt in the surroundings of the Gulf of Finland. Geological Survey of Finland, Special Paper 21: 21-57

Koivisto M, Korhonen JV (1997) Lappajärvi impact crater rocks and minerals in Quaternary sediment. In: Autio S (ed) Geological Survey of Finland, Current Research 1995-1996. Special Paper 23: 79-88

Korja A, Korja T, Luosto U, Heikkinen P (1993) Seismic and geoelectric evidence for collisional and extensional events in the Fennoscandian Shield - implications for Precambrian crustal evolution. Tectonophysics 219: 129-152

Kraus E, Meyer R, Wegener A (1928) Untersuchungen über den Krater von Sall auf Ösel (in German). Kurländische Beiträge zur Geofysic 20: 312-378

Lahtinen R, Huhma H (1997) Isotopic and geochemical constraints on the evolution of the 1.93 - 1.79 Ga Svecofennian crust and mantle in Finland. Precambrian Research 82: 13-34

Lehtinen M, Pesonen LJ, Puranen R, Deutsch A (1996) Karikkoselkä – A new impact strucure in Finland [abs.]. Lunar and Planetary Science 27: 739-740

Lehtinen M, Pesonen LJ, Stehlik H, Kuulusa M (2002) The Suvasvesi South structure, central Finland: New evidences for impact [abs.]. Lunar and Planetary Science 33, abs. #1188 (CD-ROM)

Lehtovaara JJ (1982) Stratigraphical section through Lower Cambrian at Söderfjärden, Vaasa, western Finland. Bulletin of the Geological Society of Finland 54: 36-43

Lidmar-Bergström K (1996) Long term morphotectonic evolution in Sweden. Geomorphology 16: 33-59

Lindström M, Flodén T, Grahn Y, Kathol B (1994) Post-impact deposits in Tvären, a marine Middle Ordovician crater south of Stockholm, Sweden. Geological Magazine 131: 91-103

Lindström M, Sturkell EFF, Törnberg R, Ormö J (1996) The marine impact crater at Lockne, central Sweden. GFF 118: 193-206

Lindström M, Flodén T, Grahn Y, Hagenfeldt S, Ormö J, Sturkell EFF, Törnberg R (1999) The Lower Palaeozoic of the probable impact crater of Hummeln, Sweden. GFF 121: 243-252

Mänttäri I, Koivisto M (2001) Ion microprobe uranium-lead dating of zircons from the Lappajärvi impact crater, western Finland. Meteoritics and Planetary Science 36: 1087-1095

Masaitis VL (1975) Astroblemes in the USSR. International Geology Review 18: 1249-1258

Masaitis VL (1999) Impact structures of northeastern Eurasia: The territories of Russia and adjacent countries. Meteoritics and Planetary Science 34: 691-711

Motuza G (1994) Impact craters in Lithuania [abs.]. In: Törnberg R (ed) 2nd International ESF workshop: The Identification and Characterization of Impacts, Lockne, Sweden, 1 p

Motuza G, Gailius R (1978) Supposed astroblemes in Lithuania (in Russian). In: Suveizdis PI (ed) Local structures in Belarus and Baltic Region. Geological Research Institute, Vilnius, Lithuania, pp 91-94

Müller N, Hartung JB, Jessberger EK, Reimold WU (1990) 40Ar-39Ar ages of Dellen, Jänisjärvi, and Sääksjärvi impact craters. Meteoritics 25: 1-10.

Nikishin AM, Ziegler PA, Stephenson RA, Cloetingh SAPL, Furne AV, Fokin PA, Ershov AV, Bolotov SN, Korotaev MV, Alekseev AS, Gorbachev VI, Shipilov EV, Lankreijer A, Bembinova EYu, Shalimov IV (1996) Late Precambrian to Triassic history of the East European Craton: dynamics of sedimentary basin evolution. Tectonophysics 268: 22-63

Öhman T, Badjukov D, Raitala J, Petrova T, Stehlik H (2003) Impactites of the Paasselkä and Suvasvesi S craters, Finland [abs.]. Lunar and Planetary Science 34: #1571, 2 pp (CD-ROM)

Ormö J (1994) The pre-impact Ordovician stratigraphy of the Tvären Bay impact structure, SE Sweden. GFF 116: 139-144

Ormö J, Blomqvist G (1996) Magnetic modelling as a tool in the evaluation of impact structures, with special reference to the Tväran Bay impact crater, SE Sweden. Tectonophysics 262: 291-300

Ormö J, Miyamoto H (2002) Computer modelling of the water resurge at a marine impact: the Lockne crater, Sweden. Deep-Sea Research II 49: 983-994

Ormö J, Sturkell E, Blomqvist G, Törnberg R (1999) Mutually constrained geophysical data for the evaluation of a proposed impact structure: Lake Hummeln, Sweden. Tectonophysics 311: 155-177

Ormö J, Shuvalov VV, Lindström M (2002) Numerical modelling for target water depth estimation of marine-target impact craters. Journal of Geophysical Research 107 (E12): 5120

Papunen H (1969) Possible impact metamorphic textures in the erratics of the lake Sääksjärvi area in southwestern Finland. Bulletin of the Geological Society of Finland 41: 151-155

Pesonen LJ (1996) The impact cratering record of Fennoscandia. Earth, Moon and Planets 72: 377-393

Pesonen LJ, Lehtinen M, Deutsch A, Elo S, Lukkarinen H (1996) New geophysical and petrographic results of the Suvasvesi N impact structure, Finland [abs.]. Lunar and Planetary Science 27: 1021-1022

Pesonen LJ, Elo S, Lehtinen M, Jokinen M, Puranen R, Kivekäs L (1999a) The Lake Karikkoselkä impact structure, central Finland - new geophysical and petrographic results. In: Dressler BO, Sharpton VL (eds) Large Meteorite Impacts and Planetary Evolution II. Geological Society of America Special Paper 339: 131-139

Pesonen LJ, Kuivasaari T, Lehtinen M. Elo S (1999b) Paasselkä - A new meteorite impact structure in eastern Finland [abs.]. Meteoritics and Planetary Science 34: A-90-A91

Pesonen LJ, Abels A, Lehtinen M, Plado J (2000) Meteorite impact cratering - Implications for the Fennoscandian lihosphere [abs.]. In: Pesonen LJ, Korja A, Hjelt S-E (eds) LITOSPHERE 2000. A Symposium on the Structure, Composition and Evolution of the Lithosphere in Finland, Espoo Oct. 4-5, 2000. Programme and Extended Abstracts. Institute of Seismology, University of Helsinki, Report S-41: 113-119

Pipping F, Lehtinen M (1992) Geology, stratigraphy and structure of the Lappajärvi meteorite crater, western Finland: preliminary results of deep drilling. Tectonophysics 216: 91-97

Plado J, Pesonen LJ, Elo S, Puura V, Suuroja K (1996) Geophysical research on the Kärdla impact structure, Hiiumaa Island, Estonia. Meteoritics and Planetary Science 31: 289-298

Plado J, Pesonen LJ, Puura V (1999) Effect of erosion on gravity and magnetic signatures of complex impact structures: Geophysical modeling and applications. In: Dressler BO, Sharpton VL (eds) Large Meteorite Impacts and Planetary Evolution II. Geological Society of America Special Paper 339: 229-240

Puura V, Flodén T (1997) The Baltic Sea drainage area - a model of a Cenozoic morphostructure reflecting the early Precambrian crustal pattern. In: Cato I, Klinberg F (eds) Proceedings of the Fourth Marine Geological Conference - the Bartic, Uppsala 1995. Sveriges Geologiska Undersökning Ca 86: 131-137

Puura V, Flodén T (2000) Rapakivi-related basement structures in the Baltic sea: a regional approach. GFF 122: 257-272

Puura V, Suuroja K (1992) Ordovician impact crater at Kärdla. Hiiumaa Island. Estonia. Tectonophysics 216: 143-156

Puura VA, Mens KA, Männil RM, Pirrus EA, Heinsalu HN (1987) Paleotectonics and Cambrian facies of the Baltic basin before the phosphorite and kukersite deposition (in Russian). In: Garetzki RG, Suveizdis PI (eds) Tectonics, formations and facies of the western East European Platform. Nauka i Tekhnika: 86-95

Puura V, Lindström M, Flodén T, Pipping F, Motuza G, Lehtinen M, Suuroja K, Murnieks A (1994) Structure and stratigraphy of meteorite craters in Fennoscandia and Baltic region: a first outlook. Proceedings of Estonian Academy of Sciences. Geology 43: 93-108

Puura V, Laitakari I, Amantov A (1996) Latest events affecting the Precambrian basement, Gulf of Finland and surrounding areas. In: Koistinen T (ed) Explanation to the Map of Precambrian basement of the Gulf of Finland and surrounding area 1 : 1 million. Geological Survey of Finland Special Paper 21: 115-126

Puura V, Flodén T, Mokrik R, Bjerkeus M, Monkevicius A, Vaher R (2000) The tectonic history of the Baltic sedimentary basin and its heredit'ary, the modern Baltic Sea [abs.]. The Baltic Sea. The Sixth Marine Geological Conference May 7-9 2000, Hirtshals, Denmark: pp 71-72

Raitala J. Halkoaho T (1992) Mineral chemistry of the shock-metamorphosed schists of the Lake Jänisjärvi impact structure, Karelia. Tectonophysics 216: 187-194

Raukas A, Tiirmaa R, Kaup E, Kimmel K (2001) The age of the Ilumetsa meteorite craters in South-East Estonia. Meteoritics and Planetary Science 36: 1507-1514

Reinwald I (1928) Bericht über geologische Untersuchungen am Kaalijärv (Krater von Sall) auf Ösel (in German). Publications of Institute of Geology, University of Tartu 11: 30-70

Reinwald I (1938) The finding of meteorite iron in Estonian craters. A long search richly awarded. The Sky Magazine of Cosmos News 2: 6-7

Schmitz B, Tassinari M, Peucker-Ehrenbrink B (2001) A rain of ordinary chondritic meteorites in the early Ordovician. Earth and Planetary Science Letters 194: 1-15

Shmayenok AI, Tikhomirov SN (1974) Mishinogorskaya explosive structure in the vicinity of Lake Tshudskoye (in Russian). Doklady Akademii Nauk SSSR Geologiya 219: 701-703

Suuroja K, Suuroja S (2000) Neugrund structure - the newly discovered submarine Early Cambrian impact crater. In: Gilmour I, Koeberl C (eds) Impacts and the Early Earth. Lecture Notes in Earth Sciences 91, Springer Verlag, Berlin-Heidelberg 91, pp 389-416

Suuroja K, Suuroja S, All T, Floden T (2002) Kärdla (Hiiumaa Island, Estonia) – the buried and well-preserved Ordovician marine impact structure. In: Gersonde R, Deutsch A, Ivanov BA, Kyte FT (eds) Oceanic Impacts: Mechanisms and Environment Perturbations. Deep-sea Research II 49: 1121-1144

Sturkell EFF (1998) Resurge morphology of the marine Lockne impact crater, Jämtland, central Sweden. Geological Magazine 135: 121-127

Therriault A, Lindström M (1995) Planar deformation features in quartz grains from the resurge deposit of the Lockne structure, Sweden. Meteoritics 30: 700-703

Torsvik TH, Van der Voo R, Meert JG, Mosar J, Walderhaug HJ (2001) Reconstructions of the continents around the North Atlantic at about the 60th parallel. Earth and Planetary Science Letters 187, 55-69

Tynni R and Uutela A (1984) Microfossils from the Precambrian Muhos formation in Western Finland. Geological Survey of Finland, Bulletin 330, 38 pp

Veretennikov NV, Ilkevitsh GI, Makhnach AS (1978) Logoisk sub-surface depression - An ancient meteorite crater (in Russian). Proceedings of the Academy of Sciences of Bielorussian SSR. Geology XXIII: 156-159

Veski S, Heinsalu A, Kirsimäe K, Poska A, Saarse L (2001) Ecological catastrophe in connection with the impact of the Kaali meteorite about 800 - 400 BC on the island of Saaremaa, Estonia. Meteoritics and Planetary Science 36: 1367-1376

Werner SC, Plado J, Pesonen LJ, Janle P, Elo S (2002) Potential fields and subsurface models of Suvasvesi North impact structure, Finland. Physics and Chemistry of the Earth 27: 1237-1245

Winterhalter B (1982) The bedrock geology of Lumparn Bay, Åland. In: Paleozoic sediments in the Rapakivi area of the Åland Islands. Geological Survey of Finland Bulletin 317: 35-114

Geophysical Investigations of the Siljan Impact Structure – A Short Review

Herbert Henkel[1] and Sven Aaro[2]

[1]Department of Land and Water Resources Engineering, Royal Institute of Technology, S-10044 Stockholm, Sweden (herbert@kth.se)

[2]Geological Survey of Sweden, Division of Geophysics, Box 670, S-75128 Uppsala, Sweden (sven.aaro@sgu.se)

Abstract. Siljan in southwest Sweden is the largest impact structure in western Europe, with a present topographic diameter of ca. 75 km. Recent age determinations indicate an age of 377 Ma. The bedrock geology of the region has recently been re-mapped by the Geological Survey of Sweden in the scale of 1:50 000. There is now complete coverage with airborne geophysics. New maps of the geophysical data have been prepared for this review. In connection with the Deep Gas Project, further geophysical studies were made and two drill holes were sunk to over 6 km depth in the central uplift of the structure. The Deep Gas Project produced a large number of reports and publications, which are listed in the summary report of Juhlin (1991). Some of the results are compiled and shortly summarized here. Digital elevation data are available with 50 m spatial resolution, and a gray tone map has been prepared with the regional trend removed. A profile of these data shows that the peak ring of the structure is still visible in the morphology.

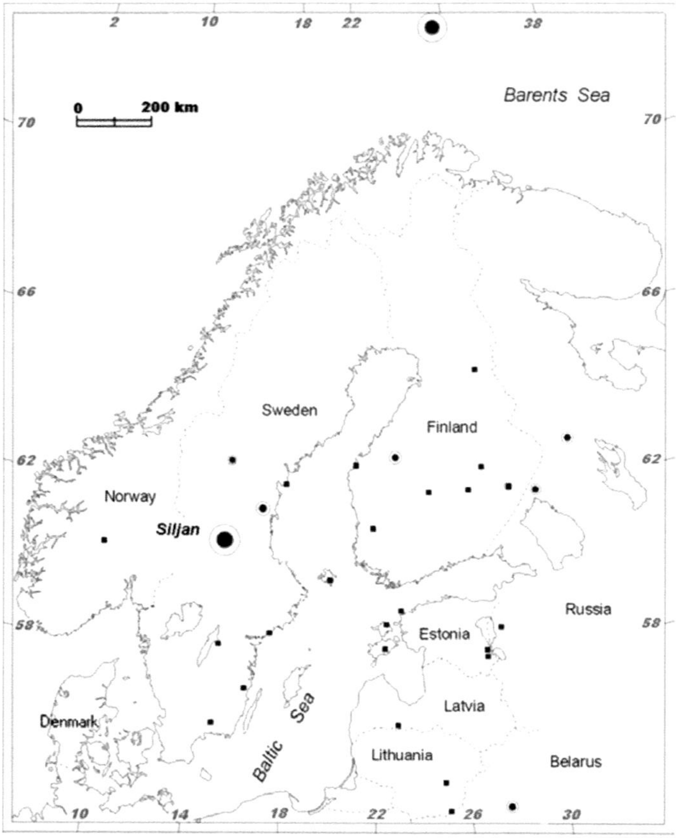

Fig. 1. Location of the Siljan Impact Structure. Impact structures smaller than 20 km in diameter are shown with square symbols. Modified from Abels et al. (2000).

1 Background

The occurrence of a dome of Proterozoic granites of the Trans-Scandinavian Igneous Belt (TIB) surrounded by a ring of Paleozoic sediments, as mapped by Stolpe (1872) and Hjelmquist (1966), posed an enigma never really discussed until the Siljan ring was suggested as an impact structure by Wickman et al. (1963) and Fredriksson and Wickman (1963). This suggestion was based on morphologic features, and it was ignored for many years until shock metamorphic evidence like planar deformation features (PDF) and shatter cones were found (Svensson 1973). Figure 1 shows the location of the Siljan structure together with other impact craters in Fennoscandia. Table 1 shows the stratigraphic column for the Siljan region and Fig. 2 gives an overview of the regional geology of the ring syncline and the central uplift. A view from the NW edge of the structure (Fig. 3) shows the still impressive crater in the landscape.

In a sector from west to north to northeast, the basement to the Paleozoic lithologies consists of mainly Proterozoic metavolcanic rocks intercalated with

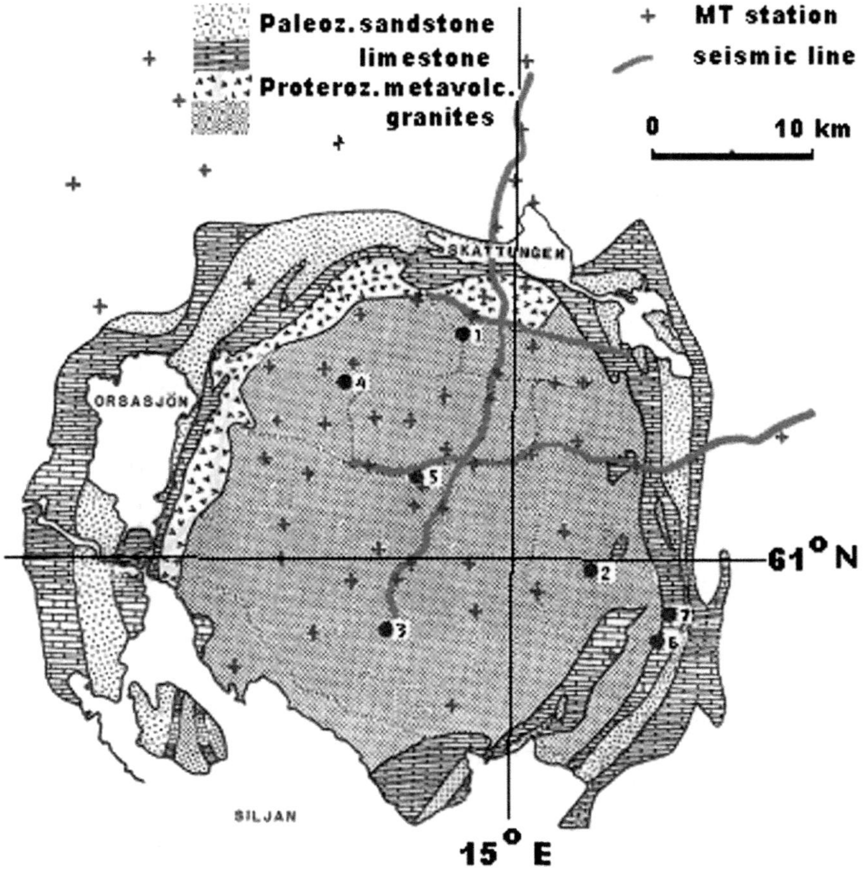

Fig. 2. Geological overview. The pre-impact cover sequence is subdivided into late Proterozoic (>), Paleozoic limestone (brick pattern) and sandstone (coarse spaced dots). The crystalline core is marked with dense dots. Location of the drillings (black dots and numbers 1 – 7), MT measurement stations (marked +) and reflection seismic lines (gray lines) made for the Deep Gas Project. The deep wells *Gravberg-1* and *Stenberg-1* are located at (1) and (5), respectively.

Fig. 3. View of the Siljan structure from the NW slope towards the south. The ring depression with lakes is followed by the gently sloping central rise and the southern crater margin forms the horizon.

minor occurrences of sandstones. The central uplift contains different types of felsic intrusives within the Trans-Scandinavian Igneous Belt (TIB) and some minor mafic masses and dykes. The Järna and Siljan granite types are predominant and are part of a large batholith within the TIB extending for 300 km to the NW (Gaal and Gorbatschev 1987). In the crystalline core of the Siljan structure, all lithologies are overprinted by characteristic penetrative lineations. These lineations have been interpreted as shatter cone features in the geologic map of Kresten et al. (1991). Small-scale shatter cones occur in both felsic and mafic lithologies.

At three localities within the crystalline central uplift spatially limited xenoliths of Proterozoic sedimentary rocks were mapped. At one of these localities, close to Hättberg (close to locality 4 in Fig. 2) in the center of the uplift, a large clast of late Proterozoic sandstone is overprinted with complex striation patterns in several orientations.

Re-appraisal of the impact idea came with the launching of the *Deep Gas Project* by a consortium in 1982. A set of bedrock geological maps by Kresten et al. (1991) at the scale of 1:50 000 is based on detailed mapping of the sparse outcrops that occur within the impact structure, and water wells and local boulder finds. Also at that time detailed low altitude (terrain clearance 30 m) airborne geophysical surveys covered the Siljan region.

The Deep Gas Project aimed at finding potential abiogenic gas emanating from the mantle and accumulated in fractured crystalline rocks within an impact structure supposed to be sufficiently large to have penetrated the entire crust (Gold and Soter 1980).

The Deep Gas Project resulted in large amounts of data from 7 shallow (up to ca 800 m) and two deep drillings (Gravberg-1 with 6.8 km, and Stenberg-1 with 6.5 km vertical depth). The drill hole at Gravberg is located within the peak ring, whereas the drill hole at Stenberg is located at the center of the uplift. Much of these data are summarized in Juhlin (1991).

Table 1. Stratigraphy of the Siljan region. The Cambrian sequences are not represented in the area.

Age		Event	Lithology
Quarternary		Glaciations	Till
Paleozoic	Devonian	Siljan impact 377 Ma	Impact melt and breccias
	Silurian		sandstone
			Limestone and siltstone
	Ordovician		Limestone and shale
			Limestone
		Discontinuity	
Proterozoic	Upper (Vendian / Riphean /Jotnian)		Mafic sills and dykes
			Sandstone
		Discontinuity	
	Middle		Felsic – intermediate volcanics
			Felsic intrusives
		Svecofennian orogeny	
			Meta supracrustal gneisses

The previous age determinations by ^{40}Ar-^{39}Ar dating were based on a few samples of suggested impact melt and ranged from 342 to 368 Ma (Bottomley et al. 1978). A collection of recently found impact melt breccias from 5 localities within the central uplift have been dated with laser-argon and argon step heating methods. A new, older age for the Siljan impact event was established at 377±1.7 Ma (Reimold et al. 2004). The chances to find more impact melt is very restricted, as the area within the central portion of the structure is basically devoid of outcrops.

The complex mass flow in connection with the transient crater collapse is discussed in Henkel and Reimold (1997 and 1998) for the Vredefort structure. In the numerical models of impact rater collapse in Melosh and Ivanov (1999), the complexity of this mass redistribution is envisaged as thinning of marker layers in the ring depression and thickening towards the central rise. The collapse flow results in radial inward mass concentration of the exterior, near-surface lithologies, and a radial outward dispersion of the central uplift lithologies. The collision of these opposed radial mass flows causes the up- and overturning of lithologies as observed in the Vredefort and Siljan structures. The collision of radial mass flows also results in the imbrication and repetition of lithologies that were located at the excavated crater edge and the central uplift, respectively. In the Siljan structure, repetitions of marker horizons of Vendian sandstone have been reported from along the central crystalline core in the northeastern part of the central uplift (Thorslund and Collini (1980). It can also be seen on the bedrock map by Kresten et al. (1991) by the distribution of Paleozoic mega-blocks in the southeastern and northwestern part of the central uplift. In the outer part of the crystalline central uplift, seven Paleozoic mega-blocks have been mapped, ranging in size from hundreds of meters to several km. In many of these occurrences limestone has been quarried for cement production. These quarries are excellent study sites for the tectonic overprint caused by the collapse flow. The lithologies of the sedimentary ring are locally strongly tectonised and broken into clasts with sizes up to several tens of meters, and are occasionally surrounded by micro-brecciated rock material.

In four places along the ring syncline, the Proterozoic basement rocks are in a high structural position and the Peleozoic lithologies appear to be missing. Kenkmann and von Dalwigk (2000) have suggested, as a working hypothesis, that these sectors represent radial transpression ridges that have formed by mass concentration by the radial inflow from the collapsing crater wall. Within individual clasts, the stratigraphic succession is often remarkably well preserved and several occurrences of Paleozoic pre-impact cover rocks have been reported as stratigraphic type localities. The occurrence large blocks of cover rocks immersed in brecciated crystalline basement have, for example, also been observed in one of the deep drill holes 3 km outside the peak ring of the Popigai structure, some 400 m below the allogenic breccia level (Masaitis et al. 1998)

The collapse of large craters has been modeled numerically for a series of structures with increasing diameters (Melosh and Ivanov 1999). In these models, transient craters with diameters >30 km collapse into peak ring structures. The near-surface structures from beyond the edge of the excavated crater are thickened in the ring basin and up- and overturned by the collapsed peak ring.

Despite huge efforts that have been put into investigations of the Siljan structure, several key issues are still unresolved. The original size of the structure, the vertical and lateral extent of different types of impact overprint (shock, tectonic, thermal, magnetic), the structure of the ring basin (depth, tectonic, stratigraphic), and the occurrence and nature of impact melt rocks, all remain to be further investigated.

In this short review emphasis is on the morphology as seen in digital elevation data and on the geophysical data from airborne, surface and drillhole measurements. The most characteristic result is that the very large and morphologically well-expressed Siljan crater largely lacks dramatic geophysical anomalies (as opposed to many smaller structures were both distinct gravity and magnetic anomalies occur).

This review concentrates on the morphological, geophysical and petrophysical aspects of the Siljan impact structure.

2 Morphology and Crater Diameter

The Siljan structure, centered at 14.9° E and 61.1° N (Fig. 1), is one of the larger impact structures on earth. It is clearly visible in the morphology, as demonstrated by digital elevation data (Fig. 4). An interpretation of these data with respect to tangentially oriented steps in the terrain is given in Fig. 5. The structure has been subjected to post-impact differential erosion due to tectonic activity of regional crustal blocks and tilting of the entire region. Especially the block to the NW of the Siljan structure is more elevated, although it still contains late Proterozoic (Riphean/Jotnian/Vendian) cover rocks (sandstone). In Fig. 4 this regional tilt has been removed by subtraction of a planar surface from the elevation data. The residual relative elevation shows the crater structure more clearly. The terrain profile based on elevation data with 50 m spacing (see section on ground-based geophysics, Fig. 10) shows that the Siljan structure has a peak ring of ca. 20 km diameter.

From the present morphology, the erosional crater diameter can be estimated to ca. 75 km based on the prominent escarpments marked in Fig. 5 and the highest topography shown in the profile in Fig. 10. This erosional

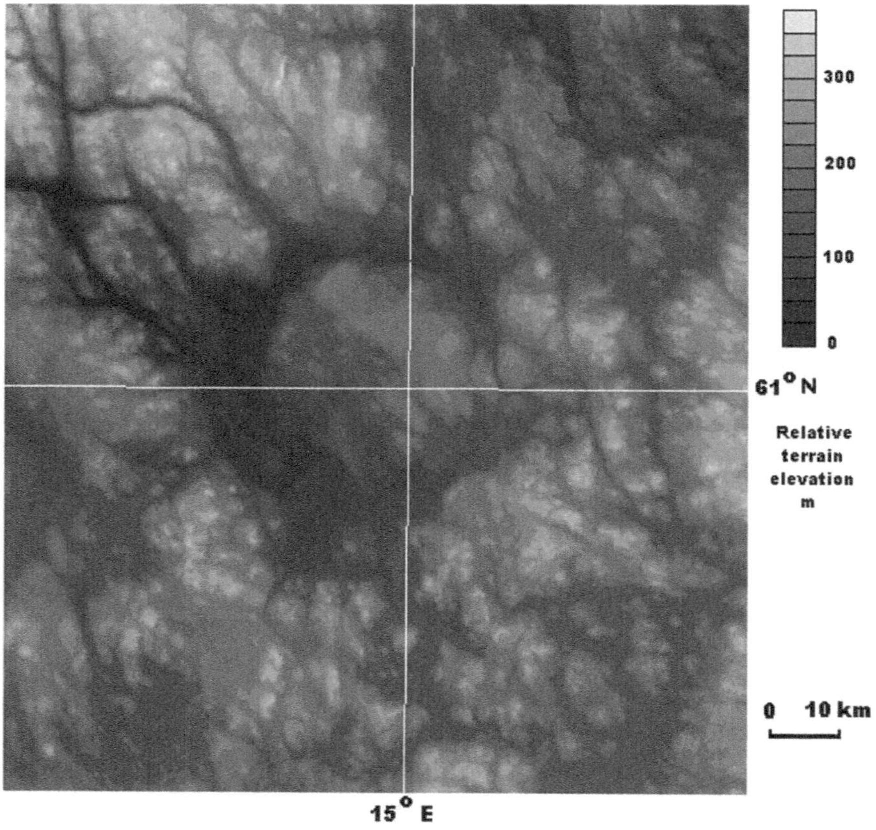

Fig. 4. The Siljan impact crater seen in digital elevation data with 50 m spacing, from Lantmäteriet (National Land Survey of Sweden). A linear regional terrain gradient has been removed.

edge, as demonstrated in Fig. 5, is highly irregular as opposed to the more circular outline of the central uplift of granitic rocks and pre-impact cover rocks. This uplift of crystalline rocks has a diameter of ca. 30 km at the present erosion level, and it has still, a partial peak ring shape, where elevations are up to 110 m higher than in the interior (Fig. 5 and 10). The height of the ring uplift is up to 265 m above the sedimentary ring basin, which, in turn, lies up to 415m below the exterior terrain. This ring basin is dominated by a series of lakes, which are wide in the south (Siljan), narrow in the north (Skattungen), and almost absent in the east (a few very small lakes). The elevation of the large lake is 162 m above the present sea level.

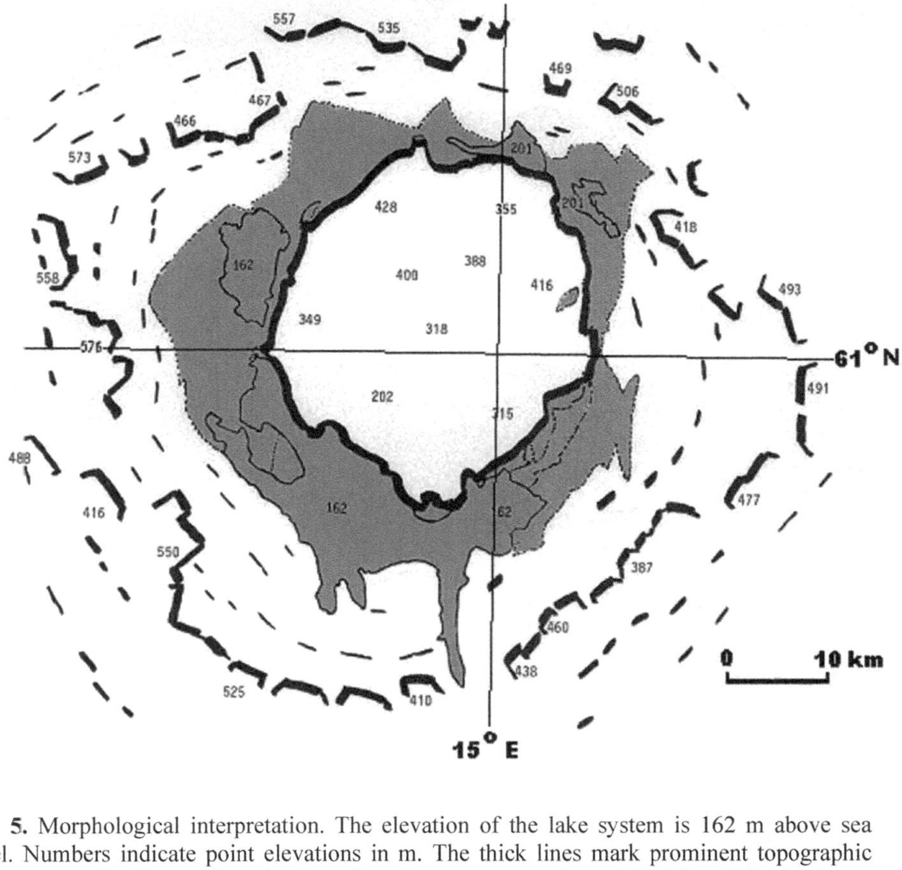

Fig. 5. Morphological interpretation. The elevation of the lake system is 162 m above sea level. Numbers indicate point elevations in m. The thick lines mark prominent topographic escarpments. Shading outlines the ring of pre-impact cover rocks and the lakes (thin lines).

Previous estimates of the crater diameter range from 52 km (Grieve 1988), i.e., just including the sedimentary ring basin, to 65 km (Kenkmann and von Dalwigk 2000) based on the occurrence of pseudotachylitic dykes assumed to be related to the impact event.

The Paleozoic sedimentary rocks are completely eroded in the surroundings of the Siljan structure. In the northwestern part of the structure, late Proterozoic rocks were also part of the pre-impact cover sequences. To the west, outside the Paleozoic ring, Proterozoic sedimentary and volcanic rocks are preserved over large areas. The minimum of erosion was, therefore, estimated to be the assumed thickness of the Cambro-Silurian sediments, i.e. 400 to 500 m (Rondot 1975). In addition, an unknown amount of younger Paleozoic cover rocks could also have been removed. Another estimate of

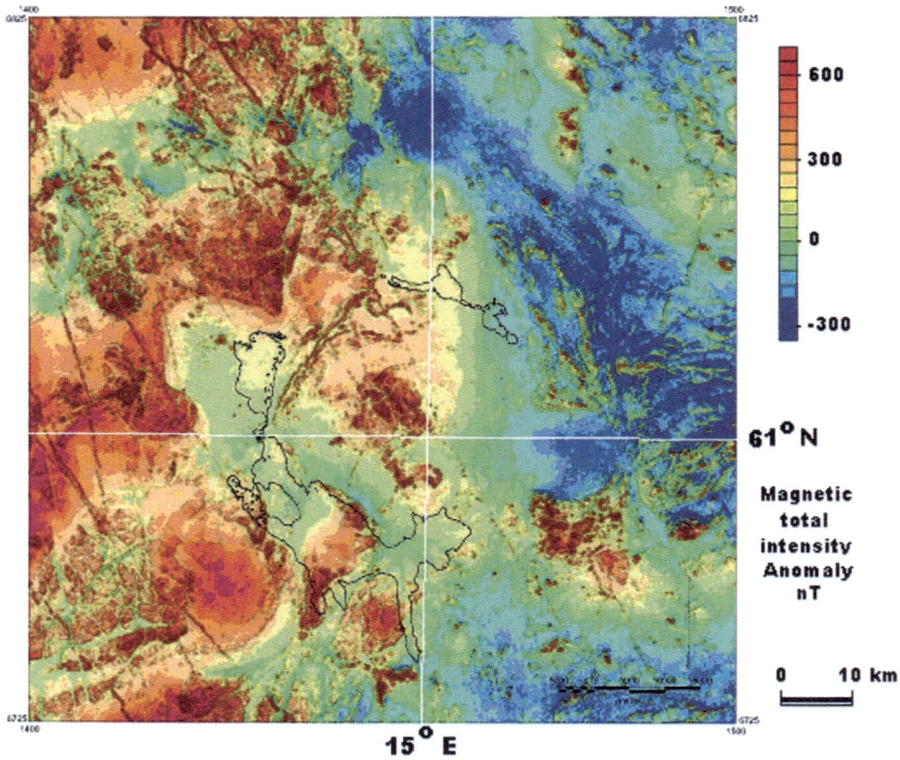

Fig. 6. Aeromagnetic total intensity anomaly map of the Siljan region, compiled by the Geological Survey of Sweden.

1 to 2 km is given in Collini (1988) based on the complete erosion of an assumed impact melt thickness of ca. 200 m in addition to the removal of the exterior Paleozoic cover rocks.

3 Geophysical Data

In this section a short summary is given of the existing geophysical data and their characteristics. Comments on their interpretation are given in a following section.

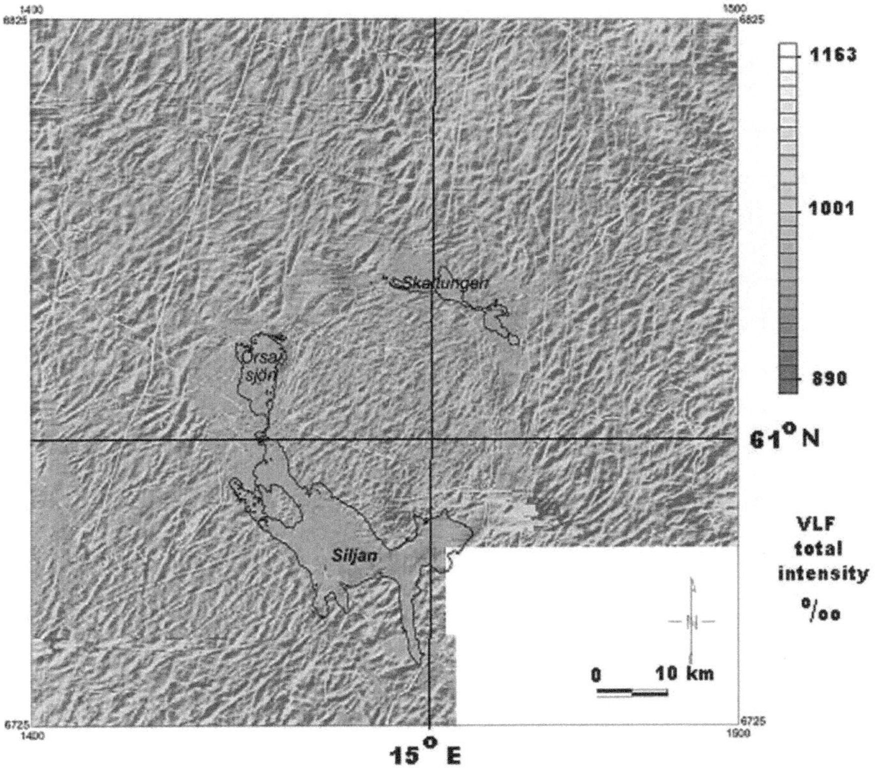

Fig. 7. VLF total intensity map of the Siljan region compiled by the Geological Survey of Sweden. The anomalies are given of the induced secondary field in $^{o}/_{oo}$ of the primary field. The transmitter is located to the SW (in southern England).

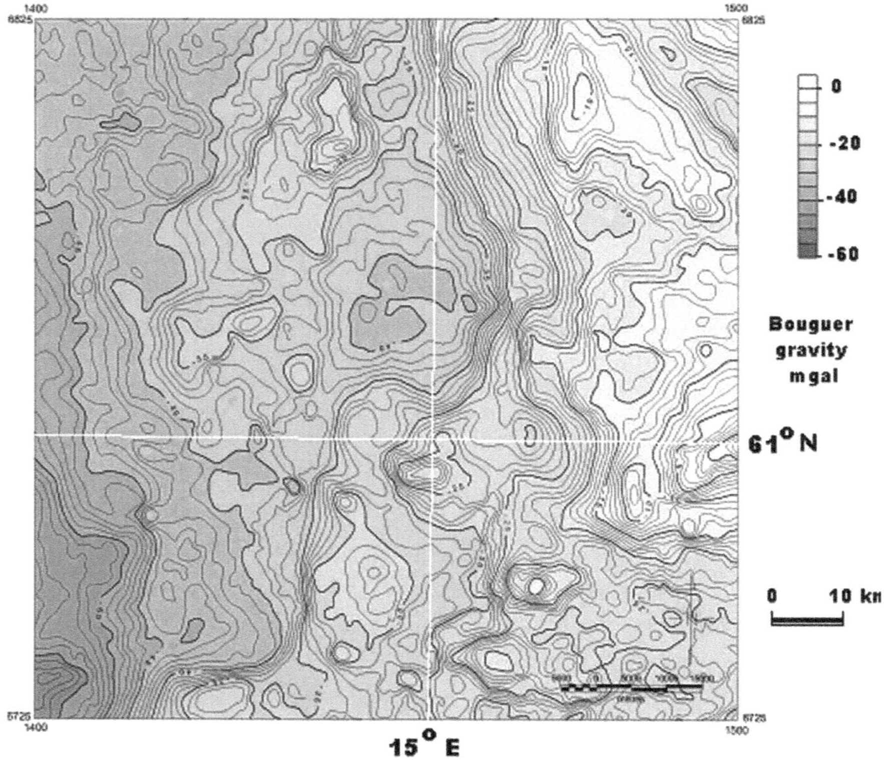

Fig. 8. Bouguer gravity anomaly map of the Siljan region compiled by the Geological Survey of Sweden. The contour interval is 1 mgal (10 gu) and the gray tone shading represents 5 mgal intervals.

3.1 Airborne Geophysics

Airborne geophysical measurements were made as part of the systematic mapping of Sweden. The ground clearance was 30 m, flight line spacing 200 m, and the spacing of measurements along flight lines 20 (modern data) to 40 m (older data). Measurements were made of the earth *magnetic field* (total

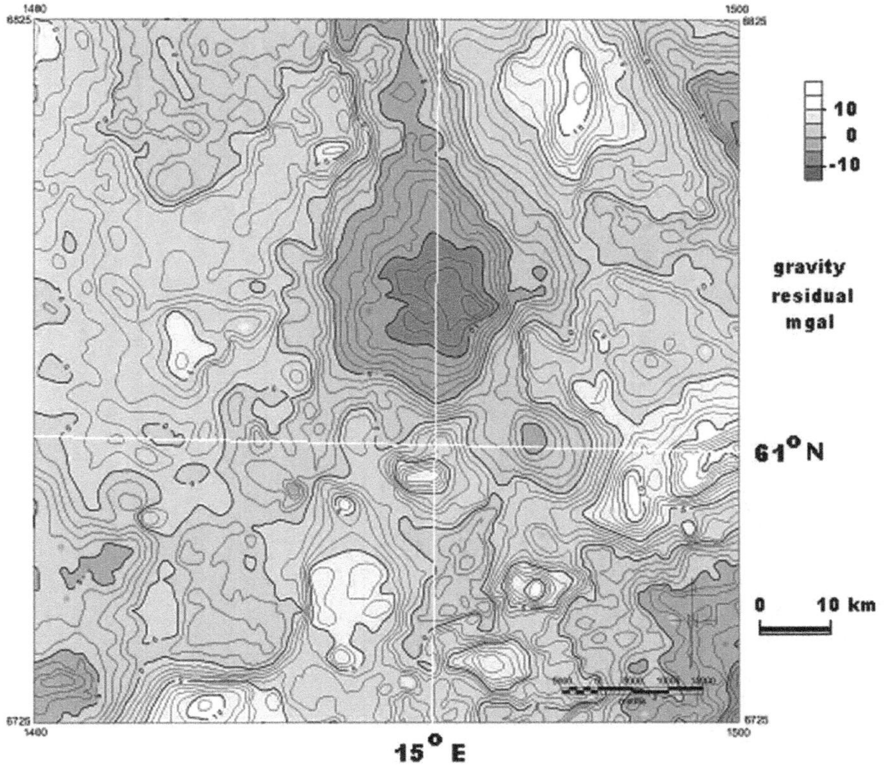

Fig. 9. Residual gravity anomaly map of the Siljan region compiled by the Geological Survey of Sweden. The contour interval is 1 mgal (10 gu) and the gray tone shading represents 5 mgal intervals.

intensity), *gamma radiation* (of U, Th, and K), and the components of induced *electromagnetic* radiation at 16 kHz frequency (VLF). These data are available from the Geological Survey of Sweden. Figure 6 shows the aeromagnetic measurements, and Fig. 7 the airborne VLF measurements. In Henkel (1992) a color map of the VLF anomaly is presented as an illustration of the increased fracture frequency seen in the central uplift region.

In contrast to the distinct features in the morphology, the magnetic anomaly pattern does not reveal any immediately clear impact related features. In the central part of the peak ring, a positive magnetic anomaly occurs that can be

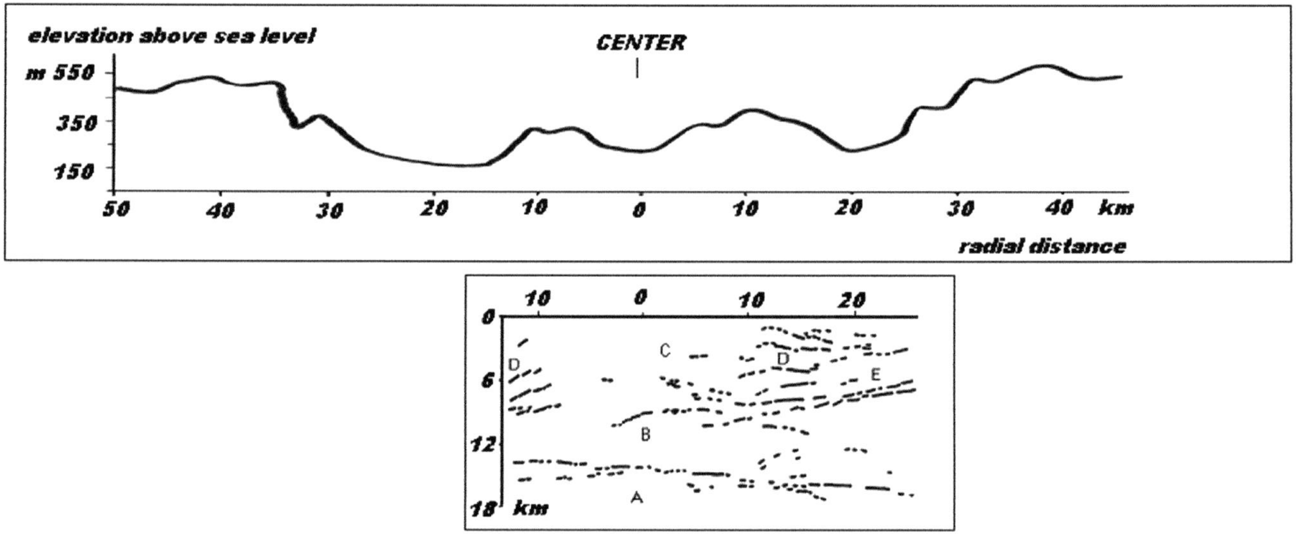

Fig. 10. Elevation profile (top) and interpretation of reflectors found in the reflection seismic surveys (below), combined from separate sections in Juhlin and Pedersen (1987), extending from west over the center to the north. (The vertical scale is approximate, as two-way-travel times have been converted to depth using 6 kms^{-1} as velocity for the whole depth range). The seismic reflection pattern indicates 5 regions, marked A-E in the figure. Region A appears to be below the level of uplift (although the length of the profile is too small to really estimate any occurring small uplift). Region B is the uplifted portion of the central rise where some reflectors still are visible. Region C is the uplift portion with steeply upturned structures resulting in no reflections. Regions D are at the flank of the uplift and show up-bent reflectors. Region E is the ring synform where structures are down-bent.

related to the impact geometry (see section on the interpretation of geophysical data).

The VLF anomaly pattern is dominated by the direction to the radio transmitter (and a few power lines). It is, however, clearly seen that the spacing of VLF anomalies differs in the exterior areas compared to the crystalline central uplift.

3.2 Ground-based Geophysics

Gravity measurements have been made in a coarse, irregular net, by the Geological Survey of Sweden, the Geodetic Survey, and in context with the Deep Gas Project by Uppsala University. All data have been collected in the national database and can be requested from the Geological Survey of Sweden. Figure 8 shows the compiled gravity with 1 mgal contours and gray tone shading in 5 mgal intervals. There is no clear gravity anomaly associated with the impact structure. Instead, local and regional differences in density of both near-surface and deep lithologies dominate the anomaly pattern. In Fig. 9, the regional field has been removed, which brings out the uppermost crustal features in the residual anomaly and suppresses the effects from deep and large structures.

The Deep Gas Project also provided 80 km of *reflection seismic* data traversing the central part of the structure and the northern and eastern parts of the sedimentary ring basin (Juhlin and Pedersen, 1987). Some of the seismic reflectors recorded within the central uplift were penetrated by the deep drilling and were found to be related to the occurrence of dolerite sills. The location of seismic lines is shown in Fig. 2. In Fig. 10, the seismic reflectors interpreted by Juhlin and Pedersen (1987 and 1993) are shown together with a topographic profile extending from west over the center to the north. The spatial characteristics of the reflective patterns have been summarized for this review.

Magnetotelluric (MT) measurements were made at 65 stations, mainly within the central uplift, traversing the ring and the nearest exterior areas to the NW and N. The magnetotelluric data were published in Zhang et al. (1989). The correlation with the impact structure is unclear, as the main low resistive feature that was found is off- center to the northwest with respect to the center of the impact structure. The location of MT stations is indicated in Fig. 2.

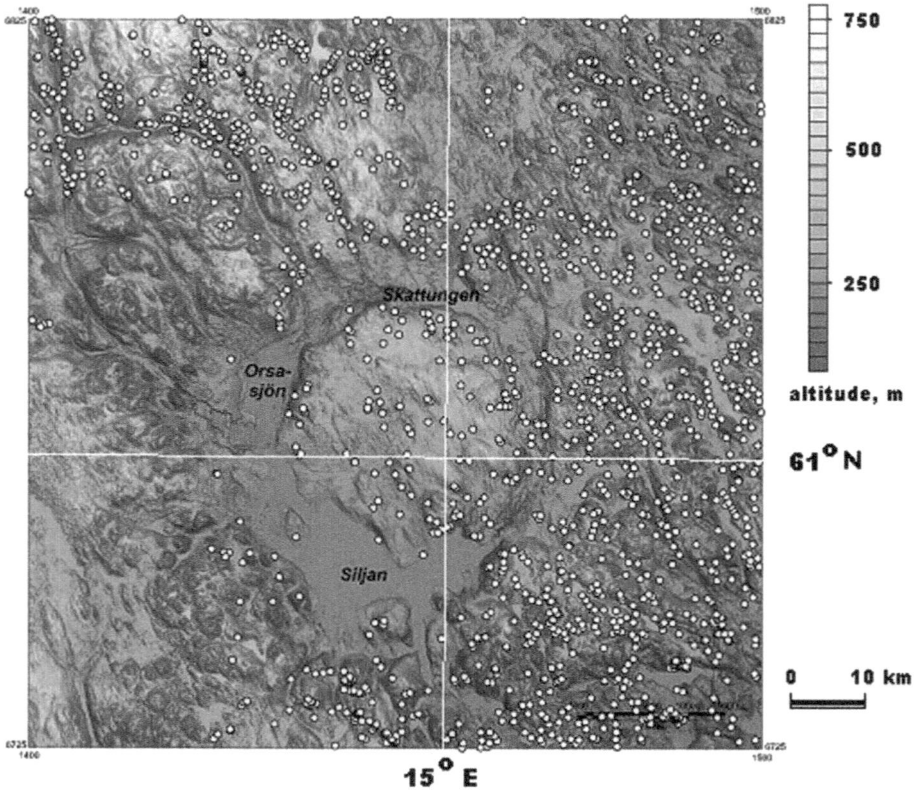

Fig. 11. Location of sampling sites for measurements of rock physical properties. The background map is based on digital elevation data with 50 m spacing from Lantmäteriet (National Land Survey of Sweden).

Rock *physical properties* have been measured on samples from the centrally located 610 m deep drill hole (No 5 in Fig. 2) and from outcrops. The unpublished data are available on request from the national database at the Geological Survey of Sweden. The measurements comprise *density*, *magnetic susceptibility* and *remanent magnetization*. In addition, the Geological Survey of Sweden has a regional data set of such petrophysical measurements containing 2392 samples. For locations see Fig. 11. The data have been compiled in standard diagrams with some of the major lithologies related to the Siljan structure outlined (Fig. 12). Some results are summarized in Table 2.

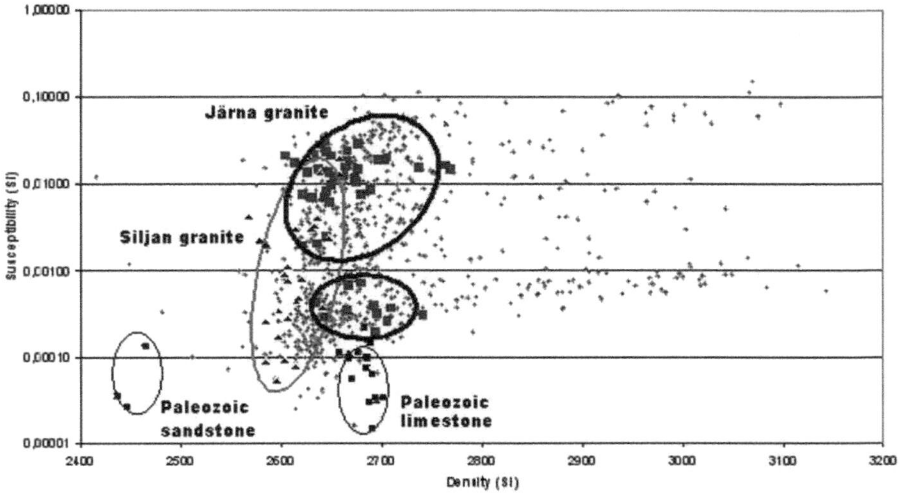

Fig. 12. Magnetic susceptibility – density plot of rock samples from the Siljan region. The variation field of some key lithologies is outlined with ellipses. The symbols refer to: Small squares - Paleozoic sandstone and limestone; Triangles: Siljan granite; Large squares: Järna granite. All other lithologies are marked with small + signs. Units are in SI (dimensionless and kgm^{-3}, respectively).

A distinct density and magnetic contrast exists between different members of the TIB intrusives that occur within the central uplift of the Siljan structure (the Järna and Siljan types, Hjelmquist 1966). There is a rather small density contrast between the Järna granite and the Paleozoic limestones. Both features complicate the interpretation of the gravity measurements with regard to structural changes caused by the impact event.

In-situ *electric resistivity* measurements have been made in a coarse net mainly in the central part of the structure. These data were evaluated by Henkel (1992). A relation between impact-induced porosity (due to more intense fracturing) and the electric resistivity could be established and seems to be a characteristic feature of brecciated crystalline rocks, and the central uplift region of complex impact structures in particular.

Table 2. Compilation of petrophysical properties for rocks from the Siljan region. The total number of measured samples is 2392. The q-ratio is the ratio between remanent and induced magnetization.

Rock type	No. of samples	Density (kgm^{-3}) mean	st.dev.	Magnetic suscept. (10^{-5} SI)	q-ratio
All rocks (except dolerites and ores)	2212	2719	122	835	1.23
TIB granite (Järna type)	65	2675	43	1188	0.31
TIB granite (Siljan type)	27	2618	24	361	0.35
Paleozoic limestone	20	2681	16	8	1.42

Table 3. Comparison of rock physical properties of Järna granite from the exterior and central uplift Samples from the outer part of the central uplift are from the drilling at locality 3 and samples from the central part of the central uplift are from drill hole 5 in Fig. 2, respectively. The q-ratio is the ratio between remanent and induced magnetization.

Source	Depth (m) below surface	Density (kgm^{-3})	Susceptibility (10^{-2} SI)	q-ratio
Exterior to impact structure	0	2712	1.4	0.21
	0	2736	3.3	0.09
Outer part of central uplift	205	2695	0.15	0.37
	225	2691	0.10	0.42
Central part of central uplift	110	2725	4.1	0.84
	170	2715	5.0	0.65
	290	2709	4.0	1.9
	410	2662	3.2	1.3

3.3
Drill Hole Geophysics

Two of the shallow drill holes in Siljan granite were logged (localities 1 and 4, see Fig. 2). Porosity up to 9 % was found and is associated with increased micro fracturing. In drill holes 3, 4, and 5 hydrological measurements were performed, showing very low hydraulic conductivity in the fractured rocks, ranging from 2.5×10^{-10} to 3.2×10^{-14} ms^{-1}.

The Gravberg-1 well was wire-line logged to 5.7 km depth. These loggings included *gamma ray spectroscopy, litho-density, geochemical, sonic, electric, electromagnetic, thermal, gravity* and *televiewer* tools. Descriptions of these logging data and their interpretation are given in Rissler-Åkesson (as sector report, see references in Juhlin 1991).

4
Interpretations Based on Geophysical Data

The following chapters report results from a large amount of geophysical surveys and their interpretation. In the discussion section some of these results are further evaluated.

4.1
Gravity

In connection with the Deep Gas Project, a preliminary *gravity model* of the Siljan structure was made by Dyrelius (1990). In this context, the erroneous assumption was made that the large gravity low in the northern part of the structure (Fig. 9) was the result of impact-induced brecciation (porosity). From density data of the different granitoids that occur in the region, it is however obvious that this anomaly is related to the occurrence of a low-density pluton (Siljan granite with average density 2618 kgm^{-3}). This pluton occurs within the upper 5 km of the crust of the TIB batholith, where the majority of granitoids have a comparatively higher density (Järna granite, 2675 kgm^{-3}). The low-density pluton extends some 30 km northwards from the center of the Siljan structure. The gravity map does not show any immediately visible features that are conform to the impact structure.

4.2 Magnetism

On the aeromagnetic map (Fig. 6), one conspicuous feature is the occurrence of distinct anomalies from an extensive SSE-NNW oriented dyke swarm (extending over nearly 800 km from southern Sweden up to and under the Caledonian thrust front in central Norway). These dykes have continuous anomalies outside the structure, disappear in the ring basin (as they are covered by the sedimentary rocks brought into the ring by the collapse flow), and seem to re-appear fragmented in the central uplift. Their general strike trend is, however, unchanged.

The only larger scale magnetic feature that is symmetric with respect to the impact structure is a centrally located positive anomaly of ca 500 nT amplitude with a diameter of ca. 10 km. A description of the aeromagnetic features is found in Pedersen et al. (1990). The absence of north-facing negative contact anomalies (and their positive counterparts on the south side), indicates either an antiform shape of the source structure or the presence of a remanent magnetization that neutralizes the asymmetry related to the inclination of the geomagnetic field. From measurements of the remanent magnetization of core samples from the shallow drill hole at the center of the central uplift (No 5 in Fig. 2, Table 3) it is obvious that this anomaly is caused by additional remanence as indicated by a high ratio of remanent to induced magnetization (q-ratios ≥ 1). Such high q-ratios do not normally occur in other parts of the TIB granitoid basement (which typically have q-ratios of 0.3 or less, see Table 2 and 3). In the residual magnetic map in Pedersen et al. (1990), several small-scale features, indicative of narrow sills or dykes, have a crater symmetric pattern. They have, however, been interpreted as dyke features without any relation to the impact.

4.3 VLF Measurements

The airborne electromagnetic very low frequency (VLF) measurements (Fig. 7) indicate a distinctly higher frequency within the central uplift of linear electric conductors, interpreted as fracture zones. The average distance between parallel fractures indicated by VLF anomalies is ca. 300 m in the central uplift, but is typically about 1.1 km outside the ring basin (Henkel 1992).

4.4 Seismics

The reflection seismic records have been interpreted with respect to discontinuities of the reflecting structures and the potential uplift of the central rise. Fig. 10 shows the reflection seismic interpretation by Juhlin and Pedersen (1987 and 1993), positioned together with a profile of the terrain surface. Several sub-horizontal reflectors occur and were targeted by the deep drilling at Gravberg; they turned out to be related to dolerite sills. The seismic reflectors appear uplifted outside ca. 11 km radius from the center of the impact structure, where they terminate (the two regions D in Fig. 10). Inside this region, reflectors are almost absent to ca. 6 km depth (region C). Below this depth, the reflectors appear in a slightly convex pattern with downward decreasing rise (region B). At the depth of ca. 14 km, the reflectors appear again sub horizontal (with a slight northward dip) and appear undisturbed by the impact event (region A). The termination of exterior reflectors has been interpreted as the onset of the transient cavity (Juhlin and Pedersen 1987), and the vertical offsets have been related to the terrace regime of the impact structure. Fault steps have been estimated from offsets of reflectors and range from 30 m to almost 1 km. In a re-evaluation of the reflection seismic data, the uplift in the center of the structure was estimated to be 2 km at 9 km depth (Juhlin and Pedersen 1993). The termination of continuous reflectors near the surface was interpreted as delineation of “the present day excavated crater” with a diameter of about 22 km. The discontinuous reflectors at larger depth within the central uplift were still related to the extent of the transient cavity (Juhlin and Pedersen 1993).

Attenuation studies on p-wave first arrivals in the borehole showed the upper 1.2 km of the crust as highly attenuating, which was interpreted to be caused by a higher fracture frequency. This is, however, an observation made also in other areas and seems to represent a general behavior of the uppermost crystalline crust in this region (Åström and Lund 1994).

4.5 Magnetotellurics

The magnetotelluric (MT) measurements showed a large region in the uppermost crust with lower than normal electric resistivity of 1 kΩm just northwest of the crater center. This low resistivity region has a 20 km diameter and a vertical extent between 6 and 20 km depth (Zhang et al. 1988). The low resistivity has been attributed to increased brecciation (porosity) hosting a relatively larger amount of saline groundwater compared to the less fractured surrounding rock volume, having a resistivity of 10 kΩm.

4.6 Electric Resistivity Logging

Resistivity logging in the Gravberg 1 well shows an increase in electric resistivity within the top 1.2 km. It then stabilizes at relatively high values around 30 kΩm. These loggings also indicate the occurrence of larger fracture zones, with a vertical spacing of 90 to 150 m throughout the drill hole. This is comparable to the results from airborne VLF data that indicate a similar lateral fracture spacing of 150 to 450 m within the whole central uplift region (Henkel 1992).

Resistivity logging in the Stenberg-1 well shows significantly more fracturing as recorded in the Gravberg-1 well. Porosity calculated from sonic and resistivity data shows a typical block size of alternating high and low porosity of 350 m. The top 1.5 km are heavily fractured throughout the central uplift with porosities of 2 to 4 %, occasionally approaching 20 %. Relatively fresh water with a salinity of 1 gl^{-1} was found down to 4.7 km depth, from where the salinity increased to 5 gl^{-1} at 5.4 km depth.

5 Petrological and Geochemical Features

Petrological and geochemical features of the Siljan impact structure were investigated by Aldahan (1990, 1991) and Ramseyer et al. (1992). The igneous rocks used for these investigations include samples from outcrops, several shallow (down to 600 m depth) and the two deep bore holes Gravberg-1 and

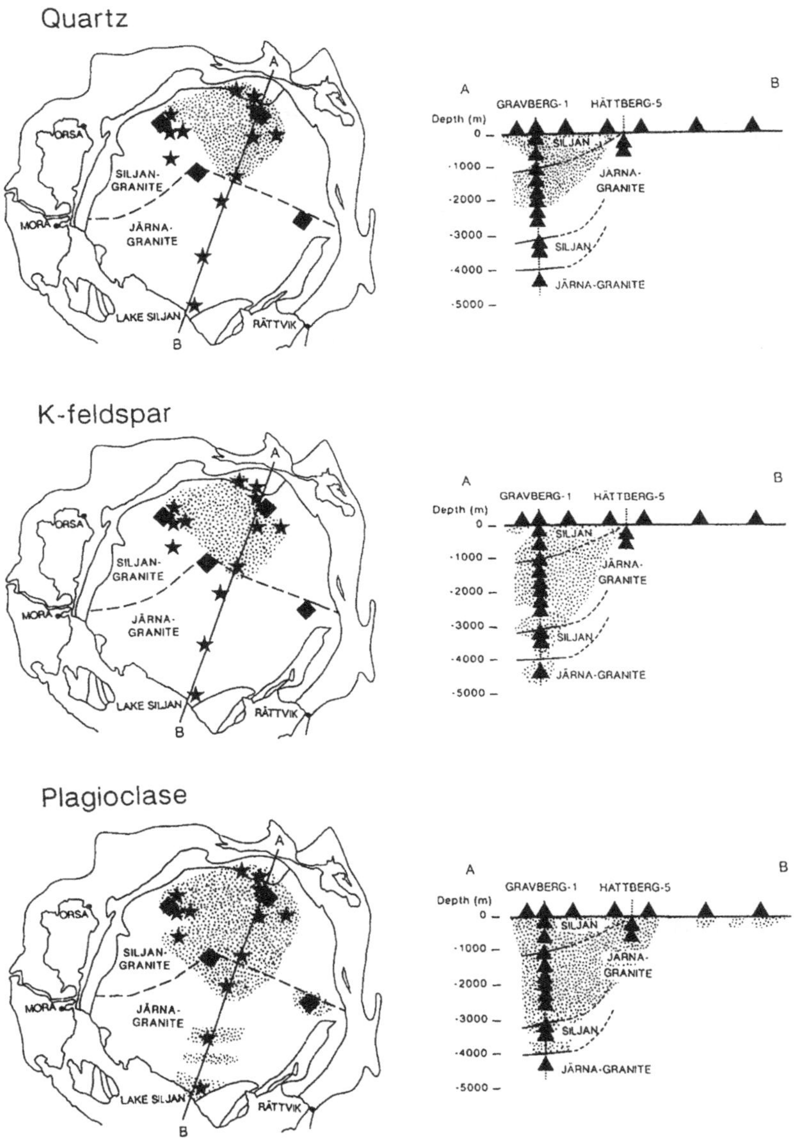

Fig. 13. Maps of observed alterations (left) of mineral cathodoluminescence. Drill hole locations are marked with diamonds. Stars denote outcrop samples. Profile across the Siljan central uplift dome (right). Triangles mark observations projected into the profile. Dotted areas: Alteration of quartz, K-feldspar and plagioclase, respectively. Dashed lines show the approximate boundary between Siljan and Järna granite types. After Ramseyer et al. (1992).

Stenberg-1 (to depth of about 6.5 km). The majority of the data from the deep bore holes concern the Siljan and Järna granite types that dominate the central uplift, and some minor dolerite.

Three major types of deformation and alteration are observed in the granites that have been related to the meteorite impact, intrusion of dolerite, and regional tectonic movements, respectively. The impact related alterations are estimated to extend down to 3 km depth in the Gravberg-1 borehole and to about 4.5 km depth in the more centrally located Stenberg-1 borehole. The impact-related alterations include 1) pseudotachylitic breccia veins, 2) planar deformation features (PDF) in quartz grains, 3) abundant fractures, 4) kinking and fragmentation of feldspars, 5) extensive saussuritisation of feldspars and chloritisation of micas.

Cathodoluminescence and microprobe analysis have been performed on quartz, plagioclase, and K-feldspar. Anomalous secondary brownish luminescence was found in albite and reddish luminescence in quartz. Primary magmatic quartz, plagioclase, and K-feldspar have blue, yellowish, and bluish luminescence colors, respectively. The change in luminescence is independent of the granite types. The depths to which anomalous cathodoluminescence is observed in the Gravberg-1 drill hole are for quartz 2.2 km, for K-feldspar 4.6 km and for plagioclase 4 km (Fig. 13). The change in luminescence of quartz is interpreted to be caused by the shock induced heating and subsequent high temperature in the granites. Most of the alteration of the feldspar occurred during the subsequent cooling period, by the introduction of meteoric water into the fractured rocks (Ramseyer et al. 1992).

6 Shallow and Deep Drillings

The Deep Gas Project started off with a series of seven shallow core drillings to characterize the uppermost part of the impact structure (down to ca. 800 m). Their locations are indicated in Fig. 2. Five drill holes were located on the central rise and two in the eastern sedimentary ring basin. From these investigations information was obtained of the porosity of the involved rocks and their physical properties (density, magnetic susceptibility and remanent magnetization). From rock magnetic measurements on drill cores from the centrally located test hole, it was found that below 240 m depth, the remanent magnetization made a relatively larger contribution to the magnetization below

240 m depth as compared to normal crystalline rocks, where it generally is very low. In Table 3 q-values are given for Järna granite from the exterior, the outer part of the central uplift and from its central part, respectively. The most shock affected crystalline rocks are located in this central part of the uplift.

The first deep drilling was located at *Gravberg* in the northern part of the central uplift (No 1 in Fig. 2) and within a large gravity low that was thought to be related to the impact. The drilling reached a depth of 6957 m, where it was terminated due to increasing technical difficulties. The Gravberg-1 hole is, thus, among the deepest drillings made in Europe. The second drill hole was more centrally located at *Stenberg* (No 5 in Fig. 2) and reached a similar depth (6529 m). Information about the Stenberg well was reported in Aldahan et al. (1993).

The two deep drillings aimed at, among others, to identify the nature of 3 seismic reflectors. It was found that they are related to the occurrence of swarms of dolerite sills with thicknesses ranging from 4 to 60 m. The sonic character is similar of deeper, not drilled reflectors, which then also can be interpreted as dolerite sills. These sills are important reference structures for the estimation of the amount of uplift at the two localities (the ring rise and the central part of the structure, respectively). From drill cuttings, about 15 such sills could be identified, however only the larger ones would be recognizable in the sonic data.

The borehole gravity data indicated a porosity of 2 to 3 % in the upper part of the central core.

Relatively fresh water was found down to at least 4 km depth and highly saline fluid was recovered below 5.5 km depth. The low resistivity modeled from magneto-telluric data indicates that a similar porosity related to fracturing of the rock also occurs below 6 km depth, which correlates well with the estimates made from the sonic log.

Only occasionally deflection cores were taken and most of the remaining rock material is a huge amount of drill chips. However a large body of data exists from measurements made in the Gravberg-1 drill hole. They are reported in the scientific summary report by Juhlin (1991). The geothermal gradient could be established to 16.1 Kkm^{-1} from the surface to 5.2 km depth. Shock effects (planar deformation features in quartz) were reported down to almost 3 km depth. Highly fractured rocks occur at the ring rise down to ca 1.2 km depth. A characteristic fracture zone spacing of 90 – 150 m prevails in the entire central uplift area as indicated by airborne VLF data (Henkel 1992) and by resistivity logging of the Gravberg-1 well (Juhlin 1991).

7 Discussion

There is no doubt that the Siljan Ring is an impact crater. The mere existence of an almost circular ring syncline with down-faulted cover rocks not anymore occurring in the surroundings of the structure is a strong argument. The finds of shock metamorphic minerals finally decided the issue after a long period of silence from the geo-science establishment. The large amount of data collected in connection with the Deep Gas Project has not yet been fully evaluated with regard to its potential to better describe the Siljan structure. Several issues are still unclear. The crater size and the depth of erosion are not well constrained. Substantial amounts of melt that would be expected for a large complex crater have so far not been found.

7.1 Crater Structure from Gravity and Magnetic Data

The deep gas idea was based on the misunderstanding of the concept of the transient cavity – it was thought to be a real opening within the entire crust. Furthermore, the general idea that impact effects generally are related to gravity lows, lead to the misinterpretation of the lithologically caused off-center gravity low in the NE part of the Siljan structure as representing impact fracturing and potential porosity in which mantle-derived gas could accumulate.

The gravity map does not show any features that, on a regional scale, conform to the impact structure. This implies that only minor density contrasts exist between the different structural units (the central uplift, the sedimentary ring, and the exterior). Thus, the average density of the sedimentary fill is similar to the density of the granitoids. It also implies that there is no major change in density of the plutonic rocks with depth in the central uplift, where lower, potentially denser rocks could have been uplifted as seen in the Vredefort structure (Henkel and Reimold 1998).

The magnetic data contain several vertical marker structures (the NNW-SSE striking dyke system) outside the Siljan structure. Similar (in strike direction) anomalies seem to re-occur, fragmented into shorter segments, also within the crystalline core. According to the modeling results illustrated in Melosh and Ivanov (1999), all pre-impact structures are rotated within the central uplift.

This rotation is marginal at large depth, is about 90° at intermediate depth and results in overturned, previously horizontal structures, at shallow levels. Previous vertical dykes would thus appear fragmented and horizontal (and provide seismic reflectors) in strike-perpendicular sectors and remain vertical but rotated in strike-parallel sectors (not providing seismic reflectors). Previous horizontal sills would appear as fragmented ring-shaped structures within the central uplift above the level of no vertical distortion.

Some of the radial and tangential magnetic dykes may however represent impact melt rocks similar to those seen in the crystalline core of the Vredefort dome (Henkel and Reimold 2002).

The nature of the increased remanence found in drill core measurements from the central uplift is still not settled. This increased remanence could be caused by shock dissociation of Fe-Mg-silicates (biotite in this case) as discussed in Feldman (1994) and overprinted with a thermo-remanence during the cooling after the impact. A study of these effects was made by Henkel and Reimold (2002) regarding the Vredefort structure. As this remanence appears first below ca. 240 m depth, the uppermost section may have suffered oxidation due to circulating meteoric water in the brecciated rocks of the central uplift, thereby changing magnetite to hematite, as observed in fracture zones in magnetic rocks (Henkel and Guzmán 1977).

7.2 Crater Structure – Comparison with Numerical Models

In the numerical modeling of Melosh and Ivanov (1999), a series of models of collapsed craters with increasing diameter are shown. The transition to peak ring structure of the central uplift seems to occur at ca. 30 km diameter. These models of collapsed complex craters can serve for an approximate scaling of eroded crater structures by using the structural markers that are included in their models. With these markers, scaling factors can be calculated for characteristic dimensions of features seen at depth in relation to the final collapsed crater (FC) and its radius (FR). Some of these features are:

1. Under the edge of the final (collapsed) crater a small but significant uplift remains from the uplifted rim of the original (excavated) crater down to considerable depth.
2. Ring subsidence starts well inside the edge of the FC and reaches a maximum under the ring syncline. This subsidence is gradually reduced with depth and is very small at a depth of ca. 0.5 FR. Most important is the

systematic inward dip of structures prevailing over large volumes in the outer part of the ring syncline. The synclinal axis is located at a similar radius over a rather large depth interval before it moves towards the crater center with depth. The maximum subsidence is ca. 0.2 FR.
3. The approximate level of no uplift occurs at a similar depth where the subsidence fades out. The uplift increases upwards under the center of the crater and reaches a height relative to the ring syncline of ca. 0.25 FR. The width of uplifted material is close to 0.5 FR at the base of the uplift and increases upwards to ca. 0.8 FR where the ring syncline is deepest. This width prevails over a rather large depth interval.
4. The peak ring rises above overturned structures and has collapsed over the uppermost stratigraphic levels from the excavated crater rim (including ejecta and pre-existing cover rocks). These lithologies are thickened, faulted and tangentially folded within the ring syncline resulting in at least one ring anticline in its upper parts. Under the entire central uplift a large region has steep (>45°) up-tilted, vertical and over-tilted structures.
5. The excavation reached down to ca. 0.12 FR from the original surface, roughly in the position of the peak ring of the collapsed crater.
6. The depth of the final morphologic crater is ca. 0.08 FR. The crater might become further flattened in later stages of its collapse (the models provide an illustration at a certain time of the crater evolution before its long term stability has been reached). Erosion must, therefore, cut deep into the surroundings before it reaches the ring syncline and the central uplift.

7.3 Crater Structure from Reflection Seismic Data

The reflection seismic data represent the at present best available information about the crater structure at depth. The data collected in the Siljan structure resolve reflectors down to ca. 18 km depth, and they extend from within the crystalline central core and beyond the Paleozoic ring to the north and east, but not into the undisturbed exterior of the crater (lower part of Fig. 10).

The reflection seismic patterns give a number of clues to both the crater size and the possible erosion level when compared to the above structural relations seen in numerical models of large collapsed craters. The results from reflection seismic surveys of the Siljan structure reveal the following relations:

1. Reflectors under and outside the ring syncline dip inward down to at least 8 km depth.

2. Reflectors under the outer part of the peak ring are upturned down to a depth ranging from 8 (in the west) to 6 km (in the north).
3. Reflectors are missing in a large portion of the central uplift.
4. At a depth of ca. 14 km below the central uplift, reflectors re-appear and seem to be least disturbed but tilted (up to the west and down to the north).
5. The width of the base of uptilted structures under the ring syncline is roughly 30 to 35 km.

The interpreted pattern of reflectors matches the patterns of marker structures obtained by modeling. Assuming the reflectors were originally horizontal, these data indicate a post impact tilt down to the NE. As the ring syncline has steep structures in this sector, and gentler outward dips are seen in the opposite direction (to the SW), oblique erosion has cut deeper into the structure to the SW (i.e,. below the level of vertically up-tilted structures). The width of the base of uplifted seismic structures corresponds roughly to a FC width of 75 to 85 km. This would result in a depth to no uplift of 18 to 21 km. The difference between the observed depth of no uplift of ca. 14 km and the corresponding model depth is 4 to 7 km and could represent a rough estimate of the possible erosion depth of the Siljan structure. The occurrence of a minimum of preserved cover rocks in a vertical position should be within 7 to 8 km depth.

The erosion with respect to undisturbed surroundings of the Siljan structure may thus amount to at least 4 km. The erosion over the central part of the crater is, however, less, ca. 3 km, as the newly formed crater also was a significant depression. The erosion level is most likely below the level of excavation.

7.4 Fracturing

The intensity of fracturing has been addressed in the interpretation of MT, airborne VLF measurements, and drill hole loggings and direct observations on outcrops. In all these data more intense fracturing was found within the crystalline central uplift. In the case of MT data, a decreased electric resistivity was interpreted as increased fracturing. In the case of airborne VLF data, the pattern of steep conductive zones (fractures) is distinctly different in the crystalline core compared to surrounding crystalline areas. The spacing between conductive zones decreases from > 1 km to ca. 300 m. In the drill hole loggings, wide zones with increased fracturing are attributed to low resistivity zones. From these data block dimensions ranging from 90 to 150 m can be derived.

In the data about fracture orientations presented in Hode et al. (2002), a tangential component in the fracture pattern can be seen in crystalline areas outside the sedimentary ring.

Vertical offsets of similar seismic reflectors in the ring syncline have been calculated in Juhlin and Pedersen (1987) and range from 30 m to almost 1 km. They have been interpreted as inward down-faulting related to the collapse of the crater wall. Some of these features are however positioned at the flank of the central uplift and are results of the collapse of the central rise. The extent of fracturing associated with the shock- and rarefraction wave in impact cratering in an isotropic medium is thought to follow a roughly hemispherical symmetry with respect to the explosion center. To this pattern the fracturing related to the collapse flow is added. Taken together, the collapsed crater represents an easier eroded volume. Fracturing had occurred also before the impact and certainly after the formation of the crater over 350 million years ago, as evidenced by the present regional distribution of lithologies.

In Kenkmann and von Dalwigk (2000), two localities with suggested radial transpression ridges limited by radial normal faults are described in the NW and SW sectors of the ring syncline, based on the bedrock geological map by Kresten et al. (1991). Due to the lack of outcrops, the evidence for raised crystalline basement in the sedimentary ring is based on finds of cuttings from wells. Together with these suggested transpression ridges, many localities around the inner part of the sedimentary ring with repeated lithologies indicate that stratigraphic thickening is associated with the collapse flow of the crater wall and the central rise.

The fracture system that developed in the Siljan crater is in parts invaded by hydrothermal deposits characterized by quartz and epidote veins in the crystalline basement and, calcite, fluorite and galena in the Paleozoic rocks (Hode et al. 2002).

8 Conclusions and Suggestions

There has bee a large gap in research activities about the Siljan structure after the closure of the Deep Gas Project in 1991. Traces of gas were found, but not of any economic interest. The large impact structure has recently again become of interest for study, this time for geothermal energy resources and for research focused on impact tectonics. This review has been prepared to encourage further studies of this extraordinary geological structure, which is easily accessible and where large amounts of data have been gathered by the efforts of the Deep Gas project and by the systematic mapping provided by the Geological Survey of Sweden.

8.1 Geophysical Data

The large geophysical database of the Siljan region contains a wealth of information about the impact structure and can be the basis for further studies of the effect of impact on the physical properties of the involved rocks. The documentation and reports of geoscientific data accumulated by the Deep Gas Project are available at Vattenfall (The Swedish State Power Board) Support AB in Stockholm. In the scientific summary report by Juhlin (1991), 17 sector reports and 41 publications concerning the results from the Deep Gas Project are listed. (Reports after 1991 have not been summarized in any publication). The airborne geophysical, gravity and petrophysical data are available from the Geological Survey of Sweden.

8.2 Geophysical Interpretations

The lack of sub-horizontal reflectors within the core of the central uplift (region C, Fig. 10) is due to the rotation into vertical structures, where a still existing reflective pattern not can be detected by vertical incidence measurements.

The termination of seismic reflectors can neither be related to the transient crater nor to the excavated crater. It is the result of strong upturning and dismembering of lithologies caused by the collapse flow.

The collapse flow is thus the likely reason for the observed patterns of seismic reflectors seen in the sections down to ca. 14 km depth. At this level reflectors appear undisturbed. The seismic surveys do however not extent far enough outside the structure to enable a comparison with conditions in the nearby undisturbed uppermost crust.

Fig. 14. Examples of lineations in granites of the central rise. The lineation is penetrative through the normally massive granite (**a** and **b**). The size of samples is ca. 10 cm.

a. In addition to the lineation also slightly stained curved surfaces are developed. These lineations have been interpreted in the geological map by Kresten et al. (1991) as shatter cone features.

b. An alternative interpretation is the formation of slickensides by the collapse flow. The two examples are from the NW sector of the peak ring.

c. Example of shatter cone developed in Siljan granite from the northern part of the central uplift.

Fig. 15. Fragmentation into large clasts of the sedimentary formations of the ring. Location is at the limestone quarry Kallholen in the NW part of the ring.

The extent at depth of mega-clasts of Paleozoic cover rocks and their orientation is of considerable interest to study. In this case more detailed gravity measurements may be needed. If these clasts mainly contain limestone they would have a generally higher density compared to the surrounding crystalline basement. Where sedimentary rocks occur, MT measurements or vertical electrical sounding (VES) measurements can be used as a depth constraint for the gravity interpretation

Additional MT measurements have the potential to better map the extent of impact induced fracturing at depth and should be extended over the entire structure and a significant part of its surroundings.

The extent of fracturing is of particular interest for geothermal energy prospecting. If the fractured rock volume at depth can be exploited as a heat exchange medium, very large thermal energy resources could be made available (Henkel 2003).

8.3
Further Magnetic Studies

From the aeromagnetic and rock magnetic data gathered so far, three interesting magnetic features are suggested to be studied further:

1. The fragmented NNW-SSE trending dyke swarm within the central uplift dome is a very interesting set of marker structures for the analysis of the mass flow related to the formation of the central rise. With magnetic modeling, the orientation of these dyke fragments can be assessed. This requires, however, more data about their magnetic susceptibility and remanent magnetization and excavations to obtain oriented samples of individual fragments.
2. The centrally located more magnetic granitoids should be studied to investigate the extent and character of impact induced remagnetization. Especially changes in the magnetization pattern and the occurrence of impact-induced production of magnetic carrier phases (like magnetite from shock dissociation of Fe-Mg-silicates) are of interest for further study.
3. Impact melt bodies can be expected to occur along the remains of the excavated crater interface along the outer part of the crystalline uplift and as injected dykes within the central uplift. The dyke- and sill-like magnetic anomalies that have a crater symmetric occurrence (radial and tangential, respectively), could thus represent impact melt rock dykes, like the one recently found near the center of the structure and similar those known from the core of the Vredefort structure (see Henkel and Reimold 1998 and references therein). This requires detailed localization with ground magnetic measurements and excavations at some localities, as there are neither boulders nor outcrops known that can be associated with these magnetic features.

8.4
Crater Diameter

From morphological studies, the present topographic diameter can be estimated at ca. 75 km. As the amount of erosion is only vaguely estimated, the original diameter cannot be constrained by surface data. The diameter of the crystalline core is substantial, about 30 km, and indicates, together with the seismic patterns seen in the central uplift, an even larger final crater of 75 to 85 km diameter.

8.5 Impact Tectonics

The characteristic penetrative lineations observed in the crystalline rocks of the central uplift (Fig. 14) have been interpreted as shatter cone features thought to be formed during the passage of the shock and rarefraction waves. Alternative interpretations are the mass flow during the excavation stage or the collapse flow during the modification stage of the crater. The lineations have a distinct orientation (with the exception of the centrally located multi-striated Proterozoic sandstone fragment) with frequent systematic offsets and could be used for the mapping of flow directions and the relative sense of displacement. The limestone quarries within the sedimentary ring are excellent places for the study of the final stages of the collapse flow. An example is the Kallholen quarry (Fig. 15) in the NW part of the ring, where large (several tens of meters) blocks of limestone are assembled in a mega-breccia. Between the large clasts, fractures are occasionally filled with microbrecciated sediments.

Acknowledgements

The permission by the Geological Survey of Sweden to publish the geophysical maps is acknowledged. The authors are grateful for the many critical and constructive comments by the reviewers.

References

Abels A, Plado J, Pesonen LJ, Lehtinen M (2002) The impact cratering record of Fennoscandia - a close look at the database. In: Plado J, Pesonen LJ (eds) Impacts in Precambrian shields. Berlin: Springer 1-58

Aldahan A (1990) Alteration and mass transfer in cataclasites and mylonites in 6.6 km of granitic crust at the Siljan impact structure, central Sweden. Contributions to Mineralogy and Petrology 105: 662-676

Aldahan A (1991) Albitized feldspars in granite and their bearing on the recognition of diagenetic vs detrital origin of albitized feldspar in sandstone. Bulletin of the Geological Institute. Uppsala 16: 47-60

Aldahan A, Norell B, Collini B (1993) Stenberg-1 deep well – Geology, Geochemistry and Geophysics. Dala Djupgas Production AB, 200 pp

Bottomley RJ, York D, Grieve RAF (1978) ^{40}Ar-^{39}Ar ages of Scandinavian impact structures. Mien and Siljan. Contributions to Mineralogy and Petrology 68: 79-84

Collini B (1988) Geological setting of the Siljan ring structure. In Bodén A, Eriksson K (eds) Deep Drilling in Crystalline Bedrock Vol 1 The Deep Gas Drilling in the Siljan Impact Structure, Sweden and Astroblemes, pp349-354, Springer Berlin

Dyrelius D (1990) Gravity field analysis in the Siljan ring area. Report, Vattenfall U(G) 1990/48

Fredriksson K, Wickman FE (1963) Meteorites. Svensk Naturvetenkap 1963, 121-157

Feldman VI (1994) The conditions of shock metamorphism. In: Dressler, BO, Grieve, RAF, Sharpton VL (eds), Large Meteorite Impacts and Planetary Evolution. Geological Society of America Special Paper 293: 121-132

Gaal G, Gorbatschev R (1987) An outline of the Precambrian evolution of the Baltic Shield. Precambrian Research 35:15-52.

Gold T, Soter S (1980) The deep gas hypothesis. Scientific American 242:154-161

Grieve RAF (1988) The formation of large impact structures and constraints on the nature of Siljan. In: Bodén A, Eriksson K (eds): Deep Drilling in Crystalline Bedrock, Volume 1: The Deep Gas Drilling in the Siljan Impact Structure, Sweden and Astroblemes, pp 328-348, Springer, Berlin

Henkel H (1992) Geophysical aspects of meteorite impact craters in eroded shield environment, with special emphasis on electric resistivity. Tectonophysics 216: 63-89

Henkel H (2003) The Björkö geothermal energy project: Norges Geologiske Undersökelse Bulletin 439: 45-50

Henkel H, Guzmán M (1977) Magnetic features of fracture zones. Geoexploration 15: 173-181

Henkel H, Reimold WU (1997) Integrated gravity and magnetic modeling of the Vredefort impact structure – reinterpretation of the Witwatersrand basin as the erosional remnant of an impact basin. Department of Geodesy and Photogrammetry, Royal Institute of Technology, Stockholm. TRITA GEOFOTO 1997: 14, 90 pp

Henkel H, Reimold WU (1998) Integrated geophysical modeling of a giant, complex impact structure: Anatomy of the Vredefort Structure, South Africa. Tectonophysics 287: 1-20

Henkel H, Reimold WU (2002) Magnetic model of the central uplift of the Vredefort impact structure, South Africa. Journal of Applied Geophysics 51: 43-62

Hjelmquist S (1966) Beskrivning till berggrundskartan över Kopparbergs län. Geological Survey of Sweden Ca 40

Hode T, von Dalwigk I, Broman C (2002) A hydrothermal system associated with the Siljan impact structure, Sweden – Implications for the search for fossil life on Mars. Astrobiology 3: 271-289

Juhlin C (1991) Scientific summary report of the Deep Gas Drilling Project in the Siljan ring impact structure. Vattenfall U(G)1991/4, 257 pp

Juhlin C, Pedersen LB (1987) Reflection seismic investigations of the Siljan impact structure, Sweden. Journal of Geophysical Research 92:14113-14122

Juhlin C, Pedersen LB (1993) Further constraints on the formation of the Siljan impact crater from seismic reflection studies. Geologiska Föreningens i Stockholm Förhandlingar 115: 151-158

Kenkmann T, von Dalwigk I (2000) Radial transpression ridges: A new structural feature of complex impact craters. Meteoritics and Planetary Science 35: 1189-1202

Kresten P, Aaro S, Karis L (1991) Berggrundskartorna Ai Nr 46, 48, 50, and 51. Sveriges Geologiska Undersökning

Masaitis VL, Maschak MS, Raikhlin AI, Selinovskaya TV, Shafranovsky GI (eds) (1998) Diamond bearing impactites of the Popigai astrobleme. Karpinsky All-Russia Geological Research Institute VSEGEI Press, 179 pp

Melosh HJ, Ivanov B (1999) Impact crater collapse. Annual Reviews of Earth and Planetary Science 27: 385-415

Pedersen LB, Rasmussen TM, Dyrelius D (1990) Construction of component maps from aeromagnetic total field anomaly maps, Geophysical Prospecting 38: 795-804

Ramseyer K, Aldahan A, Collini B, Landström O (1992) Petrological modifications in granitic rocks from the Siljan impact structure: evidence from cathodoluminescence. Tectonophysics 216: 195-204

Reimold U, Kelley SP, Sherlock S, Henkel H, Koeberl C (2004) Laser argon dating of melt breccias from the Siljan impact structure, Sweden – Implications for possible relationship to late Devonian extinction events. [abs.] Lunar and Planetary Science 35 abs. 1480, CD-ROM

Rondot J (1975) Comparaison entre les astroblémes de Siljan, Suède, et de Charlevoix, Quebec. Uppsala Universitet Geologiska Institutionens Bulletin 6: 85-92

Svenson NB (1973) Shatter cones from the Siljan structure, central Sweden. Geologiska Föreningens i Stockholm Förhandlingar 110: 418-419

Stolpe M (1872) Om Siljanstraktens sandstenar. Geologiska Föreningens i Stockholm Förhandlingar 1 (2)

Thorslund P, Collini B (1980) An Eocambrian sequence in the Siljan Ring structure. Geologiska Föreningens i Stockholm Förhandlingar 102: 188

Wickman FE, Blomqvist NG, Geijer P, Parwel A, von Ubisch H, Welin E (1962) Isotopic constitution of ore lead in Sweden. Arkiv för Mineralogi och Geologi 3:11

Zang P, Rasmussen TM, Pedersen LB (1988) Electric resistivity structure of the Siljan impact region. Journal of Geophysical Research 93:6485-6501

Åström K, Lund C-E (1994) Thin superficial layer and lateral heterogeneities in southern Sweden using short-period Raleigh wave dispersion. Geophysical Journal International 118:231-244

Mjølnir Crater as a Result of Oblique Impact: Asymmetry Evidence Constrains Impact Direction and Angle

Filippos Tsikalas

Department of Geosciences, University of Oslo, P.O. Box 1047 Blindern, N-0316 Oslo, Norway (filippos.tsikalas@geo.uio.no)

Abstract. The 40-km-diameter Mjølnir crater is proposed to have resulted from an oblique impact from the south/southwest direction and at a ~45° (possibly 30°-45°) angle from the horizontal. This is substantiated by several diagnostic structural and geophysical signatures related to obliquity and revealed through detailed re-assessment of Mjølnir's well-established structure, morphology, and gravity and seismic velocity anomalies. The diagnostic signatures include: (1) a dominant N-S/NNE-SSW crater diameter elongation, (2) a consistent northward asymmetry both in the crater radius and the shallow part of the impact-induced seismic disturbance, (3) a peak ring breached towards the N/NE, (4) an annular gravity low with a horseshoe-shape open towards the NE, (5) a transient cavity maximum-depth offset of 2 to 2.5 km towards south-southwest from the geometric crater center, combined with a similar structural uplift lateral offset towards the south, a similar central gravity high offset towards southwest, and an elongated traveltime central anomaly offset towards WSW. The oblique Mjølnir impact most probably generated a down-range sector/corridor of thicker ejecta deposits and faster travelling tsunami-waves, triggering short-term regional perturbations that are probably intensified within and adjacent to this sector.

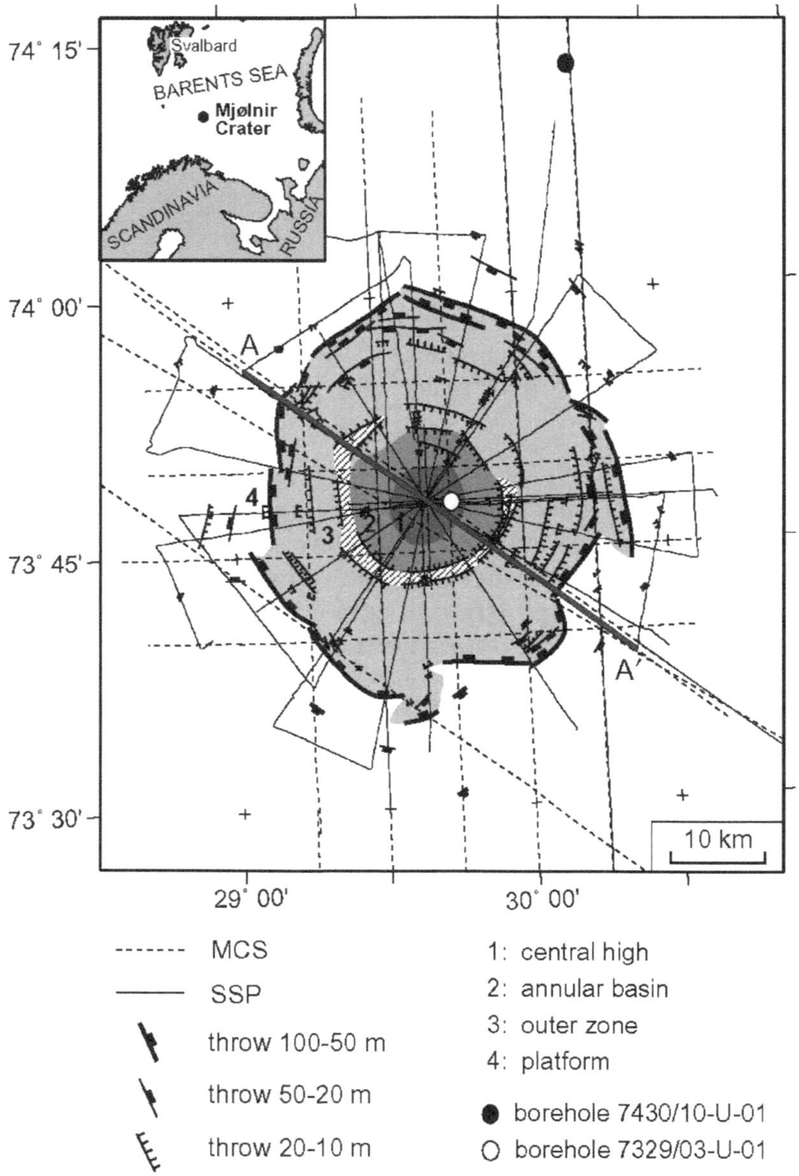

Fig. 1. Seismic reflection profiles superimposed on the Mjølnir structure defined by its radial zonation boundaries and the impact-induced structural elements. MCS, conventional multichannel profiles; SSP, shallow multichannel and shallow high-resolution single-channel profiles. Hatched-raster denotes clear definition of the raised peak ring.

1 Introduction

The 40-km-diameter Mjølnir crater is a well-established marine impact crater in the central Barents Sea (Fig. 1) (e.g., Tsikalas et al. 1999). Both geophysical and geological data unequivocally substantiate a meteorite bolide impact at ~142 Ma into an epicontinental basin with 300-500 m paleo-water depth (Gudlaugsson 1993; Dypvik et al. 1996; Smelror et al. 2001; Tsikalas et al. 2002a). In particular, a total of ~2100 km of seismic reflection profiles (Fig. 1) clearly image the impact-related and post-impact structure and stratigraphy (Fig. 2) (Tsikalas et al. 1998a, b). In addition, free-air gravity and seismic velocity anomalies exhibit a close correspondence to the impact-

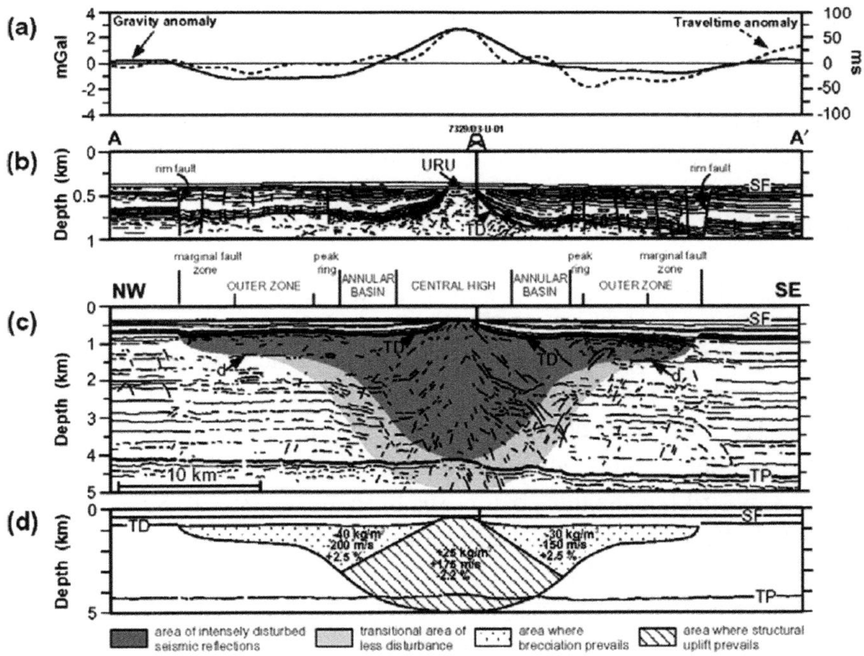

Fig. 2. Geophysical type section along profile AA′ in Fig. 1. (**a**) observed free-air gravity and seismic traveltime anomalies, (**b**) interpreted high-resolution single-channel profile, (**c**) interpreted multi-channel profile, and (**d**) impact crater model with calculated physical property distribution. SF, sea floor; URU, late Cenozoic upper regional unconformity; TD (impact horizon), the first continuous reflector above the disturbed seismic reflections; TP, Top Permian; (**d**) low-angle décollement. The crater model geometry in (**d**) is corrected for regional tilt, and the modelled density contrasts are given in kg/m^3, the seismic velocities in m/s, and the porosity anomalies in %.

induced structure and physical property distribution (Fig. 2) (Tsikalas et al. 1998c). Two shallow boreholes, one near the center and another ~30 km from the crater periphery (Fig. 1) offer a detailed seismic stratigraphic correlation and confirmed the impact origin of the structure. The boreholes revealed brecciated sediments, including shocked quartz containing abundant planar fractures/deformation features (Sandbakken 2002), and a prominent ejecta layer with an iridium abundance peak of about 1,000 ppt, about 15 times above the background level, and shocked quartz grains (Dypvik et al. 1996; Dypvik and Attrep 1999).

The physical impact resulted in an extensive disturbance both in the sedimentary target and the water column, as: (1) a 850-1400 km^3 seismically disturbed volume of target rocks at the impact-site (Fig. 2) is directly-related to the impact-driven processes of brecciation, excavation, structural uplift, gravitational collapse, and infilling (Tsikalas et al. 1998a); (2) a 180-230 km^3 excavated/ejected material volume was displaced from the impact-site and re-deposited in the near-field vicinity (Tsikalas et al. 1998a; Shuvalov et al. 2002); and (3) collapse of the impact-induced water-cavity is expected to have given rise to large-amplitude tsunami waves (Tsikalas et al. 1998a; Shuvalov et al. 2002). Dissipation of the energy released during the Mjølnir impact is directly related to the ejected material volume and the tsunami wave generation, which are, in turn, connected to possible short-term, near field perturbations affecting the Barents Sea region and, possibly, adjacent areas in the Arctic.

Although most qualitative and quantitative impact models assume, for simplicity, axial symmetry created by a near-vertical collision, no meteorite is expected to strike a planetary surface exactly vertically as the probability for near-vertical (85°-90°) and grazing impacts (0°-5°) is almost zero (Shoemaker 1962). The most probable angle of impact of a randomly incident projectile is 45° (Shoemaker 1962; Shoemaker et al. 1990), with an estimated 25% of impactors striking at angles below 30° from the horizontal (Schultz and Gault 1990). Following the above estimates, the Mjølnir impact did not, most probably, occur at vertical incidence. In this study, based on the established Mjølnir structure, morphology, and gravity and seismic velocity signatures a search for evidence revealing the impact direction and angle is carried out. Such well-constrained parameters are important for setting boundary conditions to models of ejecta and tsunami-wave distribution, and thus refining the geographic distribution of possible impact-induced regional perturbations and environmental stress.

2
Mjølnir as an Oblique Impact Event

Numerical simulations of meteorite impacts (e.g., O'Keefe and Ahrens 1985; Pierazzo and Melosh 1999) have shown that the shock wave produced in an oblique impact is roughly hemispherical in spite of the obliquity, although it is weaker than the shock wave produced by a vertical impact at the same velocity. This is in accordance with the cratering record, which reveals that the circularity of craters on different planets is a first-order morphological characteristic (Spudis 1993; Alexopoulos and McKinnon 1994; Grieve and Pesonen 1996). The circularity underscores the first-order symmetry created by large impacts down to very low angles of incidence as a result of near-symmetry in the propagation of the shock wave front and symmetry of gravitational collapse during the modification stage (e.g., Melosh 1989), although the far-field shock stress distribution may retain asymmetry in terms of intensity (Dahl and Schultz 2000, 2001). Only recently, however, have a number of detailed diagnostic structural and geophysical signatures of oblique terrestrial and planetary impacts, based on interior crater features, been appreciated; asymmetries on crater morphology and shape, on central high/peak ring morphology and placement, and on nature and distribution of ejecta (e.g., Manson crater, Schultz and Anderson 1996; Chicxulub crater, Schultz and D'Hondt 1996; Toms Canyon crater, Poag and Poppe 1998; cratering record on Venus, Schultz 1992).

Here, a re-assessment of the established structural and geophysical Mjølnir signatures is carried out searching for evidence that point out to an impact obliquity.

2.1
Elongated Crater Diameter

The seismic reflection database has revealed in detail Mjølnir's distinct radial zonation pattern of the impact-induced relief (Fig. 1 and 2) composed of a 12 km-wide complex outer zone, including a marginal fault zone and a modestly elevated peak ring, a 4-km-wide annular depression, and an uplifted central high, 8 km in diameter. Sharp boundary faults form a 100 to 150 m-high, near-circular rim wall and separate highly deformed strata within the crater from intact platform strata (Fig. 1 and 2). The most representative diameter for the near-circular Mjølnir crater periphery was found to be 40 km (Tsikalas et al. 1998b).

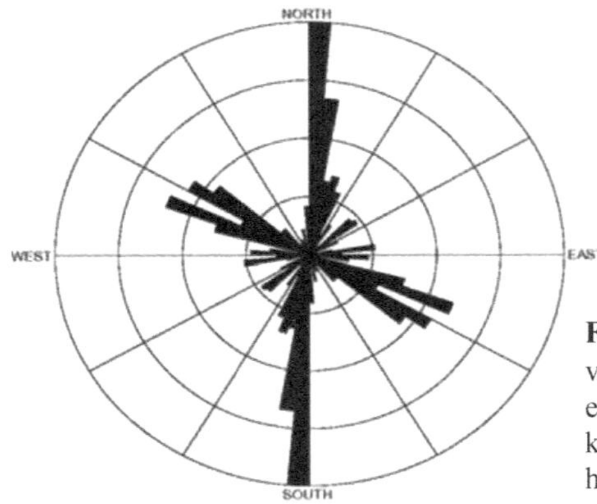

Fig. 3. Histogram of the crater diameter versus azimuth at 1° intervals around the entire Mjølnir periphery normalized to the 40 km average crater diameter. The normalized histogram lengths range from 0 to 1.

The well-established Mjølnir crater structure and morphology (Fig. 1) is further used as the basis for detailed measurements on the crater diameter. In particular, the crater diameter is estimated at 1° intervals for a 360° circle at Mjølnir's periphery, and after normalization to the 40 km average crater diameter the residuals are plotted versus the azimuth direction. The resulting histogram (Fig. 3) reveals that there is a dominant elongation of the structure in a N-S/NNE-SSW direction, and a second, smaller, elongation trend at an ESE-WNW direction.

2.2 Seismic Disturbance Asymmetry

The regional grid of multi-channel seismic profiles combined with the fan-like geometry of the single-channel seismic surveys provide a dense coverage of the structure at all directions (Fig. 1). Figure 4 shows three deep multi-channel seismic profiles stacked with three shallow single-channel profiles. All profiles cross the central high and are aligned along the geometric crater center. Note that the latter is defined as the center position where a 40 km diameter circle best fits the near-circular Mjølnir crater periphery. For each profile the ratio of the two radii, on the left and right side, between the crater center and the rim faults is estimated (Fig. 4, Table 1). If the crater radii on each profile were symmetrical, a ratio of ~1.0 is to be expected. The analysis provides a crater radius asymmetry factor for each profile and reveals a consistent asymmetry towards the northward direction (NW/N/NE) in the order of ~1.16, with a range of 1.33-1.07 (Table 1).

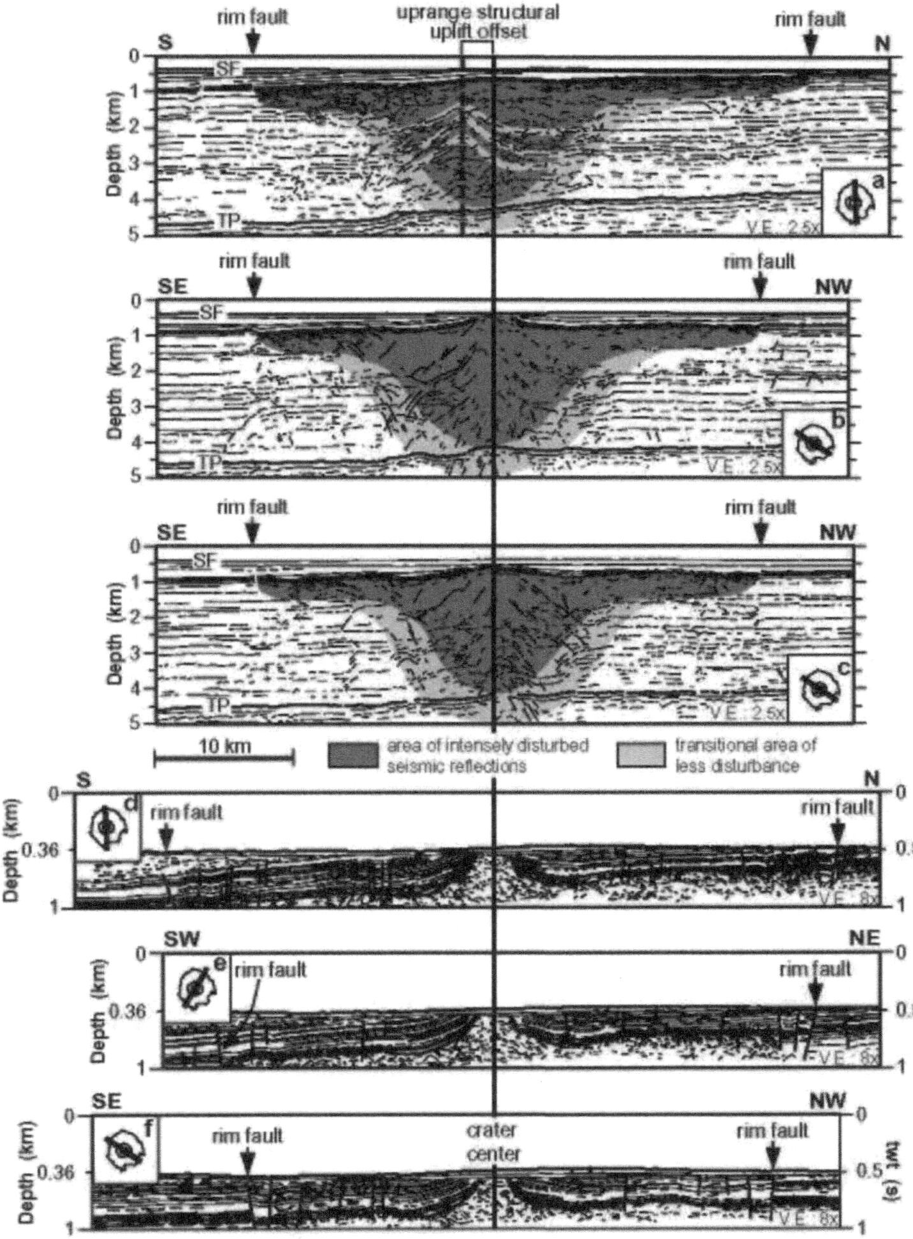

Fig. 4. Interpreted multi-channel (top three) and single-channel (bottom three) seismic reflection profiles stacked along the geometric crater center. In profile (a) selected reflector segments are connected to visualize the clear up-range structural uplift offset.

Table 1. Crater radii asymmetry factors (dimensionless) defined as the ratio of the two radii, on the left and right side, between the geometric crater center and the rim faults of each of the profiles (a-f) illustrated on **Fig. 4**

Profile (Fig. 4)	Asymmetry factor (direction)
a	1.33 (N)
b	1.12 (NW)
c	1.14 (NW)
d	1.07 (N)
e	1.17 (NE)
f	1.15 (NW)

The impact-induced seismic disturbance at Mjølnir has a parabolic bowl-shape at the center of the structure and turns into a shallow broad-brim towards the periphery (Fig. 2 and 4). This distinct shape evolved during gravitational collapse of the transient cavity wall and progressive outward expansion of the crater by inwards dipping fault-blocks floored on low-angle décollement surfaces (Tsikalas et al. 1998a). The crater radius asymmetry (Table 1) is directly translated to a similar asymmetry in the lateral extend of the shallow broad-brim part of the impact-induced seismic disturbance (Fig. 4). It also appears that the shallow part of the seismic disturbance is not only elongated but also slightly shallower in the same northward-direction (Fig. 4, profile a). Furthermore, the decompacted-thickness contour map of the intensely disturbed zone in Figure 5 is directly related to the transient cavity bounded approximately at the perimeter of the annular basin and reaches ~5 km in depth. It appears that the transient cavity maximum depth is offset by 2-2.5 km to the south-southwest relative to the geometric crater center (Fig. 5).

2.3 Peak-ring Character

The outer perimeter of the annular basin, thought to represent the maximum possible diameter of the short-lived transient cavity during the excavation and modification stages (Tsikalas et al. 1998a; Shuvalov et al. 2002), as well as the subfloor compression crater limit (Schultz 1992; Schultz and Gault 1992), exhibits a slightly raised relief (Fig. 1 and 2). Although irregular in shape, and varying in width from 1 to 3 km, the raised relief gives the impression of a

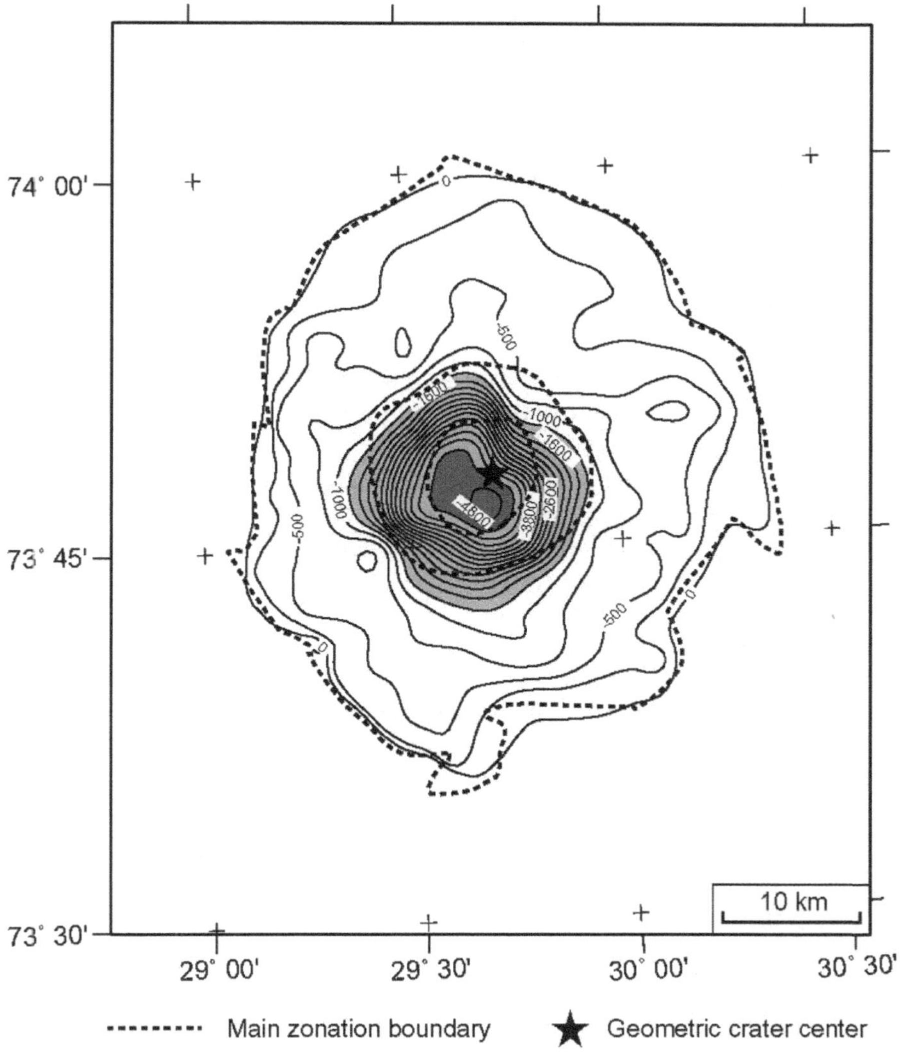

Fig. 5. Morphological characteristics of the decompacted impact-induced intensely disturbed zone (cfr. **Fig. 2** and **4**; dark-grey raster) derived through mapping of the entire seismic reflection dataset. Light-shading denotes the central bowl-shape part of the disturbance approximately at the limit of the annular basin; dark-shading denotes the transient cavity maximum depth. Contour interval in meters, but not uniform.

subdued ring structure, similar to those typically found in peak ring craters, and was thus referred to as a peak ring feature by Tsikalas et al. (1998b).

In the greatest detail, the peak ring at Mjølnir is clearly defined as a raised near-arcuate feature delineated by opposite dipping faults with 10-30 m throws (Fig. 1). This characteristic shape becomes less clear in the N- and

NE-directions where the raised relief is breached and the peak ring remains open, being replaced by faults facing the crater center (Fig. 1).

2.4 Offsets in Brecciation and Structural Uplift

Impact craters on sedimentary targets have the advantage, in comparison with similar structures on crystalline targets, that the regular, pre-impact stratification of these targets provides reference horizons against which the impact-induced structures can be identified and mapped by seismic reflection studies. Such studies have provided an effective means of mapping the large-scale geometrical structure at depth with a high-degree of horizontal and vertical resolution (e.g., Morgan and Warner 1999).

At Mjølnir, seismic mapping and analysis of the deeper structure levels in combination with gravity and seismic-velocity modelling, and with detailed numerical simulations, have provided greater insight into several cratering processes, such as brecciation and excavation, gravitational collapse of the transient crater, and structural uplift (Tsikalas et al. 1998a-c; Shuvalov et al. 2002). The integrated analysis supports the lateral differentiation of the Mjølnir seismic disturbance into a central uplift and a peripheral region (Fig. 2). The primary effect of the impact event is an impact-induced porosity increase due to extensive fracturing and brecciation (Pilkington and Grieve 1992; Tsikalas et al. 2002b). Subsequent modification of the density field takes place as a result of mass transport during gravitational collapse and structural uplift of the crater floor displacing deep, denser strata to shallower levels beneath the central structure (Fig. 2) (Tsikalas et al. 1998a, c; Shuvalov et al. 2002).

The models for Mjølnir are directly supported by the observed free-air gravity and seismic velocity anomalies. In particular, the residual free-air gravity field exhibits a circular anomaly over the structure (Fig. 6a). The anomaly is divided into an annular low, with an outer diameter of 45 km, attaining minimum values of -1.5 mGal over the periphery, and a central 14 km-wide gravity high, with a maximum value of +2.5 mGal (Tsikalas et al. 1998c). It appears that the 0-mGal gravity anomaly contour exhibits a very distinct elongated-shape in the SW-NE direction (Fig. 6a). In addition, the annular gravity low (<-0.5 mGal), which is directly connected with the region of most intense fracturing and brecciation, closely resembles a U-shaped central pit open to the northeast (Fig. 6a).

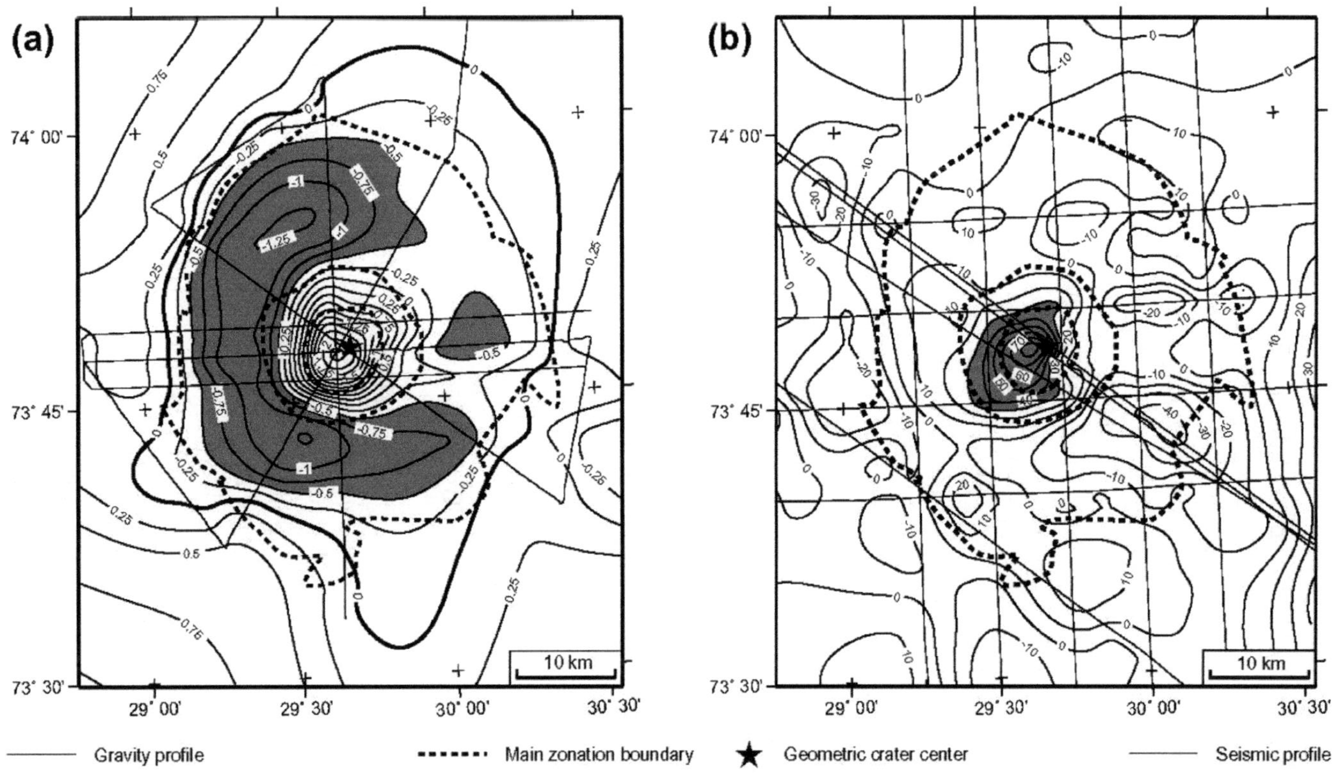

Fig. 6. (**a,** left) Residual free-air gravity anomaly map with 0.25 mGal contour interval, (**b,** right) Residual traveltime anomaly map at the Top Permian reflector (cfr. **Fig. 2**) with 10 ms contour interval (modified extensively from Tsikalas et al. 1998c).

The deep seismic profiles provide evidence of upward bending reflector segments beneath the central high and the annular basin, indicating elevation of deep strata to shallower levels (Fig. 4). By measuring the difference in depth between the extrapolated top of selected upward bending reflector segments beneath the central high and their most likely equivalent subhorizontal interfaces the structural uplift was estimated to be 1.0 -. 1.5 km and when decompacted ~1.5-2.0 km, fitting the theoretical expectations for the Mjølnir dimensions (Fig. 4) (Tsikalas et al. 1998a). The north-south profile in Figure 4 (profile a) clearly shows a maximum structural uplift lateral offset of 2 - 2.5 km towards the south from the geometric crater center. Similarly, the gravity central peak which corresponds to the maximum structural uplift (Tsikalas et al. 1998c) is laterally offset by ~1.5 - 2 km to the southwest from the geometric crater center (Fig. 6a). Furthermore, fracturing and brecciation are expected to induce changes in the seismic velocity expressed as pull-up or pull-down effects on continuous reflectors below the impact structure. Indeed, the seismic profiles reveal a small pull-up of the high-amplitude, originally planar Top Permian reflector beneath the structure (Fig. 2). The mapped traveltime anomaly is 16 km in diameter and rises to +80 ms beneath the central crater (Fig. 6b), corresponding to a +175 m/s seismic velocity anomaly (Fig. 2) (Tsikalas et al. 1998c). It appears that the central traveltime anomaly has a slightly elongated shape in the SW-NE direction and its top is slightly offset by ~2 km to the WSW from the geometric crater center (Fig. 6b).

3 Impact Direction and Angle

The Mjølnir crater lies on the Bjarmeland Platform in the central Barents Sea, which represents a stable sedimentary platform since Late Paleozoic times without evidence for considerable tectonism during its evolution (Faleide et al. 1993; Gudlaugsson et al. 1998). At the time of impact, ~142 Ma (Jurassic-Cretaceous transition), the region contained upper Paleozoic strata, mainly carbonates and evaporites, overlain by 4-5 km thick Mesozoic siliciclastic marine sediments (Worsley et al. 1988; Gabrielsen et al. 1990). As indicated by the regional sequence thicknesses of Triassic and Jurassic strata (Fig. 4) the Bjarmeland Platform at the time of impact exhibited a minor dip of 0.5° towards the south, and the pre-impact Mesozoic successions appear sub-horizontal without presence of any angular strata or angular unconformities (Gabrielsen et al. 1990; Gudlaugsson et al. 1998). Furthermore, the post-impact sedimentary succession initially filled the subtle impact-generated crater relief on the sea bottom and subsequently deposited uniformly above the entire structure (Tsikalas et al. 1998b). Therefore, I exclude the possibility that the observed asymmetric structural and geophysical patterns may be caused by any pre-impact irregularities in the regional sea-bottom topography or sequence stratigraphic relations of older strata, as well as by post-impact deformation.

The detailed analysis in this study has revealed several evidence that, when combined with the established diagnostic structural and geophysical asymmetries of oblique impacts (e.g, Schultz 1992; Schultz and Anderson 1996; Sugita and Schultz 2001), can be directly related to impact direction and angle estimates for the Mjølnir impact. First, the N-S/NNE-SSW elongated crater diameter translates into impact from a similar northward or southward direction without, however, definition of the explicit azimuth. The same impact direction is further supported by the observed slight elongation of crater diameter in the ESE-WNW direction as such elongation, approximately traverse to the impact trajectory, has been reported for both the Manson (Schultz and Anderson 1996) and Chicxulub (Schultz and D'Hondt 1996) craters. A similar northward or southward impact direction is also supported by the elongated shape of the 0-mGal gravity anomaly contour. Second, the asymmetry and elongation in both the crater radius and the shallow broad-brim part of the impact-induced seismic disturbance towards a northward direction is the first evidence for an impact from the southward direction. Third, the breached and open towards N/NE peak ring in combination with the horseshoe shape of the annular gravity low, which remains open in NE, further support a southwestern impact direction. The shape of the annular gravity low is connected with intense up-range brecciation and a down-range shallower excavation. Fourth, the transient cavity maximum-depth lateral offset by ~2 to 2.5 km towards the south-southwest from the geometric crater center, combined with a similar structural uplift lateral offset towards the south, a similar central gravity high offset towards southwest, and an elongated traveltime central anomaly offset towards WSW, all demonstrate an up-range highest elevation offset and, thus, support an impact direction from south-southwest.

Two different approaches, a quantitative and a qualitative one, were followed in order to find the angle, measured from the horizontal, of the Mjølnir impact. The quantitative approach is based on laboratory experiments and crater studies on Venus relating the effect of impact angle on the ratio of crater diameter to central high/peak ring diameter (Schultz and Gault 1990; Schultz 1992), as well as insights from structural and shock asymmetries (Schultz and Anderson 1996; Dahl and Schultz 2000, 2001). A possible impact angle ranging 30°-45°, is found to be representative for the Mjølnir dimensions. The qualitative approach uses the results obtained by the laboratory experiments of Gault and Wedekind (1978) and the numerical computations of O'Keefe and Ahrens (1985) for oblique impacts at various angles. For all resulting crater geometries, crater radii asymmetry factors are calculated by fitting a parabolic-shaped transient cavity and the results are extrapolated to the full range of impact angles. A possible impact angle of ~45° was estimated, ranging 30°-60°, based on the crater radius/seismic disturbance asymmetry factors for Mjølnir (Table 1).

Planetary crater studies and laboratory experiments showed that the oblique penetration phase of the impactor appears to be at least in part preserved within the central crater: the up-range offset of transient cavity, structural uplift and central gravity peak, and the down-range breach of the peak-ring zone are consistent with the extended region of energy transfer created during the early penetration stages by an oblique impact (Schultz and Gault 1990; Schultz 1992). A similar down-range shallower excavation and up-range offset of the central gravity high connected with oblique impacts are revealed at detailed gravity studies on lunar complex impact craters (Zuber et al. 1995). However, based on a comprehensive statistical approach for Venusian oblique impact craters, Ekholm and Melosh (2001) argued recently that the preserved transient cavity asymmetry in the final crater, which is observed in simple craters, does not necessarily hold in complex craters. Although the clarity of this issue is still debatable, the various lines of evidence presented in this study for Mjølnir all point towards a similar impact direction from the south/southwest.

4
Mjølnir Impact Obliquity Constrains Models for Near-Field Perturbations

Numerical simulations and experimental analogues have shown that obliquity is accompanied by less energy transfer from the projectile to the target (e.g., Gault and Wedekind 1978; Hayhurst et al. 1995; Schultz 1996; Burchell and Mackay 1998; Ivanov and Artemieva 2002). The Mjølnir energy release estimates of Tsikalas et al. (1998a) were made considering an elevation impact angle of 45° based, at that time, on well-known probability arguments (Shoemaker 1962; Shoemaker et al. 1990). The energy release was estimated to be in the order of 16×10^{20} J (range of $2.4\text{-}53 \times 10^{20}$ J), translating into 3.8×10^{5} megatons TNT equivalent (range of 5.7×10^{4} - 1.2×10^{6}) (Tsikalas et al. 1998a). An oblique impact at a ~45° (possibly 30°-45°) angle, as estimated in this study, is expected to have resulted in very similar energy release.

Energy release dissipation at the proposed trajectory and angle for the Mjølnir impact may have a direct consequence on the distribution of proximal ejecta and tsunami-waves following the cessation of the impact-related processes at the impact-site. This is because the oblique impact most probably have created a down-range sector/corridor of thicker ejecta deposits and greater water column disturbance (Fig. 7). Such a sector/corridor may have been responsible for a geographic variation of short-term perturbations/environmental stress magnitude on the Barents Sea and adjacent regions, as it may have intensified the stress at a specific location and left the others almost unaffected.

4.1 Nature and Distribution of Proximal Ejecta

Theoretically, the volume of material displaced from the crater equals the volume of excavated cavity (Croft 1985; Melosh 1989). Geophysical observations constrain the volume of a parabolic excavated cavity to 180 km^3 (Tsikalas et al. 1998a), whereas numerical simulations indicate a volume of ~230 km^3 (Shuvalov et al. 2002). The ejecta layer is expected to be thickest close to the crater rim, decreasing rapidly with distance from the crater center (Tsikalas et al. 1998a; Shuvalov et al. 2002). Accepting an oblique Mjølnir impact from the south/southwest, ejecta iso-thickness contours will probably not be circular around the crater site, but rather elongated towards the north/northeastern direction (Fig. 7).

Borehole evidence for the Manson crater case (e.g., Anderson et al. 1996) substantiate that ejecta deposits are thinnest in the up-range direction and that only the top target layers are ejected due to shallower excavation as a result of oblique impact (Schultz and Anderson 1996). At Mjølnir, possible shallower penetration and excavation may be the reason for the small discrepancies in the estimated transient cavity depth of 4-5 km versus 6 km and the excavated/ejected volume of 180 km^3 versus 230 km^3, based on geophysical observations and numerical simulations, respectively (Tsikalas et al. 1998a; Shuvalov et al. 2002). Note that numerical simulations have for simplicity considered a vertical incidence for the Mjølnir impact (Shuvalov et al. 2002). Furthermore, the fact that the shocked quartz grains and the iridium anomaly peak are located at the base and top, respectively, of the 80-cm-thick ejecta deposit at borehole 7430/10-U-01 (Fig. 1) (Dypvik et al. 1996) may have

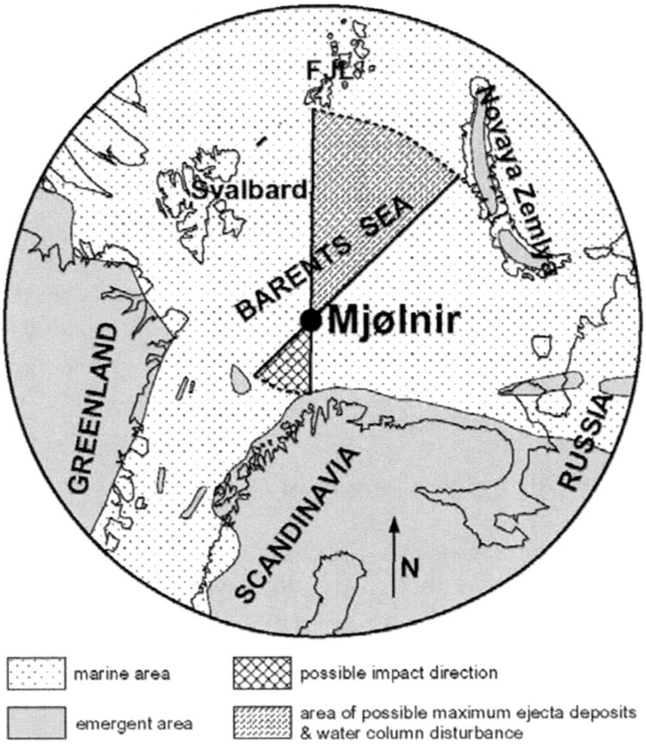

Fig. 7. Mjølnir impact location with possible range of impact direction azimuth and downrange area of possible maximum ejecta deposits and water column disturbance, shown at a ~142 Ma plate reconstruction based on Lawver et al. (1999) and overlaid on a simplified paleogeographic synthesis based on Brekke et al. (2001) approximately at the time of impact. FJL, Franz Josef Land.

possibly resulted from a multistage ejecta emplacement, similar to those attributed to oblique impacts as revealed by laboratory experiments and planetary impact crater studies (Schultz and Gault 1990; Schultz 1992; Schultz and D'Hondt 1996).

Magnetic modelling has indicated that only low quantities of dispersed-character melts localized in the crater periphery may have been produced during the water-covered, sedimentary target Mjølnir impact (Tsikalas et al. 1998c). A similar absence of considerable impact glass and melts at Manson crater has been also attributed to the sedimentary target and, more importantly, to the obliquity of the impact as this results in shallower target penetration and less direct energy transfer from the projectile to the target (Izett et al. 1993; Schultz and Anderson 1996). Recent geochemical analyses of samples from the Mjølnir central crater core (borehole 7329/03-U-01, Fig. 1) have shown absence of (Cr, Co, Ni) or weak (Ir) siderophile-element anomalies (Sandbakken 2002). This translates into a low abundance or total absence of projectile material in the crater itself, being consistent with oblique impact models, where a large fraction of the projectile material retains a net down-range motion and fragments of it may survive the impact, due to higher ejection velocity and lower shock compression, and may be deposited outside the crater proper (Schultz 1996; Pierazzo and Melosh 2000; Artemieva and Shuvalov 2001).

4.2 Tsunami-wave Distribution

The growing crater rim and ejecta curtain following the Mjølnir impact form a water surge that eventually breaks up and causes the formation of several waves that, in turn, together with reflected waves, will generate tsunamis. Tsunami wave heights resulting from a vertical incidence Mjølnir impact at various radii from the impact site have been calculated based on different approaches (Tsikalas et al. 1998a; Shuvalov et al. 2002). Although the impact tsunami theory for vertical incidence impacts is well understood (e.g., Ward and Asphaug 2002), there is an almost total absence of computational experiments of tsunami-wave distribution resulting from oblique impacts. Due to the relatively shallow water-target depth, I visualize an oblique Mjølnir impact (south/southwest azimuth at ~45° angle, possibly 30°-45°) to have generated a greater down-range water column disturbance, probably giving rise to faster travelling tsunami-waves at the down-range rather than the up-range region (Fig. 8). Such a scenario is better approximated with the non-axis-symmetrical propagation of tsunamis resulting from submarine slides (Ward 2001).

The ~80-cm-thick ejecta layer at borehole 7430/10-U-01 (Fig. 1) is the thickest Mjølnir ejecta detected so far. The minor thickness ejecta detected on Svalbard (≤1 cm; H. Dypvik, pers. comm.) and the absence (undetected so far despite the intense efforts) of tsunami-deposit signatures on NE Greenland (H. Dypvik, per. comm.), are additional evidence for the obliquity of the Mjølnir impact and possible geographic selectivity in ejecta and tsunami-waves distribution patterns. The proposed conceptual model (Fig. 7, 8) envisages thickest ejecta distribution and faster travelling (thus most devastating) tsunami waves concentrated in the area between Svalbard and

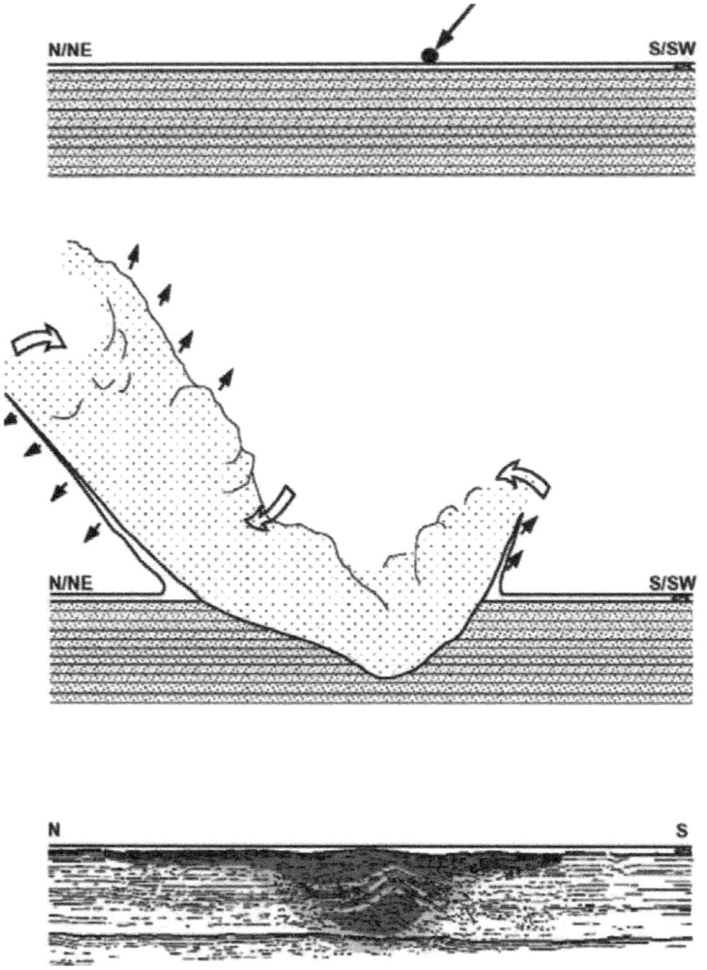

Fig. 8. Schematic cross-section diagrams showing the proposed formation sequence of the Mjølnir crater, with focus on the ejecta distribution and water-column disturbance, resulting from an oblique impact from south/southwest. Detailed hydrocode simulations of a vertical incidence Mjølnir impact are provided by Shuvalov et al. (2002).

Novaya Zemlya. The analysis clearly shows the importance of impact direction and angle in the distribution pattern of ejecta and tsunamis, and further research must, therefore, focus on the proposed down-range region (Fig. 7).

5
Conclusions

The Mjølnir impact, at ~142 Ma, took place at a stable sedimentary platform with thick, sub-horizontal siliciclastic successions. Detailed re-assessment of the 40-km-diameter Mjølnir crater in terms of its well-established structure, morphology, and gravity and seismic velocity anomalies has brought up several diagnostic structural and geophysical signatures related to oblique impacts. These include: (1) a dominant N-S/NNE-SSW crater diameter elongation, (2) a consistent northward asymmetry both in the crater radius and the shallow broad-brim part of the impact-induced seismic disturbance, (3) a peak ring breached towards the N/NE, (4) an annular gravity low with a horseshoe-shape open towards the NE, (5) a transient cavity maximum-depth offset of ~2-2.5 km towards the south-southwest from the geometric crater center, combined with a similar structural uplift lateral offset towards the south, a similar central gravity high offset towards southwest, and an elongated traveltime central anomaly offset towards the WSW.

Within the framework of established diagnostic structural and geophysical asymmetries resulting from oblique-impacts, the compelling evidence for Mjølnir clearly reveal that the bolide travelled from the south/southwest direction at a ~45° (possibly 30°-45°) angle from the horizontal. A conceptual model is proposed, visualizing a north/northeast down-range sector/corridor of thicker ejecta deposits and faster travelling (thus most devastating) tsunami-waves concentrated in the area between Svalbard and Novaya Zemlya. Such a distribution may have selectively intensified the short-term perturbations/environmental stress magnitude in this region.

Acknowledgments

I thank J.I. Faleide for discussions. I gratefully acknowledge the Norwegian Defense Research Establishment, the Norwegian Petroleum Directorate, and SINTEF Petroleum Research for providing the data this study is based on. I also thank P.H. Schultz and N.A. Artemieva for helpful comments and suggestions as reviewers. The study is part of the "Mjølnir Project" financed by the Norwegian Research Council. Statoil is also acknowledged for providing financial support.

References

Alexopoulos JS, McKinnon WB (1994) Large impact craters and basins on Venus, with implications for ring mechanics on the terrestrial planets. In: Dressler BO, Grieve RAF, Sharpton VL (eds) Large Meteorite Impacts and Planetary Evolution. Geological Society of America Special Paper 293: 29-50

Anderson RR, Witzke BJ, Roddy DJ (1996) The drilling of the 1991-1992 Geological Survey Bureau and U.S. Geological Survey Manson impact structure research cores. In: Koeberl C, Anderson RR (eds) The Manson Impact Structure, Iowa: Anatomy of an Impact Crater. Geological Society of America Special Paper 302: 45-88

Artemieva NA, Shuvalov VV (2001) Extraterrestrial material deposition after the impacts into continental and oceanic sites. In: Buffetaut E, Koeberl C (eds) Impact Studies (Geological and biological effects of impact events), Springer Verlag, Berlin-Heidelberg, pp 249-263

Brekke H, Sjulstad HI, Magnus C, Williams RW (2001) Sedimentary environments offshore Norway - an overview. In: Martinsen OJ, Dreyer T (eds) Sedimentary Environments Offshore Norway - Paleozoic to Recent. Norwegian Petroleum Society Special Publication 10: 7-37

Burchell MJ, Mackay NG (1998) Crater ellipticity in hypervelocity impact on metals. Journal of Geophysical Research 103: 22761-22774

Croft SK (1985) The scaling of complex craters. Proceedings of the 15th Lunar and Planetary Science Conference, part 2: Journal of Geophysical Research 90 (supplement): C828-C842.

Dahl JM, Schultz PH (2000) Strain rate measurements in vertical and oblique projectile impact experiments [abs.]. Lunar and Planetary Science XXXI (Lunar and Planetary Institute, Houston, Texas): abs. #1901 (CD-ROM)

Dahl JM, Schultz PH (2001) Measurements of stress wave asymmetries in hypervelocity projectile impact experiments. International Journal of Impact Engineering 26: 145-155

Dypvik H, Attrep M Jr (1999) Geochemical signals of the late Jurassic, marine Mjølnir impact. Meteoritics and Planetary Science 34: 393-406

Dypvik H, Gudlaugsson ST, Tsikalas F, Attrep M Jr, Ferrell RE Jr, Krinsley DH, Mørk A, Faleide JI, Nagy J (1996) Mjølnir structure: An impact crater in the Barents Sea. Geology 24: 779-782

Ekholm AG, Melosh HJ (2001) Crater features diagnostic of oblique impacts; the size and position of the central peak. Geophysical Research Letters 28: 623-626

Faleide JI, Vågnes E, Gudlaugsson ST (1993) Late Mesozoic-Cenozoic evolution of the south-western Barents Sea in a regional rift-shear tectonic setting. Marine and Petroleum Geology 10: 186-214

Gabrielsen RH, Færseth RB, Jensen LN, Kalheim JE, Riis F (1990) Structural elements of the Norwegian continental shelf. Part I: The Barents Sea region. Norwegian Petroleum Directorate Bulletin No. 6, 33 pp

Gault DE, Wedekind JA (1978) Experimental studies of oblique impact. Proceedings of the 9th Lunar and Planetary Science Conference, pp 3843-3875

Grieve RAF, Pesonen LJ (1996) Terrestrial impact craters: their spatial and temporal distribution and impacting bodies. Earth, Moon and Planets 72: 357-376

Gudlaugsson ST (1993) Large impact crater in the Barents Sea. Geology 21: 291-294

Gudlaugsson ST, Faleide JI, Johansen SE, Breivik AJ (1998) Late Paleozoic structural development of the south-western Barents Sea. Marine and Petroleum Geology 15: 73-102

Hayhurst CJ, Ranson HJ, Gardner DJ, Birnbaum NK (1995) Modelling of microparticle hypervelocity oblique impacts on thick targets. International Journal of Impact Engineering 17: 375-386

Ivanov BA, Artemieva NA (2002) Numerical modelling of the formation of large impact craters. In: Koeberl C, MacLeod KG (eds) Catastrophic Events and Mass Extinctions: Impacts and Beyond. Geological Society of America Special Paper 356: 619-630

Izett GA, Cobban WA, Obradovich JD, Kunk MD (1993) The Manson impact structure: $^{40}Ar/^{39}Ar$ age and its distal impact ejecta in the Pierre Shale in southeastern South Dakota. Science 262: 729-732

Lawver LA, Gahagan LM, Campbell DA, Brozena JM, Childers V (1999) Mid-Jurassic to Recent tectonic evolution of the Arctic region (*Powerpoint* animation, using the *PLATES'* animation software) [abs]. In: Lawver LA, Brozena JM, Kovacs LC, Childers V Compilations in the Canada Basin, Aerogeophysical Anomalies. Eos, Transactions, American Geophysical Union, Fall Meeting 1999, San Francisco, 80 (46): 1000

Melosh HJ (1989) Impact cratering - A geologic process. Oxford University Press, New York: 245

Morgan J, Warner M (1999) Chicxulub: The third dimension of a multi-ring impact basin. Geology 27: 407-410

O'Keefe JD, Ahrens TJ (1985) Sampling of planetary material by oblique impact jet entrainment [abs]. Lunar and Planetary Science XVI (Lunar and Planetary Institute, Houston, Texas): 629-630

Pierazzo E, Melosh HJ (1999) Hydrocode modeling of Chicxulub as an oblique impact event. Earth and Planetary Science Letters 165: 163-176

Pierazzo E, Melosh HJ (2000) Understanding oblique impacts from experiments, observations, and modeling. Annual Reviews of Earth and Planetary Sciences 28: 141-167

Pilkington M, Grieve RAF (1992) The geophysical signature of terrestrial impact craters. Reviews of Geophysics 30: 161-181

Poag CW, Poppe LJ (1998) The Toms Canyon structure, New Jersey outer continental shelf: A possible late Eocene impact crater. Marine Geology 145: 23-60

Sandbakken PT (2002) A geological investigation of the Mjølnir Crater core (7329/03-U-01), with emphasis on shock metamorphosed quartz. Masters Thesis, University of Oslo: 142

Schultz PH (1992) Atmospheric effects on ejecta emplacement and crater formation on Venus from Magellan. Journal of Geophysical Research 97: 16,183-16,248.

Schultz PH (1996) Effect of impact angle on vaporization. Journal of Geophysical Research 100: 21,117-21,135.

Schultz PH, Anderson RR (1996) Asymmetry of the Manson impact structure: Evidence for impact angle and direction. In: Koeberl C, Anderson RR (eds) The Manson Impact Structure, Iowa: Anatomy of an Impact Crater. Geological Society of America Special Paper 302: 397-417

Schultz PH, D'Hondt S (1996) Cretaceous-Tertiary (Chicxulub) impact angle and its consequences. Geology 24: 963-967

Schultz PH, Gault DE (1990) Prolonged global catastrophes from oblique impacts. In: Sharpton VL, Ward PD (eds) Global catastrophes in earth history; An interdisciplinary conference on impacts, volcanism and mass mortality. Geological Society of America Special Paper 247: 239-261

Schultz PH, Gault DE (1992) Recognizing impact signatures in the planetary record [abs]. International Conference on Large Meteorite Impacts and Planetary Evolution (Sudbury, Canada), LPI (Lunar and Planetary Institute, Houston, Texas) Contribution 790: 64-65

Shoemaker EM (1962) Interpretation of lunar craters. In: Kopal Z (ed) Physics and Astronomy of the Moon. Academic Press, New York, pp 283-351

Shoemaker EM, Wolfe RF, Shoemaker CS (1990) Asteroid and comet flux in the neighborhood of Earth. In: Sharpton VL, Ward PD (eds) Global Catastrophes in Earth History; An interdisciplinary conference on impacts, volcanism and mass mortality. Geological Society of America Special Paper 247: 155-170

Shuvalov V, Dypvik H, Tsikalas F (2002) Numerical simulations of the Mjølnir marine impact crater. Journal of Geophysical Research (Planets) 107: d.o.i: 10.1029/2001JE001698

Smelror M, Kelly SRA, Dypvik H, Mørk A, Nagy J, Tsikalas F (2001) Mjølnir (Barents Sea) meteorite impact ejecta offers a Volgian-Ryazanian boundary marker. Newsletter on Stratigraphy 38: 129-140

Spudis PD (1993) The geology of multi-ring impact basins. Cambridge, United Kingdom, Cambridge University Press, 263 pp

Sugita S, Schultz PH (2002) Initiation of run-out flows on Venus by oblique impacts. Icarus 155: 265-284

Tsikalas F, Gudlaugsson ST, Faleide JI (1998a) The anatomy of a buried complex impact structure: the Mjølnir Structure, Barents Sea. Journal of Geophysical Research 103: 30,469-30,484

Tsikalas F, Gudlaugsson ST, Faleide JI (1998b) Collapse, infilling, and post-impact deformation at the Mjølnir impact structure, Barents Sea. Geological Society of America Bulletin 110: 537-552

Tsikalas F, Gudlaugsson ST, Eldholm O, Faleide JI (1998c) Integrated geophysical analysis supporting the impact origin of the Mjølnir Structure, Barents Sea. Tectonophysics 289: 257-280

Tsikalas F, Gudlaugsson ST, Faleide JI, Eldholm O (1999) Mjølnir Structure, Barents Sea: a marine impact crater laboratory. In: Dressler B, Sharpton VL (eds) Large Meteorite Impacts and Planetary Evolution II. Geological Society of America Special Paper 339: 193-204

Tsikalas F, Faleide JI, Eldholm O, Dypvik, H (2002a) Seismic correlation of the Mjølnir marine impact crater to shallow boreholes. In: Plado J, Pesonen LJ (eds) Impacts in Precambrian Shields, Impact Studies vol. 2, Springer Verlag, Berlin-Heidelberg, pp 307-321

Tsikalas F, Gudlaugsson ST, Faleide JI, Eldholm O (2002b) The Mjølnir marine impact crater porosity anomaly. Deep-Sea Research Part II 49: 1103-1120

Ward SN (2001) Landslide tsunami. Journal of Geophysical Research 106: 11,201-11,215

Ward SN, Asphaug E (2002) Impact tsunami - Eltanin. Deep-Sea Research Part II 49: 1073-1079

Worsley D, Johansen R, Kristensen SE (1988) The Mesozoic and Cenozoic succession of Tromsøflaket. In: Dalland A, Worsley D, Ofstad K (eds) A lithostratigraphic scheme for the Mesozoic and Cenozoic succession offshore mid- and northern Norway. Norwegian Petroleum Directorate Bulletin No. 4, 42-65

Zuber M, Smith DE, Lemoine FG, Neumann GA (1995) The shape and internal structure of the Moon from the Clementine Mission. Science 266: 1839-1843

The Duobblon Structure - A Small Segment of a Large Precambrian Impact Structure?

Robert Lilljequist[1] and Ulla Preeden[2]

[1]Ecominas, Calle Horno 9, Estepuma, Spain (robertlilljequist@yahoo.se)
[2]Institute of Geology, University of Tartu, Vanemuise 46, EE-51014 Tartu, Estonia (ullap@ut.ee)

Abstract. The Duobblon Structure is located in northern Sweden, near the Scandinavian mountain range in the county of Västerbotten. The structure constitutes a 30 km long formation, slightly concave to the NW. The Revsund Granite (1.87 Ga) makes up most of the basement, and grades into a 100 m thick autochthonous basal breccia, overlain by a chert-like, clast-bearing melt rock and a polymict breccia. A fine-grained volcanic-like rock makes up the top of the series. Caledonian nappes of Cambro-Silurian age covers the structure to the W. The possible presence of planar deformation features in quartz and the similarity to the stratigraphy of proven complex impact structures support a possible impact origin. The age of the assumed impact event might correspond to a U-Pb zircon age of about 1.80 Ga, which is the indicated age of a chert-like rock in-between the autochthonous and the polymict breccia. Assuming that the arcuate breccia formations form part of a once circular structure, a crater diameter of approximately 80 km can be inferred, of which 1/7 of the outer margin is preserved.

1 Introduction

The Duobblon structure is located in northern Sweden, at 65°35′N and 17°10′ E, near the Scandinavian mountain range in the county of Västerbotten (Fig. 1). The Duobblon area has been geologically mapped (Einarsson 1979) and investigated in detail, as part of a uranium exploration campaign by the Geological Survey of Sweden in 1974-1979 (Lindroos and

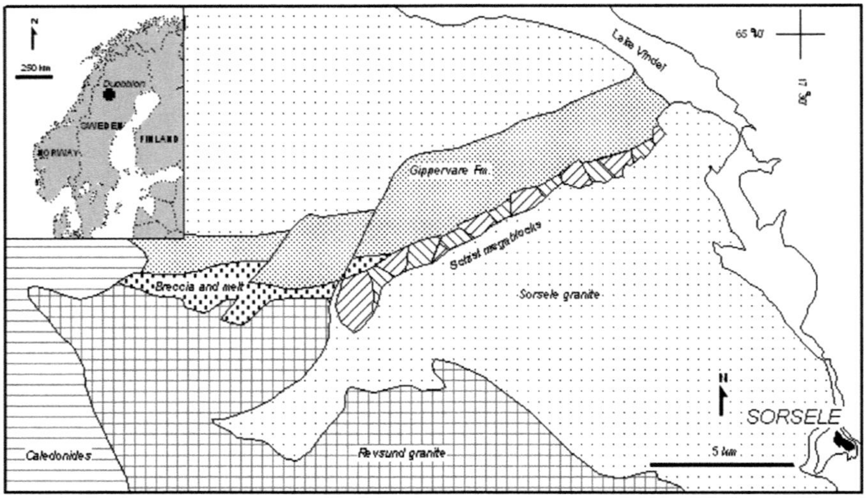

Fig. 1. The Duobblon structural sequence can be followed for 30 km in an approximately E-W direction, with a gentle dip (about 20°) to the north.

Smellie 1979). Wickman (1988) proposed that the supracrustal rocks form part of an impact-generated rock sequence, on the basis of the similarity of the stratigraphy found in other impact structures. In 1997, the author had the opportunity, on behalf of North Atlantic Natural Resources, to investigate the formation to follow up on the suggestion by Wickman.

2 Previous Geological Interpretation

The geology of the Duobblon area was first described briefly by Högbom (1937), and Gavelin and Kulling (1955). In the 1970s Einarsson published a more detailed geological map of the Duobblon area (Einarsson 1979). He introduced the term "Duobblon Group", which is divided into a lower Björnknösen Formation with "conglomerates, lithophysae-bearing volcanics and tuffitic sandstone", and an upper Gippervare Formation with "acid volcanics". The Duobblon Group is the name given to the supracrustal rocks on top of brecciated basement. A detailed description of the lithophysae can be found in Lindroos and Smellie (1979) and Smellie (1982). They range in size from 5 mm to 15 cm in diameter and vary in shape from spheroidal, ellipsoidal, to lenticular. Perlitic structures occur in the vicinity of the lithophysae, which often coalesce into amoeboid shape. The mineral assemblage within the lithophysae is dominated by quartz with subordinate feldspar, commonly exhibiting a concentric layering within the lithophysae.

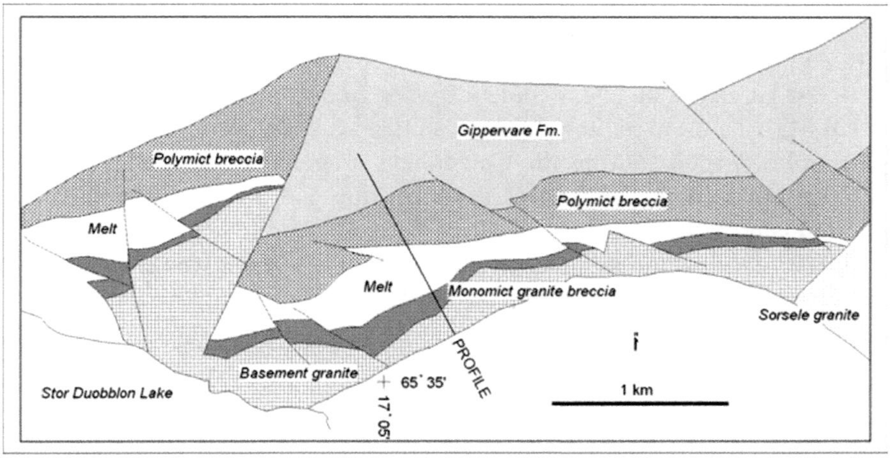

Fig. 2. Geological plan map over the central parts of the Duobblon Structure.

The lower unit is about 450 m thick and the overlying Gippervare rocks are over 1 km thick (Einarsson 1979; Lindroos and Smellie 1979).

Sixty-one drill-holes (11 252 m) penetrate rock sequence down to the brecciated granite rock. The drill-cores allow detailed lithological and mineralogical studies.

According to Einarsson (1979), a major unconformity separates the Duobblon Group from the basement, which is composed of granite and metamorphic pelitic and psammitic metasediments. The greywacke type

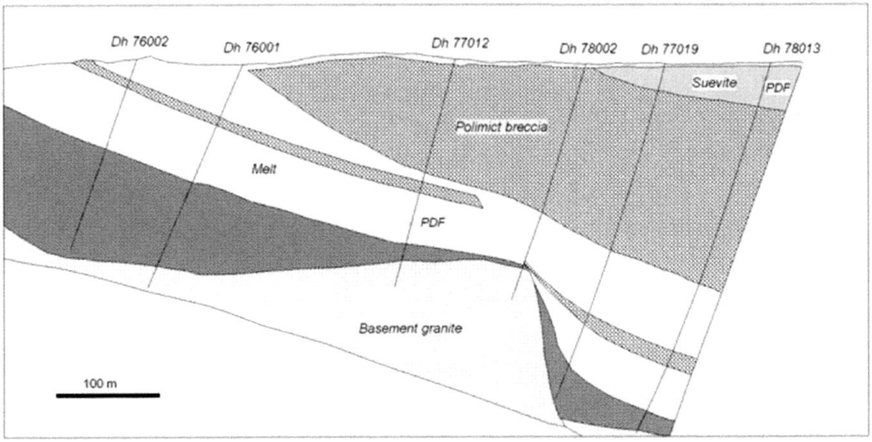

Fig3. Geological profile. The position of the profile is indicated in **Fig. 2.**

metasediments form the Paksjo Group. The Sorsele Granite (1791±22Ma; Skiöld 1988) intrudes both the basement rocks and the Duobblon Group. The last Precambrian event is the intrusion of dolerite dykes (younger than 1626 Ma). To the west, the Caledonian nappes have been thrusted over the Precambrian rocks during the Caledonian orogeny at about 450 Ma ago. The general geology of the area is shown in Fig. 1 and Fig. 2; and a cross section is presented in Fig. 3.

Lindroos and Smellie (1979) and Smellie (1982) in general support the stratigraphy introduced by Einarsson (1979) with only minor alterations.

3 The Impact Concept

The author has studied only a few of the drill-cores, and a preliminary attempt to re-interpret the geology from the concept of an impact event is presented. The interpretation, suggesting an impact stratigraphy (first proposed by F. E. Wickman in 1988) is only tentative, and more in-depth studies are needed before different formations can be more accurately interpreted. As evidenced from the shock metamorphism (described in the following paragraph), the general sequence of rock units and the appearance of the breccia formations an impact model would best explain the Duobblon Structure. The following description and discussion assumes that the structure formed by an impact, even though there is not yet final confirmation for such a proposal.

3.1 Shock metamorphic textures

Shock metamorphic features have been observed in three thin sections from three different drill sites. PDFs in 5 quartz grains have been measured with an universal stage with the following results:

2 grains with Miller index $\{1\ 0\ \bar{1}\ 3\}$ with the symbol ω,
1 grain with (0 0 0 1) with the symbol c,
1 grain with $\{5\ 1\ \bar{6}\ 1\}$ with the symbol x, and
1 grain with $\{1\ 0\ \bar{1}\ 3\}$ and $\{1\ 1\ \bar{2}\ 1\}$ with the symbols ω and s, respectively.

Typically only one set of PDF is seen in quartz grains. Two sets of PDFs in quartz have been found in a thin section from Dh 78013 at 245.6 m (Fig 4). The PDFs in quartz are mainly non-decorated or very weakly decorated.

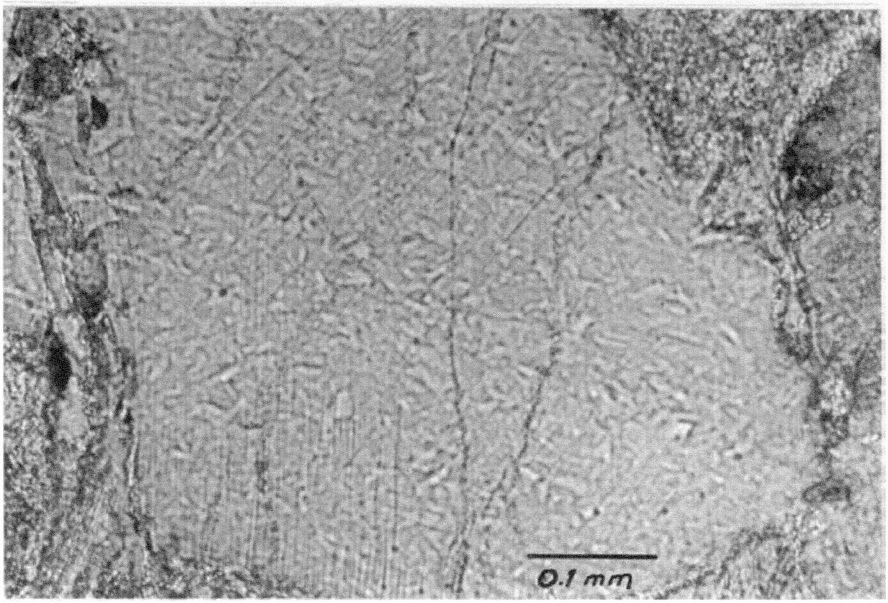

Fig. 4. Two sets of PDFs in quartz. Drill-hole 78 013, 245.6 m. Crossed nichols.

The crystallographic orientation of the PDFs against quartz optical axis has been obtained by U-stage studies, resulting in five forms of features parallel to: $\{1\ 0\ \bar{1}\ 3\}$, $\{1\ 1\ \bar{2}\ 1\}$, $\{5\ 1\ \bar{6}\ \}$), (0 0 0 1), and $\{2\ 1\ \bar{3}\ 1\}$. The main orientation is $\{1\ 0\ \bar{1}\ 3\}$, which is typical for impact-affected rocks. PDFs in feldspar (usually in K-feldspar) have been found in thin sections from three different sites (Dh 85003 at 197.7 m, Dh 78013 at 245.6 m and Dh 7872). The obtained data are however not sufficient to deduce the formation pressure.

Partly circular to semicircular textures in quartz grains have been noted in two thin sections and is interpreted as ballen quartz (Grieve et al 1996).

3.2 Target Rocks

The target rocks are granite, meta-greywacke and graphitic schist. The granite has been given an age of 1.87 Ga (based on U/Pb analyses of zircons, Skiöld 1988).

In the westernmost part of the Duobblon Structure, a sparse network of cm-wide breccia veins intersects the basement granite, which upward grades into a matrix- supported autochthonous breccia. Einarsson (1979) interpreted the veins as "fractures filled with mud and weathered residue containing larger quartz grains and granite clasts in a dense greyish green matrix". To the east, the basement consists of greywacke overlain by autochthonous schist breccia derived from the same sediments. In drill-cores pesudotachylitic-like veinlets have been observed both in the granite basement and in brecciated schist. A more detailed investigation and description is, however, needed before a correct classification of these veins can be presented.

3.3 Autochthonous Breccia

A matrix-supported autochthonous breccia with granite clasts overlies the granitic basement, whereas the schist (the Paksjö Schist) is overlain by a monomict schist breccia. In the eastern part the Sorsele granite is in intrusive contact with rotated blocks of basement greywacke with bedding planes striking more or less perpendicular to the contact and with way-up directions to both the west and the east (Fig. 3). The thickness of the autochthonous breccia varies considerably and can reach 100 m.

3.4 Assumed Impact Melt Rock

The monomict breccia grades upward into what Einarsson (1979) describe as lithophysae-bearing acid volcanic rocks. In the following this unit is interpreted as an impact melt rock. The assumption is based on planar deformation features in quartz, the fine-grade cherty-like appearance, perlitic textures (indicating that the rock was formed as glass) and the inclusion of more or less melted granite and schist clasts.

The rock ranges in thickness, from 20 m in the western parts (Dh 78001) to 140 m in the central parts of the structure (Fig. 4). As a whole, the rock is heterogeneous on a smaller, local scale (due to varying number of clasts). In some drill-holes of the structure it is more homogeneous, where the rock unit normally has a chert-like appearance, due to recrystallisation and hydrothermal alteration. The rock is, however, not completely recrystallised and in place exhibits perlitic textures (e.g., Einarsson 1979, Fig 44, and observations in Dh 78301 at 68.8 m).

In the lower part and at different levels higher up, it contains more of less melted granite/and or schist clasts (Fig. 5a). The clasts occur at different levels and in varying stages of preservation (Fig. 5b). The more altered granite grades into clasts of chert-like appearance (Fig. 5c). It is, however, not possible to exclude or to prove that these are of volcanic origin and existed in the assumed target rocks. Earlier authors (Einarsson 1979; Lindroos and Smellie 1979) described the material as dominated by fine-to medium grained clasts of granite and volcanic rocks with very variable appearance and consisting of rock types, that are not represented within the Duobblon area.

Feldspar and quartz phenocrysts are common throughout the rock (Fig. 5e). The feldspars (0.3-5 mm in diameter) have mainly albite composition, and generally are sericitised. The quartz phenocrysts show resorption and embayment features. Occasionally the rock has a pseudo-layering with alternating green and red colours. Welded particles are commonly present, and the matrix consists of a devitrified felsitic mass with grain size <1 mm. Accessory minerals include titanomagnetite, ilmenite, magnetite, apatite, titanite, epidote, calcite, pyrite, and zircon (Smellie 1982).

The most spectacular textures are spherulitic textures and cavities, which have attracted the attention of earlier investigators. Spherulites are sporadic and can reach 4 mm in size, exhibit radial growth structures, and are nucleated around quartz fragments (Smellie 1982). The lithophysae are spherical in shape or sometimes elliptical. The diameter is usually less than 5 cm (good illustrations in Einarsson 1979).

The lithophysae are cryptocrystalline in the outer margin with the same mineralogical composition as the matrix. Some of the lithophysae have a concentric growth pattern. In the inner parts of the lithophysae appear flattened, rather similar to ignimbritic textures.

In the microscope the matrix of the lithophysic rock is sericitised and has often well preserved perlitic texture, but mostly a devitrified microfelsitic groundmass is seen. In the matrix a few mineral clasts of potassium feldspar and quartz can be seen.

The zircons in the rock are commonly zoned and, in addition to their common association with opaque minerals, occur sporadically throughout the rock (Smellie 1982). Major element analyses indicate that the rock is calc-alkaline in composition with only minor variation along its vertical and its horizontal extension.

Fig. 5 A. Rounded granite clast and flattened melt clasts. Drill-hole 85 003. 185.5 m. **B.** Irregular formed clasts with reddish rim and mineral clasts in a fine-grained matrix. Drill-hole 78 013, 325.7 m. **C.** Irregular, densely packed clasts with a chert-like appearance. Drill-hole 78 013, 327.9 m. **D.** Fine-grained, partly banded rock, consisting of feldspar grains and a few quartz grains (one with 2 sets of PDFs, see Fig. 2). **E.** Rock with <1 cm size granite clasts and diffuse chert-like clasts. From this section, one quartz grain with one set of PDF has been recorded. Drill-hole 85 003, 194.1 m.

The lithophyse-bearing melt rock is up to 15 m thick in the field. At one outcrop lithophysae have been seen to occur in two layers separated by breccia (Einarsson 1979). At another locality, a 3-m-thick schist breccia grades upwards into 5 m of polymict breccia with mainly granitic clasts, and uppermost follows a 5-m-thick greyish green lithophysae-rich rock. This unit in between breccia formations, was earlier interpreted as an ignimbrite, and has given an U/Pb zircon age of 1803±15 Ma age (Skiöld 1988).

3.5 Suevitic Ash?

The lithophysae-bearing volcanic rocks are overlain by what has been interpreted as a grey to red tuffitic sandstone (Fig. 5d). Larger fragments of quartz, microcline-perthite and sericitised plagioclase is found in a very fine-grained sericitised matrix. The finest fraction consists of small angular quartz and potash feldspar fragments in a dense matrix of sericite and calcite with varying amount of epidote. It is here interpreted as a fine-grained suevitic, based on two sets of PDFs in quartz. The rock is sometimes difficult to separate from the lower fine-grained melt unit.

3.6 Polymict Breccia

An upper polymict breccia overlies the rock, above interpreted as a suevitic ash, without a distinct contact. In the central part of the rocks exposed in outcrop the thickness of the polymict breccia is approximately 160-200 m.

The lower part of the polymict breccia is interpreted as a coarse-grained suevite breccia with clasts of granite, schist and "porphyries" (which either are melted granite clasts or have a volcanic origin).

Part of the upper part of the polymict breccia has more rounded clasts and could represent a re-deposited breccia. A more in depth study might be able to separate different units.

3.7 The Gippervare Formation

The Gippervare Formation concordantly overlies the polymict breccia. According to Einarsson (1979) and Lindroos and Smellie (1979), the formation is about 1000 m thick and extends over an area of 30 km with a maximum width of 2 km. The rocks have an overall rhyolitic composition (Einarsson 1979). Two main types can be distinguished: in the lower a part grey dense rock rich in quartz phenocrysts and higher up a coarser grained rock with feldspar phenocrysts. Coarse, fragment-bearing rocks with clasts up to 30 cm in size occur in the central part of the sequence. Einarsson (1979) suggested that these rocks could largely be of ignimbritic origin, forming relatively thin units with large areal extent. The position on top of the polymict breccia indicates that the formation could have originated as a fine-grained suevite. No investigation of the presence of shock metamorphic minerals has yet been performed. For a more detailed description of the Gippervare Formation the reader is referred to the publication of Einarsson (1979).

4 Tectonic Structures

On a detailed scale NNE and NW striking faults displace the Duobblon Structure (Fig. 1 and 3). On the regional scale a NNE shear zone is cutting the structure towards the southeast (Fig 6). The Caledonian Front covers the structure towards the west (Fig. 1), and might have caused repetitions

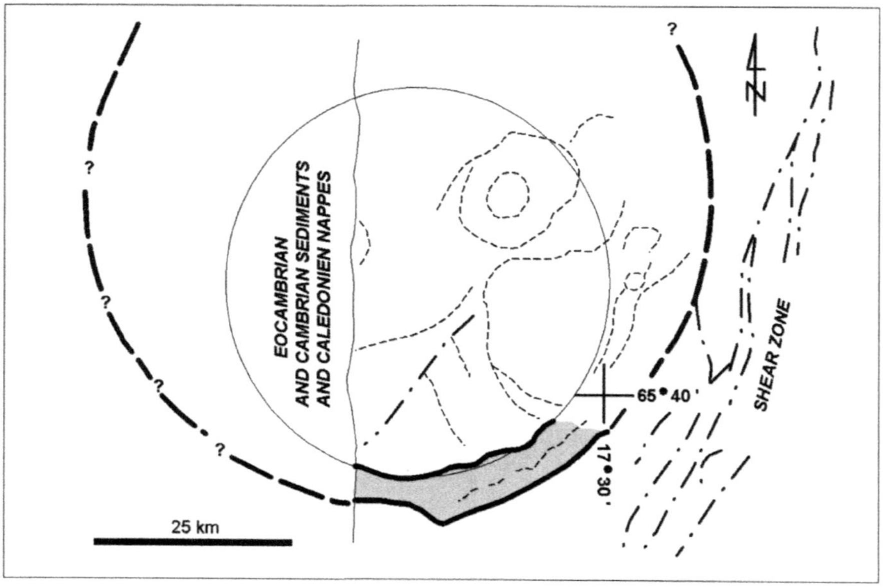

Fig. 6. Lineaments interpreted from aeromagnetic measurements and two possible interpreted sizes of the structure. The exposed part of the Duobblon structure is marked with gray shading.

of individual sections by thrust movements, although this has not been verified. Several kilometres of autochthonous Palaeozoic sediments and thrust units of the Caledonian Orogeny may have covered the structure but were subsequently removed due to erosion. This has contributed to the lithification of the various rock units.

The continuation of the Duobblon structure towards the E to ENE is not well established. On the geological map the structure continues about 10 km East of Lake Vindeln (Fig. 1), where it disappears in a granite terrain, and terminates against faults striking NNW-SSE (Fig. 6).

5 Mineralizations

The melt rock contains relatively high uranium concentrations with an average of 200-300 ppm U. The uranium is considered to be "syngenetic" and related to the devitrification (Smellie 1982). The uranium occurs as fine granular pitchblende around Fe-Ti-Mn oxide aggregates and is associated with increased amounts of Pb, V and Mo (for a more detailed description the reader is referred to Lindroos and Smellie 1979 and Smellie 1982). Very fine-grained disseminated chalcopyrite, pyrite and arsenopyrite, and less commonly sphalerite, occur sporadically within fractures.

In zones of strong fracturing and brecciation, hydrocarbons are present, almost certainly of hydrothermal origin. Shrinkage cracks are a common texture in the hydrocarbons (Lindroos 1979). The mineralised zone is rather thin (< 1 m) and runs in NNW-SSE. The radioactivity within the mineralised zone is 300-1000 mR/h, whereas the background value is 25 to 30 mR/h.

A gold-bearing sulphide mineralization occurs in the westernmost part of the autochthonous breccia (at Duobblonbäcken). The sulphides are pyrrhotite, pyrite, chalcopyrite, sphalerite, and arsenopyrite, with traces of molybdenite and gold. The gold content in the best drill-hole intersection is 0.52 g/t over 37 meters. The mineralization can be followed over a length of at least 200 m and with a maximum width of 50 m.

6 Discussion: Reinterpretation of the Duobblon Stratigraphy as a Sequence of Impactites

The geology of the rocks comprising the Duobblon structure displays a stratigraphy in outcrops and drill-holes that is similar to that in other proven impact structures. The possible presence of PDFs in quartz and feldspar, described above, suggests an impact origin for the structure. On this basis it is assumed that the Duobblon Structure originated as a complex impact crater.

It is known that each proven or suspected impact crater has its own particular characteristics, and the description of each new impact site continues to open up new interesting windows into the geological history. This is also the case with the Duobblon structure. Only a part of an original ring structure (about 50^{0} of the concave outer ring) is exposed close to, and to a large part covered by the Caledonian overthrusted formations. In this aspect it is similar to the Lockne impact structure (Lindström et al. 1996).

The supposed impact melt body at Duobblon is located both on top of the authigenic granite breccia (e.g., drill-holes Nos. 77025; 78013, 78313 and at Duobblonbäcken) and on suevite (e.g., drill-holes 76001, 76002). The thickness of the melt ranges from a few meters in the westernmost parts (Duobblonbäcken) to 140 m in the central part of the exposed structure. The Caledonian front is situated close to the structure, and overthrust structures might have affected the rock sequence. The Caledonian front with Palaeozoic sediments have most likely covered the whole part of the now exposed part of the Duobblon structure, and were probably removed by erosion. If the presently preserved impact rocks are part of a marginal formation, the melt thickness might be considerably increased in the areas closer to the central uplift, which is located under the Caledonian thrust sheets.

At the Manicuagan impact structure (100 km diameter), the clast content in the melt decreases and the mineral grain size increases upward in the melt sheet (Floran et al. 1978). A similar variation has been observed in the melt sheet of West Clearwater. Mineral clasts in melt rocks are predominantly quartz and feldspar. Obviously the reaction of hot melt with less shocked debris resulted in preferred dissolution of femic minerals and an enrichment of the most refractory quartz in the clast component. The melt in Duobblon is clast-rich in its lower part, but a clast-rich layer is also found higher up in the melt sheet.

Spherulites and lithophysae occur in the Duobblon melt. Similar textures, although not characteristic, are also reported from other impact structures. In the Mistastin Crater (28 km diameter), vesicles and many of the country-rock inclusions are drawn out and define a crude subhorizontal foliation (Grieve 1975). Spheroids occur here as discrete structures in the matrix. Spheroids are also reported in glassy melt rocks at Charlevoix (Rondot 1971), Manicouagan (Floran et al. 1978) and West Clearwater (Simonds et al. 1978) craters.

A notable characteristic of fresh melt rocks from some large, coherent impact melt sheets is their chemical homogeneity, even when the target rocks have diverse composition (Grieve et al. 1987, Engelhardt 1984).

This is also so in the melt sheet of the Duobblon structure where analyses of the melt from different parts are rather similar. Melts are however, texturally heterogeneous over distances from mm to meters, mostly due to incomplete melting of clasts (Engelhardt 1984; Grieve et al 1987).

It is a remote possibility that the granite - north of the Gippervare formation - could represent a slow cooling melt sheet and thus forms part of the impact-generated rocks. The 1791±22 Ma age given by U/Pb analyses of zircons (Skiöld 1988) is not very far from the age given for the melt (1803±15 Ma). Isotope studies might prove or disregard such an explanation.

The volumes of coherent melt masses at Brent, Mistasin, West Clearwater and Manicouagan amount to 1-5% of the excavated cavity (Engelhardt 1984). As only part of the Duobblon Structure is preserved, it is not possible to use this estimation to indicate the melt volume and the original size of the structure. The erosional edge of the Popigai structure dips 30° toward the centre. At about 6 km from the margin, the melt sheet dips 20° but in generally the melt follows the topography of the underlying allogenic breccia (Figure 2.12 in Masaitis 1999). The dip of the different units of the Duobblon structure is 20° towards the NNW, but a post-impact tilting might have occurred. If we assume that the preserved Duobblon structure is part of a circular configuration, a hypothetical crater diameter of 80 km can be inferred (Fig 6). This is not an unrealistic assumption, comparing the thickness of preserved impact formations from other parts of the world. The preserved part of the hypothetical 80 km diameter Duobblon structure represents a 50 degree sector or about 1/7 part of the outer margin. The edge of the structure might originally have been several kilometres further out from the present erosional edge.

The gold sulphide mineralization at Duobblonbäcken lies in the continuation of the so-called "Gold Line" (Andersson and Lax 2000), along which several gold deposits have been found. The "Gold Line" is interpreted as a deep crustal zone that has been re-activated at several occasions with the generation of hydrothermal fluids. The monomict breccias covering the basement granite at Duobblonbäcken was a suitable trap for post-impact hydrothermal solutions, maybe generated by the heat energy created by the impact event.

The uranium mineralization in the central part of the Duobblon structure is confined to the fine-grained melt. The uranium deposition is described as syngenetic (Smellie 1982), and could either be formed as a separation of uranium metal during the melting of the target rocks or formed short afterwards by hydrothermal fluids driven by the heat generated in the rocks by the impact event.

7 Conclusions

The presence of single and multiple sets of PDFs in quartz and feldspar and the relative position and composition of the rock units strongly indicate that the Duobblon structure was created by a Precambrian impact. The age of the supposed impact event might correspond to a U-Pb zircon age of about 1.80 Ga, which is the indicated age of the lithophysae-rich vitreous melt in-between the monomict and the polymict breccia.

Only part of the ring structure is exposed, and even this is, to a large part, covered by the Caledonian thrusts. The assumed size of the original impact structure diameter is 80 km (Fig. 6).

The "impact melt" is mainly situated between the brecciated target rocks and an allogenic polymict breccia.

Two interesting types of mineralizations have attracted the attention of exploration organisations in the past. A uranium deposit is confined to the melt sheet and a gold occurrence in the monomict autochthonous breccia. The uranium mineralization in the impact melt is interpreted as formed by hydrothermal solutions short after or during the solidification of the melt, and derived from the uranium content of the target rocks. The gold-bearing sulphide mineralization in the monomict granite breccia at Dobblonbäcken most likely formed at a later event during the re-activation of a regional deep-reaching structure, along which several other deposits are found in different geological units.

A total of 61 drill-holes penetrate the preserved crater formation down to the brecciated target rock. Most of the drill-cores are available at the Mineral Office of the Swedish Geological Survey in Malå, Västerbotten County. This offers opportunities for future, more detailed studies (e.g., the search for iridium anomalies and impact diamonds).

Acknowledgements

Herbert Henkel has creatively read, discussed and corrected the manuscript. The reviewers B. Dressler and Uwe Reimold have made constructive improvements and Christian Koeberl contributed useful remarks. Zdenka Ivanic have improved the drawings and Ted Posey, manager of North Atlantic Natural Resources, kindly gave his support to the publication. Ulla Preeden did the U-stage measurements under the supervision of Juhu Kirs and the kind participation of Väino Puura at Tartu University in Estonia.

References

Andersson M, Lax K (2000) Geokemiska kartan. Swedish Geological Survey. Report SGU-GK: 2, ISSN: 14004-3157 (in Swedish), 49 pp

Billström K, Weihed P (1996) Age and Provenance of host rocks and ores in the Paleoproterozoic Skellefte District, Northern Sweden. Economic Geology 91: 1054-1072

Dence MR (1971) Impact melts. Journal of Geophysical Research 76: 5525-5565

Einarsson Ö (1979) The Precambrian rocks of the Duobblon area, Västerbotten County (in Swedish with English summary). Swedish Geological Survey Ser C 748, 123 pp

Engelhardt Wv (1984) Melt products from terrestrial impact structures. Proceedings of the 27th International Geological Congress 19: 149-163

Floran RJ, Grieve RAF, Phinney WC, Warner JL, Simonds CH, Blanchard DP, Dence MR (1978) Manicouagan impact melt, Quebec, 1. Stratigraphy, petrology and chemistry. Journal of Geophysical Research 83: 2737-2759

Gavelin S, Kulling O (1955) Beskrivning till berggrundskarta över Västerbottens län (Description to the bedrock map over the county of Västerbotten. With English summary). Swedish Geological Survey Ca 37, 295 pp

Grieve RAF (1975) Petrology and chemistry of the impact melt at Mistastin Lake crater, Labrador. Geological Society of American Bulletin 86: 1617-1629

Grieve RAF, Reny G, Gurov EP, Ryabenko VA (1987) The melt rocks of the Boltysh impact crater, Ukraina, USSR. Contributions to Mineralogy and Petrology 96: 56-62

Grieve R (1987) Terrestrial impact structures. Annual Reviews of Earth and Planetary Science 15: 245-270

Grieve RAF, Langenhorst F, Stöffler D (1996) Shock metamorphism of quartz in nature and experiment: II Significance in geoscience. Meteoritics and Planetary Science 31: 6-35

Högbom A (1937) Skelleftefältet (in Swedish with English summary). Swedish Geological Survey C 389: 1-122

Lilljequist R (2002) The Duobblon structure in northern Sweden. [abs.]. In: 8th Workshop of the European Science Foundation on Impact Tectonism, Mora, Sweden, p 39

Lilljequist R, Reimold WU (1999) The Duobblon Structure - a suspected impact formation [abs.]. Meteoritics and Planetary Science 34: A74

Lindroos H (1979) Geological description of the uranium deposit at Brånaberget. Unpublished report in Swedish No 1979:20. Available at the Mineral Office of the Swedish Geological Survey

Lindroos H, Smellie J (1979) A stratabound uranium occurrence within Middle Precambrian ignimbrites at Duobblon, Northern Sweden. Economic Geology 74: 1118-1130

Lindström M, Sturkell EFF, Törnberg R, Ormö J (1996) The marine impact crater at Lockne, central Sweden. Geologiska Föreningens Förhandlingar 118: 193-206

Masaitis VL (1984) Impact reworking of the Earth's crust. Comparative Planetology, VNU Science Press. Proceedings of the 27th International Geological Congress 19: 135-147

Masaitis VL (1999) Impact structures of northeastern Eurasia: the territories of Russia and adjacent countries. Meteoritics and Planetary Science 34: 691-711

Rondot J (1971) Impactite of the Charlevoix structure, Quebec, Canada. Journal of Geophysical Research 76: 5414-5423

Skiöld T (1988) Implications of new U-Pb zircon chronology to Early Proterozoic crustal accretation in northern Sweden. Precambrian Research 38: 147-164

Smellie J (1982) The mineralogy and genesis of uranium in rhyolite ignimbrites of Precambrian age from Duobblon, Sweden. Mineralogical Magazine 46: 187-199

Wickman FE (1988) Deep drilling in crystalline bedrock. 1. In: Bodén A, Eriksson KG (eds) The Deep Gas Drilling in Siljan Impact Structure, Sweden and Astroblemes. Exploration of the deep continental crust, vol 1. Springer-Verlag, Berlin Heidelberg New York, pp 298-327

Åvike Bay – a 10 km Diameter Possible Impact Structure at the Bothnian Sea Coast of Central Sweden.

Herbert Henkel[1], Väino Puura[2], Tom Flodén[3], Juho Kirs[2], Mare Konsa[4], Ulla Preeden[2], Robert Lilljequist[5] and Joanne Fernlund[1]

[1]Royal Institute of Technology, Department of Land and Water Resources Engineering, S-10044 Stockholm, Sweden (herbert@kth.se)

[2]Institute of Geology, University of Tartu, Vanemuise 46, EE-51014 Tartu, Estonia (puura@ut.ee)

[3]Department of Geology and Geochemistry, Stockholm University, S-10691 Stockholm, Sweden (tom@geo.su.se)

[4]Institute of Geology, Tallinn Technical University, Estonia pst.7, EE-10143 Tallinn, Estonia (mare@gi.ee)

[5]Ecominas, Calle Horno 9, Estepuma, Spain (robertlilljequist@yahoo.se)

Abstract. Åvike Bay is a 270° degree wide near-circular, 114 m deep bay on the Swedish coast of the Bothnian Sea, northeast of Sundsvall. The structure has a diameter of about 10 km. It was classified as a probable impact structure because of its extraordinary circular topography in the overwiew of impact structures in Fennoscandia. Recent studies lend further support to this interpretation. The structure has a submarine central mound, which is elevated some 40 m above the adjacent sea floor. It has a very distinct tangential and radial on-shore fracture pattern as seen in the topographic map. Along the southwestern shore of the Bay, an enigmatic quartzite breccia of unknown age occurs as part of a larger outcrop of polymict breccia with clasts of crystalline rocks and quartzite of unknown age. In thin section, planar fractures can be observed in quartz and feldspar grains. A detailed investigation showed that in a few cases the quartz grains contained microdeformation features closely resembling PDFs.

1 Introduction

The semi-circular Åvike Bay at the fast rising coast of the Bothnian Sea has previously been investigated to show its formation as a caldera related to Alnö intrusives and for prospecting of mineral deposits. In this presentation the Åvike Bay is suggested to be caused by an impact event.

Studies have been made of the rock physical properties based on in-situ measurements and laboratory measurements of rock samples. The shore and some of the islands have been searched several times for finds of potential impactites and a team of divers has collected samples from the shallow sea floor. At one outcrop, an enigmatic breccia was found that further strengthened the suspected impact connection of the structure. Aeromagnetic measurements indicate an anomalous pattern of magnetic structures within the bay. Gravity measurements across the bay on ice have been planned for several winter seasons, but still await an occasion with sufficient thickness of the ice. This report gives an acccount of the information so far collected and its interpretation with respect to a possible impact origin of the Åvike Bay structure.

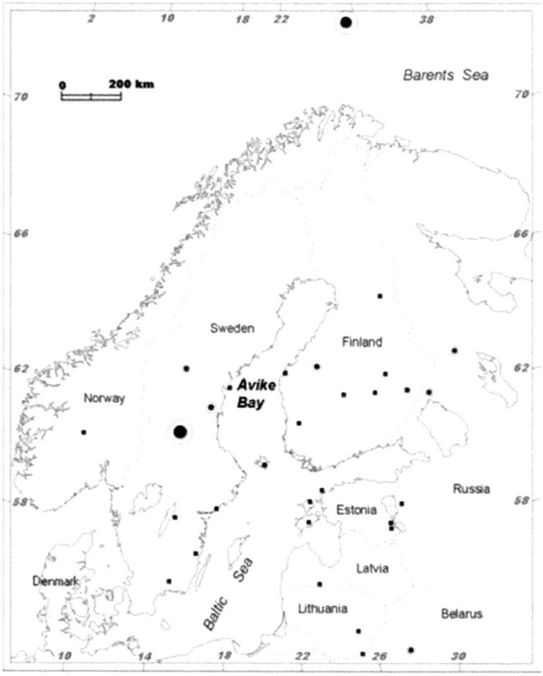

Fig. 1. Impact structures in Fennoscandia. The Åvike Bay structure is located at the western shore of the Bothnian Sea. Modified from Abels et al. (2002).

2 Background

The circular shape of the Åvike Bay, centered at 62.5° N and 17.8° E, (Fig. 1), first drew attention of scientists during detailed mapping of the nearby Alnö intrusive complex in the 1960s. A study of the dikes typical for the Alnö complex showed that several dykes extended to the shores of the Åvike Bay. Söderström (1966), therefore, suggested that the bay was a caldera structure related to the Alnö complex. The find of an enigmatic quartzite breccia at the SW shore of the bay was reported by Lundquist et al. (1990), who regarded it as a tectonic breccia related to an unknown fault zone. Boulders of alnöitic and carbonatitic breccias containing clasts of similar quartzite and also a fragment of sövite have been reported from the shore of the bay and were considered to emanate from within the bay. The composition of this sövite is, however, different from that of other parts of the Alnö region (Lundquist et al. 1990).

Recent studies have determined the physical properties, especially the electric resistivity, of rocks from this area. The enigmatic breccia has been studied in more detail. At several shallow areas in the bay, boulders of Paleozoic limestone and brecciated dolerite were recovered from the seafloor.

An aeromagnetic survey of the bay region was performed by the Boliden Mining Co.; later the region was re-measured by the Geological Survey of Sweden as part of a regional geophysical mapping program. Gravity measurements are so far restricted to land areas. Rock physical properties have been measured both in-situ and on rock samples to facilitate modeling of the magnetic anomalies observed in the aeromagnetic data and potential gravity anomalies.

The coastal region around Åvike Bay is presently emerging from the Bothnian Sea at a rate of almost 8 mm a^{-1} (Sveriges National Atlas 1994). Wave erosion has removed un-consolidated Quaternary glacial sediments from the exposed parts of all near coastal hills. Due to extensive erosion in connection with glaciations, also the weathered top part of the bedrock has been removed; the rock outcrops are therefore extremely smooth and clean. The predominant ice flow direction is 130° (towards SE) as evidenced by glacial striae on rock outcrops on several small and thus the boulder transport is in a similar direction. The ice flow divide was located near the present western coast of the Bothnian Sea, but it is not known with sufficient precision to facilitate knowledge of the source region for erratic boulders within the bay.

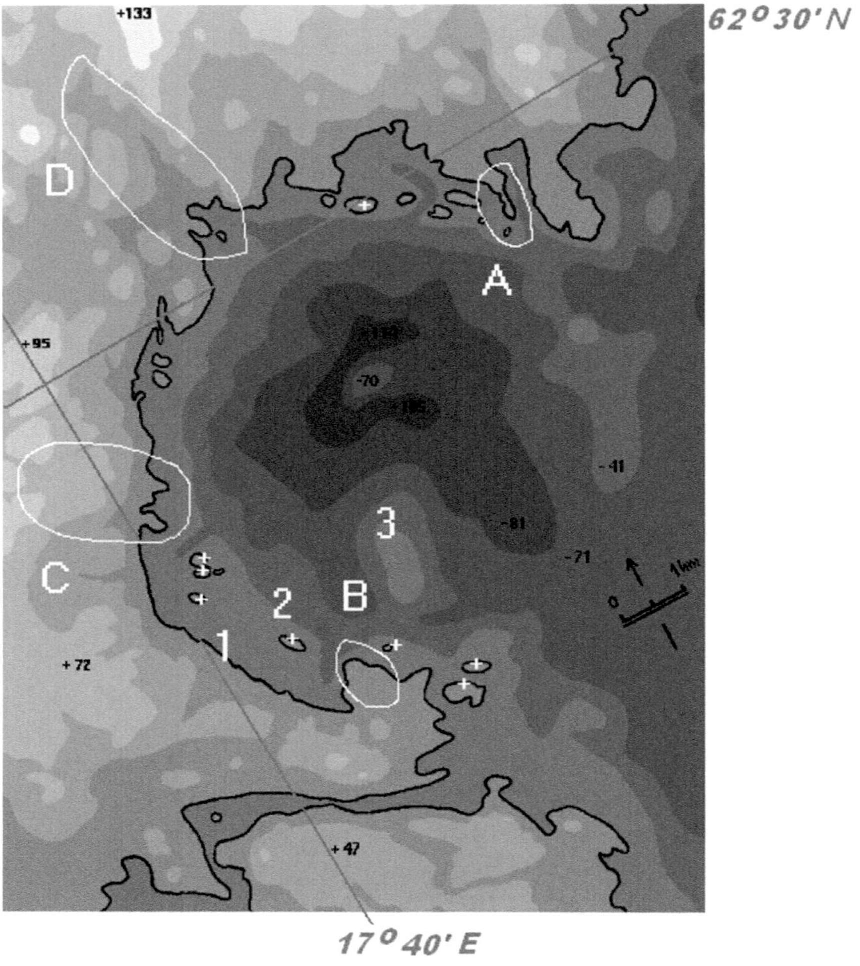

Fig. 2. Simplified elevation and bathymetric map, showing the general morphology of the Åvike Bay structure. The gray level intervals are 25 m and the present shore line has been marked together with a few elevations and depths, respectively. The structure is ca. 150 m deep with a ca. 40 m high and ca. 1 km wide submarine central mound. Sampling was made in the marked regions (Density and magnetic properties within A and B, in-situ electric resistivity measurements within C, D, and at scattered localities marked with +). Localities 1, 2, and 3 are mentioned in the text.

The Åvike Bay structure was first suggested as an impact structure by Henkel and Lilljequist (2001), based on the extraordinarily circular topography and the presence of polymict breccia.

3 Topography and Fracture Pattern

At the time of the first geological investigations by Söderström (1966), detailed topographic maps with elevation contours did not exist. Modern maps clearly indicate that the circular structure extends several kilometer on shore, and has a diameter of ca. 9.5 km. From the edge of the structure, where the terrain is ca. 100 m above sea level, the elevation decreases stepwise to the shoreline along apparent tangential scarps. Just offshore in the SW part of the bay, there are several scattered islands that mark an inner ring with ca. 6.5 km diameter. Bathymetric information from navigation charts has been combined with onshore elevation data (Fig. 2). The circular structure is even more clearly depicted by a submarine ridge located in the NE part of the bay. In the onshore and submarine morphology, thus all of 330° of a circular structure are evident. A central submarine mound, ca. 1 km wide, rises 40 m above the surrounding sea floor, which constitutes a deep ring basin with a near coastal unusually large depths of over 110 m. In addition to the series of tangential morphological features, numerous radial topographic scarps extend from the coast up to 5 km inland (Fig. 3). On one of the islands, at locality 2 (Fig. 2), the gneiss basement is strongly fractured and striated in a pattern somewhat reminescent of shatter cones (Fig. 4d). A boulder of Åsby type dolerite recovered from an escarpment at the sea floor 2.5 km off shore (locality 3), was strongly sheared and brecciated, unlike the otherwise abundant well preserved and undeformed boulders of this rock type.

4 Geology

The rocks of the Åvike Bay region have been mapped and described by Lundquist et al. (1990). Gneisses of meta-sedimentary rocks predominate, with ages older than 1.9 Ga. The gneiss is intruded by small local massifs of granite of 1.8 Ga age (Claesson 1987). Gneisses are predominant over the local granite bodies. These Svecofennian rocks are intruded by a large north-south striking and eastwards dipping Mesoproterozoic dolerite sill of

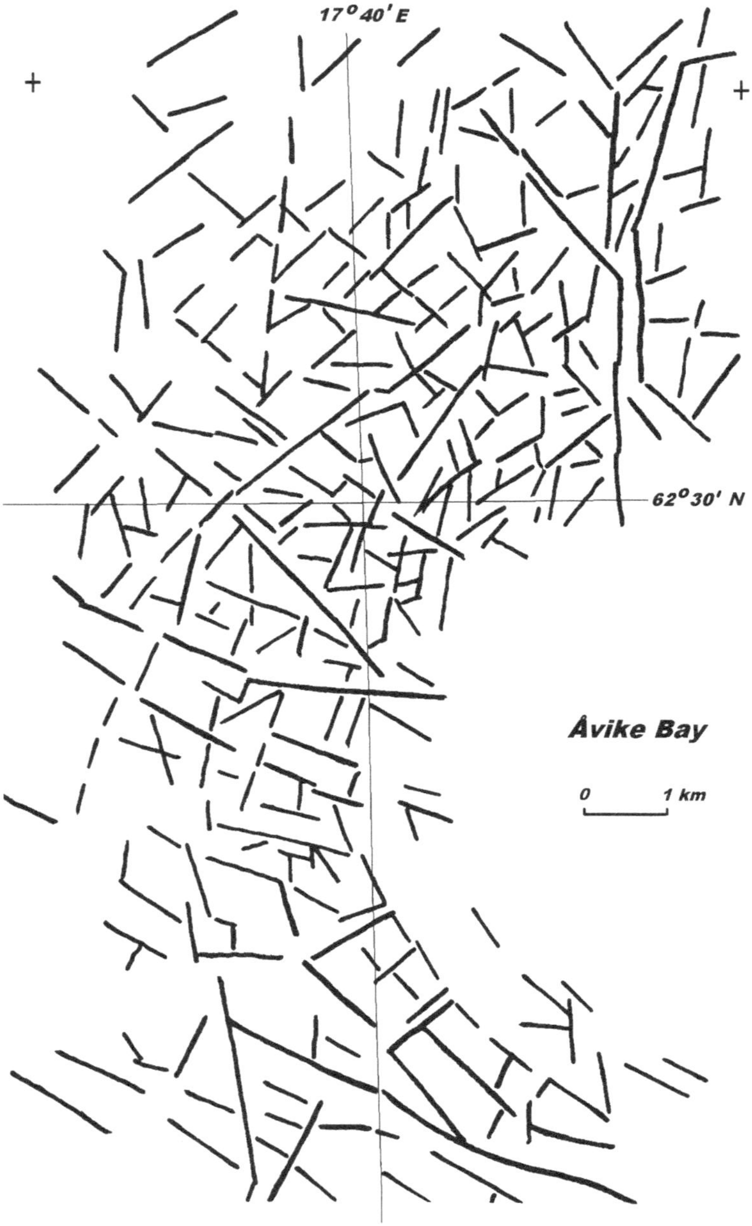

Fig. 3. Interpretation of fracture patterns based on the detailed standard topographic map in scale 1:50 000 with 5 m contour interval.

so called Åsby type. These dolerite intrusions are widespread in central Fennoscandia and are characterized by plagioclase, augit and olivine as their major mineral constituents, and a coarse grained ophitic texture (Lundquist 1979). Dolorites intrude the Jotnian sandstone and have been given an age of 1.2 Ga (Welin 1979). Boulders (Fig. 4a) of dolerite have been recovered from shallow water areas within the bay at locality 3 (Fig. 2), indicating that the sill extends several km into the bay.

The circular Åvike Bay structure clearly cuts the dolerite sill. Along the shore, at locality 1 (Fig. 2), an enigmatic quartzite breccia occurs (Fig 4b). The stage of metamorphism of this quartzite is much lower than that of the migmatized Svecofennian gneiss and granite complex. The quartzite is assumed to be of Jotnian (Riphean) age due to its similarity with known occurrences around the Bothnian Sea and the Bothnian Bay. The closest location of outcrops of Jotnian sandstone at the coast is ca. 50 km north of Åvike Bay (Lundquist et al. 1990). The brecciation pattern of the quartzite is identical with patterns observed in ejected clasts of sedimentary target rocks in the Ries and Steinheim impact structures (Stöffler and Ostertag 1983), in terms of the variation in clast size, rotation of clasts and a matrix of very fine grained fragments. At low waterstand in 2000 and 2001, the breccia was observed to be part of a larger occurrence of polymict breccia with clasts of quartzite and local crystalline rocks, cemented with fine grained calcitic breccia of mineral fragments (Fig. 4c).

Summarizing the results of field observations and thin section studies from 2000 to 2003, the characteristics of different types of breccias at the shore of Åvike Bay are as follows: (1) Breccia dikes are derived from Svecofennian rocks and distributed inside Svecofennian basement fracture zones. The macroclasts and brecciated micromatrix of the dykes are cemented with carbonate. (2) Meter-sized blocks of so called "enigmatic" brecciated quartzite are also distributed within fracture zones. (3) The regolith of the polymict breccia is composed of both Svecofennian and Jotnian (?) clasts, scattered as erratic cobbles and boulders along the sea shore near the zone of brecciated bedrock. These erratics are derived from within the bay, where further outcrops could be observed at low waterstand. A Jotnian age of the low-metamorphic quartzite is suggested based on the occurrence of Jotnian sediments covering the Svecofennian basement of the Sea and Gulf of Bothnia, including the area just east of Åvike Bay. The Jotnian sediments were eroded from the western coast of the gulf as late as in Phanerozoic time (Puura and Plado, this volume). (4) Small dikes of alnöitic intrusives occur scattered in the region and have ages from 210 to 600 Ma (Brückner and Rex 1980).

Fig. 4. Pictures of the deformed rocks. For locations see Fig. 2.

a. Boulder of brecciated dolerite recovered from shallow water. The scale has cm divisions.

b. Outcrop of quartzite breccia. The compass is 6 cm wide.

c. Polymict breccia with fine grained breccia matrix containing calcite healed fractures, deformed mica, melted quartz grains and recrystallised rims of mineral grains. The scale divisions are 1 cm wide.

d. Strongly fractured gneiss with shatter cone like lineations.

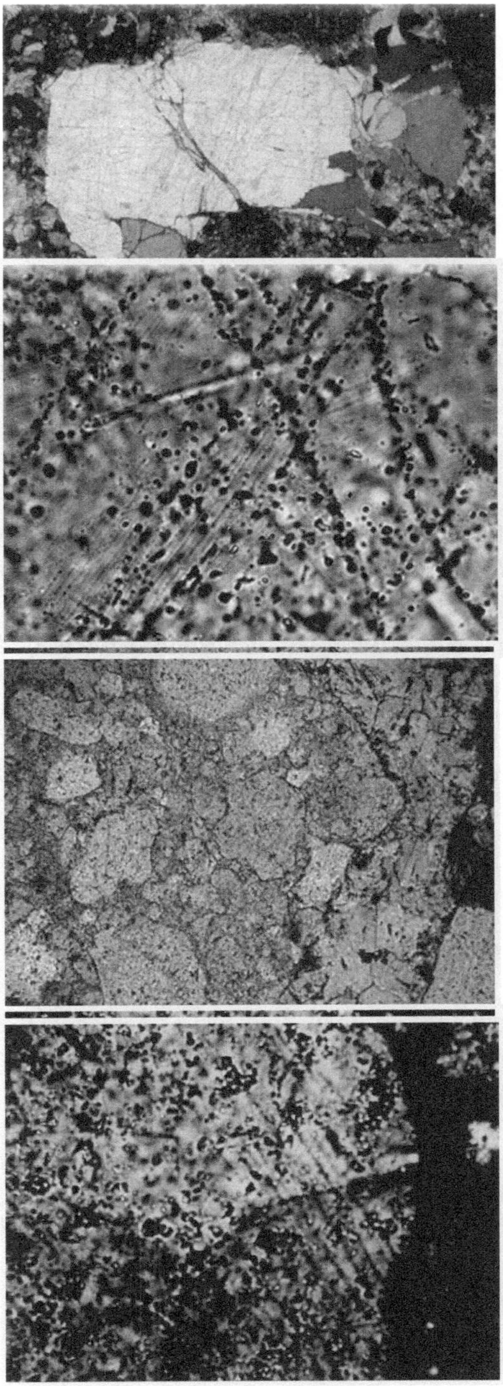

Fig. 5. Planar features in mineral grains from polymict breccia at locality 1; for location see Fig. 2.

a. Sheared quartz fragments with fluid inclusion trails in polymict breccia from thin section Av-5. The length of the largest quartz fragment is 1.4 mm. Oblique polarizers at ca. 30° angle.

b. Interior (SW marginal part) of the quartz grain from Fig. 5a with two systems of microdeformation features. The spacing is 3-8 μm. Oblique polarizers. The width of the picture is 0.1 mm.

c. Clast of Jotnian (?) sandstone in a carbonate dominated breccia matrix. The quartz grain with microdeformation features is in the center of the image. Oblique polarizers. The width of the picture is 0.5 mm.

d. The quartz grain with microdeformation features seen in the center of Fig. 5c with one system with spacing of 7 μm. The width of the grain is 0.28 mm.

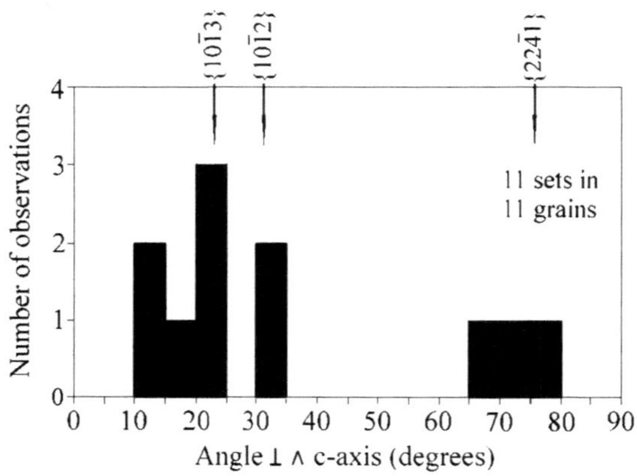

Fig. 6. Distribution of micro-deformation orientations in quartz from polymict breccia. The histogram shows the angle of observed planes with the c-axis. The class width is 5°.

Along the shore of Åvike Bay, numerous boulders of brecciated alnöitic material (beforsite etc.) have been observed (Lundquist et al. 1990). Also a few boulders of unfractured, orthoceratitic, brown limestone of Ordovician age were recovered from the shore of the bay and from shallow water within the bay, at locality 3 (Fig. 2). Ordovician limestone is interpreted to occur in a large basin just east of the distinct offshore escarpment marking the west coast of the Bothnian Sea (Ahlberg 1986). This escarpment is located right at the eastern margin of the circular bay and truncates a sector of the structure. To the NW, i.e. in the direction from which the ice flowed, the nearest preserved occurrence of this limestone is in Jämtland, ca. 150 km from the bay.

5 Occurrence of Microdeformation Features in Quartz

Thin sections were prepared from the polymict breccia. As a whole, the polymict breccia consists of large, angular, broken clasts of quartz, plagioclase and rare microcline in carbonate-rich, fine-grained clastic matrix. In some fractures, what appears to be idiomorphic dolomite rombohedra have grown. The plagioclase clasts are rich in carbonate and occasionally contain patches of microcline and rounded, semi-opaque, brownish-grey earthy aggregates. In places, the breccia contains rounded and altered "alnöitic nodules" with quartz. The rounded nodules are probably related to the Alnö alkaline igneous complex, which is widely developed in the region in the form of dikes.

In the calcite-dominated matrix of the polymict breccia, mm-sized lithic clasts of quartz-rich Svecofennian rock (Fig. 5a and b) and Jotnian quartzite (Fig. 5c and d) occur. The clasts are intersected by cracks with jagged edges. The cracks are filled with carbonate or carbonate-microcline mass, in places together with kink-banded muscovite. Well-developed decorated planar fractures occur in several samples. Their spacing is around 15 μm.

Microdeformation features in quartz (of the quartz grains shown in Fig. 5a and b) are demonstrated in Fig. 5b and 5d. In three out of 25 studied thin sections, planar microdeformation features in quartz were observed in quartz-rich rocks. The spacing of the two non-decorated systems measured in one 0.2-mm-sized aggregate of quartz fragments varies from 5 to 8 μm.

In thin section Av-02-51A, in a Jotnian (?) quartzite clast, a 0.2-mm rounded quartz grain with 7 μm-spaced microdeformations was observed (Fig. 5d). The crystallographic orientation of these microdeformation features was studied with an universal stage in seven quartz clasts in three thin sections. Poles of the planes of microdeformation features that have angles of 11°, 14°, 24°, 31°, 32°, 71°, 72° against the quartz c-axes were detected. The orientations $\{1\ 0\ \bar{1}\ 3\}$, $\{1\ 0\ \bar{1}\ 2\}$, and $\{2\ 2\ \bar{4}\ 1\}$ are well represented (Fig. 6). The orientation $\{1\ 0\ \bar{1}\ 3\}$ is most frequent and typically occurs in impact-affected rocks.

6 Rock Physical Properties

In-situ measurements of magnetic susceptibility and electric resistivity (Fig. 7, for location see Fig. 2) were made on outcrops of gneiss and dolerite. The gneisses have low susceptibility of ca $1.8 \cdot 10^{-4}$ SI and the dolerite has >100 times higher susceptibility, $3 \cdot 10^{-2}$ SI (Fig. 7a). Rock samples were measured with respect to density, magnetic susceptibility, and remanent magnetization. The gneisses and granites are generally low-magnetic, in contrast to the high-magnetic dolerite, whose magnetic anomalies can be followed under the bay from outside the shore to the north and south. The dolerite has a typical remanent magnetization with declination 40° and inclination ranging from +10° to –10°, acquired during its formation in the Jotnian (Riphean) period (Fig. 8). The ratio of remanent to induced magnetization is 1.35, resulting in a total magnetization with low inclination (ca. +25°) and northeast directed declination (ca. 30°).

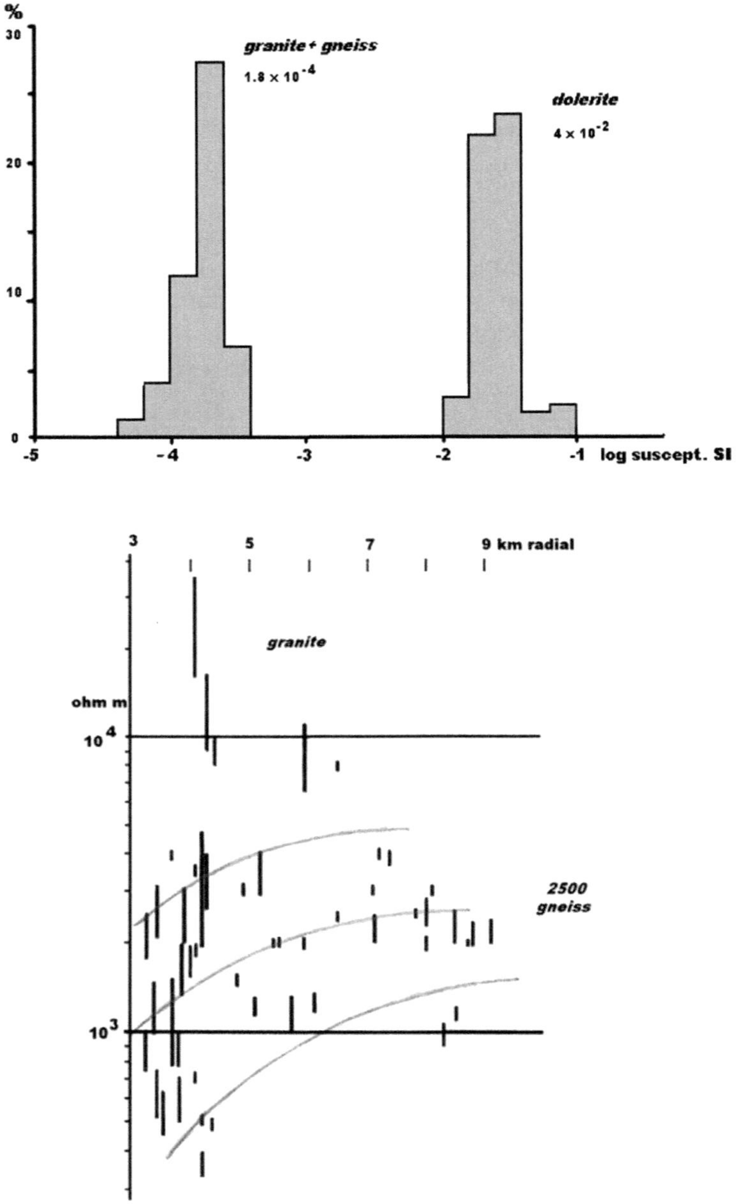

Fig. 7. Rock physical properties. Numbers indicate type values. **a.** (top) Magnetic susceptibility of gneiss + granite, and dolerite, respectively, from regions A and B in Fig. 2. **b.** (bottom) Radial variation of *in situ* electrical resistivity. For location see Fig. 2.

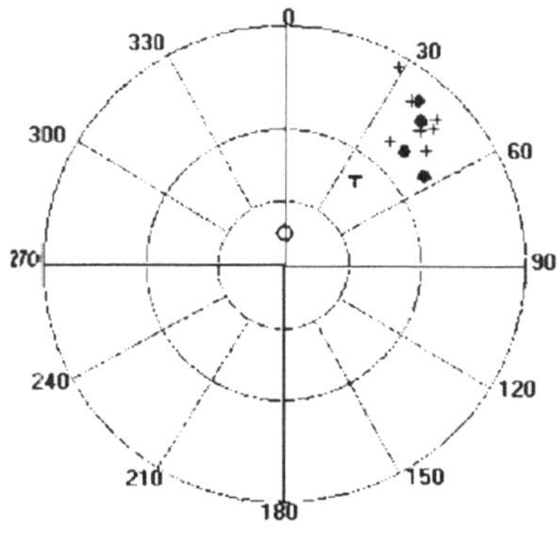

Fig. 8. Directions of remanent magnetization in dolerite from oriented rock samples in regions A and B. The present geomagnetic field is shown within the small circle. Negative inclination is marked with + and positive with dots, respectively. The total magnetization vector (induced + remanent magnetization) would be located at T in the diagram. Radial variation of electric resistivity collected in regions C and D and at scattered localities shown with + in **Fig. 2**. On average a reduction with a factor 2.5 is observed towards the center of the structure. The lines represent the uncertainty interval due to unisotropy.

The electric resistivity (Fig. 7b), is generally low for the gneisses with a normal resistivity of 2.5 kΩm outside the Åvike Bay structure and decreasing towards 1 kΩm at the shore of the Bay and on nearby islands. The low resistivity of the gneiss is related to the content of conductive minerals (potentially graphite and pyrrhotite). An additional decrease in resistivity is related to an increased fracturing, similar to that observed in large impact structures (Henkel 1992 and Bäckström, this volume).

7 Magnetic Anomaly

The aeromagnetic map (Fig. 9) shows the extent and structure of the magnetic dolerite sill (the irregular features) and the effect of two deep-seated magnetic sources, giving low-gradient, rounded anomalies to the west and south. The southern anomaly is spatially associated with the Alnö alkaline intrusives. Outside to the north and south of the Åvike Bay structure, high gradient northeast-skewed anomalies reflect the low inclination angle of the total magnetization vector and the shallow dip of the sill. Within the bay, the anomalies are subdued and have northwest skewing, indicating a larger depth and a different direction of the total magnetization vector, respectively. Roughly coincident with the present shoreline, the magnetic anomaly is discontinuous, marking the onset of significant subsidence. There are several options for the interpretation of the interior magnetic features of the bay (see discussion).

The presently available gravity data indicate a weak gravity low over the bay area and a predominant high to the west of the bay, which is also seen in the aeromagnetic map.

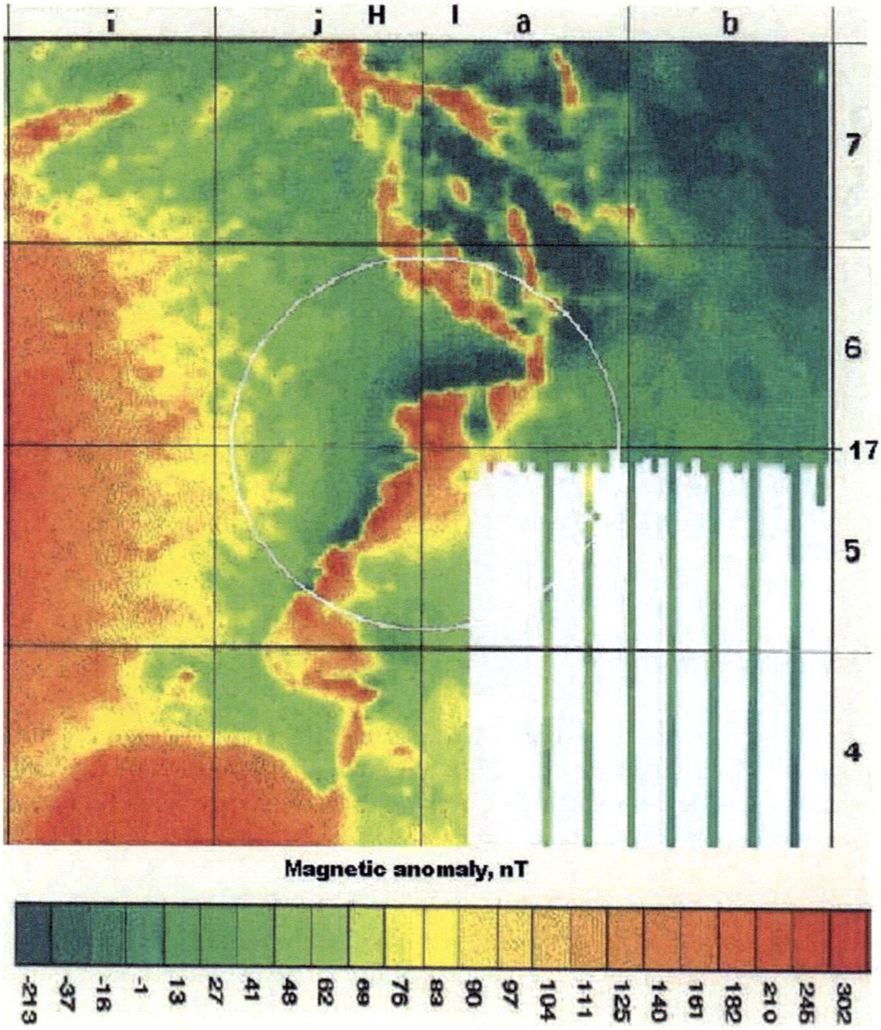

Fig. 9. Aeromagnetic anomaly map of the region around the Åvike Bay structure. The grid has 5 km spacing and the Åvike Bay structure is outlined by the white circle. The measurements were made by the Geological Survey of Sweden (SGU) at low altitude and with 200 m line spacing. Within the edge of the Åvike Bay structure the magnetic anomaly of the dolerite sill has a different asymmetry as compared to the outside appearance. From the aeromagnetic map 17H-I 4-7 i-b; with permission by the Geological Survey of Sweden (SGU).

8
Age of the Åvike Bay Structure

Based on the appearance of the magnetic anomaly associated with one of the sills of Åsby-type dolerite (occurring at the shore to the south and to the north of the bay and continuing displaced into the bay) the Åvike Bay structure seems to be younger than this dolerite, i.e., < 1.2 Ga. The occurrence of brecciated dykes of Alnö type (Lundquist et al. 1990) may place the age younger than the 600 to 210 Ma interval of ages for the Alnö magmatism. Dykes of Alnö type contain both angular and somewhat rounded fragments similar to clasts found in impact related breccias. If the inferred impact predates the Alnö dykes, its age would then be between 1.2 and 0.2 Ga.

The occurrence of Paleozoic limestone boulders, in which no impact signature has been reported, may have several bedrock sources. Quaternary glacial transport from Jämtland is one option. Late marine ice rafted transport from the Bothnian Sea another, and a third alternative could be locally derived Paleozoic rocks within the structure. In the latter case, the limestone would constitute a down-faulted part of sedimentary cover rocks or rocks formed within the impact crater. The change in the magnetic anomaly pattern, with an overprinted remanence in the central uplift area, may result in a paleomagnetic age estimate after detailed modeling of the exterior and interior magnetic anomalies associated with the dolerite sill.

9
Discussion and Conclusions

There are several indications for a possible impact origin of the Åvike Bay structure. The extraordinary circularity of the Åvike Bay, the on-shore fracture pattern as seen in the topography, and the bathymetry of the Bay with a ring basin and a central mound, are all consistent with the morphometry of an impact structure.

The occurrence of a polymict breccia, unrelated to any known fault zone, containing fragments of both Svecofennian rocks and Jotnian sandstone is another indication of a possible impact origin.

The Quaternary ice flow has eroded and transported material from the structure further off-shore into the Bothnian Sea and thus the chance to find boulders from within the structure are extremely small.

The brecciation pattern seen in the sandstone part of the outcrop resembles very closely the patterns observed in ejecta of sedimentary target rocks seen at impact structures (e.g., like the Ries and Steinheim structures; Stöffler and Ostertag 1983).

Within an outcrop of brecciated rocks at the shoreline, microdeformation features were observed in several quartz grains this lends further support to the suggestion that the Åvike Bay might be an impact structure. The main types of the Precambrian rocks in the region – Svecofennian quartz-rich crystalline rocks and Jotnian (?) sandstones - were both subjected to this microdeformation. The quartz grains in the Jotnian (?) sandstone can, however, have widely varying sources. The 1.2 Ga sill of Åsby dolerite, as well as other Precambrian rocks are locally strongly fractured and brecciated. Although the number of thin sections and the number of quartz grains with distinct microdeformation features is fairly low, the common impression of a strong deforming overprint of the different samples is evident. The crystallographic orientation of microdeformation features is at angles from 11° to 72° against the c-axis of quartz. Besides the microdeformation features in quartz, similar patterns in feldspars have been observed together with kink banded biotite. The grain with microdeformation features (Fig. 5b) that comes closest to planar deformation features (PDFs) indicates that the clasts of the rocks in studied samples could have been derived from a relatively low-pressure impact environment and were possibly subjected to pressures of 8-20 GPa.

The age of the structure must be younger than the post-Jotnian dolerite sill, i.e., <1.2 Ga, which is the only regional structure presently known to be clearly cut by the Åvike Bay structure. Rocks from the youngest phase of the Alnö intrusive could mark a younger isotopic age limit of 600 - 210 Ma. The source for the Ordovician limestone boulders cannot be determined with certainty and limestone clasts are not observed in the breccia. If they are of local origin, their age would represent a minimum age of the impact event.

The geophysical data are also consistent with the Åvike Bay being an impact structure. The electric resistivity decreases radially inward from normal levels to values below 1 kΩm due to increased fracturing typically observed along the edge of impact structures (Henkel 1992) like Dellen or Lockne (Bäckström, this issue), or in the central uplifted basement of impact structures like Siljan.

The magnetic anomaly related to the dolerite sill is discontinuous across the shore line. Within the bay the anomaly is subdued due to the increased water depth and has a different skewing. The dolerite sill is subsided in the ring part of the bay and may be absent in the central mound region.

The cause of the anomalous magnetic feature in the central mound can either be re-magnetised gneiss basement, which however requires the formation of ferrimagnetic minerals in the otherwise low magnetic gneiss, or a remagnetized part of the existing ferrimagnetic minerals in the dolerite. Magnetite has been shown to form by shock dissociation of Fe-Mg-silicates (Feldman 1994) and re-magnetization of ferrimagnetric rocks is a characteristic feature in the central parts of large impact structures (Henkel and Reimold 2002). By detailed modeling of the magnetic anomalies outside and inside the bay, combined with gravity modeling, this aspect could be further clarified. It also gives the option to determine the most likely remanent magnetization vector that can explain the observed anomalies and serve as a paleomagnetic age constraint.

In order to outline the 3D-structure of the Åvike Bay gravity measurements on ice are needed within the bay area. This can only be done when there is a thick ice cover. Combined with the knowledge of the magnetisation and density of the involved rocks, a rather detailed modeling can then be made.

The previous interpretation of the structure is a caldera related to the Alnö intrusives. The occurrence of dykes from the Alnö complex is characteristic for the entire region around the Alnö structure, extending far to the north and south of Alnön, and well exposed along the western shores of the Bothnian Sea, including the Åvike Bay area. There are, however, no similar rounded morphologic structures associated with the Alnö complex in the entire region.

The various observations of anomalous features cited for the Åvike Bay are not by themselves proofs of an impact origin of the bay, however when taken together they constitute a good argument in favour of an impact structure. It is suggested that the Åvike Bay structure is possible eroded impact crater that was formed after the intrusion of the Åsby dolerite sill (< 1.2 Ga old). The topographic diameter of the structure is defined to ca 9.5 km at the present erosion level. The ca. 1 km wide central mound and the surrounding ring basin are two features that are chracteristic for complex impact craters. Geophysical data of the magnetic anomaly pattern and the radial decrease of electric resistivity lend further support to the impact hypothesis as well as the radial and tangential fracture pattern seen in the topographic map. The occurrence of polymict breccia and finds of microdeformation features in quartz grains within this breccia also point towards a possible impact cause of the Åvike Bay. The present natural conditions (the deep bay) and the recent glacial action (that transported unconsolidated material offshore) unfortunately prevents additional geological observations without future drilling.

Acknowledgements

Comments and suggestions from two reviewers are appreciated, as well as the kind assistance offered by the *Sundsvall Divers Club* to collect rock samples from the shallow bay floor.

References

Abels A, Plado J, Pesonen LJ, Lehtinen M (2002) The impact cratering record of Fennoscandia - a close look at the database. In: Plado J, Pesonen LJ (eds) Impacts in Precambrian Shields. Berlin: Springer, 1-58

Ahlberg P (1986) Den svenska kontinentalsockelns berggrund. Sammanställning av tillgängliga geologiska undersökningar. Geological Survey of Sweden. Rapporter och Meddelanden 47, 101 pp

Brückner HK, Rex DC (1980) K-Ar and Rb-Sr geochronology and Sr isotopic study of the Alnö alkaline complex, northeastern Sweden. Lithos 13: 111-119

Claesson S (1987) Nd isotope data on 1.9 – 1.2 Ga old basic rocks and metasediments from the Bothnian basin, Central Sweden. Precambrian Research 35: 115-126

Feldman VI (1994) The conditions of shock metamorphism. In: Dressler BO, Grieve RAF, Sharpton VL (eds), Large Meteorite Impacts and Planetary Evolution. Geological Society of America Special Paper 293: 121-132

Henkel H (1992) Geophysical aspects of meteorite impact craters in eroded shield environment, with special emphasis on electric resistivity. Tectonophysics 216: 63-89

Henkel H, Lilljequist R (2001) The Åvike Bay structure. In: Martinez-Ruiz F, Ortega-Huertas M, Palomo I (eds) Abstract Volume 6th ESF – IMPACT Workshop "Impact Markers in the Stratigraphic Record", Granada, Spain p 47

Henkel H, Reimold WU (2002) Magnetic model of the central uplift of the Vredefort impact structure, South Africa. Journal of Applied Geophysics 51: 43-62

Lundquist T (1979) The Precambrian of Sweden. Geological Survey of Sweden C 768, 87 pp

Lundquist T, Gee D, Kumpulainen R, Karis L, Kresten P (1990) Beskrivning till berggrundskartan över Västernorrlands län. Geological Survey of Sweden Ba 31, 285 pp

Sveriges NationalAtlas (1994) Jord och Berg, ISBN 91-87760-27-4, 208 pp

Söderström L (1966) The kimberlites of Åvike Bay, on the Bothnian coast of Sweden. Geologiska Föreningens i Stockholm Förhandlingar 88: 351-360

Stöffler D, Ostertag R (1983) The Ries impact crater. Fortschritte der Mineralogie 61: 71-116

Welin E (1979) Tabulation of recalculated radiometric ages published 1960-1979 for rocks and minerals in Sweden. Geologiska Föreningen i Stockholm Förhandlingar 101: 309-320

The Structure and Age of the Kaali Main Crater, Island of Saaremaa, Estonia

Anto Raukas[1], J.-M. Punning[2], T. Moora[3], Ü. Kestlane[1], A. Kraut[4]

[1]Institute of Geology at Tallinn Technical University, 7 Estonia Avenue, EE-10143 Tallinn, Estonia (raukas@gi.ee)

[2]Institute of Ecology at Tallinn Pedagogical University, 2 Kevade St., EE-10137 Tallinn, Estonia

[3]Institute of History, 6 Rüütli St., EE-10130 Tallinn, Estonia

[4]National Heritage Board, 18 Uus St., EE-10111 Tallinn, Estonia

Abstract. We have studied in detail the main crater at Kaali with 98 borings and several diggings, carefully levelled. As a result, the first detailed morphologic map of the crater is presented. The diameter of the meteorite, which formed the main crater, was probably not more than 3 m and the weight about 40-50 t. Bronšten and Stanyukovich (1963) maintained that at the time of impact, its weight was probably 20-80 t. According to Pokrovski (1963), the diameter of the Kaali meteorite was evidently 4.8 m and its mass 450 t. Thus, the explosive energy released by the event was not powerful enough to induce remarkable environmental consequences. Pollen stratigraphy and radiocarbon dates up to 3390±35 BP from the bottom sediments in the main crater suggest that the Kaali crater group is more than 4000 years old. Based on an iridium-rich layer in the Piila Bog, 8 km NW of Kaali, Rasmussen et al. (2000) and Veski et al (2001 2001a) concluded that this event took place about 800 to –400 BC. Our budget calculations of the spatial-temporal distribution of mineral impurities in the peat layers do not support the hypothesis about direct extraterrestrial origin of Ir in the Piila Bog. The changes in the iridium content in the vertical profile of the peat section are due to natural geochemical processes and have no connection with the Kaali impact.

1
Introduction

At least nine impact craters are located in an area of one square kilometre within the Kaali crater field (58°24'N, 22°40'E), 19 km NE of the town of Kuressaare on Saaremaa Island (Fig. 1). This paper deals with the main crater, which has a diameter of 105–110 m and is about 22 m deep (Fig. 2). For the first time the crater was described 175 years ago (Luce 1827); however, up to our recent studies there was no authentic information available on its morphology. Reinwald* (Reinvaldt 1933) having performed a few drillings in the main crater reached the conclusion that the crater had a flat and symmetrical bottom. But in later publications the depression was depicted as an asymmetrical hollow with a deepened central part (Saarse et al. 1991). Different opinions have been expressed as to the energy needed for the formation of the crater. For truthful calculations of explosive energy new information on the crater morphology was urgently needed. Budget calculations for iridium distribution around the crater field were undertaken to find out the age of the main crater.

2
Early Descriptions of the Main Crater Morphology

Johann Wilhelm Ludwig von Luce (1756–1842) is usually acknowledged as the discoverer of the Kaali craters. From the description of the main crater he compiled and published in 1827 (Luce 1827) appears that he had mapped the lake as early as 1780. However, it is obvious that islanders were aware of the existence of the exotic hollow much earlier. The morphology of the crater and uplifted dolomitic rocks gave rise to several legends and tales.

It was believed that the lake had no bottom and its waters were hiding an entrance to the hell. This belief was partially triggered by statements of some lake researchers. For instance, Luce wrote in his work of 1827: "*The water in the lake is always clear and fresh and it is very deep. When I fifty years ago exerting all my strength threw an about-4-m-long pole into the lake, and it flew out of the water in a few seconds, and I grabbed it, I couldn't find any sign suggesting that the tip of the pole had touched the*

---

* In the scientific literature the name Reinwald is written in different ways: Reinwald, Reinwaldt, Reinvaldt, Reinvald. In this paper the first variant is used.

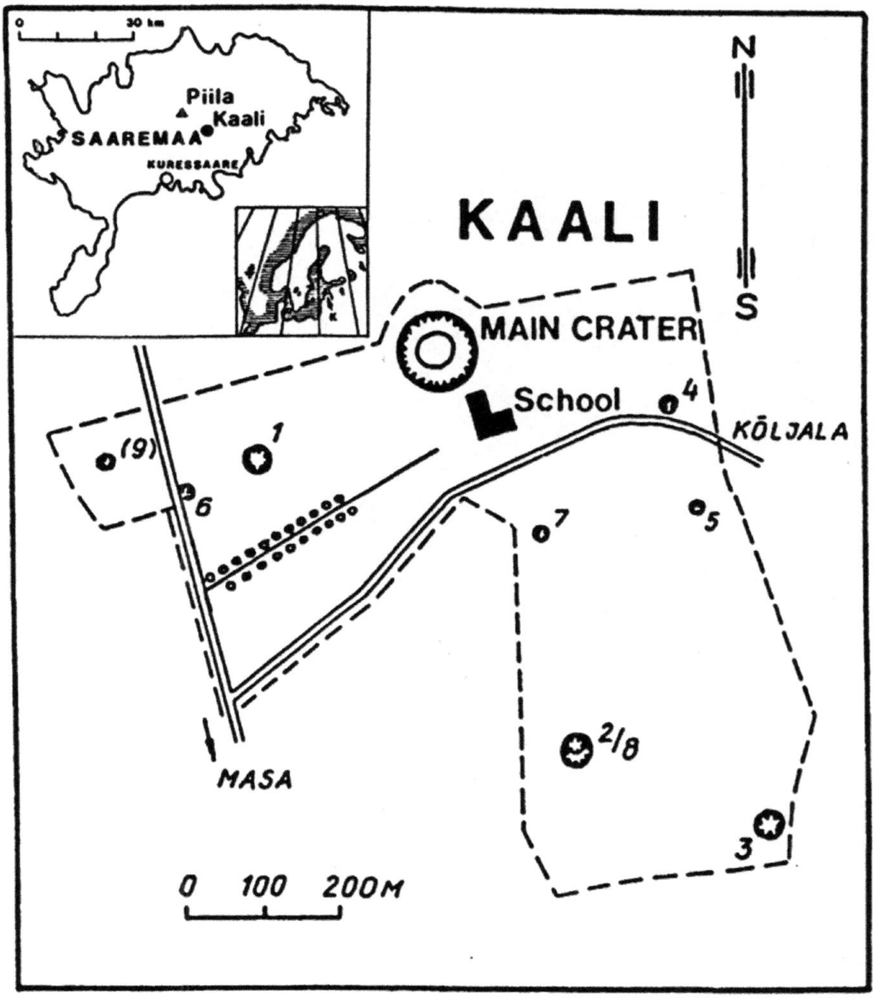

Fig. 1. Location of the Kaali impact craters and Piila Bog in the central part of the Island of Saaremaa. Numbers on the map indicate the location of small craters.

mud on the lake bottom". In reality, the water depth in the lake is 5 to 6 metres during high-water periods. During low-water periods the basin is almost dry (Fig. 2). The most rapid lake-level rise was recorded in 1927 when, as a result of heavy rainfall that lasted from 6 to13 September, the water rose by nearly 2 m (Lõugas 1996).

Hofman was the first to describe the structure of the Kaali main crater in 1837. He noted that the Kaali depression bore an astonishing resemblance to maars – volcanic funnels on the Eifel plateau – and the tilt of the

Fig.2. Lake Kaali in the main crater during the low stand of water on October 10, 2000. A peat sample from a depth of 1.18–1.25 m in the dighole in the northeastern side of the crater (the right bank on the photo) was dated as 2958±51 BP. Photo by T. Moora.

uplifted dolomite layers on the crater slopes indicated an explosion the force of which had been directed from bottom upwards (Hofman 1837). In 1839, during an archaeological expedition on the Island of Saaremaa, Kruse, professor of general history at Tartu University, compiled a schematic plan of Lake Kaali, showing also the location of park roads and surrounding buildings. The plan was published three years later (Kruse 1842).

Hofman's theory suggesting an abrupt eruption of water, steam, gas and mud in the Kaali crater area was supported by Wangenheim von Quelen (1849, a.o.), who presented a visual plan and assumable cross-section of the Kaali main crater. Linstow (1919), a supporter of gas eruption, presented also a scheme of the main crater, four smaller craters and the surroundings of the crater. As a source of gas eruption he considered the organic-rich Dictyonema argillite rocks well known in northern Estonia. In the area under consideration, their bedding depth is about 200 m.

In the autumn of 1927, following the order of the Mining Department of the Ministry of Trade and Industry of Estonia, Reinwald explored the solid bed in the craters by digging and boring (Reinwald 1928). One borehole was put down at the base of the main crater from outside to a depth of 63.14 m, and two smaller ones were sunk inside the wall, the form and character of the lakebed were explored and water-level variations were studied. Reinwald established that the bottom of the crater is approximately horizontal and has a somewhat curved form. A funnel-shaped depression with a prolonged form was discovered in the hard bottom. Ground magnetic measurements (Aaloe et al. 1976) and several borings (Saarse et al. 1991) showed that the main crater is a bit asymmetric, being about 0.4 m deeper at the northwestern margin than in the central part. As another peculiarity of the crater, Reinwald mentioned huge dolomitic blocks tilted upwards from inside at an angle of 30–40 degrees (Fig. 3). After clearing the slope on the north side of the crater, it became evident that the general thickness of the tilted dolomites was 8 m, and they were absolutely identical with the upper part of the dolomitic strata at the exterior base of the crater where the aforementioned borehole had been put down. Beneath these tilted dolomite blocks, white or slightly brownish powder with soft pieces of whitish dolomite was discovered. In the centre of the bottom large stones had been heaped up for a monument to a deceased owner of the estate. The stones originated partly from the interior of the crater itself, and partly from elsewhere.

Based on the above facts, Reinwald reached the conclusion that the craters were of meteoritic origin (Reinwald 1928). The same hypothesis had already been advanced in 1919 by Kalkun-Kaljuvee (in Kaljuvee 1933). In September 1927, Meyer and Kraus from Riga visited the Kaali

Fig. 3. Dolomites uplifted and destroyed during the Kaali impact. Photo by R. Tiirmaa.

area under the guidance of Reinwald. They were accompanied by Alfred Lothar Wegener (1880–1930), founder of the theory of continental drift. As a result of the short five-day field work Kraus, Meyer and Wegener (Kraus et al. 1928) more or less simultaneously with Reinwald persisted the idea of meteoritic origin of the Kaali craters. The same opinion was expressed by Hinks (1933), Fisher (1936, 1938) and Kranz (1937), but the last doubt about the genesis of the Kaali craters disappeared only in 1937 when Reinwald (1938) found meteoritic pieces containing nickel and iron (8.5 and 91.5%, respectively). Thus, at that time the Kaali craters were the second terrestrial object in the world – after the Diablo/Arizona crater in the USA– which could have been generated by extraterrestrial forces. With the decision of the Nature Conservation Council of Estonia the Kaali craters were taken under nature protection in November 1937.

3 Formation of the Crater

According to Reinwald (1933), the meteorite had pierced the dolomite massif by 2.5 m. The mighty impact was accompanied by heat which turned the water in the rocks instantaneously into steam and heated the gases, which had formed, so far that they caused burning of the rock not only at the level of explosion, but also at the depths they reached through a dense net of impact-generated cracks. At the same time, these huge quantities of instantaneously formed gases caused an acute explosion which shattered the surrounding rock thereby crushing it partly into rock-flour, and throwing it upwards out of the formed crater. The focus of the explosion was presumably situated in the place of the greatest pressure, i.e., directly under the solid body.

According to Reinwald, the meteorite that had produced the main crater had a diameter less than 3 m. His calculations showed that one of the small craters – crater 4 – with a diameter of 20 m was produced by a meteorite body the diameter of which was about 0.5 m, i.e. approximately 1/40 of the crater's diameter. In size, the meteorite generating the Kaali main crater is comparable to the Hoba meteorite in Namibia (length and width 2.7 m, height from the ground 0.9 m, mass ca 70 t) and Ahninghito meteorite (mass 34 t). However, compared to the Kaali meteorite, they hit the ground under a different angle and did not produce a crater.

With the death of Reinwald in 1941, the study of the Kaali craters was interrupted for a long time. It was resumed in 1955 on the initiative of Karl Orviku, head of the Commission on Meteoritics of the Estonian Academy of Sciences. The studies were carried out by researchers of the Institute of

Geology of the Academy of Sciences of the Estonian SSR, mainly by Ago Aaloe. Main attention focused on small craters, but in 1974 large-scale geophysical investigations were undertaken in the main crater in co-operation with researchers of Moscow State University (Aaloe et al. 1976). Using seismic and electrometric methods, the contours of the zone of shattered rocks surrounding the crater were determined. The zone surpasses the dimensions of the crater twice. The zone of shattered rocks has a specific contour, the axis symmetry of which is directed from east to west.

Opinions differ as to the direction from south to northeast and angle of incidence of the Kaali meteorite. On the basis of the size of the craters, Reinwald (1937a, 1937b) came to the conclusion that the direction of movement was from the east-southeast to the west-northwest. This idea was supported by Krinov (1962). It is known that a meteorite with a reasonable mass maintains its initial cosmic velocity for a longer time than smaller ones. Therefore, the larger one should be in front of the dispersal ellipse, and the smaller ones behind it (Fig. 1). This shows that the meteorite fell from the southeast (Raukas 2002). Based primarily on the study of impact traces at the bottom of craters 4 and 5, Aaloe (1958) maintained that the possible angle of incidence had been 35–40^0 relative to the horizon.

The energy needed for the formation of the Kaali main crater is estimated as 4×10^{12} J (or 10^{19} erg), and approximately two orders of magnitude less for the formation of small craters (Bronšten and Stanyukovich 1963). The initial velocity of the meteorite with an initial mass of 400–10,000 t (most probably ~1000 t upon entering the atmosphere) is estimated as 15–45 km/s. At the time of impact, its weight was probably 20–80 t and its velocity was 10–20 km/s (Bronšten and Stanyukovich 1963). According to Pokrovski (1963), the diameter of the Kaali meteorite was probably 4.8 m, its mass 450 t, and its impact velocity 21 km/s. Koval (1974) suggested a somewhat lower velocity (~13 km/s) of the Kaali meteorite at impact. His calculations showed that the mass of the meteorite, which produced the main crater, must have been about 40–50 t. According to Koval, the meteorites that formed the small craters may have weighed between 1 and 6 t. Thus, the force of the Kaali meteorite was too small to induce thus great environmental consequences as maintained by Veski et al. (2001a, 2001b). To our mind, its explosion did not cause any serious ecological catastrophe in the surroundings.

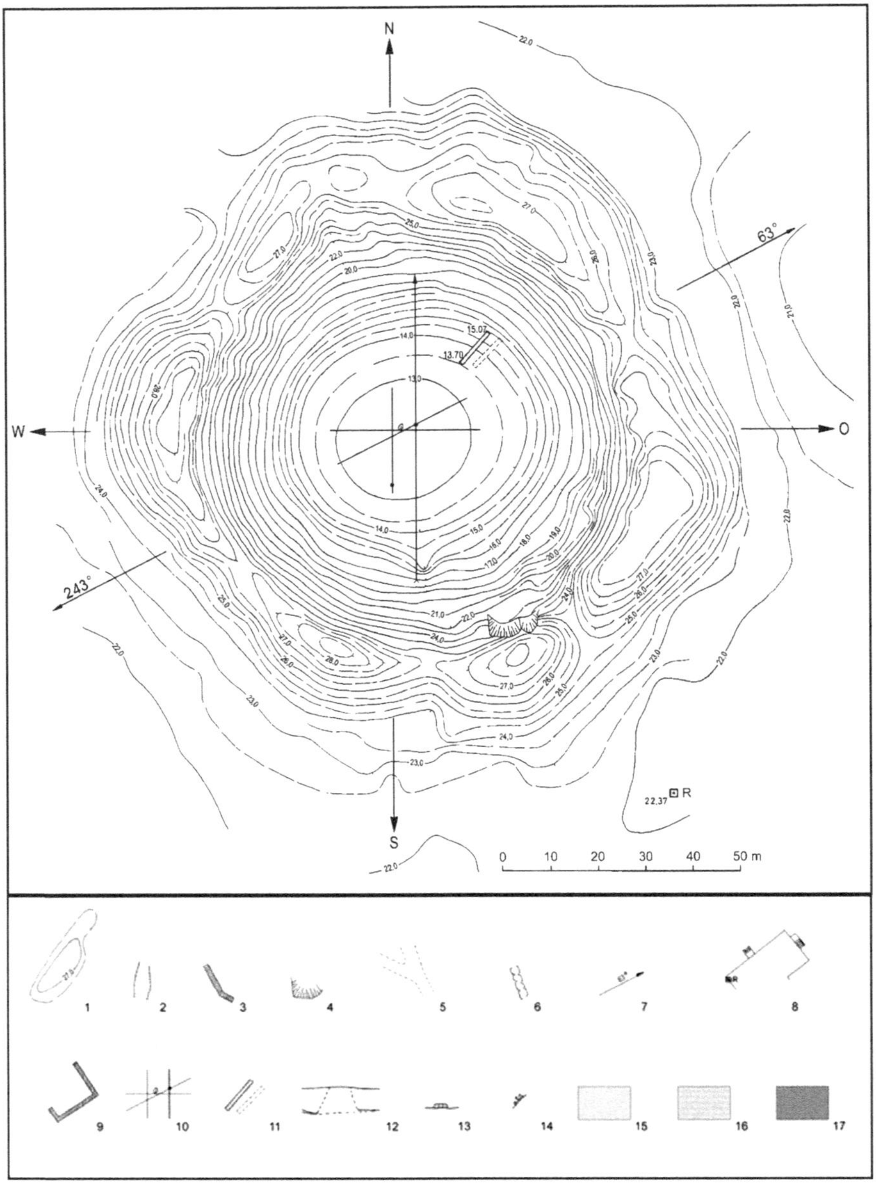

Fig. 4. Morphology map of the Kaali main crater and the profiles of the crater in different directions. Abbreviations: R – bench mark (22.37 m a.s.l.); ABSO – absolute zero level of the Baltic System; K – digging; T – trench. Legend: 1– isohypses; 2 – old diggings; 3 – steps; 4 – steep slope; 5 – footpaths and small roads; 6 – stone wall; 7 – profiles; 8 – schoolhouse with bench mark (R); 9– old foundation; 10 – crossing point of profiles on top of the heap of stones; 11 – digholes made in 2001 (see Fig. 2); 12 – heap of stones in the centre of the lake; 13 – cross section of the stone wall; 14 – stones on the crater's slope. On the profiles: 15 – water; 16 – lake sediments; 17 – stones on lake bottom.

4
Recent Data on the Crater Structure

When Reinwald studied the geology of Lake Kaali in 1937, he tried to determine the thickness of lake sediments. However, none of the boreholes penetrated the mud layers, which were full of stones and tree trunks. The aim was achieved not until the autumn of 1978 when the water was partly pumped out from the lake depression. Paap and Kessel penetrated the sediment layer by means of two boreholes.

During the expeditions of 1989–1990, the working group under the guidance of Saarse laid out 6 boreholes in the crater. In 2000 and 2001, some more boreholes were made by Veski. However, the data obtained this way was insufficient for elucidating the morphology of the lake floor. In 1999–2000, Kestlane and Moora made 98 boreholes and several diggings in the crater and studied the crater floor structure in particular detail. The results are presented in Fig. 4.

Regretfully, human activities have caused great changes in both the Kaali crater field and the main crater. For the buildings erected northeast of the main crater and in the Kaali manor yard material was taken from the inner slope of the main crater and its surroundings (Reinwald 1928). The top of the crater mound was flattened for a hiking path. A wide passage was dug into the mound from the side of the schoolyard to offer a convenient access to the lake. Next to the passage, on the inner slope of the mound, there is a flattened plot with a bench. Great load of visitors – some 10,000 people annually – promotes erosion of the crater slope and accelerates filling of the lake depression with sediments.

5
On the Age of the Impact

Traditionally, in Estonia the age of the impact is placed at some 4000 yr BP. Based on the investigation of microimpactites (mainly glassy spherules formed by melting and vaporization of meteoritic matter and target rocks), Raukas (2000; Raukas et al. 1995, 1999) reached the conclusion that the formation of the craters took place ca. 7500 yr BP.

On the basis of very scanty and contradictory archaeological evidence obtained from the burning of ancient strongholds at Asva and Ridala on the Island of Saaremaa, some archaeologists reached the conclusion that the Kaali meteorite could not have fallen before the turn of the seventh to the eighth centuries BC, i.e. about 2600 years ago (Lõugas 1995). The idea was supported by the geologist Aaloe (Aaloe et al. 1975; Aaloe 1981), who based his conclusion on radiocarbon dates of charcoal from the bottom of the twin craters 2 and 8 (2530±130 BP, TA-19 and 2660±250 BP, TA-22), and from the bottom of crater 4 (2920±240 BP, TA-769). According to Aaloe, the age of the craters is 2800±100 BP. The analysis of trace elements in peat cores taken from the Piila Bog ($58^0$25'N, $22^0$36'E), 8 km northwest of the craters (Fig. 1), suggested an age of 2400–2370 BP for the impact (Rasmussen et al. 2000). This weakly grounded idea was recently supported by Veski et al. (2001a, 2001b). In these papers the authors assumed that the impact caused serious geochemical (Rasmussen et al. 2000) and ecological changes (Veski et al. 2001a, 2001b) in the surrounding environment. In their analyses the authors of the mentioned papers have made some conclusions that are in contradiction to the geochemical and stratigraphical data presented as well as to the accepted palaeoecological knowledge about the development of natural processes.

The main argument in both papers is that the increased content of mineral matter and iridium in peat layers at a depth of 172–183.5 cm (dated by ^{14}C as 2307±22 to 3030±85 BP, see Rasmussen et al. 2000, Fig. 2 and Table 2) is a direct evidence of the eject of extraterrestrial matter and that this event is synchronous with the Kaali meteorite fall. Using the data presented in the above-mentioned papers, it is possible to calculate the minimum total amount of dispersed mineral matter and Ir in the surroundings of the Kaali impact.

As Rasmussen et al. (2000) mentioned, the Ir concentration in the samples in the peat section from the Surusoo mire some 25 km away from Lake Kaali was below the Ir detection limit (10–60 pg). Also it is known that all mineral matter (and naturally iridium) on bog peat is allochthonous. Thus, we can make very simplified calculations of ejected matter assuming that in the case of extraterrestrial origin Ir was distributed evenly on the circle with a radius of 8 km around the impact craters. Naturally, it is the minimum value, because the distribution of atmospherically dispersed matter has a logarithmic character decreasing from the center to the periphery, and we do not know the spatial limit of the distribution of Ir (its content in the peat was nearly zero at a distance of 25 km from the epicenter).

Using these simplifications it is possible to calculate the total influx of Ir on the surface. Figure 2, of Rasmussen et al. (2000) shows an elevated distribution of Ir in the layers from a depth of 172 to at least 182 cm. The authors explain the vertical distribution of Ir over sediments formed during some 700 years with mechanical falling of particles through peat surface or by percolating rainwater. For budget calculations we can use the average Ir influx value of some 400 pg/g in the 10-cm-thick peat layer with a mean density 0.2 g/cm^3. Thus, the average Ir spatial input would be ca 8x10^{-10} g/g cm^2. Now we can calculate the total amount of extraterrestrial Ir influx over the area of a circle with R=8 km (that is ca 2x10^{11} cm^2). Thus, the total amount of dispersed Ir over the area from the Kaali crater up to the Piila reference site is not less than 1600 g. Rasmussen et al. (2000) took the Ir concentration in the Kaali meteorite after Kracher et al. (1980) equal to 2.8 µg/g (see also Czegka and Tiirmaa 1998). Thus, for 1600 g Ir at least 600 t of meteoritic matter must be dispersed. According to different authors, the maximum mass of the meteorite was 400–450 t at entry to the atmosphere and, naturally, not all of it was dispersed.

Analogous mineral matter calculations show that at least 28x10^4 t was dispersed around the crater. The sources of mineral matter are the meteorite itself and the mass of matter ejected from the crater by explosion. As the mass of the meteorite was less than 500 t, the mass of the ejected matter can be calculated from the measures of the crater (R=50 m, h=15 m; nearly conic in shape, Fig. 4) assuming the mean density of the matter to be 2.6 g/cm^3. The maximum amount of the ejected mineral matter is less than 100 000 t (not subtracting the rim around the crater), which is much less than geochemical data of the peat sequence indicate (280 000 t). As mentioned before, because of simplified presumptions, all calculated values are manifold lower than the real data.

Summarising the above, it is obvious that the spatial-temporal distribution of mineral matter and iridium in the Piila Bog is not directly connected with the impact of the Kaali meteorite. The increase in mineral matter and, thus, the iridium in the peat layers are most probably related to the input of the deflated matter from surroundings. A very strong signal confirming this is changes in the lithology of the peat sequence at a depth of 171 cm, where fen peat is replaced by bog peat (Veski et al. 2001a). Naturally, this extremely strong geochemical barrier caused changes in both chemical and mechanical migration as seen vividly from the vertical distribution of Ir. The changes in the lithological composition of the peat sequence testify also very clearly that like in many areas of northern Europe, climate became drier around 2500–2300 years ago. Naturally, such deep changes in the climate seriously affected the landscapes and vegetation. Degradation of vegetation and changes in land cover caused increasing deflation in areas surrounding the mire and an influx of mineral matter to the peat surface. The mean concentration of Ir in the total amount of dispersed mineral matter is approximately 0.4 ng/g, which is a rather typical background level of the Earth's crust (e.g., Alvarez et al. 1980). Even the value 1 ppb is sometimes given (Greenwood and Earnshaw 1998). An increase in the proportion of mineral matter transported by wind from the surroundings is obvious also from the paper by Veski et al. (2001b) where much attention is paid to forest fires. As is well known (Huttunen 1980), in the boreal zone regular forest fires generally occur after every hundred years. Increasing dryness, fixed in the Piila peat sequence, might be the reason of forest fires that increased greatly the activity of deflation.

Therefore, the speculations about the ecological catastrophe caused by the Kaali meteorite on Saaremaa Island have no scientific basis. Rich palaeoecological material in the paper by Veski et al. (2001b) describes, in our opinion, the natural process and has no connection with the Kaali impact.

Already the palynological analyses by Kessel (1981) showed that the basal sediments in the Kaali main crater are at least 3500 years old. Saarse et al. (1991) obtained a ^{14}C age of 3390±35 BP: Tln-1353 for the bottom sediments in the main Kaali crater. However, there is no proof that the sediments studied actually originated from the base of the section. The time interval between crater formation and the age of the lowermost radiocarbon-dated sample from the crater lake deposit is also uncertain. Undoubtedly, the heavily crushed crater bottom was for a long time dry and the sediments which accumulated during a high stand of water were washed away via cracks like in typical karst hollows. In the dighole in the flank of the crater (Fig. 2), much higher from the bottommost part of the crater, we recently dated peat (2958±51 BP: Tln-2576). Considering all the above, there is no doubt that the Kaali craters are older than 4000 years.

6 Conclusions

We have compiled a detailed morphological map of the Kaali crater, which allows to establish the formation energy of the crater. The results obtained through direct dating of organic matter in the main crater and palynological studies show that the craters are at least 3390±35 years old; however, actually they should be much older. The crater-producing impact was rather weak and did not cause any ecological catastrophe in the environment.

Acknowledgements

The authors thank Mrs. Helle Kukk, Mrs. Reet Tiirmaa and Dr. Rein Vaher for assistance with the manuscript. The financial support of European Union Culture 2000 Project "European Pathways to Cultural Landscapes" is appreciated.

References cited

Aaloe A (1958) Kaalijärve meteoriidikraatri nr. 5 uurimised 1955. aastal [Studies of Kaali meteorite crater no.5 in 1955]. ENSV TA Geoloogia Instituudi Uurimused 2: 105–117 (in Estonian, with English and Russian summaries)

Aaloe A (1981) Erinevused Kaali kraatrite vanuse määrangutes [Discrepancies in dating the Kaali meteorite craters]. Eesti Loodus 4: 236–237 (in Estonian, with English and Russian summaries)

Aaloe A, Eelsalu H, Liiva A, Lõugas V (1975) Võimalusi Kaali kraatrite vanuse täpsustamiseks [On the correction of the age of the Kaali meteorite craters]. Eesti Loodus 12: 706–709 (in Estonian, with English and Russian summaries)

Aaloe A, Dabizha B, Karnaukh B, Starodubtsev V (1976) Geophysical investigations of Kaali crater. Eesti NSV Teaduste Akadeemia Geoloogia Instituudi Toimetised, Keemia, Geoloogia 25/1:58–65 (in Russian, with English and Estonian summaries)

Alvarez LW, Alvarez W, Asaro F, Michel HV (1980) Extraterrestrial cause for the Cretaceous-Tertiary boundary extinction. Science 271: 1095–1108

Bronšten V, Stanyukovich K (1963) On the fall of the Kaali meteorite. Eesti NSV Teaduste Akadeemia Geoloogia Instituudi Uurimused 11: 78–83 (in Russian, with English and Estonian summaries)

Czegka W, Tiirmaa R (1998) Das holozäne Meteoritenkraterfeld von Kaali auf Saaremaa (Ösel), Estland. Aufschluss 49: 233–252

Fisher C (1936) The meteor crater in Estonia. Natural History 38/4: 292–299

Fisher C (1938) Exploring Estonian meteor craters. Sky Magazine of Cosmic News 2/5:8–9

Greenwood NN, Earnshaw A (1998) Chemistry of the Elements, 2nd edn. Oxford, Auckland, Boston, Butterworth-Heinemann

Hinks AR (1933) Kurznotiz oder Überschrift. Geographical Journal 82: 375

Hofman E (1837) Geognostische Beobachtung auf einer Reise von Dorpat bis Abo. Dorpat

Huttunen P (1980) Early land use, especially the slash-and-burn cultivation in the commune of Lammi, Southern Finland, interpreted mainly using pollen and charcoal analyses. Acta Botanica Fennica 113: 1–45

Kaljuvee J (1933) Die Grossprobleme der Geologie. Tallinn (Reval) F. Wassermann

Kessel H (1981) Kui vanad on Kaali järviku põhjasetted? [How old are the bottom deposits of Lake Kaali]. Eesti Loodus 4: 150–155 (in Estonian, with English and Russian summaries)

Koval VI (1974) About the mass and composition of the Kaali meteorite. Astronomicheskiye vesti 8/3: 169–176 (in Russian)

Kracher A, Willis, J, Wasson JT (1980) Chemical classification of iron meteorites – IX. A new group (II F), revision of I AB and III CD, and data on 57 additional irons. Geochimica et Cosmochimica Acta 44: 773–787

Kranz W (1937) "Krater von Sall" auf Ösel Wahrscheinlich "Meteoritenkrater". Gerlands Beiträge zur Geophysik 51: 50–55

Kraus E, Meyer R, Wegener A (1928) Untersuchungen über den Krater von Sall auf Ösel. Gerlands Beiträge zur Geophysik 20: 312–378

Krinov EL (1962) Meteorite craters on the surface of the Earth. Meteoritika 22:3–30 (in Russian)

Kruse F (1842) Necrolivonica oder Alterthümer Liv-, Ehst- und Curlands. Dorpat

Linstow O (1919) Der Krater von Sall auf Oesel. Zentralblatt für Mineralogie und Geologie 21/22: 326–339

Luce JW von (1827) Wahrheit und Muthmassung Beytrag zur ältesten Geschichte der Insel Ösel. Pernau 1827, 20–22

Lõugas V (1995) Must auk Saaremaa ajaloos [Black hole in the history of Saaremaa]. Luup 2: 52–56 (in Estonian)

Lõugas V (1996) Kaali kraatriväljal Phaethonit otsimas [Searching for Phaethon in the Kaali crater field]. Tallinn Eesti Entsüklopeediakirjastus (in Estonian)

Pokrovski G (1963) Computation of the parameters of a meteorite according to the crater caused by its fall. Eesti NSV Teaduste Akadeemia Geoloogia Instituudi Uurimused 11: 61–71 (in Russian, with English and Estonian summaries)

Rasmussen LK, Aaby B, Gwozdz R (2000) The age of the Kaalijärv meteorite craters. Meteoritics and Planetary Science 35: 1067–1071

Raukas A (2000) Investigation of impact spherules – a new promising method for the correlation of Quaternary deposits. Quaternary International 68–71: 241–242

Raukas A (2002) Postglacial impact events in Estonia and their influence on people and the environment. In Koeberl C, MacLeod KG (eds) Catastrophic Events and Mass Extinctions: Impacts and Beyond. Boulder, Colorado, Geological Society of America Special Paper 356: 563–569

Raukas A, Pirrus R, Rajamäe R, Tiirmaa R (1995) On the age of meteorite craters at Kaali (Saaremaa Island, Estonia). Proceedings of the Estonian Academy of Sciences, Geology 44: 177–183

Raukas A, Pirrus R, Rajamäe R, Tiirmaa R (1999) Tracing the age of the catastrophic impact event in sedimentary sequences around the Kaali meteorite craters on the Island of Saaremaa, Estonia. Journal of the European Network of Scientific and Technical Cooperation for the Cultural Heritage (PACT) 57: 435–453

Reinwald I (1928) Bericht über geologische Untersuchungen am Kaalijärv (Krater von Sall) auf Ösel. Loodusuurijate Seltsi Aruanded 35: 30–70

Reinwald I (1933) Kaali järv – the Meteorite Craters on the Island of Ösel (Estonia). Loodusuurijate Seltsi Aruanded 39:183–202

Reinwald I (1937a) Meteoorkraatrid Saaremaal [Meteorite craters in Saaremaa]. Looduskaitse 1: 118–131 (in Estonian)

Reinwald I (1937b) Kaali järve meteoorkraatrite väli [The Kaalijärv field of meteorite craters]. Loodusvaatleja 4: 97–102 (in Estonian)

Reinwald I (1938) The finding of meteoritic iron in Estonian craters. A long search richly awarded. The Sky Magazine of Cosmic News 2/6: 6–7

Saarse L, Rajamäe R, Heinsalu A, Vassiljev J (1991) The biostratigraphy of sediments deposited in the Lake Kaali meteorite impact structure, Saaremaa Island, Estonia. Bulletin of the Geological Society of Finland 63: 129–139

Veski S, Heinsalu H, Kirsimäe K, Poska A, Saarse L (2001a) Kaali meteorite impact induced ecological catastrophe 800–400 BC, as revealed by pollen, XRD, LOI and ^{14}C analyses of peat containing impact ejecta. In 6th ESF–IMPACT Workshop Impact Markers in the Stratigraphic Record. Abstract Book. Granada (Spain) May 19–25, 2001, pp 132–135

Veski S, Heinsalu A, Kirsimäe K, Poska A, Saarse L (2001b) Ecological catastrophe in connection with the impact of the Kaali meteorite about 800–400 B.C. on Saaremaa Island, Estonia. Meteoritics and Planetary Science 36:1367–1375

Wangenheim von Qualen F (1849) Der Krater bei Sall auf der Insel Oesel. Korrespondentzblatt des Naturforscher–Vereins zu Riga 3: 48–68

The Lockne Crater: Revision and Reassessment of Structure and Impact Stratigraphy

Maurits Lindström[1], Jens Ormö[2], Erik Sturkell[3], and Ilka von Dalwigk[4]

[1]Department of Geology and Geochemistry, Stockholm University, S-10691 Stockholm, Sweden (Maurits.Lindstrom@geo.su.se)

[2]Centro de Astrobiología (CSIC/INTA), Ctra de Torrejón y Ajalvir, km 4, E-28850 Torrejón de Ardoz, Madrid, Spain (ormo@inta.es)

[3]The Icelandic Meteorological Office, Bustadarvegur 9, IS-150 Reykjavik, Iceland (erik@vedur.is)

[4]Department of Geology and Geochemistry, Stockholm University, S-10691 Stockholm, Sweden (ilka@geo.su.se)

Abstract. The Lockne impact crater in central Sweden has features characterizing a relatively deep marine environment. Recently, improved outcrop (road construction, lumbering, etc.) has favoured examination of important features, resulting in greater precision and understanding. Under a water depth of at least 500 m, the target seabed consisted of 80 m Cambrian and Ordovician sediments. About one-half of this succession was lithified limestone; the other half was soft claystone and semi-lithified limestone. This succession rested on a peneplain cut into crystalline basement. The crater in the basement is just over 7 km wide. The sub-Cambrian peneplain can be followed all the way up to this crater, which thereby is shown to lack any significantly raised rim. The peneplain next to the crater was stripped of most sediment by the expanding water crater. Instead, there is a roughly 50 m thick brim of ejected crystalline rock, resting on the peneplain and on some remaining sediment. The brim is 2.5 km wide in the west, but less than 1 km in the east. Crystalline ejecta bodies, each with a volume of thousands of cubic meters, are scattered towards westerly directions as far as 9 km from the crater centre. An oblique impact from an easterly direction is indicated (Shuvalov et al., this volume). Radial furrows through the brim ("resurge gullies") might have originated during ejection but were utilized by resurging seawater. Target carbonates in the wider surroundings were brecciated during the ejection and excavation stage. During the resurge stage, another breccia was deposited including both limestone and redeposited clasts of crystalline ejecta.

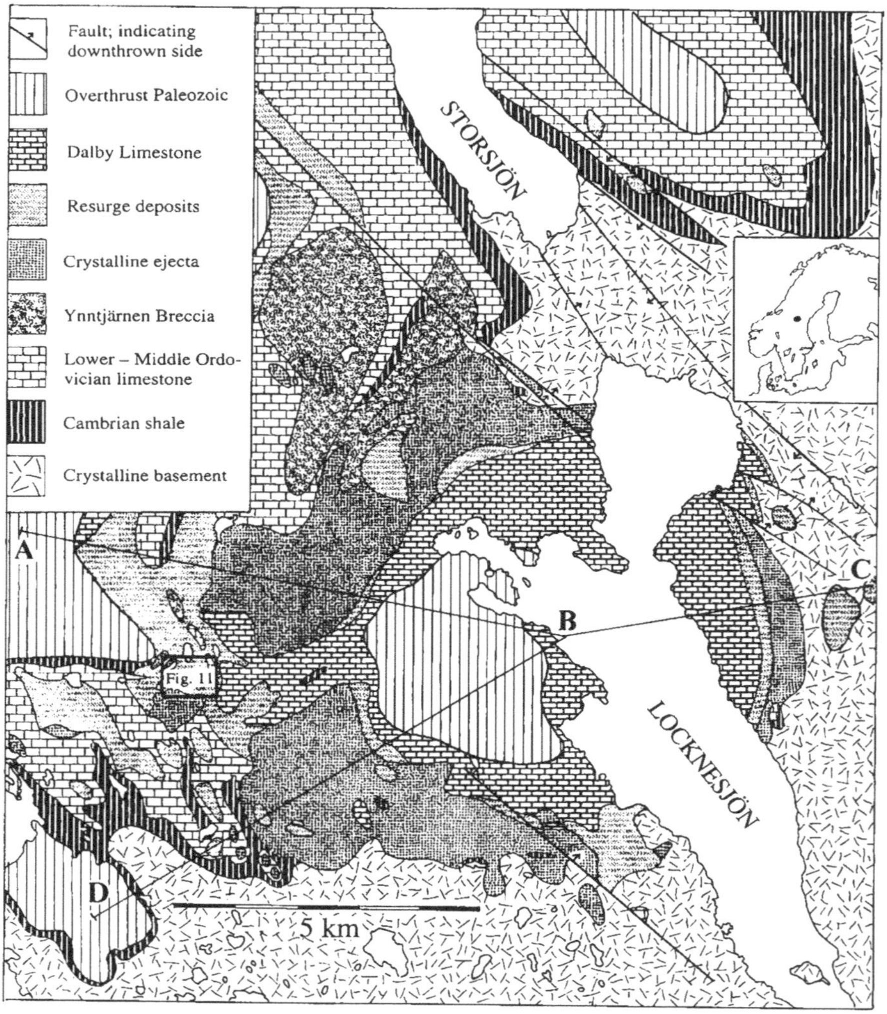

Fig. 1. Geology of the Lockne area. The corresponding topographic information is given on **Fig. 2**. *Insets*: Right: Position of Lockne on the Scandinavian Peninsula. Lower left: Position of the Ynntjärnen area (**Fig. 11**). *A-B, B-C, D-B*: Sections shown in **Fig. 3.**

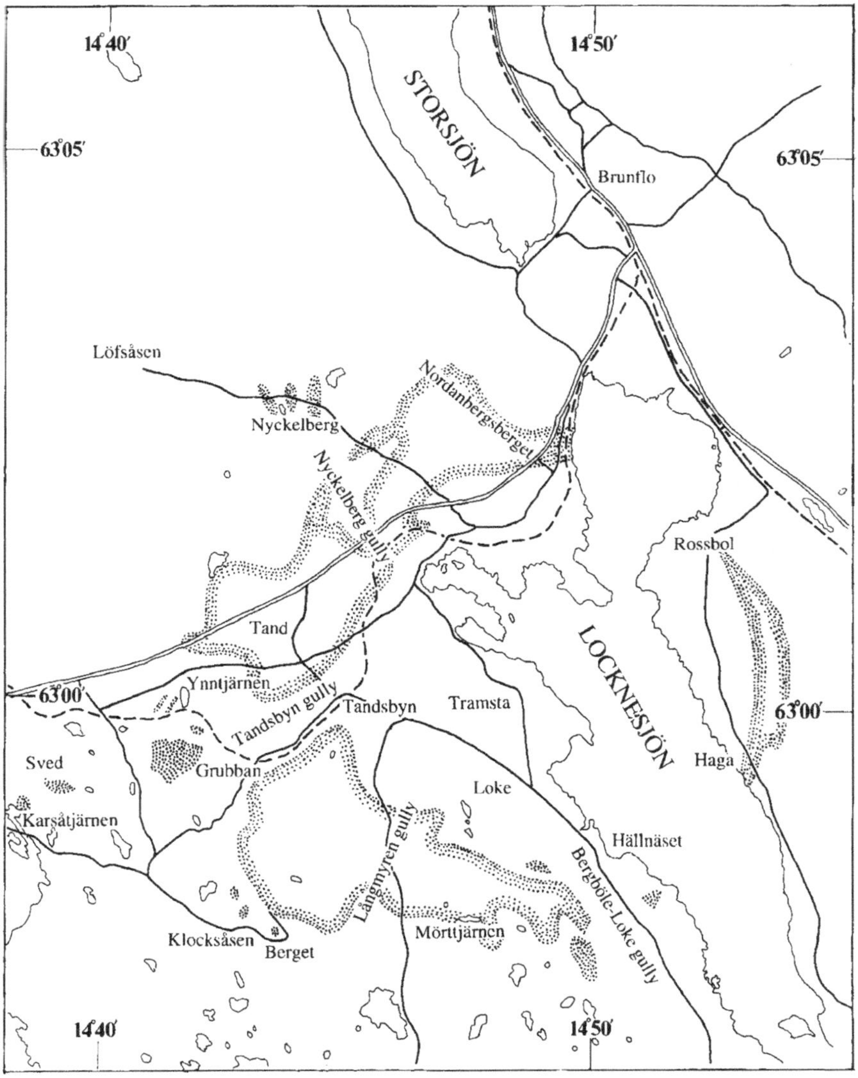

Fig. 2. Topographic information relating to **Fig. 1**. The road net (solid lines) and railroads (broken lines) are shown. For orientation relative to the geology the principal bodies of crystalline ejecta are indicated by stippling.

1 Introduction

The Lockne crater in central Sweden (N 63°00'20", E 14°49'30") (Fig. 1, 2 and 3) has become established as a reasonably well preserved and displayed marine-impact structure with fair accessibility. It is situated at the present Caledonian front, to the east of which Caledonian overthrust nappes have been stripped away by erosion. The crater owes its preservation to a cover of nappes that existed in the past, and its exposure to the removal of these nappes by erosion. Caledonian and maybe also later deformation influenced the present structure of the crater to some minor extent that we shall discuss.

Essential data concerning the structure were summarized by Lindström and Sturkell (1992), Lindström et al. (1996), and Sturkell (1998a). Since 2000, important new data have come to light owing to large-scale lumbering and the progress of quarrying. Furthermore, numerical modeling (Ormö and Miyamoto 2002, Ormö et al. 2002, Shuvalov et al., this volume) has highlighted the hitherto neglected significance of some of the earlier observations. Hitherto unrecognized features of the crater can be regarded as connected with the depth of the target sea that might have been as much as 800 m. The purpose of this paper is to identify and discuss such features.

2 Terminology

The *inner crater* (Fig. 4) is a just over 7 km wide, almost circular depression floored by crushed, crystalline basement rocks at depths mostly over 100 m below the surface, and filled with sediments laid down either as resurge deposits within minutes after the impact or as secular deposits during the following epoch. The inner crater appears in the landscape as a relatively smooth, low terrain, much of which is surrounded by hills that are several tens of meters higher. The *rim* of this inner crater is a narrow (not over tens of meters wide) zone in which crushed crystalline basement comes to the surface and meets the filling referred to above. In most cases this happens in the lower slope of the surrounding hills. Thus defined, the rim is not appreciably raised.

The *brim* is an essentially flat and as much as 2.5 km wide zone of more or less crushed basement rocks. It surrounds the inner crater, from which is delimited by the rim. The term brim alludes to the resemblance to the brim

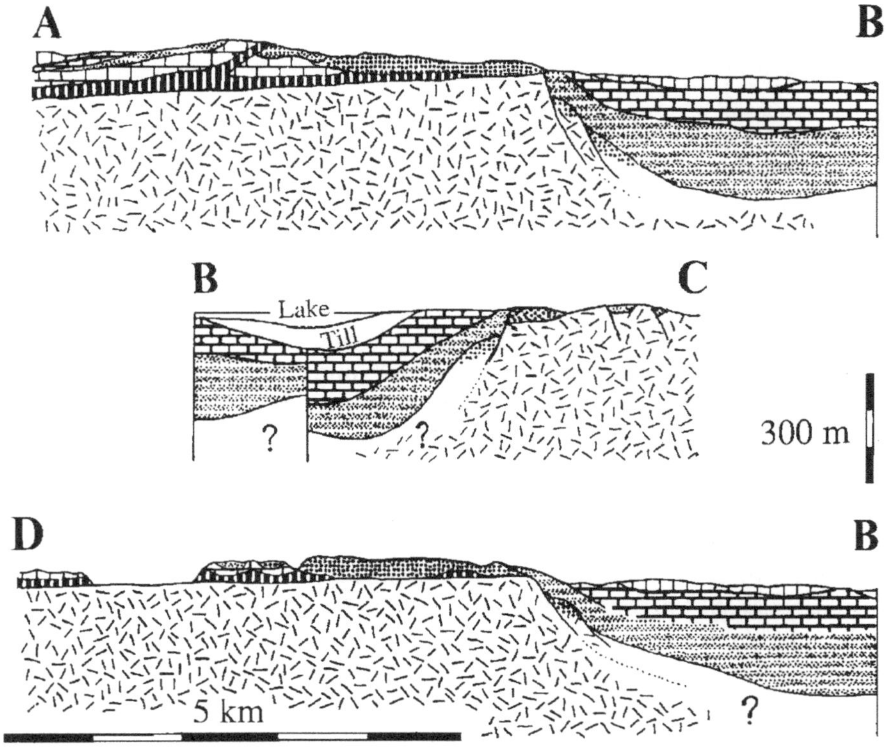

Fig. 3. Geological sections through the Lockne crater. The sections are located as shown in **Fig. 1**. The orientation of the sections is with west to the left. *B* is at the centre of the inner crater. Symbols are the same as in **Fig. 1**.

of a conventional-style plate, the deeper, inner part of which corresponds to the inner crater. For a comparison with the brim of a hat, one must think of the hat as turned upside down.

Resurge gullies cut radially through the brim. They are tens of meters deep, and can be filled with resurge deposits and post-impact secular deposits.

The twelve-figure references given for different localities identify their positions on the Swedish National Grid (=modified UTM) coordinates that are used on Swedish ordnance survey maps, such as the 1:50,000 topographic maps in common use in the country. They give the position to the nearest ten meters.

3
Research History

Lockne was known since 1900 for its unique, coarse-clastic facies of Baltoscandian Ordovician rocks which were interpreted as the products of a high-energy coast (Wiman 1900, Thorslund 1940). According to Lindström (1971) there was hardly any place for a high-energy coast in the Baltoscandian Ordovician, and Lindström et al. (1983) regarded the Ordovician coarse clastics of Lockne as debris flows, to which the name Lockne Breccia was formally applied, and turbidites. A gravity-flow model was subsequently elaborated by Simon (1987a), who furthermore launched the discussion of two more strange elements of the Ordovician at Lockne. For one thing there was the regular occurrence of an abundance of decomposed fragments, apparently of volcanic ash, in arenitic beds of a turbidite-like rock that is locally known as loftarsten (in this paper we will use the anglicized version loftarstone). For the other, there is a monomictic breccia of crystalline rock, mainly granite (Tandsbyn Breccia, Lindström and Sturkell 1992) that had been interpreted by Thorslund (1940) as a product of Early Palaeozoic, terrestrial weathering. Simon admitted his failure to explain either of these occurrences from what could be reconstructed of the local geological evolution, but stated emphatically that he could not see in the monomictic crystalline breccia any evidence of weathering; instead, he saw evidence of mechanical crushing.

On evidence from Simon (1987a) Wickman (1988) postulated the existence of a major meteoritic impact structure in the area. Although impact-related rocks are concentrated about Locknesjön (Lake Lockne) and rapidly get rare and ultimately disappear away from that area, Wickman assumed the impact to have its centre about 20 km northwest of Lockne, which is out of the question in view of the distribution of any potential impact products. The shape of the inner crater at Lockne could be approximately identified subsequently, the diameter being 7-8 km (Lindström et al. 1991, Lindström and Sturkell 1992). After the impact stratigraphy had been identified at Tvären, another coeval crater formed at sea (Lindström et al. 1994), a similar stratigraphy was found to occur at Lockne (Lindström and Sturkell 1992, Lindström et al. 1996). Whereas a fallback breccia was found in the Tvären crater, this feature could not be identified at Lockne. Both craters have a thick, water-laid deposit of fining-upward clastics consisting of both ejecta and rocks eroded from the surroundings of the crater. Sea-water resurging into the crater after collapse of the transient water crater is held responsible for the deposition of this sediment. The sediment was taken to be a characteristic feature of craters excavated into the bottom of any sea that is not too shallow for a major movement of water (Ormö and Lindström 2000).

4 Stratigraphy

The bedrock outside the brim consists of Proterozoic crystalline basement that is cut by a fairly smooth, sub-Cambrian peneplain with a slight, secondary inclination towards the northwest (Karis and Strömberg 1998). On the sub-Cambrian peneplain there rests a Lower Palaeozoic succession consisting of about 30 m thick, bituminous shale of Middle to Late Cambrian age (Kläppe Shale), and about 50 m of mainly Lower to Middle Ordovician limestone (Karis and Strömberg 1998). The limestone is bedded, with argillaceous interbeds at some levels. The uppermost limestone bed formed immediately before the impact, and carbonate deposition of identical kind continued immediately afterwards. The Cambrian shale can be followed to the margin of the brim. The upper parts of the pre-impact Ordovician limestone are missing near the crater and get complete only at distances over 7 km from the centre of the impact structure. The pre-impact Ordovician is covered unconformably by polymictic resurge deposits. These, as well as other circumstances have been taken as evidence for cratering extending as far as about 12 km from the touchdown point of the impactor (Sturkell 1998b, Ormö and Lindström 2000).

Those impact products that rest unconformably on pre-impact sedimentary rocks include, in addition to resurge deposits, also large masses of crystalline ejecta (Sturkell 1998b, Lindström and von Dalwigk 2002). They have been found as far as 9 km from the centre of the impact.

The arenitic resurge deposits (loftarstone) are referred to above. The coarse clastics, most of which are sedimentary rocks, which range from gravel to megablocks, have been collectively named Lockne Breccia. The original definition of this breccia (Lindström et al. 1983) includes also large, isolated masses of crystalline ejecta that are surrounded by sedimentary breccia. The argument for the very wide scope of this lithostratigraphic unit must now be viewed as historical and cannot be upheld in the light of new observations. Rocks previously brought to the Lockne Breccia are here interpreted as belonging to three different units (Fig. 4). The first (here called the Ynntjärnen Breccia) consists of pre-impact Ordovician beds with evidence of strong stirring that was induced either by the growth of the water crater at their upper contact or by the forceful descent of crystalline ejecta upon them. The second group of rocks consists of major (tens to hundreds of meters wide) bodies of crystalline ejecta that were thought to be embedded in the Lockne Breccia, although they actually rest on the Ynntjärnen Breccia. Though the degree of cataclasis varies, much of these ejecta consists of monomictic breccia that was called Tandsbyn Breccia by Lindström and Sturkell (1992). The third group of rocks consists of coarse, polymictic clastics that were deposited mainly by resurge. The name Lockne Breccia will be reserved for this third unit.

5 The Inner Crater

The inner crater forms low ground round the northern part of Locknesjön. It is centred at Tramsta, a couple of hundred meters off the shore of the lake. The diameter is just over 7 km. In the central part of the structure there is a thin (tens of meters) outlier of the lowermost Caledonian overthrust nappe. The near-surface bedrock of the rest of the inner crater is Middle Ordovician Dalby Limestone that formed during a prolonged interval after the impact, with a maximum thickness of at least 88 m. This thickness is reduced to a matter of about 10 m near the rim of the inner crater.

The Dalby Limestone is underlain by resurge sediments that reach at least 203 m thickness in the central part of the inner crater, diminishing strongly towards the rim. These deposits are fining upwards, the lower part consisting of Lockne Breccia in the restricted sense discussed later in this paper, whereas the upper part is loftarstone, beginning with fine gravel and coarse sand, and ending upward with fine silt.

The Lockne Breccia rests on monomictic crystalline (mainly granitic) Tandsbyn Breccia, in which the deeper drill-holes ended, at 335 m in the hole that was deepest. No body of melt rock has been identified. However, this circumstance may be due to the lack of deep drill-holes in the central part of the inner crater, rather than to a genuine lack of melt rock within the crater. The abundant presence of melt fragments in the loftarstone resurge arenite demonstrates that melt was indeed produced by the Lockne impact.

Lindström et al. (1996) reported on geophysical evidence for the existence of a central uplift.

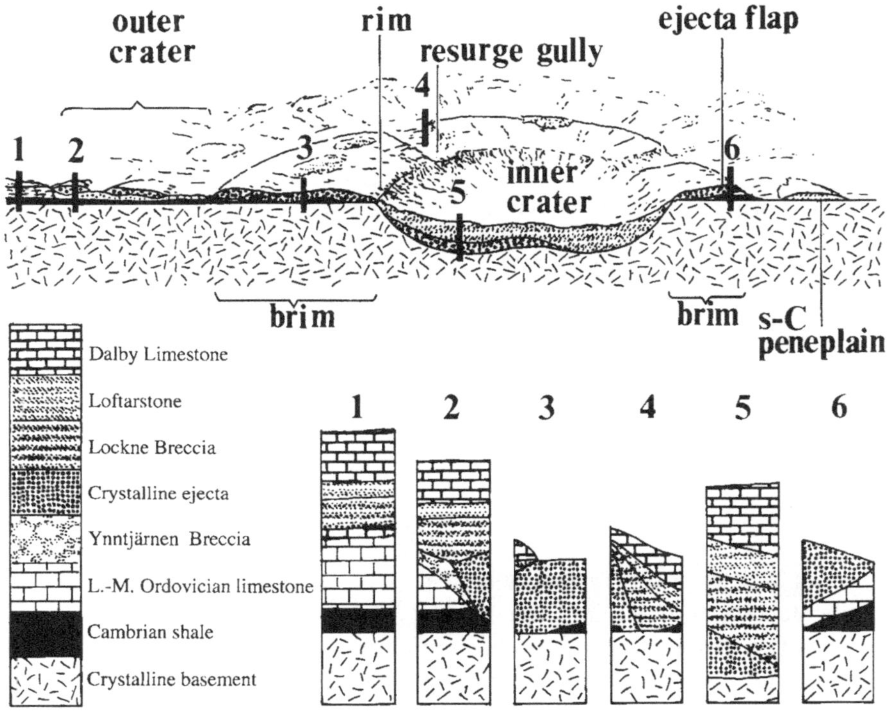

Fig. 4. Principal stratigraphy and terminology used for the Lockne crater. The sketch of the crater leaves out any tectonic complications and is intended to show only very generalized features, such as the relation of the principal structures to the sub-Cambrian peneplain. *s-C peneplain*: sub-Cambrian peneplain

6 The Brim

The brim (Fig. 4) is identified as a body of more or less strongly fractured, crystalline rock that includes monomictic breccia (Tandsbyn Breccia). Where it is preserved from erosion, it extends continuously outwards from the rim of the inner crater. At its outer margin the brim rises slightly above the adjacent terrain which consists of either the well-preserved basement rocks of the sub-Cambrian peneplain, or Cambrian and Ordovician sedimentary rocks that rest on the peneplain. In the west and north sectors, where this brim is widest, the width is about 2.5 km. In the southwest and south the brim forms outliers on the sub-Cambrian peneplain. At least some of these outliers were isolated from one another by erosion. The brim

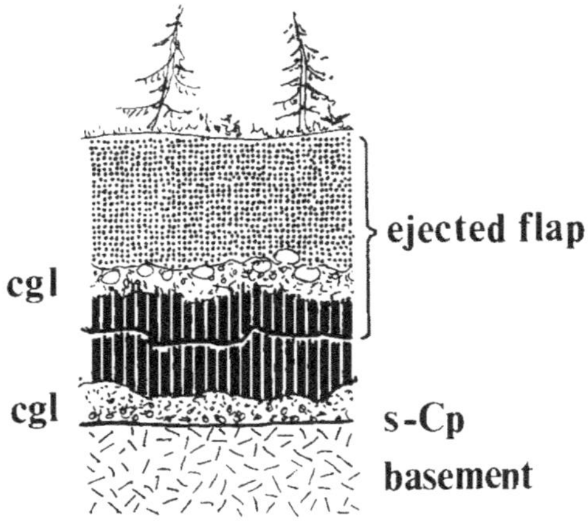

Fig. 5. Structure of the flap of crystalline ejecta at Nordanbergsberget and underlying geological units. The section is about 15 m high. The symbols are the same as in **Fig. 1,** excepting the conglomerates. *cgl*: basal conglomerate of Cambrian. *s-Cp*: sub-Cambrian peneplain. The sketch shows an inverted succession of strongly fractured, crystalline basement, Cambrian basal conglomerate, and Cambrian black shale that rests on a normal succession of (downwards) Cambrian black shale, basal conglomerate, the sub-Cambrian peneplain, and intact crystalline basement.

is less well exposed in the eastern part of the crater, but available data indicate a width less than 1 km.

The topography of the brim has numerous irregularities with amplitudes from a few meters to tens of meters, and rises at the most about 50 m above the rest of the terrain that surrounds the inner crater (Sturkell and Lindström 2004). Thus it appears as a low platform on the larger scale. It does not conform to the concept of a raised rim, as described from different terrestrial craters and summarized by Melosh (1989). Much of the crystalline rocks of the brim is brecciated to various degree, but at least as much looks fairly well preserved in outcrop. However, no blocks of apparently well-preserved rock are demonstrably larger than about 100 m. The preservation of patches of resurge deposits and post-impact marine sediments on the brim, indicates that its upper surface has not been significantly lowered by erosion after the impact.

Our previous studies of the Lockne crater (Lindström et al. 1996) have dealt with major parts of the brim as if they consisted essentially of autochthonous basement granite that was crushed to various degree by the impact. This interpretation can be regarded as an attempt to apply the concept of structural, or stratigraphic, uplift (Melosh 1989). However, we are now convinced that the brim is not autochthonous but consists of ejecta from the inner crater, and that these ejecta are underlain by the essentially intact sub-Cambrian peneplain.

The only vertical outcrop section through the brim just outside the inner crater occurs at the Skanska quarry at Nordanbergsberget (National Grid Reference 699320/144900) (Lindström and Sturkell 1992, Lindström and von Dalwigk 2002). This section (Fig. 5) exposes about 350 m of a sheet of fractured granite in a southeast-northwest direction. 150 m of it rests on Cambrian black shale in an identifiable state, including pyritized trace fossils (Simon 1987a). The basal Cambrian gravel deposit was found still in place under this shale. The gravel deposit was in its turn underlain by the granite of the sub-Cambrian peneplain that apparently extends throughout the quarry. It is a most significant feature of this section that the Cambrian basal gravel is present above as well as beneath the black shale. The upper gravel sheet has an overturned sedimentary contact against the granite that rests upon it (Lindström and Sturkell 1992, Lindström and von Dalwigk 2002). Thus, it is the very top of the basement that became overturned and ejected from the nearby outer part of the inner crater.

The section was described by Simon (1987b), who explained it as a possible instance of Caledonian overthrusting. This explanation appeared to be justified before the Lockne impact structure was identified. However, in order to be emplaced by overthrusting, the over 350X50 m^2 large and almost 20 m thick, heavily segmented sheet of granite, with weathering crust and overlying basal conglomerate, would have had to be detached along a horizontal surface. Thereafter it must have rotated 180° about a horizontal axis. All of this must have happened without much rotation of the granite segments relative to one another or other significant change of shape. Whereas the shearing-off of horizontal basement slices occurs commonly at the base of major overthrust complexes, the practically non-destructive overturning of entire slices of this structure. shape, and size is mechanically and geometrically most unlikely to occur within an overthrust. Therefore we explain the inverted basement sheet in the Skanska quarry as overturned ejecta, which normally occur at impact craters.

Cambrian shale has been reported to underlie about 50 m of granitic rock belonging to the brim in water wells at Tand 5 km southwest of the Skanska quarry and 4.7 km west of the crater centre (Fig. 2) (Lindström and von Dalwigk 2002). Another occurrence of Cambrian shale relevant to the interpretation of the brim was reported from 1 km west of Tandsbyn by Thorslund (1940). This outcrop in the midst of the Tandsbyn gully is now covered by bush and a road, so that it cannot be verified. If the Tandsbyn gully was formed by resurge flow eroding the crushed granite of an essentially autochthonous brim, this would have been the last place where one would look for an outcrop of autochthonous Cambrian shale. Because this was the interpretation of Lindström et al. (1996), the Cambrian shale reported by Thorslund (1940) was neglected as questionable, probably for no good reason. If this outcrop was correctly identified by Thorslund, as

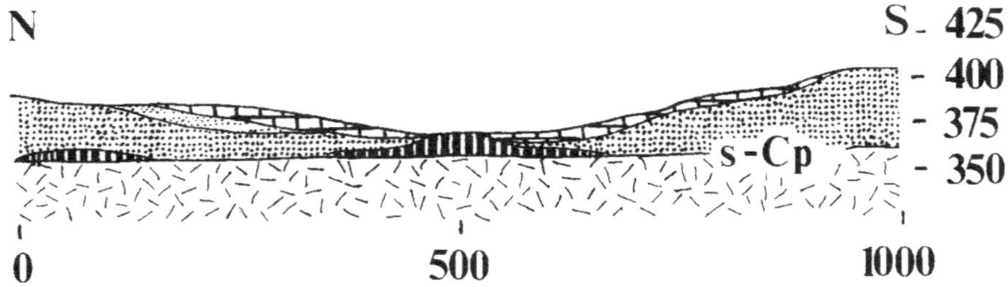

Fig. 6. North-south section through inlier of Cambrian black shale in the Tandsbyn gully west of Tandsbyn. Vertical and horizontal scales are in meters; the vertical scale is in meters above sea-level. The symbols are the same as in **Fig. 1.** *s-Cp*: sub-Cambrian peneplain. The sketch shows a sheet of crystalline Proterozoic ejecta than rests on the sub-Cambrian peneplain with patches of preserved Cambrian shale. The ejecta sheet is itself overlain by a thin cover of resurge sediment and secularly deposited, post-impact Dalby Limestone. The section roughly coincides with a section given by Thorslund (1940, Fig. 28).

we now believe, one can conclude that the bottom of the Tandsbyn gully is close to the level of the sub-Cambrian peneplain, with sporadic patches of preserved Cambrian shale, and that the brim on both sides of the gully consists of rootless ejecta that rest on the peneplain (Fig. 6).

An analogous outcrop of Cambrian shale in the Långmyren gully (Fig. 2; see below, chapter Resurge Gullies) is evidence that this gully, too, locally bottoms at the sub-Cambrian peneplain. An outcrop of fossiliferous Upper Cambrian shale at Haga (at approximately 698710/145350, east of Locknesjön), briefly referred to by Thorslund (1940), is evidence that this locality is outside the inner crater. Nearby outcrops of monomictic Tandsbyn Breccia, referred to in Thorslund's (1940) publication, but likewise nonexistent nowadays, are evidence of the ejected brim.

Cambrian shale with preserved basal conglomerate and a substratum of weathered Proterozoic dolerite occurs in an erosion window within the domain of the brim at Nordanbergsberget west of the Skanska quarry (National Grid coordinates 699305/144875). The locality is occupied by a water ditch in a golf course and is surrounded on all sides by crystalline ejecta. Other occurrences of Cambrian shale have been identified at outliers of the brim outside the southwest sector of the inner crater (Mörttjärnen and 1 km E of Mörttjärnen; Fig. 2).

Ar distances over 1.5 km outside the rim of the inner crater the substratum of the crystalline ejecta may include not only Cambrian shale, but Ordovician limestone as well. As a rule the Ordovician limestone is either strongly folded, or its bedding is disintegrated into small limestone nodules chaotically scattered in a marly matrix. Limestone affected in this way occurs for instance on both sides of the small road to Löfsåsen 500-

Fig. 7. Matrix-supported Tandsbyn Breccia (monomictic; granitic protolith). 900 m southeast of Hällnäset. Photo Jens Ormö.

800 m E of Nyckelberg (Fig. 2). The segment of the crater on the east shore of Locknesjön has Upper Cambrian and Lower to Middle Ordovician rocks within a few hundred meters from the rim of the inner crater, and apparently overlain by ejected monomictic breccia from the crystalline basement.

In accordance with the interpretation of the brim as ejecta, monomictic breccias and strongly deformed crystalline rock are prominent in its composition. Close to the margin of the inner crater the crystalline rock of the brim may, however, be apparently well-preserved and continuous within about 100 m long segments, as for instance in the Skanska quarry at Nordanbergsberget and on the south side of the Tandsbyn gully southwest of Tandsbyn.

There are three principal varieties of monomictic crystalline breccia at Lockne: *matrix-supported breccia, clast-supported breccia,* and *carbonaceous breccia.* The matrix of the matrix-supported breccia (Fig. 7) is very hard and consists of particles near and below the size that can be

Fig. 8. Clast-supported, granitic Tandsbyn Breccia. 1,300 m southwest of Hällnäset. Photo Jens Ormö.

resolved in thin sections under the light microscope, i. e., the matrix tends to appear isotropic under crossed Nicols. Macroscopically this matrix is commonly dark grey and porcellaneous in fresh specimens and light brownish in weathered outcrops. Clasts are commonly smaller than 4 cm; they frequently have irregular outlines. These breccias show microscopic evidence of shock, such as deformed feldspar lattices and micas, but no quartz with PDF has been identified so far.

Clast-supported breccias (Fig. 8) mostly have angular clasts with a size variation from less than a millimeter to many centimeters. However, the smallest clasts do not dominate to the extent that they form a continuous, low-porosity matrix. These breccias represent crushing with little evidence of shock.

In the carbonaceous breccias the crystalline clasts vary in size from less than a millimeter to several centimeters. They are separated by dark, sooty or asphalt-like matrix with an abundance of curved and striated shear-planes. The planes and striations lack any obvious preferred orientation. These breccias disintegrate easily and do not appear in natural outcrops. They probably originated from the topmost level of the crystalline basement, because the area has no other known source of carbonaceous matter than the Cambrian black shale that overlies the basement in undisturbed successions.

Among evidence of shock deformation in the brim zone is small-scale disharmonic folding of the basement/Cambrian contact. This was observed in several places in the Skanska quarry and would not occur under normal tectonic stresses at the contact between a massive, competent unit (basement) and a very weak unit (Cambrian bituminous shale). Another observed evidence is splayed crinkling of foliated inclusions in basement rocks. Thirdly, movement parallel stretching and fluting of shock-weakened basement rock occurs in movement zones (Fig. 9). Despite much search, no convincing shatter cones have been found.

7 Resurge Gullies

The brim is crosscut by radial channels (Fig. 2 and 4), at the bottom of which there are Locknee Breccia and loftarstone, deposited by the resurge from the collapsing water crater. The Nyckelberg gully to the north is neither deep nor particularly extensive. The Tandsbyn gully to the west is the best preserved and exposed. The Bergböle-Loke gully to the south is large, but its western margin is distorted tectonically. Its eastern margin is missing owing to erosion in late geologic time and, probably, tectonics as well. Roughly at mid-distance between the Tandsbyn gully and the Bergböle-Loke gully there is a less conspicuous gully that has come under observation only recently; it can be called the Långmyren gully.

Lindström et al. (1996) suggested the presence of a further similar structure, which they named the Rossbol gully, in the northeastern sector of the crater. The existence of this structure was inferred from two shallow core drillings. However, as will be discussed below, the informations from these drill-holes can be interpreted differently in the light of recent work. The Rossbol gully is therefore questionable but will be discussed at the end of this chapter.

The gullies are tens of meters deep. They either cut through to the base of the brim, in which case the sub-Cambrian peneplain forms their bottom, or they are not quite as deep and bottom in the rocks of the brim. The latter case is represented by the Nyckelberg gully. The Långmyren gully is eroded to the sub-Cambrian peneplain in one place (National Grid coordinates 698550/144690), where Cambrian shale is also exposed. Its

Fig. 9. Fluting in basal part of sheet of crystalline ejecta. North shore of Mörttjärnen. The height of the outcrop is about 1.5 m. Photo Jens Ormö

lower reach crosses ejected crystalline rock that subsided during the crater modification.

The gullies widen away from the inner crater. The Tandsbyn gully is about 1 km wide near the rim of the inner crater but widens considerably and branches to the northwest, west, and southwest about 1.5 km west of

Fig. 10. Ynntjärnen Breccia. Brunflo Church, lawn of the congregation localities. The width of the pictured surface is about 35 cm. Photo Erik Sturkell.

Tandsbyn. Its bottom is close to the sub-Cambrian peneplain and contains Cambrian shale (Fig. 6).

The existence of the gullies immediately after cratering is proven by the resurge deposits inside them. However, it is debatable, how much they owe their existence to erosion by resurge. Von Dalwigk and Ormö (2001) demonstrated that such erosion was active. The study of Shuvalov and von Dalwigk (2002) indicates that no other agent than erosion could have completed the formation of the gullies, but that it must be taken into account that the amount of erosion might be small in view of the enormous mass of water in motion. Furthermore, the preservation of soft Cambrian shale at the bottom of gullies shows that such erosion could not have cut deeply, if at all, into the autochthonous basement.

We argue that an ejected blanket of inhomogeneously disaggregated rock will develop thickness variations in the tangential direction, that is, normal to the ejection. Attenuation in this direction will appear as radially orientated tears that tend to widen outwards. If the mechanism is real, it implies that incipient, though perhaps not continuous, furrows were already there when the resurge current touched the surface of the brim. Subsequent erosion by the resurge extended these pre-existing furrows in the radial direction.

Although bedrock outcrops are much more abundant in the Lockne area than in most other areas in mainland Sweden, there is not enough of them to make it an easy task to follow the gullies beyond the brim, if they do at all exist there. The velocity of the culminating resurge in the identified gullies was demonstrably enough to transport cobbles, or coarser (Ormö and Miyamoto 2002). It was even greater than that according to Shuvalov et al. (this volume). For instance, Lockne Breccia in the Bergböle-Loke gully contains a mixture of slabs from diverse levels of the Lower-Middle Ordovician orthoceratite limestone, many slabs being over 5 m^3 (Lindström et al. 1983). Some exceptionally large clasts may, however, occur near the sites they occupied when the water cavity was excavated (Shuvalov et al., this volume). This may be the case with an inverted slab of limestone of at least 2,000 m^3 that occurs close to Grubban (Fig. 2; loc 20E, Lindström and von Dalwigk 2002)

8 Ynntjärnen Breccia

Since the publication of the paper of Lindström et al. (1983) those Middle Ordovician (lower Caradoc) rocks occurring in the Lockne area that contain isolated bodies of reorientated or physically disintegrated limestone, have collectively been called Lockne Breccia. The original assumption was that all of these rocks have a shared origin as debris flows. The idea of a common origin remained even after it was realized that debris flows, properly defined, had not been active. The discussion of von Dalwigk and Ormö (2001) opened the door to doubt, and by now it has become clear that we are dealing with two discrete units of the impact stratigraphy, with slightly different lithologies, discrete timing, and important differences as regards the mode of origin.

The whole Lockne Breccia in the earlier sense was regarded as a product of the resurge flow (Lindström and Sturkell 1992). Frequently occurring transitions to the immediately overlying, arenitic loftarstone

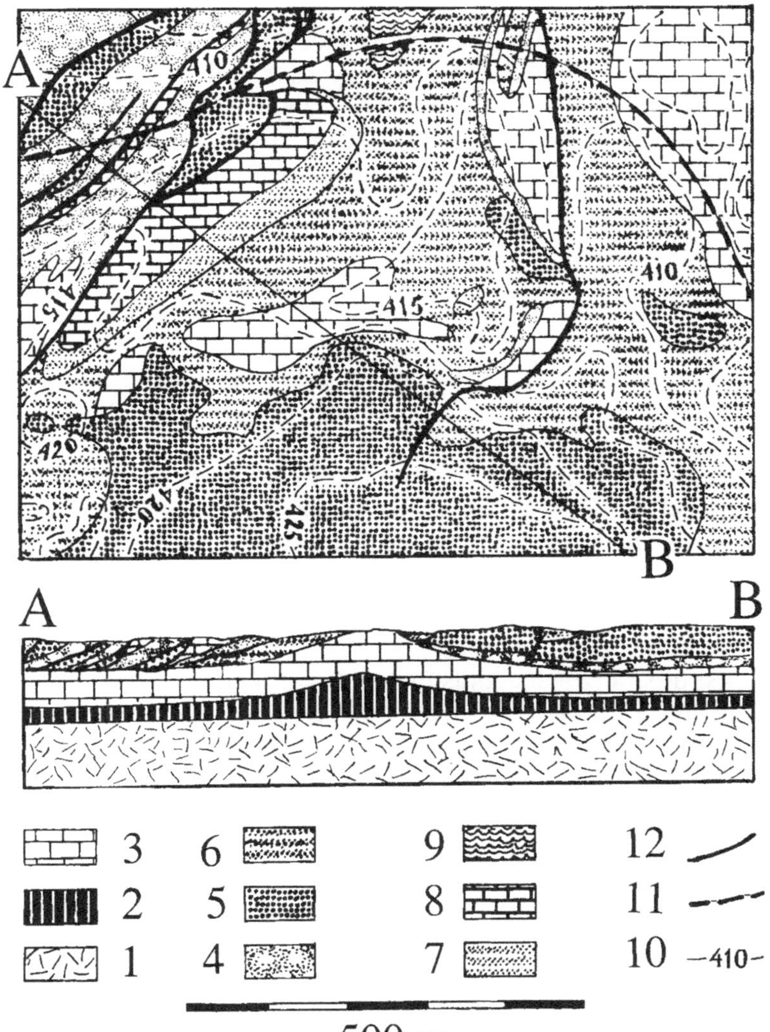

Fig. 11. Ynntjärnen area. The mapped area is located on **Fig. 1**. *1*:Crystalline basement. *2*: Cambrian black shale. *3*: Lower and Middle Ordovician, mainly limestone. *4*: Ynntjärnen Breccia. *5*: Sheet of crystalline ejecta. *6*: Lockne Breccia (resurge deposit). *7*: Loftarstone (resurge deposit). *8*: Dalby Limestone (post-impact, secular deposit). *9*: Lake Ynntjärnen (at the middle of the map's north margin). *10*: Meters above sea-level. *11*: Railroad. *12*: Thrust fault.

were observed, and the occasionally abundant occurrence of crystalline ejecta was regarded as a characteristic feature. The failure to find such ejecta in many outcrops was simply taken for scarce occurrence, but now we are convinced that these outcrops consist exclusively of fragments of local limestone and calcareous-argillaceous ("marly") matrix.

Renewed mapping and improved outcrop conditions have corroborated the conclusion that the Lockne Breccia is a polymictic sedimentary breccia with crystalline ejecta clasts. It is a widely distributed lithology, occasionally transitional upwards to the loftarstone, and should keep its name. We have also become convinced that there is an earlier formed monomictic limestone breccia which may consist of limestone disaggregated into nodules that are chaotically dispersed in a calcareous-argillaceous matrix (Fig. 10). This early-formed breccia will be referred to under the new name Ynntjärnen Breccia (Fig. 4) (although the need for this kind of local names can indeed be discussed, we consider them useful because they can apply regardless of possible fluctuations of interpretation).

The Ynntjärnen Breccia has its name from the area within less than 1.5 km west, south, and east of the small lake Ynntjärnen (Öhntjärn in Thorslund 1940) (Fig. 11). The area is characterized by a succession of southeast-vergent thrust faults with a period of 40-400 m and throws of the order of 50 m. This instance of Caledonian deformation has led to an estimated 300 m of horizontal shortening, but more important, to a repetition of stratigraphic units that makes it relatively easy to unravel the stratigraphy. The lowermost outcropping unit is well-preserved beds of Lower to Middle Ordovician orthoceratite limestone that form exposed successions of at least 10 m thickness. At the top of this unit there is Ynntjärnen Breccia consisting of disaggregated marly-nodular limestone. In the eastern outcrops this breccia is but a few meters thick at the most, but the western outcrops, mainly along the railway, have thicknesses of over 10 m (Fig. 11).

The orthoceratite limestone and the Ynntjärnen Breccia are overlain in places by bodies of ejected Tandsbyn Breccia with volumes of the order of 10,000 m^3. The next higher unit is water transported Lockne Breccia that includes a variable content of crystalline clasts of gravel to cobble size. The Lockne Breccia rests unconformably on either orthoceratite limestone, Ynntjärnen Breccia, or Tandsbyn Breccia.

We interpret the Ynntjärnen Breccia as derived from the top layers of the orthoceratite limestone. These layers took the rap from, first, cratering in the overlying water body, thereafter, the impact of crystalline ejecta, and, finally, the resurge with the full force of its initial phase. Its components were strongly stirred, but not carried by water. Its principal occurrence is in the west and northwest sectors of the area adjacent to the crater. A large area on both sides of the road between Nyckelberg and Löfsåsen, as well as northwest of Nordanbergsberget, has Ynntjärnen

Breccia occurring on top of widely distributed orthoceratite limestone. This area was earlier mapped as Lockne Breccia.

9 Large Masses of Crystalline Ejecta

Some of the most spectacular outcrops of Tandsbyn Breccia occur outside the brim at distances of 6-9 km from the centre of the crater. The Tandsbyn Breccia has resisted erosion better than the surrounding lithologies and commonly forms hillocks that are several meters high. Thorslund (1940) thought that these local rises of monomictic crystalline breccia are rooted in the basement vertically beneath them. He developed a widely accepted model of a local granitic archipelago in the Ordovician sea (Thorslund 1940).

Table 1. Major masses of crystalline ejecta. *Coordinates* Swedish National Grid (modified UTM), *Names* used in maps and text, *V* volume in estimated 10^3 m^3, *Distance* given in km from crater centre, *Direction* to centre of ejecta mass from crater centre (east is 90°, south 180°, west 270°).

Coordinates	Name	V	Distance	Direction
698425/144525	Berget	20	6.3	228°
698458/144455	Klocksåsen 1	50	6.5	232°
698525/144445		20	6.3	239°
698454/144430	Klocksåsen 2	10	6.8	233°
698740/144360	Ynntjärnen S	2,000	6.4	258°
698785/144390		50	6.0	263°
698770/144405		10	5.9	260°
698800/144350	Ynntjärnen W	50	6.5	265°
698805/144330		50	6.7	265°
698700/144150	Sved	50	8.6	259°
698630/144095	Karsåtjärnen	10	9.2	256°
699335/144550	Nyckelberg	400	6.3	320°

The volumes given in Table 1 are low estimates based on observed horizontal dimensions and average thicknesses of 10 m. At Nyckelberg the estimate is based on the assumption that a few discrete bodies lying close to one another were continuous before becoming isolated from one another by erosion following upon minor folding. The distances are given from the approximate geometric centre of the inner crater to the centres of the ejected bodies.

Fig. 12. Loftarstone resting with sharp contact on Lockne Breccia. 1 km southwest of Hällnäset. Swiss army knife for scale. Photo Erik Sturkell.

The listed bodies spread within a sector of about 90°, with the bisectrix at about 270° and the arithmetic mean at 257°. Karsåtjärnen (Fig. 2) is exceptional for the over 9 km distance of ejection, but also for its composition. It is the only isolated ejecta body that is carbonaceous. These two circumstances indicate that it was ejected from a position that was near the impact centre as well as close to the boundary between the crystalline basement and the Cambrian shale. The size and position of Ynntjärnen South (Fig. 2) suggest that it can be regarded as an outlier of the brim, although its separation from the brim existed already in the resurge phase, on the evidence of resurge deposition between it and the brim.

Fig. 13. Lockne Breccia transitional upwards to loftarstone. 1 km southwest of Hällnäset. Swiss army knife for scale. Photo Erik Sturkell.

10 The Lockne Breccia and Loftarstone

As remarked above, the definition of the Lockne Breccia should be restricted to resurge deposits that contain clasts of crystalline ejecta. The content of limestone clasts is mostly of diverse provenance. Transitions to the overlying loftarstone may be either sharp (Fig. 12) and erosional, or gradual (Fig. 13). The principal difference from the loftarstone is the smaller grain size in the latter. The sharpening of recognition criteria has led to a considerable reduction of the distribution on the map that is particularly obvious in the northwest sector.

Quartz with PDF, particles of melt rock, and anomalous concentration of iridium have so far been found only in the loftarstone arenite (Therriault and Lindström 1995, Simon 1987a, Sturkell 1998c)

In the lower part of the Lockne Breccia the limestone clasts are frequently matrix supported and can be almost a thousand cubic meters large. This part has properties that characterize a debris flow (Lindström et al. 1983). This is true for outcrops as far from the crater centre as Brunflo

(Fig. 2) and Grytan (just beyond the north edge of Fig. 2) (9-10 km from the crater centre; Lindström and von Dalwigk 2002). Crystalline ejecta at these locations are several cubic meters large, and reoriented blocks of orthoceratite limestone and Ynntjärnen Breccia reach hundreds of cubic meters.

11 The Sub-Cambrian Peneplain and the Crater Rim

When marine sedimentation began in large parts of Sweden in the Cambrian, it covered a surface that had been land for long enough time to cut extensive peneplain facets into the deeper, crystalline parts of Proterozoic orogens. A sub-Cambrian peneplain can thus be identified in most parts of the country where Paleozoic sedimentary rocks are preserved. In Jämtland this peneplain has got a north-westward slope of 1.5% through mainly Caledonian deformation of the lithosphere (Karis and Strömberg 1998, Sturkell and Lindström, in press in Meteoritics and Planetary Science). Non-deformed segments of this peneplain are flat on the kilometer scale, and deviations of several meters from a flat surface are commonly signs that tectonic deformation has occurred.

As indicated above in the chapters on the brim and the resurge gullies, the presence of the peneplain at expected levels indicates that the inner crater at Lockne lacks a significantly raised rim. There is, however, evidence of local collapse of the rim during the modification of the crater. In the Långmyren gully (Fig. 2), there is a peneplain outcrop at about 385 m a.s.l., north of which there is a drop of the peneplain of over 10 m before the gully enters the inner crater. In the Nyckelberg gully (Fig. 2), Lockne Breccia rests at 350 m on Tandsbyn Breccia that evidently belongs to the brim, which in its turn is expected to overlie the peneplain at a deeper level. The level of the peneplain in the upper reach of the gully, which is at Nyckelberg, is 380 m. A collapse of the crater rim would explain the subsidence of the lower part of the gully. In the southern part of the Bergböle-Loke gully there is a succession of terraces with 5-10 m subsidence of the peneplain toward the north at each of three east-west striking faults. These instances show that minor collapse, with amplitudes of up to a few tens of meters, has affected the rim of the Lockne inner crater.

12 Younger Tectonic Structures

The sub-Cambrian peneplain can be followed eastwards from lake Näkten to near the southwestern margin of the inner crater of Lockne. It remains a relatively flat surface that rises from about 325 m in the west to somewhat over 435 m in the east, over a distance of 9.5 km. This corresponds to the rate that is normal in the region (Sturkell and Lindström, in press in Meteoritics and Planetary Science). The 435 m-plus level is reached at the top of a hill 1.7 km southwest of Hällnäset (National Grid coordinates 698400/145040). 1.1 km to the northeast of this hilltop the level of the surface identified as the sub-Cambrian peneplain has dropped to 340 m. The drop is effected mainly along a fault that can be followed for 6.5 km to the southeast and 2.5 km to the northwest, beyond which we have not been able to follow it.

A fault zone with the same strike and about half as large throw forms the northeast flank of Nordanbergsberget (Fig. 2) at the north margin of the inner crater. In this case the peneplain can be identified at 385 m in the southwest and drops to about 340 m in the northeast. This fault zone can be followed about 4 km along the strike north of Nordanbergsberget. A continuation toward the southeast at Rossbol on the other side of Locknesjön would explain the crater-related distribution of rocks there and finds support in the topography. It would extend the fault zone at least 7 km toward southeast.

The circumstance that the sub-Cambrian peneplain occurs at somewhat below 345 m at Haga (see the preceding chapter on the sub-Cambrian peneplain) is evidence of a drop of about 100 m on the east side of lake Locknesjön. In this case the most likely position of the fault is along the northwest-southeast striking depth axis of Locknesjön. The combined effect of the three cases of vertical relative movement referred to in the preceding paragraphs is that the eastern segment of the Lockne crater has been lowered at least 100 m relative to the rest of the crater. The evidence that this tectonic subsidence has taken place is the preservation of the eastern crater rim at Haga and well-preserved resurge deposits resting on the intact, crystalline basement outside the eastern rim of the crater,

These arguments conflict with the hitherto held view (Lindström et al. 1996) that the eastern segment of the Lockne crater must have been eroded to some depth because of the general north-westward inclination of the regional structure. Having been downfaulted about 100 m, it could, however, be equally well preserved as the western part. It has been impossible to verify this possibility because of an omnipresent, if mostly

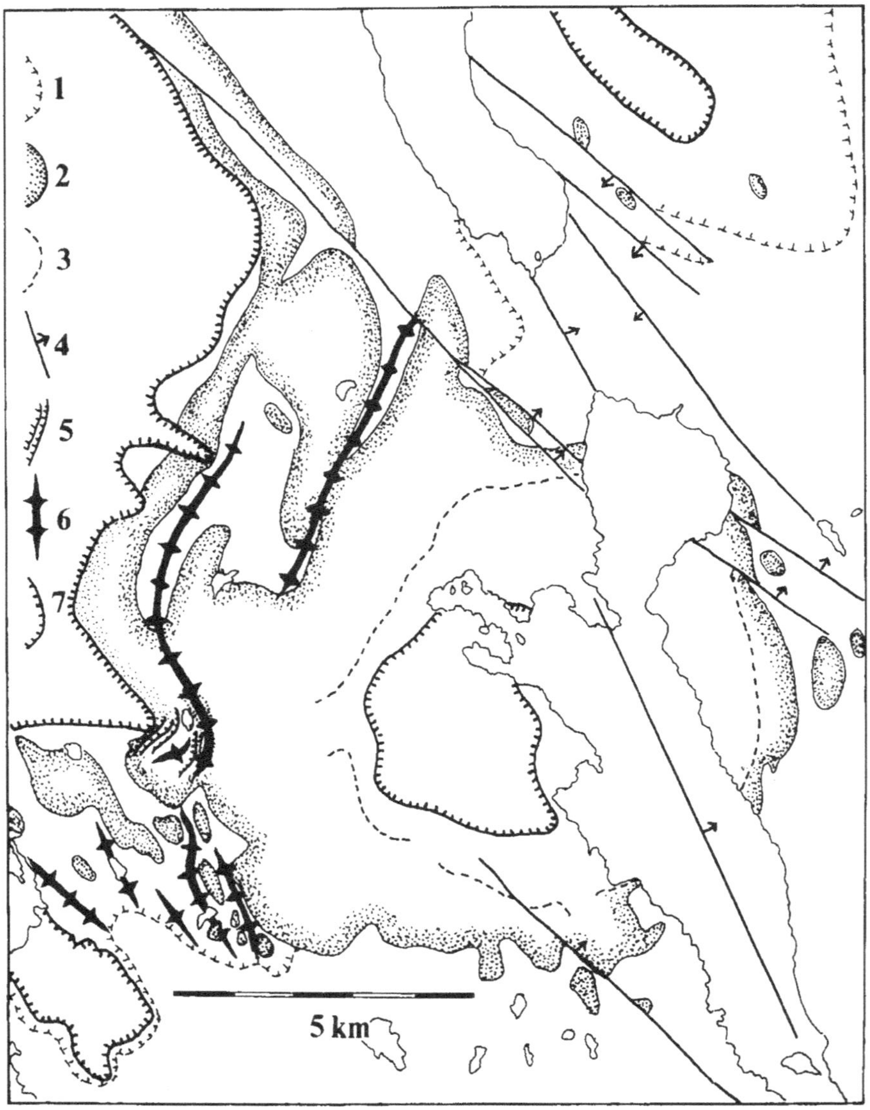

Fig. 14. Principal structures of the Lockne area. *1*: Boundary between the basement and the autochthonous Palaeozoic. *2*: Boundary of impact-related rocks. *3*: Boundary of inner crater. *4*: Fault; the arrow indicates the downthrown side. *5*: Imbricated thrust-faults. *6*: Anticline. *7*: Boundary of overthrust nappe.

thin, earth cover. The new situation is that construction of lumbering tracks has created numerous exposures of the sub-Cambrian peneplain with a sporadic cover of resurge sediments. Because the peneplain is preserved, the same should be true for the crater.

The Lockne crater is situated at the southeast front of Caledonian overthrusts. An advanced outlier of the basal overthrust is preserved in the midst of the crater. The overthrusts originally reached far to the east of the present front, and the buildup of nappes by far exceeded the thickness of those remains that have survived erosion. The nappe outlier within the crater consists of Lower and Middle Ordovician rocks that show no signs of having been stirred by the impact. Therefore its roots must be sought at least 9 km northwest of the centre of the crater.

In previous publications we have assumed that the Cambrian and Ordovician sedimentary rocks adjacent to the west side of the crater were allochthonous, although the distance they were transported eastwards might have been modest (Lindström et al. 1996). The principal reasons for this assumption were that the Cambrian black shale at the base of the succession is sheared and evidently served as a detachment zone, and that Ordovician limestones are folded, which requires movement relative to the basement. However, the observed spread of very large-sized ejecta across the concerned terrain would have been impossible if the terrain had travelled from a distance great enough to make it useful to speak of allochthony (Shuvalov et al., this volume). Instead, the mapped structures must be examined in order to find their least, arguable distance of movement, which in this case must be taken to be the most likely one.

The Ordovician terrain west of the brim exhibits five principal, northwest-southeast striking anticlines with wavelengths of 0.5-1 km and amplitudes not over 50 m (Fig. 14). The northeast flanks are steep but not overturned. In this case there is no evidence for a southwest-northeast total shortening of over 200 m. Shortening of a similar order in the northwest-southeast direction occurred normal to two anticlines that strike tangentially to the northwest margin of the brim. These anticlines are strongly southeast-vergent and pass laterally into thrust faults (Fig. 3, section A-B). Tectonic shortening of a slightly different style occurs in the sedimentary and impact-related succession near Ynntjärnen (Fig. 11). The transport involved in this case is of interest, because the tectonized succession includes major bodies of crystalline ejecta (see the chapter on the Ynntjärnen Breccia). A shortening of about 300 m along a 700 m long northwest-southeast transect can be derived for this outcrop area. The discussed deformations do not concern the basement, which was rigid at this scale (Fig. 11). The sum of these arguments is that the nappe boundary should be removed that hitherto was drawn close to the brim (Lindström et al. 1996). A thrust-plane with significant transport occurs 1-4 km farther to the west. No impact-related lithologies have been observed above this nappe boundary.

The arrangement of Caledonian fold axes to the northwest and southwest of the brim suggests that the brim presented some degree of obstacle to tectonic transport in the immediately overlying beds. This was probably because it replaces the weak Cambrian shale that is preserved and serves as level of detachment between the basement and the Ordovician limestones in areas untouched by the impact.

13 Conclusions

The just over 7 km wide crater at Lockne was excavated into crystalline basement, the upper surface of which has a present dip of 0.85° toward the west-northwest. If it had not been for a set of northwest-southeast striking faults with about 100 m subsidence to the northeast, the eastern segment of the crater would have been significantly eroded owing to the regional dip (Lindström et al. 1996). However, things being what they apparently are, we assume its preservation to equal that of the larger western part.

The inner crater is surrounded by a brim of ejecta from the crystalline basement target. This brim, which can reach a thickness of at least 50 m, is as much as 2.5 km wide in the west but less than 1 km in the east. It rests on the sub-Cambrian peneplain that forms the boundary between the basement and the normal succession of Cambrian and Ordovician cover rocks, where the succession is not disturbed. Remains of Cambrian basal conglomerate and black shale are preserved in places and thus facilitate identification of the peneplain surface below the ejecta. Distorted Ordovician sediments are present under the outer reaches of the brim. The described conditions show that the immediate surroundings of the crater were bared of the sediment blanket before the ejecta sheet descended on them. Numerical modelling (Shuvalov et al., this volume) indicates that the removal of the sediments was connected with the growth of the water transient crater.

There is no indication that the sub-Cambrian peneplain should rise toward the margin of the inner crater. The absence of a significantly raised rim is an unusual feature that differs from the normal situation in terrestrial environments (Melosh 1989). This situation might be related to the circumstance that the inner crater at Lockne is not to be equated with an apparent crater formed after impact on land but only, in a sense, with the lowermost part of such a crater. The upper, main part of the transient crater could not leave any immediate trace in the apparent crater, because it formed in water. The depth of this water roughly equalled the diameter of the bolide, that is, at least 500 m (Shuvalov et al., this volume).

The brim is crosscut by a number of radial channels ("resurge gullies") at the bottom of which resurge deposits of variable thicknesses occur. The gullies may have been initiated as radial rifts in the spreading ejecta blanket, in two cases also as tongue-shaped collapse features in connection with crater modification.

The removal of the sediment blanket from a wide zone along the rim of the inner crater during growth of the water crater would not have occurred in water shallower than about 500 m, nor is the force and volume of the resurge an expected feature of a shallow-water crater (Shuvalov et al., this volume). A depth of about 700 m is an arguable condition for the origin of these features in the case of Lockne. At such depth of water, ejecta as large as those found as far as 6-9 km to the west of the crater centre would, however, not have occurred there, if the impact had been vertical. An oblique impact is more likely in this case. The large ejecta would spread in the direction of impact (Shuvalov et al., this volume), which evidently was toward west to southwest. The polarity to be expected from an oblique impact is well shown by the gravity profiles and model sections through the Lockne crater presented by Sturkell et al. (1998).

The Lockne Breccia of previous publications (Lindström et al. 1983, Lindström et al. 1996) should be separated in two: 1) a breccia with sedimentary and most frequently monomictic clasts (Fig. 10) that formed during expansion of the water crater in the areas that surround the present inner crater, and 2) another breccia that formed in connection with the resurge throughout the area affected by the impact. The unit characterized by the former lithology is given the name Ynntjärnen Breccia. The second breccia is the Lockne Breccia proper; it has polymictic sedimentary clasts with an admixture of more or less frequent clasts of crystalline ejecta (Fig. 12 and 13). The strongly stirred calcareous rock of the Ynntjärnen Breccia shows some superficial similarity to the rock known as Gries in the Nördlingen crater (Hüttner and Schmidt-Kaler 1999), with which it shares the descriptive term "mortar-like". However, cataclasis prevails throughout the Gries lithology, but neither the limestone fragments nor the argillaceous matrix of the Ynntjärnen Breccia show penetrative cataclasis. The stirring appears to have taken place in a carbonate rock some nodule-like parts of which were lithified (the limestone fragments), whereas other parts were richer in clay minerals and therefore remained in a weak, either plastic or viscous state, from the beginning of events. Carbonate rocks with evidence of a corresponding stage in their diagenesis are common in the Ordovician of Sweden.

The Ynntjärnen Breccia may be overlain by crystalline ejecta or by Lockne Breccia, but unlike the Lockne Breccia it never has a gradational contact with the loftarstone arenite. It happens most frequently that the Lockne Breccia rests on either Ynntjärnen Breccia or orthoceratite limestone, without any body of ejected Tandsbyn Breccia in between. Big clasts of Ynntjärnen Breccia can occur either within the Lockne Breccia or possibly within the Ynntjärnen Breccia itself. Several such clasts with volumes over 100 m^3 occur just west of the church of Brunflo, at about 9.6 km from the crater centre. Regardless whether they are evidence of the dynamics of the resurge flow, or they occur within the Ynntjärnen Breccia itself, they are relevant to the interpretation of the environment and process of cratering.

Acknowledgements

We thank Boris Ivanov and Valery Shuvalov of the Russian Academy of Science, Moscow, for essential discussion of the Lockne crater in the field and in the laboratory, Lars Karis at SGU, Uppsala, for access to the field notes of Per Thorslund, and Jan Backman, Department of Geology and Geochemistry, Stockholm University, for space, facilities, and a congenial atmosphere. Risto Kumpulainen, Department of Geology and Geochemistry, Stockholm University, invested his skill and much precious time in the electronic handling and improvement of the illustrations. The work by Jens Ormö was partially supported by the Spanish Ministry for Science and Technology (Reference AYA2003-01203). We thank Henning Dypvik and David T. King, Jr., for thorough and highly useful reviews.

References

von Dalwigk I, Ormö J (2001) Formation of resurge gullies at impacts at sea: The Lockne crater, Sweden. Meteoritics and Planetary Science 36: 359-369

Hüttner R, Schmidt-Kaler H (1999) Die Geologische Karte des Rieses 1:50 000.Geologica Bavarica 104: 132 pp

Karis L, Strömberg AGB (1998) Beskrivning till berggrundskartan over Jämtlands län. Del 2: Fjälldelen. (Description to the bedrock map of the province of Jämtland. Part 2. The Caledonian part). Sveriges Geologiska Undersökning, Ca 53:2, Uppsala, 363 pp

Lindström M (1971) Vom Anfang, Hochstand und Ende eines Epikontinentalmeeres. Geologische Rundschau 60: 419-438

Lindström M, von Dalwigk I (2002) Geological guide to the Lockne and Dellen impact structures. Stockholm Contributions in Geology 44: 45 pp

Lindström M, Sturkell EFF (1992) Geology of the Early Paleozoic Lockne impact structure, central Sweden. Tectonophysics 216: 169-185

Lindström M, Simon S, Paul B, Kessler K (1983) The Ordovician and its mass movements in the Lockne area near the Caledonian margin, central Sweden. Gelogica et Palaeontologica 17: 17-27

Lindström M, Ekvall J, Hagenfeldt SE, Säwe B, Sturkell EFF (1991) A well-preserved Cambrian impact exposed in Central Sweden. Gelogische Rundschau 80: 201-204

Lindström M, Flodén T, Grahn Y, Kathol B (1994) Post-impact deposits in Tvären, a marine Ordovician crater south of Stockholm. Geological Magazine 131: 91-103

Lindström M, Sturkell EFF, Törnberg R, Ormö J (1996) The marine impact crater at Lockne, central Sweden. Geologiska Föreningens i Stockholm Förhandlingar 118: 193-206

Melosh HJ (1989) Impact Cratering. A Geologic Process. Oxford University Press, New York, 245 pp

Ormö J, Lindström M (2000) When a cosmic impact strikes the seabed. Geological Magazine 137: 67-80

Ormö J, Miyamoto H (2002) Computer modeling of the water resurge at marine impact: the Lockne crater, Sweden. Deep Sea Research II 49: 983-994

Ormö J, Shuvalov V, Lindström M (2002) Numerical modeling for target water depth estimation of marine-target impact craters [abs.]. Journal of Geophysical Research 107, E11, 1-9, olio: 10.1029/2002JE001865

Shuvalov V, von Dalwigk I (2002) Numerical modeling of resurge flow and Lockne gullies formations: preliminary results [abs.]. 8th workshop of the European Science Foundation program IMPACT, Impact Tectonism p 57

Shuvalov V, Ormö J, Lindström M (this volume) Hydrocode simulation of the Lockne marine target impact event, pp 405-422

Simon S (1987a) Stratigraphie, Petrographie und Entstehungsbedingungen von Grobklastika in der autochthonen, ordovizischen Schichtenfolge Jämtlands (Schweden) Sveriges Geologiska Undersökning C 815, 156 pp

Simon S (1987b) Caledonian deformation of basement in the Lockne area, Jämtland, central Sweden. Geologiska Föreningens i Stockholm Förhandlingar 109: 269-273

Sturkell EFF (1998a) The marine Lockne impact structure, Jämtland, Sweden. Geologische Rundschau 87: 253-267

Sturkell EFF (1998b) Resurge morphology of the marine Lockne impact crater, Jämtland, central Sweden. Geological Magazine 135: 121-127

Sturkell EFF (1998c) Impact related Ir anomaly in the Middle Ordovician Lockne impact structure, Jämtland, Sweden. Geologiska Föreningens i Stockholm Förhandlingar 120: 333-336

Sturkell E, Lindström M (2004) The target peneplain of the Lockne impact. Meteoritics and Planetary Science 39: 1721-1731

Sturkell EFF, Ekelund A, Törnberg R (1998) Gravity modeling of Lockne, a marine impact structure in Jämtland, central Sweden. Tectonophysics 296: 421-435

Therriault AM, Lindström M (1995) Planar deformation features in quartz grains from the resurge deposit of the Lockne structure, Sweden. Meteoritics and Planetary Science 30: 700-703

Thorslund P (1940) On the Chasmops Series of Jemtland and Södermanland (Tvären). Sveriges Geologiska Undersökning C 436, 191 pp
Wickman FE (1988) Possible impact structures in Sweden. In: Bodén A, Eriksson KG (eds) Deep drilling in crystalline bedrock. I. The deep gas drilling in the Siljan impact structure, Sweden and astroblemes, Springer Verlag, Berlin Heidelberg, pp 298-327
Wiman C (1900) Eine untersilurische Litoralfacies bei Locknesjön in Jemtland. Bulletin of the Geological Institution of the University of Uppsala 4: 133-151

A Study of Impact Fracturing and Electric Resistivity Related to the Lockne Impact Structure, Sweden

Ann Bäckström

Department of Land and Water Resources Engineering, Royal Institute of Technology, S-100 44 Stockholm, Sweden (annb@kth.se)

Abstract. The fracture frequency and the electric resistivity of outcrops of crystalline basement rocks at the Lockne meteorite impact site have been studied in order to investigate the extent and radial changes of impact induced fracturing. By measuring the electric resistivity and the fracture frequency at the same outcrops, the effect of fracturing on the electric properties of the rock is estimated and correlated with the fracture frequency. A negative linear correlation between the Log of fracture frequency and the Log of electric resistivity was found.

It was also found that the fracture frequency decreases in a transition zone over a distance of about 1100 m across the southern margin of the impact structure. A similar set of measurements was made across the suggested northern limit of the structure, but no change was detected. This implies that the outer limit of the Tandsbyn Breccia is further to the north. The studied area is, therefore, not likely to be the margin of the structure.

1 Background

The Lockne Area, Jämtland, Sweden, provides unique opportunities to study the interior structure of a relatively large impact crater. Relatively recent erosion has exposed many outcrops that can be studied geologically and geophysically. The purpose of this investigation is to investigate the correlation between electric resistivity and fracture frequency of crystalline rocks using the electro-magnetic, Very Low Frequency-Resistivity (VLF-

R), method. I also present the result of a modified window mapping method to estimate the fracture frequency of impact-affected crystalline rocks and the unaffected surrounding crystalline rocks. Window mapping of glacially eroded crystalline outcrops is considered to give the most reliable estimate of the fracture frequency in this area compared to scan line mapping (Priest 1993), which has an orientation bias.

1.1 Impact Fracturing

During a meteorite impact, the fractures are mainly caused by the tensile stresses in the target induced by the rarefaction wave. The rarefaction wave is the response to the compressional high-pressure shock wave generated by the impact as the shock wave meets the free surface, which is a surface of zero stress. The rarefaction wave is equal in strength to the shock wave but of opposite sign, starting at the free surface and moving downwards when the shock wave arrives. The shock wave and the rarefaction wave occur simultaneous at the free surface. Close to the free surface, the rarefaction wave reduces the stress caused by the shock wave but, at a distance further from the free surface, the difference in speed between the shock wave and the tensile rarefaction wave causes tensile stress. This stress causes fractures to develop in the target when it reaches the tensile strength of the target material. The rarefaction wave moves away from the crater center into the target, causing fracturing to a distance of several times the crater depth or diameter (Melosh 1989). Fractures are also formed during the subsequent excavation flow and the crater modification process.

The fracturing in impact crater structures has been studied from the deep drilling at the Puchezh-Katunki impact structure (Russia), where a 5374 m borehole was drilled and 3082 m of drill core were retrieved (Masaitis et al. 1999). The authigenic breccia took up the interval between 546-5374 m. In this impact crater, with a diameter of 80 km, strong brecciation occurs down to 3.5- 4 km depth. The size of the undisturbed rock-blocks does not exceed 3-5 m. Below 4 km, the core recovery was poor but, according to the acoustic properties, the diameters of the blocks increase to many tens of meters at 5 km (Masaitis et al. 1999).

The porosity of the crystalline rocks down to 5 km was measured and it was found that the decrease of porosity follows a power law. A correlation between fracture frequency, electric resistivity and porosity has not been established yet. Drilling performed at the Popigai impact crater (Russia) shows that fractured rocks in the ring structure can be allochtonous to a very large depth (Masaitis et al. 1975). In the annular ring structure, large blocks of sedimentary rocks have been found immersed in the fractured crystalline rocks. This indicates that considerable movements have also occurred within the brecciated crystalline crater basement.

The impact-generated fractures overprint and may reactivate existing fractures. The more recent fractures cannot generally be separated from older ones. The spatial variation of the fracture frequency values can, however, reveal the impact-generated overprint as being several orders higher than the fracture frequency of the unaffected surroundings.

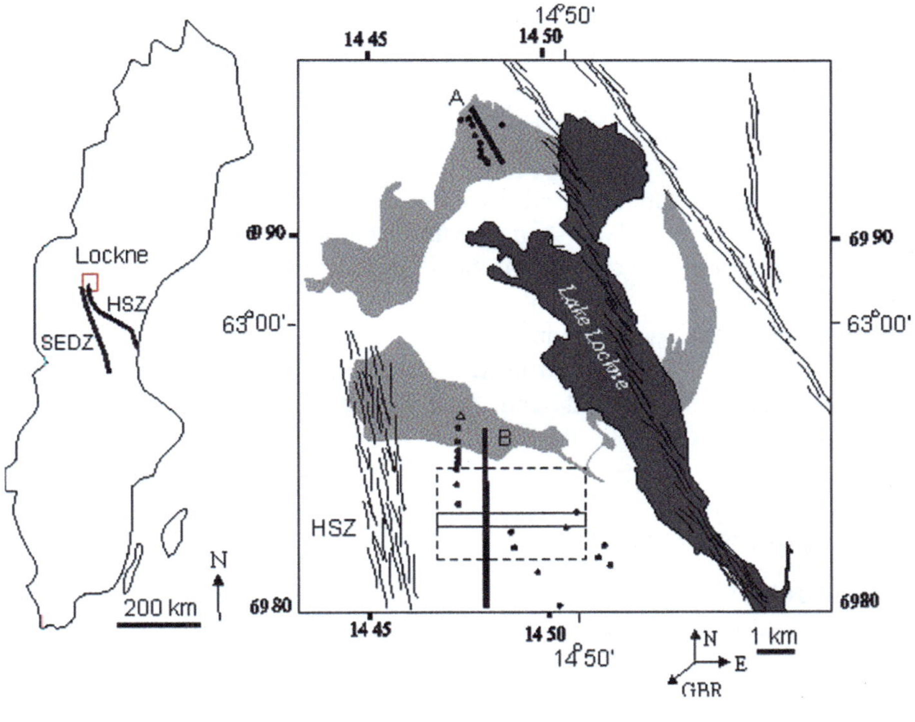

Fig 1. Location of the study area in Sweden where two major deformation zones are seen, the Storsjö Edsbyn Deformation Zone (SEDZ) and the Hassela Shear Zone (HSZ). The HSZ can be seen as thin lines in the area map (after Högdahl 2000). The light grey areas in the area map are the extent of the Tandsby Breccia (after Lindströn et al. 1996). The sites for fracture frequency and electric resistivity measurements are marked with dots. The triangle marks the location of the outcrop shown in figure 2. The large box (broken lines) at section B shows the transition zone. The step in fracture frequency / electric resistivity is marked by the narrow box. The direction to the GBR radio transmitter in southern England is SV. Coordinate numbers refer to the Swedish National grid. Thick lines indicate the sections A and B on which the measurements are projected.

1.2 Geological Setting of the Lockne Area

The study area in and around the Lockne structure is situated in Jämtland, Sweden (N 63°00'20", E 14° 49'30", Fig. 1). It is situated just east of the Caledonian thrust front in an area with autochonous Lower Palaeozoic sedimentary sequences on Proterozoic crystalline basement. The deposits resulting from the Lockne structure were suggested by Wickman et al. (1988) to be the result of a meteorite impact. The Lockne crater is an impact crater formed in a sea environment, which is now exposed on land allowing detailed studies of the impact-induced effects to be made.

The Lockne impact structure was formed in the Ordovician Sea 455 Ma ago (Grahn and Nõlvak 1993). An up-to-date description of the crater structure and relevant references is given in Lindström et al. (this volume).

The structure has been covered by Caledonian thrust nappes of at least 3000 m thickness (Karis and Strömberg 1998). Thus, the structure has been preserved from erosion for a long time. Presently, the cover rocks have been eroded so that the front is less than 1 km NW of the suggested margin of the structure. A small outlier of the lowermost Caledonian overthrust nappe called the Tramsta nappe has been preserved from erosion due to its low topographic level in the crater depression (Lindström and Sturkell 1992).

Radial erosional gullies cut the "brim" of the structure (von Dalwigk and Ormö 2001; Lindström et al. this volume). They eroded through the allochtonous to the autochthonous breccia and, at some locations, to the less fractured basement.

1.2.1 Pre-Impact Geology

Proterozoic. In the Lockne Area, the oldest rocks are represented by a sequence of metavolcanites, the Börön volcanic suite (Mansfeld et al. 1998), occurring on the east side of the southern part of Lake Lockne. After the peak metamorphism, there was an extensive magmatism of Revsund granite. The Revsund granite occupies vast areas in the western part of the Bothnian Basin (Gorbatschev 1997).

The major part of the Lockne Area consists of Revsund granite with ages from 1.82 Ga to 1.65 Ga (Gorbatschev 1980). A prominent ductile deformation zone is located to the southwest of Lake Lockne, the Hassela Shear Zone, striking NNW. Further to the southwest, another deformation zone, the Storsjön Edsbyn Deformation Zone (SEDZ), has affected the basement. The Storsjön Edsbyn Deformation Zone extends from the Bothnian Sea to Lake Storsjön. The two deformation zones experienced ductile deformation during the late stage of the Svecokarelian orogeny between 1.85 and 1.7 Ga ago (Högdahl 2000). The youngest crystalline rocks in the area are represented by Åsby dolerite sills, which are 1.2 to 1.3 Ga old (Patchett et al. 1987; Souminen 1991). After the intrusion of these dolerites and before the transgression of the Cambrian Sea, the area was exposed to weathering (Karis and Strömberg 1998).

Cambrian to Ordovician pre-impact autochthonous cover rocks. The autochthonous Lower Palaeozoic is of standard Baltoscandian type (Lindström and Sturkell 1992). Prior to the meteorite impact, the deposition of the Dalby Limestone began.

1.2.2
The Impact Event

The Lockne Impact Crater formed in a marine environment with a water depth of less than 500 m (Lindström et al. this volume), 455 million years ago (Grahn and Nõlvak 1993). The resulting structure has an inner crater diameter of 7.5 km that is surrounded by a zone, about 2.5 km wide, with a brecciated crystalline basement and a few traces of pre-impact sedimentary rock called the "brim" (Lindström et al. this volume).

Authigenic breccia. Authigenic breccia (autochtonous or parautochtonous) is an impact generated monomict breccia of the locally occurring rocks, remaining relatively coherent during the crater formation. It shows no signs of significant transport and contains no exotic fragments. The fragments are angular and resemble breccias formed by other geological mechanisms, such as volcanic explosions or tectonic movements (French 1993). Authigenic breccias are often given local names. In the Lockne impact structure, it has been named Tandsbny Breccia (Lindström and Sturkell 1992).

Tandsbyn Breccia consists of intensely fractured but otherwise little altered fragments of local Proterozoic basement rocks. The lithology varies depending on the source rocks, the most common being Revsund Granite but dolerites or metavolcanites as protoliths have also been found (Lindström and Sturkell 1992). On several locations, the Tandsbyn Breccia grades into more intact basement granite. Cataclastic zones and displacements occur as well as spaces filled with bituminous materials. The Tandsbyn Breccia has a bituminous matrix containing angular fragments of quartz and feldspar, lithic fragments and fragments of fossils and shale. The fragments of Alum shale are slightly deformed along the margins but have an intact interior. The fractures in the Tandsbyn Breccia have a random distribution (see Fig. 7 in Simon 1987). The fracture fill is similar to that of Loftarsten (see below) down to very small dimensions and cemented with chlorite, calcite, and fluorite (Simon 1987). A study to investigate the occurrence of shocked quartz has been made in thin sections of Tandsbyn Breccia and it was found to contain quartz with a lamellar structure reminiscent of shocked quartz, but it was not as distinct as examples of Planar Deformation Features (PDF) in quartz from other established impact craters (Lindström and Sturkell 1992). The PDFs found in connection with this structure were found in the allogenic breccia (Therriault and Lindström 1994).

Allogenic breccias. Breccias with rock fragments that have been incorporated from different sources form allogenic breccia deposits, usually deposited in and around the structure (French 1993). In the Lockne impact structure, the Lockne Breccia (Simon 1987) and the Loftarsten (Thorslund 1940) are resurge sediments containing ejecta formed during the impact. Orientations of sets of PDFs corresponding to those found in known impact craters have been measured by Therriault and Lindström (1994) in 25 quartz grains from the rock type Loftarstone from the Lockne impact structure.

1.2.3 Post-Impact Geology

After the impact, the sedimentation continued, with the Dalby Limestone as the lowermost unit. The youngest autochthonous deposit, still existing in the Lockne Area, is the Slandrom limestone from the upper Caradoc (458-448 Ma) identified in this area by Thorslund (1940) and Lindström et al. (1983).

Allochtonous units of the Caledonian thrust nappes. The base of the Caledonian thrust front, the lower allochton, is composed of thrusted rocks of late Precambrian to early Palaeozoic age with a very low-grade metamorphism. The main orogenic displacement occurred in the Silurian, extending over the already existing crater. A small nappe, called the Tramsta nappe, still covers an area in the center of the topographic depression. Large parts of brecciated basement of sizes from 10 to 10 000 m^3 (Lindström et al. this volume) have been found lying on top of transported sediments. These large clasts are interpreted as having been ejected from the crater to a distance of several kilometers and then transported back towards the structure with the Caledonian nappes in the general direction of the Caledonian thrusting. The thrusting direction was E-SE, with an estimated 10-40 km range of transportation distance of the units of impact age (Lindström et al. 1996).

Post Caledonian tectonic events. Shearing and fracturing of the crystalline basement in the Lockne Area unrelated to the meteorite impact occurred mainly before the Caledonian orogeny (described earlier) (e.g., Gorbatschev 1997; Högdahl 2000). However, the displacement of the thrust front and the impact structure indicate some post-Caledonian shearing and faulting, which also displaced the Åsby dolerite sills. At present, the top of the basement surface dips about 1° to the northwest (Thorslund 1940).

2 Methods

The two main methods are *in situ* electrical resistivity measurements with VLF-R technique along with fracture frequency measurements on outcrops.

2.1 Electric Resistivity Measurements with the VLF-R Technique

Electric resistivity is the inverse of conductivity, and the unit used is the ohmmeter (Ωm). The aggregate rock resistivity is, in a first approximation, the weighted sum of the resistivities of the rock components. One of the components is water (the electrolyte) residing in the pores of sedimentary rocks or in the fractures in crystalline rocks. The fracturing of rocks changes the rock physical properties by an increase in porosity and thus the electric conductivity. This effect has previously been studied for fracture zones (e.g., Eriksson 1980; Henkel 1988). It has also been observed in several impact structures like Dellen and Siljan (Henkel 1992).

The resistivity measurements were made in-situ using a signal from a distant Very Low Frequency (VLF) radio wave transmitter. Typical transmitted VLF frequencies lie in the 10-30 kHz range. The instrument, Geonics EM16R, is a VLF receiver equipped with a 10-meter reference antenna. This equipment was used to receive the signal with a frequency of 16.0 kHz from the GBR transmitter located in southern England. The signal induces secondary Electro-Magnetic (EM) fields in conductors in the ground or on the surface. The instrument EM16R (Geonics 1979) measures the ratio between the horizontal electric and magnetic field as the phase angle and the apparent resistivity (ρ_a) of the subsurface in Ωm by adjusting the resistivity dial and phase dial for signal minimum. As the readings are determined by listening to sound variations, the precision is about 5° in phase angle and about 10% in apparent resistivity. The penetration depth of the resistivity measurements depends on the ρ_a of the measured site and is inversely related to the signal frequency. The penetration depth at frequency 15 kHz is 60 m at ρ = 200 Ωm, 200 m at ρ = 2000 Ωm, and 600 m at ρ = 20 000 Ωm. The ρ measured are in the range of 300 - 10 000 Ωm, giving a penetration depth much larger than the depth to the ground water level. The rock volumes measured are thus saturated with water.

As the earth is not a perfect conductor, the electric vector of the EM-field is tilted near the earth's surface; thus it also has a horizontal component. The phase angle represents the tilt of the electric vector compared to the magnetic vector caused by the vertical anisotropy of the resistivity within the penetration depth. A two-layer model can be applied to the readings, to estimate the overburden thickness and resistivities of the overburden and bedrock. One of these three model parameters must, however, be assumed. This procedure is put into practical form in the two-layer nomograms constructed by Geonics (1979). If the subsurface has constant electrical resistivity at least down to the penetration depth, the phase angle is 45°. When applied to two-dimensional conductive structures, the two-layer model results in several typical distortions that must be accounted for. When a highly conductive vertical structure, like a fault zone, has a strike direction parallel to the VLF-R antenna, the phase angle measurements give a false and too high value over rather large distances. Above thick conductors, the resistivity can be estimated, but the depth indication becomes unreliable (Hjelt et al. 1985).

It is known that the electric resistivity in fractured crystalline rock has rather low values from studies of mapping fracture zones using an electric method, Slingram, and the electromagnetic technique VLF (e.g. Eriksson 1980; Henkel 1988). Typical values are around 2000 Ωm in fractured rocks, while normal crystalline rock resistivities are >10 000 Ωm. Even lower values are found for fault gouge, down to 30 Ωm (Henkel 1988).

2.2 Investigation of Impact-Induced Fracturing

The fracture frequency measurements were made within an 8 km wide area between the two large shear zones mentioned previously but not across them (Fig. 1). The study was made on outcrops along two forest roads, section A and B in Fig. 1. Section B follows the eroded path of one of the mentioned resurge gullies formed through the rim of the crater. Each crystalline rock outcrop was investigated using window-mapping technique to estimate the fracture frequency.

Fig. 2. A nearly vertical surface of Tandsby Breccia similar in appearance to the horizontal surfaces on which the fracture frequency was measured. The hammer length is 0.6 m and the location of the outcrop can be seen as a triangle in Fig. 1.

The outcrops used were smooth, glacially eroded, horizontal surfaces having a homogenous lithology. Figure 2 shows a nearly vertical surface of Tandsby breccia, the same rock type for which the fracture frequency was measured.

The surface was cleaned of all loose materials including lichen or moss growth. Because of the glacial scouring, the surface weathering was minor, but had often reinforced the contrast between the fractures and the edges of the clasts, thus making the counting easier. The general size of each outcrop was 10-50 m^2.

At each site, the measurements were divided into two groups, one for large fractures (with a trace length from 0.25 to 5 m) and one for small fractures (with a trace length of 0.1 to 0.25 m). The large fractures were counted over the entire outcrop, whereas the small fractures were counted within 0.25 m^2 due to the large number of fractures.

The maximum distance between large fractures was restricted to 5 m because of the size of the outcrops, which had a width of about 5 - 10 m.

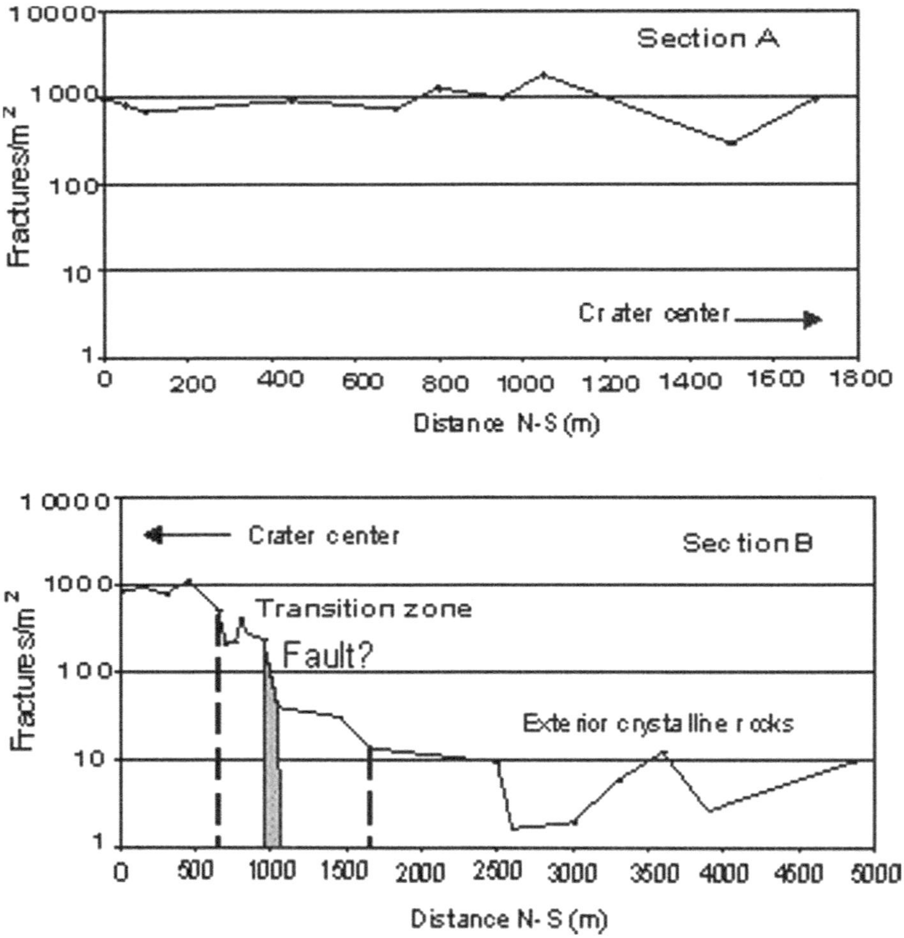

Fig. 3. Projection of the fracture frequency (number of fractures per m^2) to the northern (section A) and southern (section B) sections. For the location of sections and individual measurements, see Fig. 1.

The minimum distance between small fractures was truncated at 2 cm; fracture sets with less than 2 cm between fractures were not considered. Fractures smaller than that were too difficult to distinguish from the rock structure. The numbers of fractures were counted including the sides of individual clasts that were considered as fractures. The sum of the small fractures and the large fractures normalized to 1 m^2 is the fracture frequency presented in this study.

2.3 In-situ Resistivity Measurements on Rock Outcrops

The resistivity measurements were preferentially made over a flat part of an outcrop that contained the fracture frequency sampling area, with no steep edges around. All nearby highly conductive features, such as power lines, railroad tracks, buried electricity cables and wire fences, were avoided. Tectonic fault zones parallel to the transmitter direction, as well as steep slopes in the terrain were also avoided.

At each measurement site, the phase angle and the apparent resistivity values were measured with the Geonics 16R instrument and a short description of the rock type and its soil or moss cover was made.

3 Results

The fracture frequency observed on outcrops of the crystalline rocks is presented in section A and B (Fig. 3), with a logarithmic scale for the fracture frequency because of the large range. In section A, the fracture frequency is generally very high, close to 1000 fractures/m^2 over a distance of 1700 m. No transition from high fracture frequency to low is noticed towards the suggested northern margin of the structure. In section B a transition from high to low fracture frequency can be seen. The frequency is variable but high in the northern part of the section, over 1000 fractures

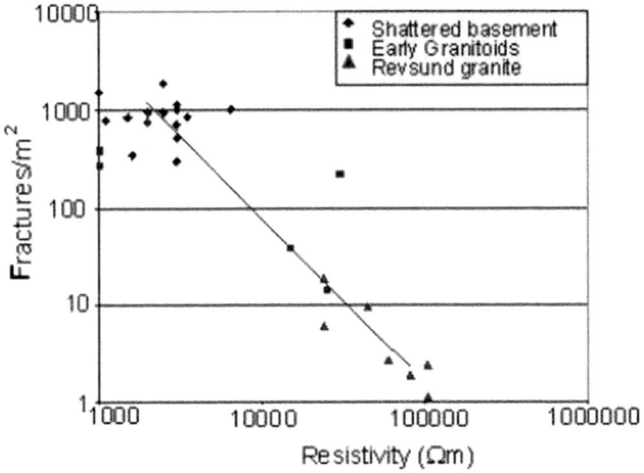

Fig. 4. Correlation between the fracture frequency and electric resistivity of crystalline rocks in the Lockne impact structure measured in two sections shown in Fig. 1.

per m^2 and thus similar to the frequency in section A. In the transitional part, the fracture frequency declines from ~ 1000 fractures/ m^2 to ~10 fractures/m^2. This transition occurs within 1100 m, from 500 to 1600 m distance from the start of the section. In the southernmost area of section B, the fracture frequency remains at 10 fractures/m^2. At 1000 m in section B there is a step in the fracture frequency over one order of magnitude, shown as a light gray area in Figure 3.

The crystalline rocks have different resistivity values depending on the amount of brecciation. The non-brecciated crystalline rocks have resistivities of ≥10 000 Ωm. The brecciated crystalline rocks have resistivities ranging from 10 000 to 300 Ωm depending on the amount of brecciation.

The correlation between the two-dimensional fracture frequency and resistivity measured on the outcrops along section A and B is shown in Fig. 4.

The correlation plots as a straight line on doubly logarithmic axes, indicating that it has a power law functional form. The equation for the correlation shown in Fig. 4 relating resistivity and fracture frequency in the Lockne structure is:

$$\text{Log F} = 9.4 - 1.9 \text{ Log R}$$

where F = fracture frequency (fractures/m^2) and R = electric resistivity (Ωm) and the constants 9.4 and -1.9 are derived from the linear reduction of the correlation.

The correlation is assumed to be linear although the distribution of data indicates two populations with a rather wide gap, as seen in Fig. 5. This is caused by the rapid decline in fracture frequency and increase in resistivity at 1000 m in section B, when moving away from the crater center. The frequency decrease of the resistivity can be seen in the histogram of resistivity values in Fig. 5, where there are no values between 7000-15000 Ωm.

4 Discussion

Measurements of the fracture frequency of both vertical and horizontal surfaces would be ideal to estimate the 3-dimensional fracture frequency. Most of the existing vertical surfaces in connection to the Lockne structure were blasted during road construction, and the artificial fractures from the blasting make them unsuitable for these kinds of studies (Priest 1993). There is, however, no visible difference in the fracture frequency of the vertical and horizontal surfaces of non-blasted rocks as in Fig. 2. It can, therefore, be assumed that an estimate of the fracture frequency of horizontal surfaces is representative for the comparable volume of rock.

In an authigenic breccia formed in-situ, the clast edges represent the fractures that formed in the target, hence the counting of the clast sides as fractures in this study. Earlier fractures in the target were erased by the formation of the authigenic breccia and cannot be identified. Fractures younger than the Tandsbyn Breccia crossing the outcrops are so few that they do not contribute to the fracture frequency. The fractures counted in the fracture frequency study are thus considered to be primarily impact-generated.

In section A (Fig. 3), the fracture frequency is high and shows no tendency to decline in either direction. A similar decline in the fracture frequency as observed in the southern part of section B would be expected in the north part of section A if this is a similar part of the structure. This indicates that the outer limit of the structure has not been reached in section A. Potential occurrences of crystalline rocks further to the north are covered by sedimentary rocks and cannot be investigated.

The systematic energy decay with distance from the impact center would create a gradient in the radial distribution of fracture frequency. The

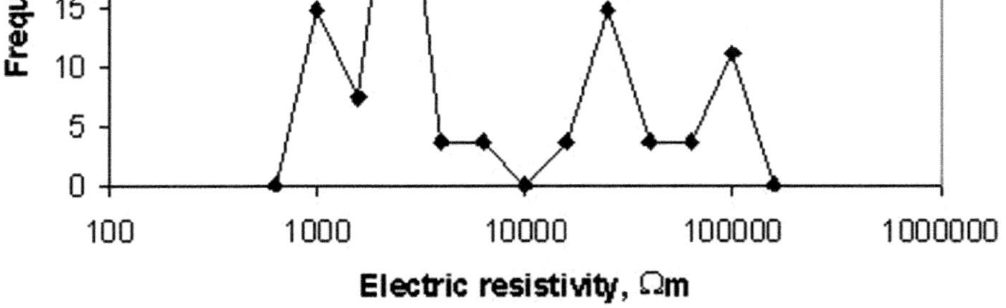

Fig. 5. Frequency of electric resistivity readings from measurements at section A and B seen in figure 1. Tandsbyn Breccia has low resistivity values (<1000 to 7000 Ωm) and the crystalline basement has high values (15000 to 100 000 Ωm).

observations however, show a rather limited transition zone of 1100 m at the southern crater margin. The relatively narrow interval where the fracture frequency decreases towards the margin of the structure, and the step at the distance 1000 m from the start of the section contradict the systematic radial trend in fracture frequency assumed earlier.

The description of the Tandsbyn Breccia in Simon (1987) indicates an allochtonous nature of the rock type. On the other hand, the observed gradual change to less fractured basement and the predominance of locally derived clasts points to an authigenic breccia type. The homogenous fracture frequency and resistivity in the present investigations favors an allogenic formation process for the Tandsbyn breccia. The transition to authigenic breccia and undisturbed basement is a narrow zone of 1100 m that is only observed at the southern margin.

The distinct step in fracture frequency, also found in the resistivity distribution at the southern margin, shows that an allogenic, generally low resistive, part of the Tandsbyn Breccia is juxtaposed with an authigenic part with higher resistivity along a fault zone. An alternative interpretation of the step in fracture frequency would be non-systematic radial stress decay.

5 Conclusions

The electric resitivity distribution of crystalline basement and impact generated Tandsbyn Breccia has been determined. The negative correlation between impact-induced fracturing and electric resistivity in crystalline rocks in the Lockne Area (Fig. 4) shows that the extent of impact brecciation in crystalline rocks can be measured with electro-magnetic or electric techniques. A radial fracture frequency or resistivity variation in Tandsby Breccia was not found; rather it was homogenous over large distances, indicating an allogenic formation process. As no transition zone similar to the southern margin could be found either in fracture frequency or in electric resistivity, the northern margin of the structure has not been reached. The gap in resistivity and fracture frequency related to the distinct step at 1000 m in section B is interpreted as a fault.

Acknowledgements

I wish to thank H. Henkel at the Royal Institute of Technology in Stockholm and M. Lindström at the Stockholm University for initializing and supervising this work at Stockholm University. I am also grateful for their advice and suggestions in connection with the preparation of this paper. I would like to extend my gratitude to J. Plado, Estonian Land Board, J. Plescia, US Geological Survey, and C. Koeberl, University of Vienna, for their helpful comments regarding this paper. Finally, I would like to thank R. Törnberg for his help and advice during the fieldwork.

References

Dalwigk I von, Ormö J (2001) Formation of resurge gullies at impacts at sea: The Lockne crater, Sweden, Meteoritics and Planetary Science 36: 359-369

Eriksson L (1980) Electric and magnetic methods to map fault zones in the bedrock. (in Swedish). Geological Survey of Sweden, Geofysiska byrån Report 8012, 79 pp

French BM (1998) Traces of catastrophe - a handbook of shock-metamorphic effects in terrestrial meteorite impact structures. LPI contribution 954, Lunar and Planetary Institute, Houston, 120 pp

Geonics (1979) Operating manual for EM16R VLF resistivity meter. Geonics Ltd, Ontario, Canada

Gorbatschev R (1980) The Precambrian development of southern Sweden. Geologiska Föreningens i Stockholm Förhandlingar 102: 129-136

Gorbatschev R (1997) Description to the bedrock map of the county of Jämtland part 1 (in Swedish). Geological Survey of Sweden Ca 53, 250 pp

Grahn Y, Nõlvak J (1993) Chitinozoan dating of Ordovician impact events in Sweden and Estonia. A preliminary note. Geologiska Föreningens i Stockholm Förhandlingar 115: 263-264

Henkel H (1988) Tectonic studies in the Lansjärv region, Swedish Nuclear Fuel and Waste Management Co. Technical Report 88-07, 66 pp

Henkel H (1992) Geophysical aspects of meteorite impact craters in eroded shield environment, with special emphasis on electric resistivity. Tectonophysics 216: 63-89

Hjelt SE, Kaikkonen P, Pietilä R (1985) On the interpretation of VLF resistivity measurements. Geoexploration 23: 171-181

Högdahl K (2000) Late-orogenic, ductile shear zones and protolithic ages in the Svecofennian Domain, central Sweden. Meddelande från Stockholms Universitets Institution för Geologi och Geokemi. Avhandling för filosofie doktorsexamen no 309, 85 pp

Karis L, Strömberg AGB (1998) Description of the bedrock map over the county of Jämtland: part 2: The Caledonian mountains (in Swedish). Geological Survey of Sweden Ca 53, 363 pp

Lindström M, Sturkell EFF (1992) Geology of the Early Palaeozoic Lockne impact structure, Central Sweden. Tectonophysics 216: 169-185

Lindström M, Simon S, Paul B, Kessler K (1983) The Ordovician and its mass movements in the Lockne Area near the Caledonian margin, central Sweden. Geologica et Palaeontologica 17: 17-27

Lindström M, Ormö J, Sturkell E, von Dalwigk I (this volume) The Lockne crater: Revision and reassessment of structure and impact stratigraphy, pp 357-388

Lindström M, Sturkell, EFF, Törnberg R, Ormö J (1996) The marine impact crater at Lockne, central Sweden. Geologiska Föreningens i Stockholm Förhandlingar 118: 193-206

Mansfeld J, Sturkell EFF, Broman C (1998) Meteorite impact influence on a Proterozoic shear zone mineralization at Gullgruvan, Jämtland, Sweden. Nordic Geological Winter Meeting 23, p 195

Masaitis VL, Pevzner LA (1999) Deep drilling in the Puchezh-Katunki impact structure. VSEGEI-Press, St. Petersburg, 392 pp

Masaitis VL, Mikhailov MS, Selivanovskaya TV (1975) The Popigai Meteorite Crater (in Russian). Nauka Press, Moscow, 124 pp

Melosh H J (1989) Impact cratering A geologic process. Oxford University Press, New York, USA, 245 pp

Patchett PJ, Todt W, Gorbatschev R (1987) Origin of continental crust of 1.9-1.7 Ga age: Nd isotopes in the Svecofennian orogenic terraines of Sweden. Precambrian Research 35: 145-160

Priest SD (1993) Discontinuity analysis for rock engineering. Chapman & Hall, London, Glasgow, New York, Tokyo, Melbourne, Madras, 473 pp

Simon S (1987) Stratigraphie, Petrographie und Entstehungsbedingungen von Grobklastika in der autochthonen, ordovizischen Schichtenfolge Jämtlands (Schweden). Geological Survey of Sweden C 815, 156 pp

Souminen V (1991) The chronostratigraphy of southwestern Finland with special reference to Postjotnian and subjotnian diabases. Geological Survey of Finland Bulletin 356, 100 pp

Therriault AM, Lindström M (1994) Planar deformation features in quartz grains from resurge deposits of the Lockne structure, Sweden. Meteoritics 30: 700-703

Thorslund P (1940) On the Chasmops Series of Jemtland and Södermanland (Tvären). Geological Survey of Sweden C 436: 1-191

Wickman FE (1988) Possible impact structures in Sweden. In Bodén A, Eriksson KG (eds) Deep Drilling in Crystalline Bedrock. 1. The Deep Gas Drilling in Siljan Impact Structure, Sweden and Astroblemes. Springer, Berlin: 298-373

Hydrocode Simulation of the Lockne Marine Target Impact Event

Valery Shuvalov[1], Jens Ormö[2], Maurits Lindström[3]

[1]Institute for Dynamics of Geospheres, Russian Academy of Sciences, 38 Leninsky pr., (build.6), ROS-119334, Moscow, Russia (shuvalov@idg.chph.ras.ru)

[2]Centro de Astrobiología (CSIC/INTA), Instituto Nacional de Técnica Aeroespacial, Ctra de Torrejón a Ajalvir, km 4, E-28850 Torrejón de Ardoz, Madrid, Spain (ormo@inta.es)

[3]Department of Geology and Geochemistry, Stockholm University, S-10691 Stockholm, Sweden (maurits.lindstrom@geo.su.se)

Abstract. In this study 2D and 3D numerical simulations are used to model the formation of the Lockne crater (centered at 63°00'20" N, 14°49'30"E) during Middle Ordovician times, about 455 Ma ago. We study a possible mechanism of shallow excavation to explain the concentric structure of the crater, as well as the interaction between basement the ejecta curtain and the water ejecta curtain, to explain the final ejecta distribution on the Earth's surface. We also consider different angles of trajectory inclination to understand how obliquity can influence the cratering flow in a marine target impact. Comparison between the results of numerical simulations and field studies allows us to estimate a water depth at the time of the impact of about 700-900 m.

1
Introduction

Geological observations and numerical simulations show that the morphology of submarine craters strongly depends on the depth of the seawater at the time of the impact. On the strength of these results it was suggested to use the crater morphology for an estimate of the target water depth (Ormö et al. 2001). To develop and discuss this idea we consider here the Lockne crater, central Sweden, which is known to have formed in a marine environment, and the morphology of which is well known.

The Lockne crater (centered at 63°00'20" N, 14°49'30"E) formed in Middle Ordovician time, about 455 Ma ago (Lindström et al. 1996; Sturkell 1998). The target consisted of an unknown, though not oceanic, depth of sea water, about 80 m thick Cambrian and Ordovician sediments, approximately half of which were well-lithified limestone, and crystalline basement that included granite, meta-volcanics, and dolerite. There is an inner, 7.5-km-wide crater, cut into the basement rocks, and a surrounding, 2-2.5 km wide, "brim" of exclusively crystalline ejecta that is tens of meters thick. These ejecta rest mostly on an apparently non-deformed peneplain that extended across the crystalline basement before the beginning of Cambrian sedimentation. Thus, with the exception of local, thin remains of the Cambrian basal sediments, the Cambrian and Ordovician strata were removed from the immediate vicinity of the inner crater before deposition of the crystalline ejecta. The distribution of crystalline ejecta gets discontinuous at distances greater than 2.5 km from the inner-crater margin. There also are a few major, radially oriented disruptions of the brim of crystalline ejecta. They have been called resurge gullies (Ormö and Lindström 2000); demonstrably, they served as channels to the resurge that resulted from the collapse of the transient water-crater. The total volume of crystalline ejecta preserved in the brim can be estimated at 0.3-0.7 cubic kilometers.

At distances greater than 1.5-2 km from the margin of the inner crater, not only Cambrian, but also Ordovician sediments of variable thickness were still present when the brim of crystalline ejecta was deposited. The Ordovician sedimentary succession preserved underneath the impact products becomes thicker and more complete with increasing distance from the crater center, until no more removal of pre-impact sediments can be identified. This condition is met at a distance of about 12-13 km from the center of the crater.

A most impressive feature at Lockne consists of debris deposited by the resurge of seawater into the newly formed crater. The resurge deposits grade upwards from extremely coarse breccia with debris-flow features, to arenites with upwards decreasing grain-size (Lindström et al. 1996; Dalwigk and Ormö 2001). The resurge deposits are characterized by greatly varying proportions of crystalline ejecta that were caught by the inward flow. Cobbles of such ejecta occur at least as far as 40 km from Lockne (Sturkell et al. 2000).

First attempts at numerical modeling of the Lockne impact were presented in (Ormö et al. 2001; Ormö et al. 2002). The purpose of those studies was to estimate the impactor size and possible water target depth at a moment of the impact from comparison between results of numerical simulations and field observations. We trie to derive a model that would fulfil the three most important constraints from the Lockne crater:

(i) correct basement crater diameter;
(ii) the so-called "brim" structure, which probably formed by the excavation of sediments and following deposition of the overturned basement flap;
(iii) the crater rim wall must have allowed a forceful resurge to enter the inner crater.

The sedimentary part of the target was not considered in those simulations, because this would require a higher spatial resolution and, therefore, would lead to a significant increase of the necessary computer time. The size of an area of probable sediment excavation was roughly estimated as the size of the transient water cavity.

The results of Ormö et al. (2002) show that the most probable radius of a spherical stony impactor was about 300 m. At 200 m water depth, a well-developed overturned flap would have formed, but the crater rim would have been too high to allow a forceful resurge. At 500 m water depth there is still a well-developed overturned flap; however, the zone of shallow excavation that took place on the surface to be covered by the overturned flap was quite narrow. At 1000 m water depth shallow excavation provides a large stripped surface around the central depression, however, the overturned flap decreases in size. In other words, at a small depth we have no shallow sediment excavation (stripping) from the region to be covered by the overturned flap; at large depth the flap itself is missing, because the basement crater is formed by material displacement within the target, not by excavation. Thus, we have a rather narrow range of possible sea depths, where both shallow excavation (at least, large water cavity) and the flap are present. For the Lockne crater the conclusion was drawn that the target water depth should have been in the range of 500 to 1000 m, with 700-900 m as the most probable numbers.

However, many problems remained unsolved. The first relates to the mechanism of shallow excavation. The assumption that the stripped surface coincides with the water transient cavity, at least, needs confirmation. To study this process in more detail it is necessary to consider an impact on a three-layered (water, sediments, crystalline basement) target. The second problem not considered in the previous numerical modeling is the interaction between basement ejecta curtain and water ejecta curtain. This interaction strongly influences the final ejecta distribution on the Earth's surface. This distribution is known from geological field observations and provides one more important constraint for the numerical model. The third open question is about how obliquity can influence the cratering flow after a marine target impact. The purpose of this paper is to achieve some progress in solving these three problems by means of direct numerical simulations.

2 Numerical Model

We used the same numerical model as Ormö et al. (2002), which is based on the SOVA multi-dimensional multi-material hydrocode (Shuvalov 1999). SOVA is an Eulerian material response code with some Lagrangian features. It allows to consider strong hydrodynamic flows with an accurate description of the boundaries between different materials (e.g., soil, gas, water, etc.) The code is similar in concept to the CTH hydrocode (McGlaun et al. 1990), which is widely used in the United States. The Eulerian forms of the governing equations are solved in two steps: a Lagrangian step and a remapping step. In the first step the Lagrangian forms of equations are solved, then the distorted cells are remapped back to initial mesh (or some new map changing in time) by the second order Van Leer scheme (Van Leer 1977). The two-step procedure is more suitable for multi-material problems.

Some interface reconstruction techniques are necessary to remap density and internal energy in the case of multi-material (mixed) cells. The SLIC scheme (Noh and Woodward, 1976) is used in SOVA for 3D simulations. The high-resolution interface tracker developed by McGlaun et al. (1990) is exploited in the case of two-dimensional geometry.

To model material strength the approach developed by Melosh and Ivanov (1999) is used. It is based on the "rigid-plastic" model (Dienes and Walsh 1970). For fractured rocks (loose materials with finite cohesion), the yield strength is defined as by Lundborg (1968) and Zamyshliaev and Evterev (1990):

$$Y=\min(Y_0+kP, Y_{max}), \tag{1}$$

where Y_0 is the cohesion, k is the coefficient of dry friction, P is the pressure, and Y_{max} is the limiting yield strength of the material at high pressure. The mechanism of acoustic fluidization in the treatment by Ivanov and Turtle (2001) is also taken into account.

In all numerical simulations the grid size increases progressively near the outer boundaries. As a result, perturbations can freely extend to high distances, and reflected perturbations do not disturb the flow we are interested in.

3 Shallow Excavation

To study the effect of shallow excavation we performed numerical simulations of an impact against a three-layered target. We used the same numerical model as in Ormö et al. (2002), but increased the spatial resolution of the numerical grid (80 cells across water column) and explicitly included a sedimentary layer (10 cells thick). Based on the results of previous simulations we considered the impact of a 300-m-radius stony asteroid into 800-m-deep water at a velocity of 20 km/s. The crystalline basement was covered by a 100-m-thick sediment layer with low strength (10^7 Pa cohesion and zero internal friction). The strength model used for the sediments is rather arbitrary. We do not know the real strength and simply consider a limiting case of low strength media. The effect of spallation resulting from the interaction of the shock wave with solid-water boundary could cause an intrusion of sediment fragments into water. This could result in mixing of the water and sediments and formation of a mud-like medium with very low strength. We used the Tillotson equation of state (Tillotson 1962) for wet tuff (with a density of 2 g/cm^3) to describe the thermodynamical properties of the sediments and the ANEOS equation of state (Thompson and Lauson 1972) for granite to describe thermodynamical properties of the impactor and crystalline basement. We consider an axisymmetric 2D problem. The nonuniform computational grid consists of 250×250 cells in vertical and horizontal directions. Near the sea floor we used rather high spatial resolution (space step equals 10 m, that is 10 cells across the sedimentary layer). The impactor was considered to be an ellipsoidally shaped granite body flattened in the vertical direction (with a diameter that is two times its height). Its mass equals the mass of a 300-m-radius spherical impactor. For land impacts this flattened shape provides a transient crater that is similar to the transient crater formed after an oblique impact (Boris Ivanov, personal communication).

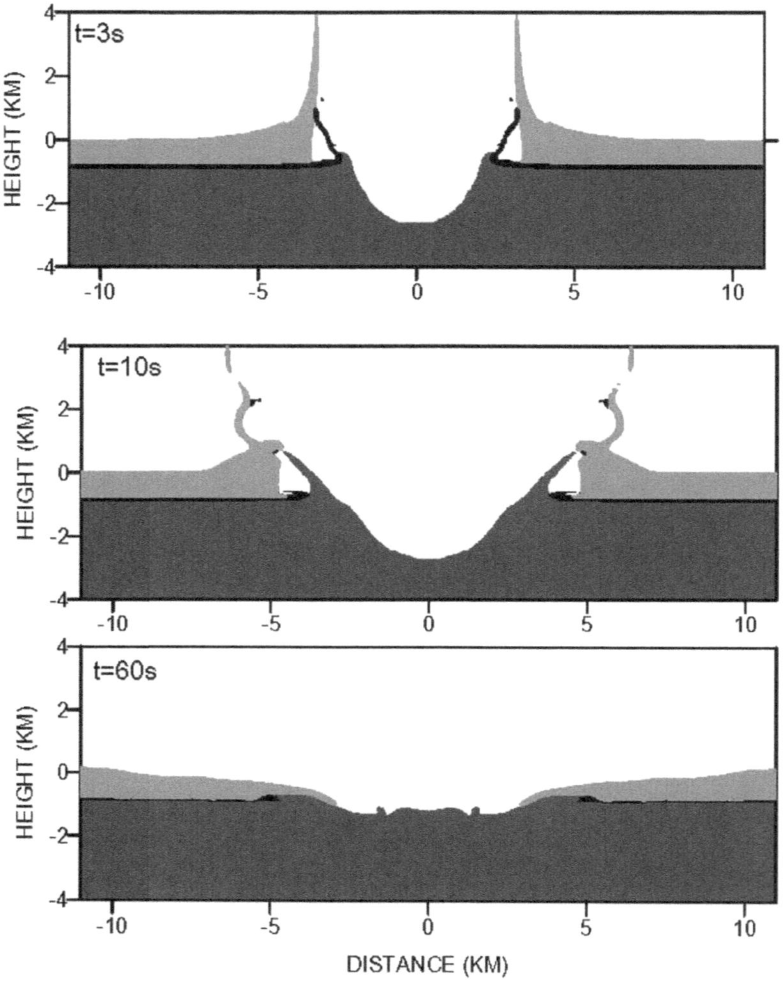

Fig. 1. Cratering flow at various times after the vertical impact of a 300-m-radius granitic asteroid into a 800 m deep sea. Black is sediments, light gray is water, dark gray is basement material. Atmospheric gas, projectile material, and ejecta with low bulk density are not shown.

The results of the numerical simulations are shown in Fig. 1. At the beginning the ejecta curtain consists of sediment fragments only. Later, a more powerful basement ejecta curtain form. We can see a considerable difference between the behavior of water and that of sediments, which is due to the higher density and strength (water has no strength) of sediments. The rather minor thickness of the sediment layer also plays some role in the different behavior of the sedimentary layer. In analogy to water we could say that we consider an impact into "shallow sediments". As a result there is no separate sediment cavity (similar to separate water cavity). However, moving outwards, the basement ejecta curtain pushes off low strength sediments (like a bulldozer). As a result an area that is cleared of sediments forms around the cavity. Similarly the strengthless water is pushed outwards after impact into shallow water (Ormö et al. 2002).

The width of the area cleared from sediments and covered by basement ejecta reaches a width of 1-1.5 km outside the basement crater rim. This roughly agrees with the field observations. Moreover, the area cleared of Ordovician strata (the upper part of the sedimentary layer) is slightly larger. The area cleared from sediments can also increase due to water erosion (during water cavity growth), which is not taken into account in these simulations. Dense, shock compressed water moves outside at a velocity of about 100 m/s and can strongly erode a sedimentary surface disturbed by shock wave and spallation. The water density does not change significantly (maximum 1.5-2 times), but pressure increases by orders of magnitude.

The erosion also can explain the wide range of available estimates of the outer crater diameter (from 12 to 24 km). The flow velocity continuously decreases with distance from the central crater, and the erosion also decreases. The erosion of the sea floor is determined by a powerful turbulent flow in the boundary layer near the sea floor (Dalwigk and Ormö 2001). When the surface is eroded by powerful gas flows, the eddy diffusion coefficient and the flux of the blown-off material are proportional to the squared velocity of the flow (Adushkin and Nemchinov 1994). This corresponds to the scenario in which a particle is broken from the surface by force, which arises as a result of the gas flow around the particle and is proportional to the squared velocity of the stream. Similar ideas are used for water flows in river engineering (Paintal 1971). If we assume that the quadratic dependence is correct also for impact-induced water-stream erosion, the thickness Δ of the soil washed away by water at distance R from the crater center can be estimated as $\Delta = k\int[u(R)]^2 dt$, where u is the flow velocity given from the numerical simulations described above and k is an unknown constant.

Figure 2 shows the dependence of the integral $\int u^2 dt$ on the radial coordinate R both for outward and inward (resurge) water flows. We can see that in the brim area (4 km$<R<$5.5 km) erosive power of the outward directed flow is comparable with that of the resurge (within a factor of two). Outside the brim most erosion of the sea floor is produced by the resurge. Note that the expression used for eroded depth is only a very

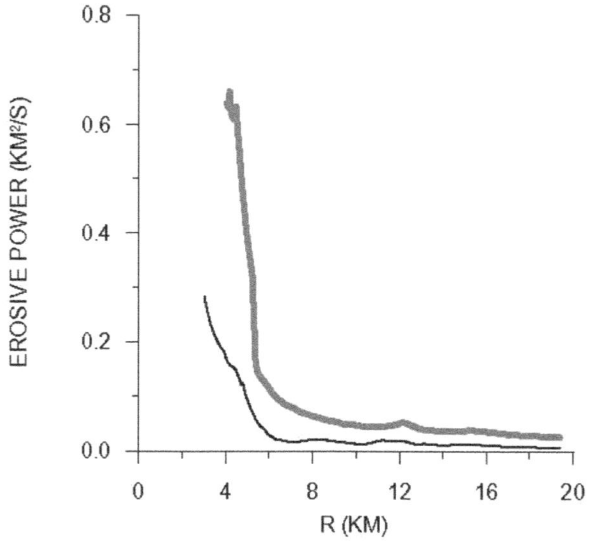

Fig. 2. Integral $\int u^2 dt$ (erosive power) versus distance R from the crater center. Black line indicates outward flow, the thick gray line resurge flow.

crude approximation. In particular, the value of k can depend on the soil density, the surface roughness, the size of the particles washed away, and so forth. Furthermore, there is a minimum velocity threshold for the stream at which soil particles of a definite size break away from the surface.

4
Ejecta Distribution

Geological field data show that, apart from the overturned flap of crystalline ejecta, which occurs within roughly 2.5 km from the boundary of the inner crater, fragments of granite ejecta with sizes from 0.1 to 10 m lie at distances r of up to 10 to 15 km from the crater center. This seems to be the norm for land impacts, where ejecta are continuously distributed over a large area around the crater with a density roughly decreasing with distance as r^{-3} (McGetchin et al. 1973). However, this is not obvious for marine target impacts. Numerical simulations (Ormö et al. 2002) and the new simulations presented here show that the flight of basement ejecta is restricted by the transient water cavity and water ejecta curtain (at least at great water depth, which results in a large water cavity).

A typical situation is shown in Fig. 1. The angle of ejection and the angle of the ejecta curtain depend on material strength (Melosh 1989). In strengthless water the ejecta curtain is almost vertical and initial ejecta trajectory angles exceed 60°. In solid targets the angle of the ejecta curtain is about 45° and initial ejecta trajectory angles are smaller than 60°. This means that ejecta curtains must intersect. At shallow water depth the mass of ejected water is small and cannot strongly influence the flight of the solid ejecta. However, a large water transient cavity (necessary for shallow excavation) evolves at high water depth where most of the excavated mass is water. Thus we can conclude that the basement ejecta are absorbed by water and its late motion is defined by water flow and gravity sedimentation. Moreover, it is known that fast ejecta (those that are ejected to great distances) form only from the top target layers. In our case the top layer is water. To consider the interaction between small ejecta fragments (smaller than cell size) and water flow we should abandon the treatment of ejecta as a continuous medium.

A simple rough estimate of solid ejecta transport by water can be done from considering water flow velocities and velocities of sedimentation for ejecta particles of different size. Typical velocity time dependences at different distances from the crater center are shown in Fig. 3. The velocity of sedimentation can be estimated from equilibrium between particle weight, buoyancy force, and drag force:

$$4/3\pi r^3(\rho_g-\rho_w)g = C_D\rho_w V^2\pi r^2, \qquad (2)$$

where r is the radius of the spherical particle, ρ_g and ρ_w are granite and water densities, g is gravity, and V is the velocity of sedimentation. For r = 0.1, 1, and 10 m, we obtain V = 3, 10, and 30 m/s, respectively. The time τ of particle transport by water can be estimated as $\tau = H/V$, where H is the sea depth. For H = 800 m we have τ = 250, 80, and 25 s for r = 0.1, 1, and 10 m correspondingly. Combining these values with data from Fig. 3 we can conclude that basement ejecta fragments could be transported to a distance of 1 to 3 km outside the water transient cavity. Figure 3 shows flow velocity averaged through a water depth. In reality, the velocity changes with water depth and in some layers it can considerably exceed the averaged value. In particular, when the water transient cavity reaches its maximum radius, water near the seabed begins to move inwards, forming the resurge, while water in the top layers and water surge continue to move outwards. This can increase the area of ejecta fragment distribution.

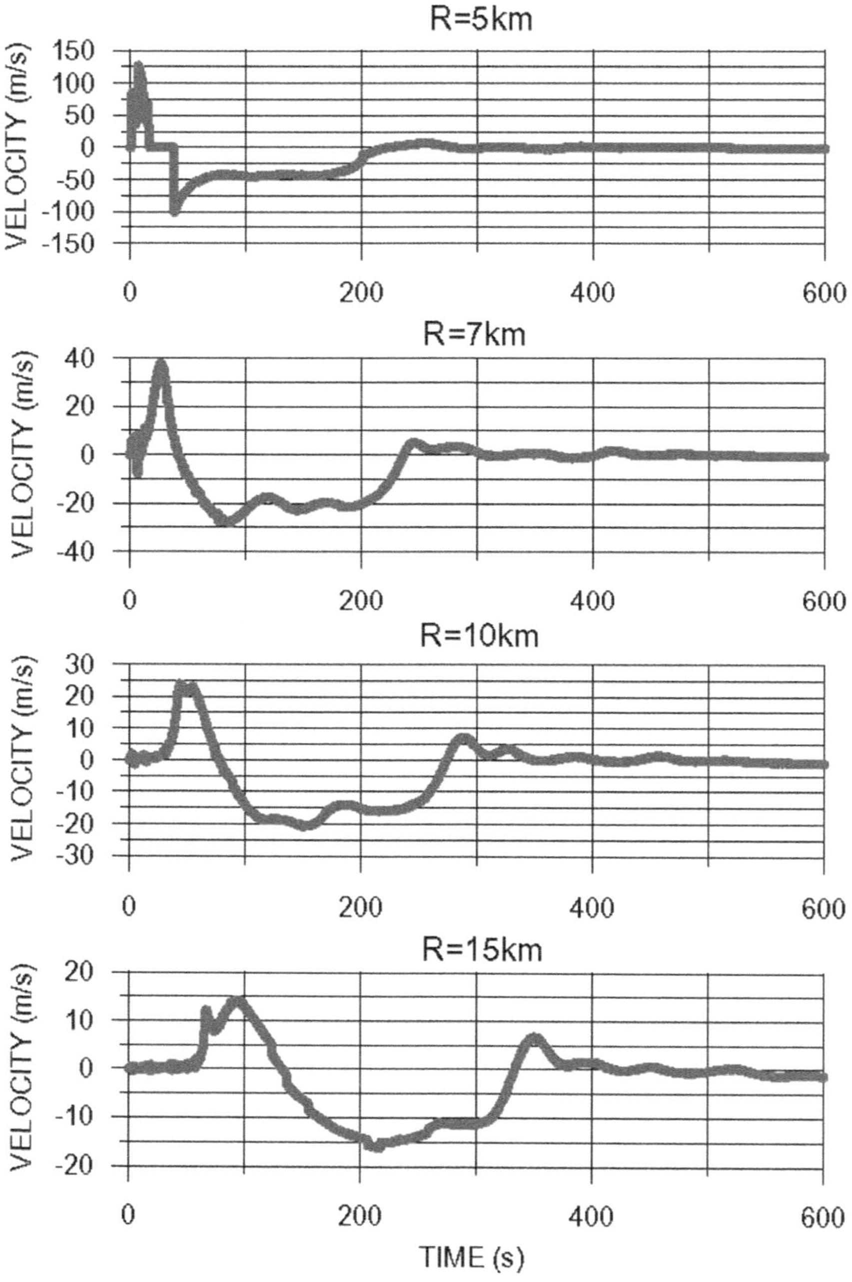

Fig. 3. Water flow velocity at different distances *R* from crater center versus time. A positive velocity indicates outward flow.

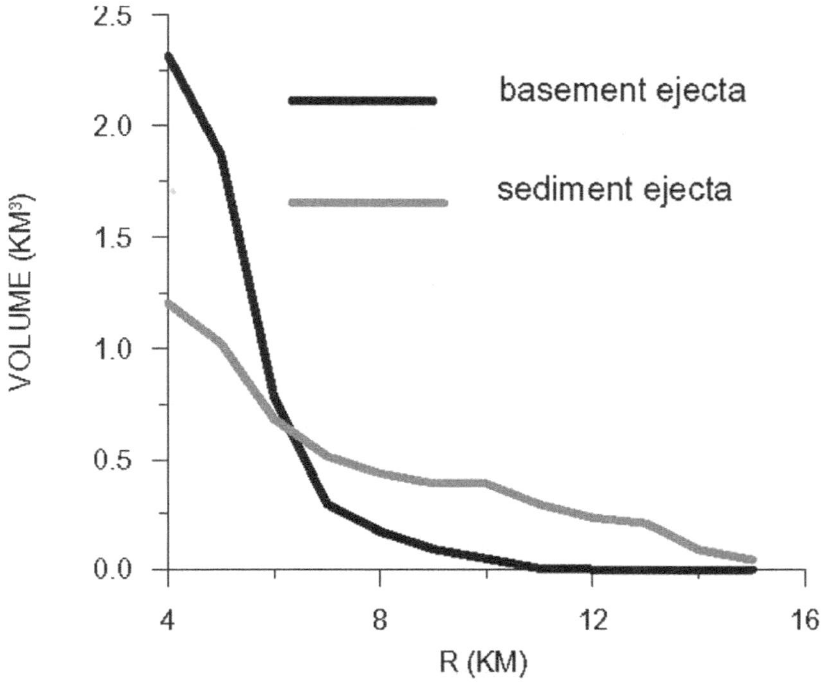

Fig. 4. Volume of basement and sedimentary rocks ejected at distances larger than *R* versus *R* (Distance from crater center).

To study this effect more thoroughly we performed numerical simulations where solid ejected material (described as continuous medium before ejection) was replaced by a set of discrete 1-m-radius particles with a density of 2.6 g/cm^3. These particles move with flow velocity (like passive tracers), but simultaneously they descend with velocity *V* defined by (1). Figure 4 shows the ejecta distribution obtained with this approach for both sediments and granite rocks. Our results show that granite particles could reach a distance of 11 km from the crater center. The boundary of sediment ejecta distribution is at least 15 km. We do not consider here the influence of the size of the particles and the trajectory angle on the ejecta distribution, which would be the subject of a further study.

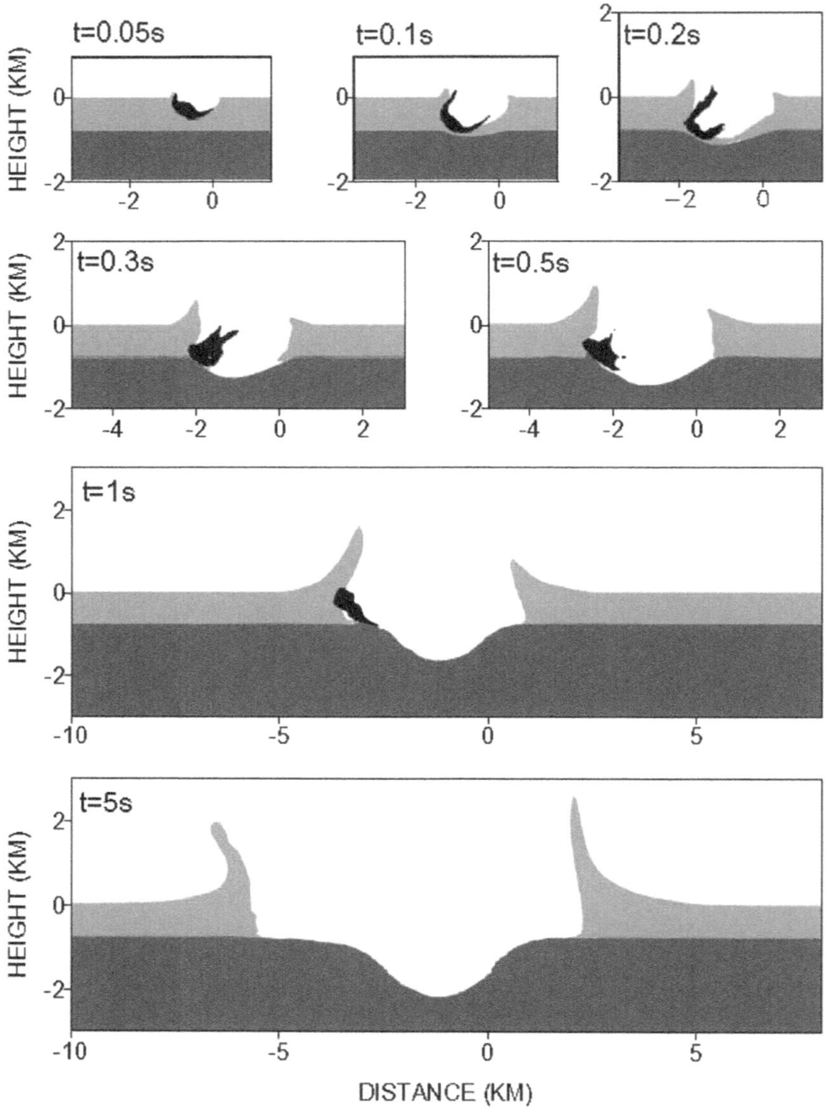

Fig. 5. Cratering flow after the oblique (45°) impact of a 300-m-radius granitic asteroid into a 800 m deep sea. Dark gray is basement material, light gray is water, black is impactor material (with bulk density exceeding 0.01 g/cm^3). Atmospheric gas and ejecta with a low bulk density are not shown.

5
Oblique Impact

To investigate the influence of the impact angle we have performed 3D numerical simulations of an oblique (45°) impact of a 300-m-radius granite projectile into a target consisting of crystalline basement overlain by a water layer that is 800 m deep. We used the same numerical model based on a 3D version of the SOVA hydrocode (Shuvalov 1999). The spatial resolution was lower than in the 2D simulations. The initial cell size in the central high-resolution region (100×50×80 cells) was 0.025 km in both vertical and horizontal direction and increases progressively far from the center. When the blast wave reached the grid boundary, the cell size (and the size of the computational region consequently) was doubled. The maximum cell size in the high-resolution region equaled 0.1 km in vertical direction and 0.2 km in horizontal direction.

The results (initial stage) are shown in Fig. 5. We can see that the basement crater is produced by a high velocity water stream, which is generated by the shock wave propagation through water. The projectile

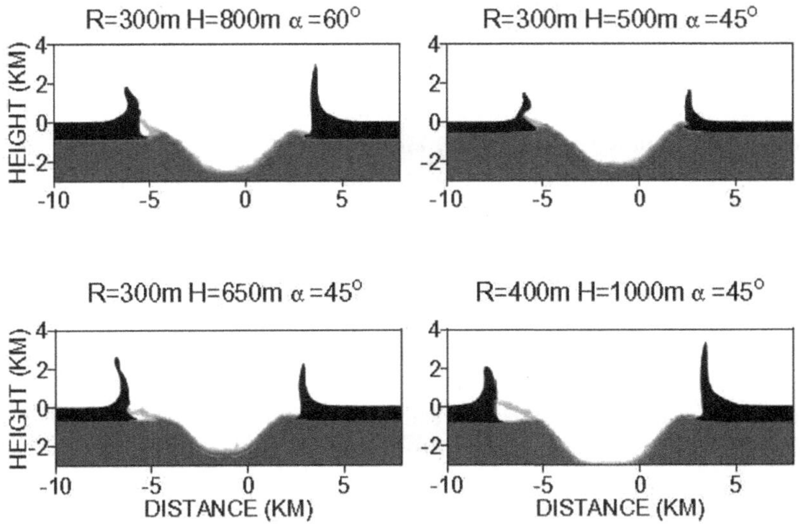

Fig. 6. Cratering flow 10 s after vertical (2D axisymmetrical and 3D simulations) and oblique impacts with different impactor radius R, sea depth H, and impact angle α. Water is shown by black, and the basement is gray. Atmospheric gas, projectile material, and ejecta with a low bulk density are not shown.

does not hit the basement crater floor and does not even penetrate into the crater. The crater center lies between a point of initial projectile-water contact and a point (near the boundary of the crater) where the projectile hits the sea floor.

The basement crater proves to be considerably smaller than in the vertical impact of the same projectile. In both cases (oblique and vertical impacts) approximately the same energy is released into the water, because in the vertical impact we considered an ellipsoidal projectile. The flattened projectile has a larger cross-section than an equivalent sphere, which compensates the longer trajectory of the oblique impact. As a result in both cases we have similarly-sized water transient cavities. However, the basement crater is produced by a water stream, and a vertically directed water stream creates a larger crater than an oblique water stream. Thus, the conclusion can be derived that obliquity increases the ratio between radii of water transient cavity and basement depression. At the same time the excavation flow from the basement crater (and, consequently, the overturned flap) decreases in an oblique impact of the same projectile size. To obtain the same basement crater we would need to increase the impactor size or decrease the water depth. However, the geological constraints for the Lockne crater (see "Introduction") and numerical simulations described below show that we can not decrease the sea depth below approximately 600 m, otherwise we will not obtain a large water crater and the resulting "shallow excavation".

Figure 6 shows the results of numerical simulations for several other values of sea depth, impact angle, and impactor radius. In particular, the results for the vertical impact obtained in both 2D (axisymmetrical) and 3D simulations are shown, as well as the results of 2D modeling with different spatial resolutions. These results demonstrate that 2D and 3D simulations of the axysymmetrical problem give the same distributions. The increase of spatial resolution (cells size lowered by a factor of two) only slightly influences the model output.

In all cases under consideration we obtain a basement crater diameter of approximately Lockne size (but with a somewhat different depth). The maximum water transient crater forms in a deeper sea (1000 m). This is also the modeled result that best fits the geological constraints for Lockne. Obliquity leads to asymmetry in the area cleared from water and to strong asymmetry of ejecta curtain. It is more extensive in the downward direction. However, the water surge is higher in the opposite direction. Thus we can expect that in the case of an oblique impact extensive distal ejecta deposits (up to 10 km away and more) should be found only in the downrange direction.

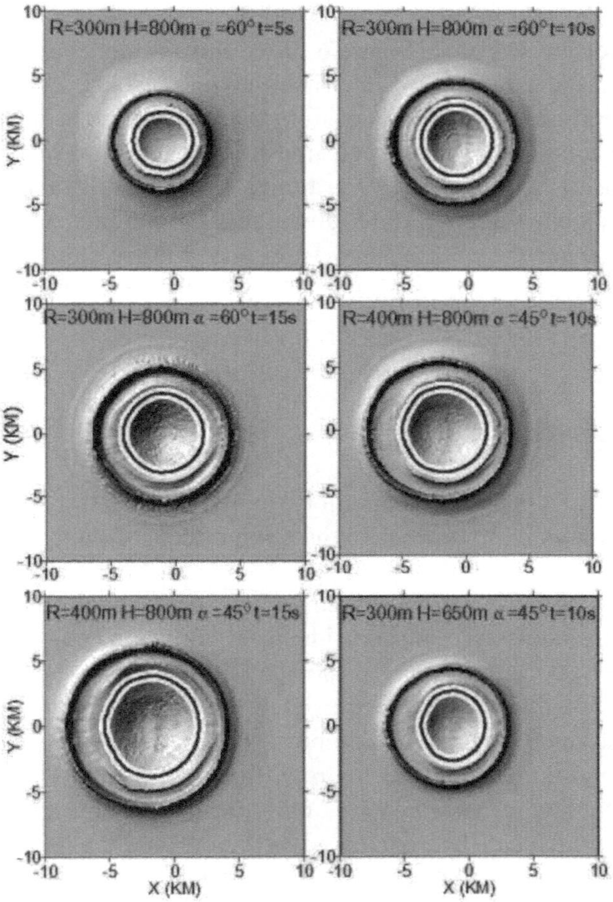

Fig. 7. This shows the zenith perspective of various craters formed by oblique impacts. In this figure the boundary solid surface is defined by the level at which the material density becomes equal to half of the normal density of condensed water (on the water-air boundary) or basement rocks (for the basement transient cavity). Therefore, only the denser part of the ejecta cone (with a high bulk density) is seen in the figure. In all cases the water transient crater is circular, the basement crater is also almost circular, however, the crater centers do not coincide. Both craters are shifted to a distance of 1 to 3 kilometers from the impact point, which has coordinates (0, 0). It is interesting to note that in the 45 degree alternative the basement crater is slightly elongated in the direction perpendicular to the direction of impactor flight.

Although distal ejecta are strongly anisotropic, the overturned flap is almost symmetric. At least, a dense part (with bulk density close to the density of basement rocks) of ejecta curtain shown in Fig. 7 is almost symmetric. This correlates with the results of laboratory experiments

(Anderson et al. 2000). The overturned flap forms due to overturning of near-surface layers, not by deposition from rarefied (with low bulk density) ejecta curtain. Figure 6 also shows more or less symmetric overturned flaps, but not symmetric ejecta curtains.

A general conclusion derived from these simulations is that the oblique impact results in a larger area cleared from water (compared to the vertical impact case), can provide overturned flap, and distal ejecta deposits, however, in one direction only (downrange of the impact point). In the uprange direction the area cleared from water is smaller, and there are no distal ejecta.

6 Conclusions

The series of numerical simulations presented in this paper allows us, at least qualitatively, to reproduce the main effects of the Lockne impact and typical features of the Lockne crater for a water depth of about 800 m. At the same time these results should be considered preliminary, because our numerical models are still very idealized, values of constants used in the models are poorly known, and the spatial resolution of numerical grids is not high enough to resolve all the details. Even typical land impacts are not well constrained (Melosh and Ivanov 1999), and the presence of water and layered solid target adds even more difficulties.

The problem of sedimentary and basement ejecta outside the water transient cavity can be considered as being qualitatively explained. The ejecta are transported to great distances outside the water transient cavity by the water surge. However, quantitative dependences presented in Fig. 4 need further elaboration. Ejecta redistribution due to resurge is also not investigated yet.

The problem of shallow excavation is far from solved. The numerical simulations presented in this paper show that with very low strength sedimentary rocks (pre-impact or resulting from spallation, mixing with water, etc.) we can obtain an area about 1-1.5 km wide, which is cleared from Ordovician sediments and is covered by a thick overturned flap. However, the model in use is only a rough approximation, as the real strength of the sedimentary rocks is unknown, and the spatial resolution is rather limited.

3D numerical simulations show that obliquity increases the ratio between the radii of the water transient cavity and the basement depression. The impactor size should also be increased or the water depth should be decreased in order to reach the same basement crater size as in the case of the vertical impact. Both water and basement craters are nearly circular, however, and the cratering flow is strongly anisotropic, because the circle centers do not coincide. The water crater extends further in the direction of impactor trajectory. The result of an oblique impact yields a crater that, down-range, has a larger area cleared from water (than in the vertical impact), overturned flap, and distal ejecta. Up-range, however, the area cleared from water is smaller, and there are no distal ejecta.

Our results confirm the previous conclusion of Ormö et al. (2002) that target water depth should have been in the range from 500 to 1000 m, with 700 to 900 m being most probable. A large water transient cavity, which is necessary for strong sea floor erosion and shallow excavation, can be obtained only in this depth range.

Acknowledgements

This study was supported by a research grant for Cooperation between Sweden and the Former Soviet Union, provided by the Royal Swedish Academy of Sciences. Facilities were provided by the Department of Geology and Geochemistry of the Stockholm University, through its director, Prof. Jan Backman, and we thank its staff for their generous assistance and cooperation. We thank the reviewers E. Pierazzo and E. Asphaug for valuable remarks.

References

Adushkin VV, Nemchinov IV (1994) Consequences of impacts of cosmic bodies on the surface of the Earth. In: Gehrels T (ed) Hazards due to Comets and Asteroids, University of Arizona Press, Tucson and London: 721-778

Anderson JLB, Schultz PH, Heineck JT (2000) A new view of ejecta curtains during oblique impacts using 3D particle imaging velocimetry [abs]. Lunar and Planetary Science 31: abs. #1749 (CD-ROM)

Dalwigk I, Ormö J (2001) Formation of resurge gullies at impacts at sea: The Lockne crater, Sweden. Meteoritics and Planetary Science 36: 359–369

Dienes JK, Walsh JM (1970) Theory of impact: Some general principles and the method of Eulerian codes. In: Kinslow R (ed) High-Velocity Impact Phenomena, Academic Press, New York, London, pp 46-104

Ivanov BA, Turtle EP (2001) Modeling impact crater collapse acoustic fluidization implemented into a hydrocode [abs.]. Lunar and Planetary Science 32: abs. #1284 (CD-ROM)

Lindström M, Sturkell EFF, Törnberg R, Ormö J (1996) The Lockne Marine Impact in the Ordovician of Central Sweden. Meteoritics and Planetary Science 31: A80

Lundborg N (1968) Strength of rock-like materials. International Journal of Rock Mechanics and Mining Sciences 5: 427–454

McGetchin TR, Settle M, Head JW (1973) Radial thickness variation in impact crater ejecta: Implications for lunar basin deposits. Earth and Planetary Science Letters 20: 226-236

McGlaun JM, Thompson SL, Elrick MG (1990) CTH: a three-dimensional shock wave physics code. International Journal of Impact Engineering 10: 351-360

Melosh HJ (1989) Impact Cratering: A Geologic Process. Oxford University Press, New York, 245 pp

Melosh HJ, Ivanov BA (1999) Impact crater collapse. Annual Review of Earth and Planetary Sciences 27: 385-425

Noh WF, Woodward P (1976) SLIC (simple line interface calculation). Proceedings of the Fifth International Conference on Numerical Methods in Fluid Dynamics. Springer Lecture Notes in Physics 59: 330-340

Ormö J, Lindström M (2000) When a cosmic impact strikes the sea bed. Geological Magazine 137: 67-80

Ormö J, Shuvalov V, Lindström M (2001) A Model for target water depth estimation at marine impact craters [abs.]. Meteoritics and Planetary Science 36 (Supplement): A154

Ormö J, Shuvalov V, Lindström M (2002) Numerical modelling for target water depth estimation of marine-target impact craters. Journal of Geophysical Research 107 (E12): 3-1–3-9 doi:10.1029/2002JE001865

Paintal AS (1971) Concept of critical shear stress in loose boundary open channels. Journal of Hydraulical Research 9: 91-113

Shuvalov VV (1999) Multi-dimensional hydrodynamic code SOVA for interfacial flows, application to thermal layer effect. Shock Waves 9: 381-390

Sturkell EFF (1998) Resurge morphology of the marine Lockne impact crater, Jamtland, Central Sweden. Geological Magazine 135: 121-127

Sturkell E, Ormö J, Nolvak J, Wallin A (2000) Distant ejecta from the Lockne marine-target impact crater, Sweden. Meteoritics and Planetary Science 35: 929-936

Thompson SL, Lauson HS (1972) Improvements in the Chart D radiation-hydrodynamic CODE III: Revised analytic equations of state. Sandia National Laboratory Report SC-RR-71 0714, Albuquerque, 119 pp

Tillotson JH (1962) Metallic equations of state for hypervelocity impact. General Atomic Report GA-3216, 137 pp

Van Leer B (1977) Towards the ultimate conservative difference scheme IV. A new approach to numerical convection. Journal of Computational Physics 23: 276-299

Zamyshliaev B V, Evterev LS (1990) Models of Dynamic Deforming and Failure for Ground Media. Nauka Press, Moscow, 215 pp (in Russian)

Numerical Simulation of Shock Propagation in Heterogeneous Solids

Jan-Martin Hertzsch[1], Boris A. Ivanov[2] and Thomas Kenkmann[3]

[1]School of Mathematics, University of Bristol, BS8 1TW, United Kingdom (jan-martin.hertzsch@bristol.ac.uk)

[2]Institute for Dynamics of Geospheres, Russian Academy of Sciences, Leninsky Prospekt, 38-6, ROS-117939 Moscow, Russia

[3]Institut fur Mineralogie, Museum fur Naturkunde, Humboldt-Universität, D-10099 Berlin, Germany

Abstract. Hypervelocity impacts of asteroids and comets on Earth and other planets lead to shock compression of the affected rocks. Heterogeneities, which are ubiquitously present in rocks, are likely to influence the dynamic behavior of geological materials under shock loading. The presence of small-scale heterogeneities like lithological interfaces, fractures, pores, etc. influence the propagation, magnitude, and geometry of shock and subsequent release waves and need to be considered if pressure and temperature achieved during an impact process are derived on the base of observed shock effects. In this study, we performed numerical simulations using the hydrocode SALE to analyze the effects of a simple planar interface on the shock wave behavior, and determine the magnitude of temperature, pressure, and density near the interface. The simulated materials, their geometries, and the scale of the model are adapted to shock recovery experiments. The ANEOS equation of state has been employed. The samples are either composed of rectangular pieces of quartzite and dunite enclosed in an iron container and impacted by an iron flyer plate, or consisting of dunite with a wedge-shaped quartzite inclusion presenting an inclined interface and impacted by a dunite slab. The computations indicate that during shock compression, pressures and temperatures are achieved in quartzite, which lead to completely reversible solid-solid phase transitions in the target material that start to develop at the interface. Shock heating alone is not sufficient for melt formation in the systems considered in the present study, but localized shear in particular at inclined boundaries between different materials results in a significant additional temperature rise of up to 400 K in addition to regular shock heating in the present model. Further mechanisms of energy dissipation are needed for a proper description of experimentally observed melting.

1 Introduction

The natural formation of shock waves in solids is restricted to hypervelocity collisions of bodies, usually of asteroids or comets with planets, asteroids or their satellites. As a shock wave progresses through a rock, it causes irreversible deformation in that particular material. Some of the produced deformation features are not exclusively restricted to shock loading (e. g., fractures), whereas others can be used as sensitive shock gauges. Among the latter are Planar Deformation Features (PDF) and high-pressure polymorphs. The term "shock metamorphism" describes the bulk of transformations that occur from when a rock is forced from an unshocked to a shocked state. Different stages of progressive shock metamorphism can be distinguished starting near the Hugoniot Elastic Limit (HEL) of a rock and ending with the complete vaporization of that rock. A variety of shock effects were calibrated with respect to pressure (and temperature) using shock recovery experiments. They were performed for the main rock forming minerals (e.g., Stöffler 1972, Reimold and Stöffler 1978, Stöffler et al. 1991, Stöffler and Langenhorst 1994, Schmitt 2000). These pressure gauges are frequently used to prove and analyze the size of terrestrial impact structures. However, components of polymineralic rocks usually exhibit different behavior in compression, making the response of inhomogeneous material to shock loading rather complicated compared to the compression of simple materials like most metals.

Inhomogeneities causing impedance contrasts are practically always present in rocks in the form of pre-existing cracks and fissures, inclusions of solid or liquid material, and pores. On the microscale, the list of inhomogeneities contains grain boundaries, crystal anisotropies, lattice defects, stacking faults or crystal impurities. They cause reflection, refraction, and interference of shock waves and may lead to localized concentrations of pressure, temperature, and deformation rate, and to phase transitions. Inhomogeneities can, therefore, influence considerably the behavior of geological materials under impact loading and need to be taken

into account if the magnitude of shock loading is to be estimated on the base of shock effects achieved during impact. An increased knowledge of shock wave propagation and interaction at interfaces is also required, for example, to understand the occurrence of highly localized melt veins (s-type pseudotachylites) or high pressure phases in the meteorites and craters. Finally, micro- and mesoscale heterogeneities may cause runaway instabilities in rocks and, hence, probably lead to considerable deviations from what is expected from bulk EOS. The presented investigation is also suited to derive improved composite EOS for macro-scale crater modeling (Crawford et al. 2003).
In this study, we will investigate the effects of a single lithological interface between two different rock types on the shock propagation and present results of numerical simulations of these shock recovery experiments. Our work has been motivated by laboratory studies on centimeter-sized samples composed of different lithologies which were enclosed in an iron container and impacted by an iron flyer plate and which showed localized melting at material interfaces (Kenkmann et al. 2000). The principal aim of our numerical studies is to illustrate the shock response of natural polymineralic rock, in particular the temperature increase at the contact of different materials. In this work we investigate the interaction of a planar shock wave with a planar lithological interface and compare the results with mineralogical findings in the corresponding experiments (Kenkmann et al. 2000).

2 Experimental Shock Compression of Rocks

We start with a brief summary of the laboratory studies, which motivated the present numerical work. In these experiments (Kenkmann et al. 2000) sized-sized two-component cylindrical rock samples were enclosed in an iron container and impacted by an iron flyer plate at 2540 m/s. The experimental setup is shown schematically in Fig. 1. The individual samples consisted of two lithologies, namely quartzite and dunite, which have been chosen because they show a strong contrast in their behavior under compression. The rock sections were fitted together along a smooth planar interface of varying orientation with respect to the plane of the incoming shock wave. Cross-sections of the samples after the experiments are shown in Fig. 2. In these experiments, gradients in deformation have been found in the two rock sections with decreasing distance from the lithological interface: Fracturing, comminution, and mosaicism are observed in the dunite section, and become more prominent near the

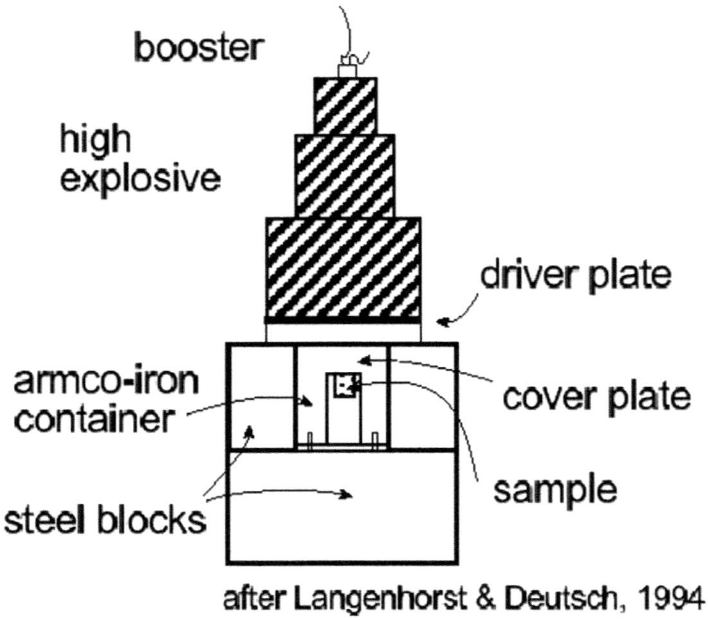

Fig. 1. Experimental setup. By means of an explosive, a driver (flyer plate) is forced onto the sample, which is enclosed in an iron container.

interface. A reduction in birefringence and multiple sets of planar deformation features are characteristic for the quartzite section, both become more intensive towards the interfaces. The occurrence of diaplectic glass, which was formed in one experiment, is restricted to the vicinity of the interface. Melts of a few micrometer width lubricated the lithological interface itself (see Fig. 3). Neutron and synchrotron radiation measurements (Walther et al. 2002, Kenkmann et al. 2003) on the shock deformed samples showed that the crystallite size in the quartzite section decreases with decreasing distance from the interface. The residual strain in olivine of the dunite section increases with decreasing distance from the interface. No high-pressure phases have been found in the recovered samples. The melt formation was attributed to a combination of shock heating and frictional heating.

The melt formation (at least part of it) is proposed by Kenkmann et al. (2000) to occur in the compressed state. Note that the melting point at high pressure is well above the melting point at normal pressure referred to by Kenkmann et al. (2000). We discuss this point briefly at the end of the paper.

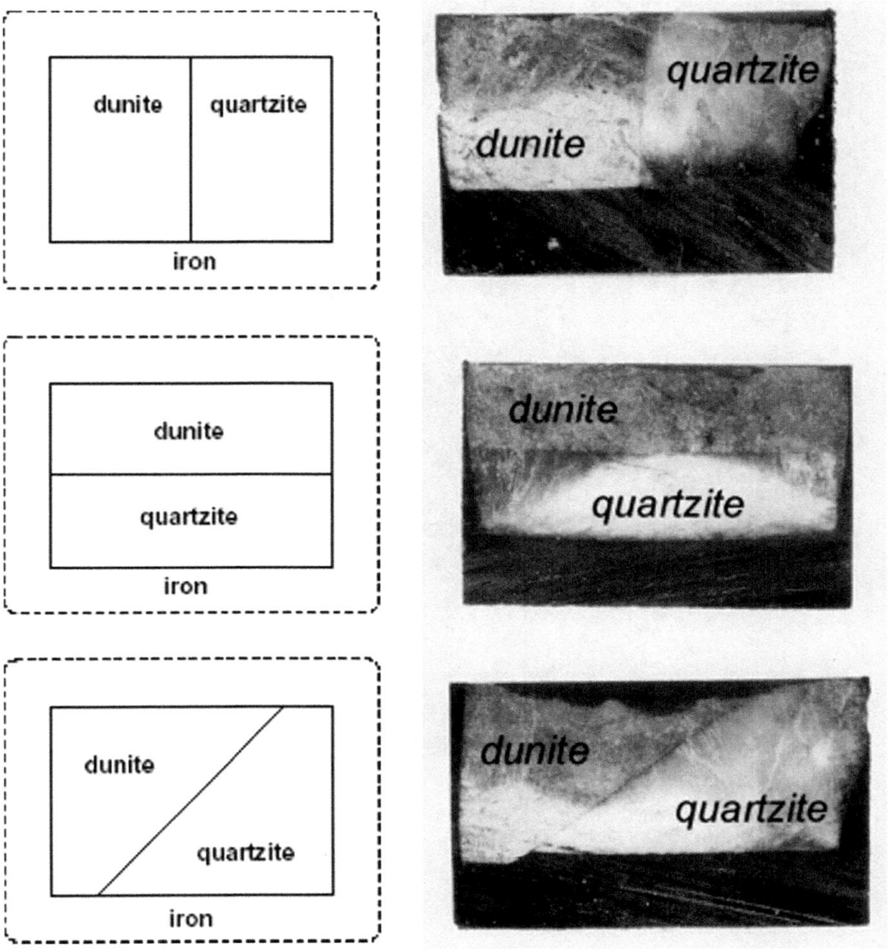

Fig. 2. Two-component rock samples after shock (Kenkmann et al. 2000), displayed after cutting the container in half. Top to bottom: vertical, horizontal, and inclined interface between quartzite and dunite. Width of the rock samples after experiment: ~17-19 mm. Note that the original sample width before shock compression was 15 mm, the original height was 10 mm. The original geometry is shown in the sketches at the left. Shear displacement at the inclined interface is 1.3 mm.

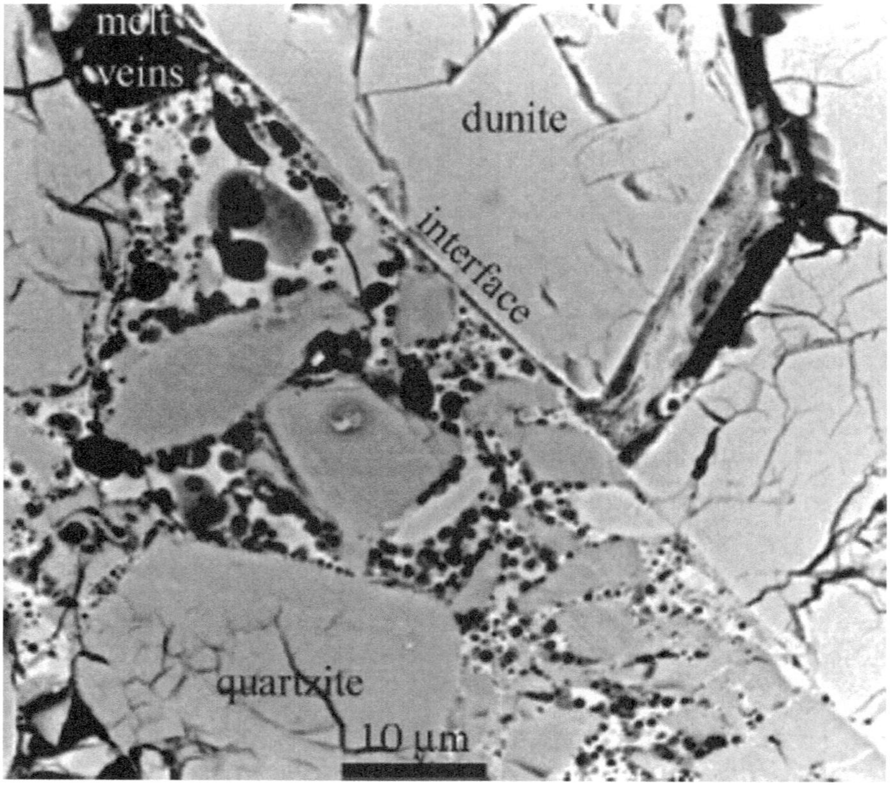

Fig. 3. Microscopic close up of inclined dunite-quartzite interface after shock experiment (Kenkmann 2000). The interface is lubricated with melt, which has intruded into cracks, dominantly of the quartzite section. The network of vesicular melt resembles natural pseudotachylite veins.

3 Numerical Modelling

In order to investigate the thermal effects at boundaries of different rock material, we have performed finite-difference hydrocode simulations of shock recovery experiments with samples consisting of quartzite and dunite. Particular attention has been paid to the effects of material boundaries on the temperatures reached during compression. These interfaces are originally perpendicular or inclined by 45° to the incident shock wave plane. The simulation technique is based on the hydrocode SALE (Amsden et al. 1980). It is a finite-difference code for the description of material flow at arbitrary speeds in which the region of interest is discretised by means of a mesh, and the continuity equations for mass, momentum, and energy are integrated over time. The behavior of the materials is treated in the framework of standard models:

gradual damage accumulation (Johnson and Cook 1983);
increase of rock shear limit with pressure (Lundborg 1968);
thermal softening of rocks (Ohnaka 1995);
thermal softening and strain hardening of iron (Johnson and Cook 1983).

Two simulation modes are used: In the Lagrangian mode, the grid moves with the material. This makes it easy to handle material interfaces and shock fronts, but one often encounters the problem of collapsing cells under high deformation. The Eulerian mode uses a grid, which is fixed, in space, and the material flows through the cells. Its advantage is that large deformations and complicated geometries can be simulated, but the handling of material interfaces is difficult. We use an extension of the code using a method developed for the hydrocode SOVA (Shuvalov 1999), which enables it to deal with mixed cells containing up to two materials plus vacuum, and in the present simulations two rock material types can be used when this mode is applied.

A complete description of the conditions in shocked material requires to relate pressure, density, and internal energy (or temperature) by means of an appropriate equation of state. Preliminary computations using the Tillotson equation of state (Tillotson 1962) have shown that pressures and temperatures occur during compression, which lead to phase transitions in the target, in particular to the formation of a high-pressure phase in quartzite. This means that one is compelled to use an equation of state, which can describe phase transitions in a thermodynamically consistent way. The Tillotson equation of state can not correctly describe the pressure and density changes in a two-phase region (Melosh 1989) because for a given internal energy, the pressure can assume any value in a wide range depending on the mass fraction of the two phases (Stevenson 1987), and such media can not be described as usual by just two thermodynamic variables. Therefore, we have instead used a computer program describing the thermodynamic relations between pressure, temperature and density in form of the ANEOS program package (Thompson and Lauson 1972, see also Benz et al. 1989) to pre-compute tables for the materials in question as an input for the hydrocode. ANEOS incorporates standard physical models for solids, liquids, gases and plasmas within their respective domain of validity and is based on a decomposition of the free energy in terms describing the behavior at zero temperature ("cold" pressure) and the thermal motion of atoms/ions and electrons (thermal pressure). Using continuous parameters, it provides a fit to the experimentally observed Hugoniot curves. Its treatment of phase transitions in the solid state, melting and evaporation is thermodynamically more correct than in the Tillotson equation of state. The phases are described in terms of the Helmholtz free energy from which the thermodynamic functions are derived. Thus, a thermodynamically consistent and complete treatment is ensured. ANEOS is a widely used tool for the purpose of simulating high-velocity impacts (see, e. g., Benz et al. 1989, Pierazzo et al. 1997, Shuvalov et al. 1999).

Although silicates have been found to undergo irreversible phase transitions under certain conditions (Ahrens and Rosenberg 1968, Swegle 1990), kinetics of phase transitions is not taken into account. In ANEOS, standard release behavior of the materials and thermal equilibrium of low-pressure and high-pressure phases (which are regarded as components of a mixture in thermodynamic equilibrium which coexist at the same Gibbs potential) is assumed. Neither phase boundaries nor delayed back transformations to the low-pressure phase, as it is the case in silica, are reproduced with this approach. This is a less serious restriction than one might assume at first, because the simulations indicate that the material was heated in the compressed state, and the differences in the release path have therefore only little influence on the results.

The material model for ARMCO iron is based on work by Johnson and Cook (1983). The plastic shear limit s is assumed to depend on the accumulated plastic strain e and the local temperature T according to

$$s = (E + F e^n)\ (1 - (\ (T-T_0)\ /\ (T_m - T_0)\ )^m$$

where the fitting parameters take the values E=175 MPa, F=280 MPa, n=0.32, m=0.55. E is actually the initial yield strength of the material at room temperature and a strain rate of 1 s^{-1}, T_m denotes the melting temperature and T_0 the room temperature as a reference temperature. As in simulations of impact cratering (Ivanov 2004), the strength of the rock components is assumed to grow with pressure, which is described in the Lundborg approximation (Lundborg 1968):

$$Y(p) = Y_0 + pk_p\ /\ (1 + pk_p\ /\ (Y_m - Y_0))$$

where Y_m is a plastic shear strength limit at high pressure, Y_0 the plastic shear strength at zero pressure, and k_p is a pressure coefficient providing "internal friction" for low pressures. The shear strength of the rock materials is assumed to decrease to zero at melt temperature (Ohnaka 1995) according to

$$Y(p,T) = k_T\ Y(p,T_0)$$

with a temperature coefficient

$$k_T = \tanh a\ (T_m/T - 1)$$

where a is a fitting parameter which has the theoretical value E^*/RT_m with an empirical constant E^*. All simulations presented here have been performed in planar geometry (two-dimensional), and heat conduction has been neglected because of the short timescale of the shock wave passage.

4 Simulations in Lagrangian Mode

4.1 Geometry

The experiments of Kenkmann et al. (2000) suggest that the different compressibility of materials in contact is the cause of considerable shear at their interface, which leads to additional heating on top of the compressive effects and to localised melting. We have, therefore, simulated the behavior of a two-component rock sample with a vertical interface, encased in an iron container and impacted by an iron flyer plate. The latter is comparatively thin because the experiments had been designed with the aim to reach the maximum pressure before the shock wave had passed the length of the rock sample and before a reflected wave arose at the interface between rock sample and container.

The dimensions of the sample correspond to a cross-section of the samples examined in the experiments (Kenkmann et al. 2000). The system is composed of rectangular pieces of quartzite and dunite of 10 mm height and 7.5 mm width enclosed in a 45 mm wide and 50 mm high iron container so that the top of the rock sample is situated 13 mm below the top of the container. It is impacted by an iron flyer plate with a thickness of 4 mm. The initial geometry is shown in the upper left image in Fig. 4, the other parts of the figure illustrate the compression of the sample.

4.2 Results

During the passage of the shock wave, the dunite part of the sample remains rather cold and reaches only a temperature of about 400 K, the quartzite part is heated up considerably more to about 700 K due to its higher compressibility compared to dunite. The quartzite section is strongly compressed, and pressure and temperature in this section show considerable peaks near the interfaces to the denser and less compressible

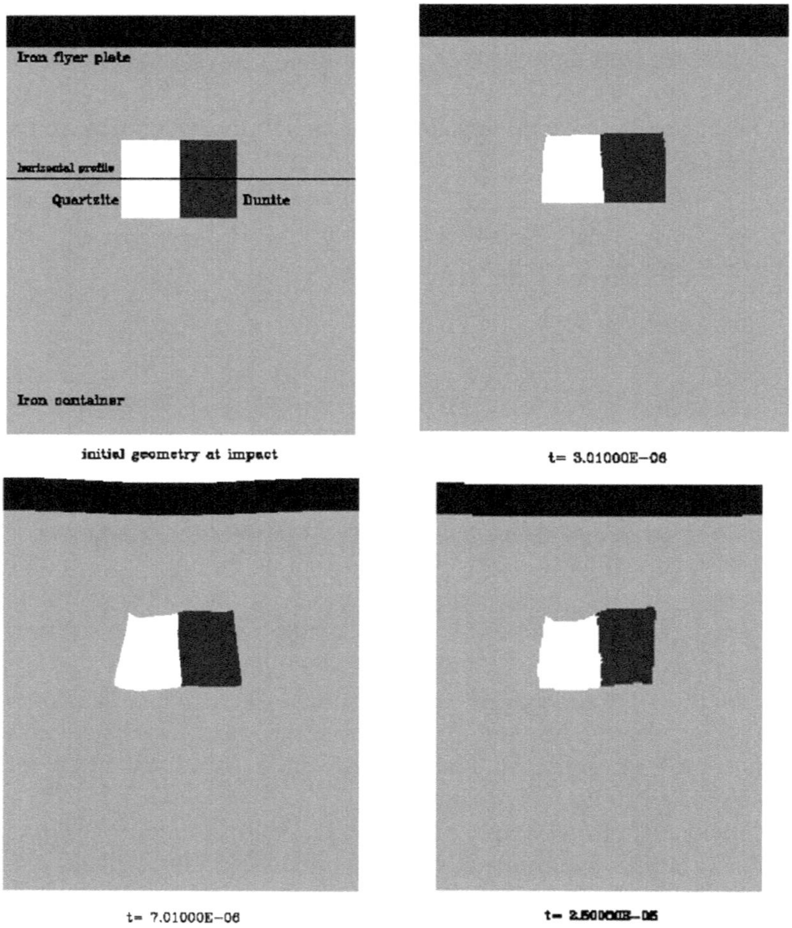

Fig. 4. Zone plots of materials showing the geometry of the sample at impact and selected later times. Width of container: 4.5 cm. Height of container: 5 cm. Thickness of flyer plate: 0.4 cm. Width of rock sample: 1.5 cm. Height of rock sample: 1 cm. Velocity of flyer plate: 2.54 km/s. Time in seconds. Spatial resoluti

materials during compression and fall earlier than in the center of the quartzite during decompression. Figure 5 shows density, pressure, and temperature changes in a horizontal cross-section cutting through the center of the rock sample 18 mm below the top of the iron container, a phase transition is observed in quartzite beginning at 3.5 microseconds, and the reverse transition occurs at 5.6 microseconds after impact of the flyer plate. Hence, quartzite stays in the high-pressure phase for about 2 microseconds. The onset of the phase transitions occurs at the material interface and subsequently affects the entire quartzite section with advancing time. From first principles of thermodynamics one can see that the adiabatic (standard) release above the lpp-hpp Hugoniot curve will result in a considerable amount of heat and correspondingly high residual temperatures. The occurrence of these phase transitions underlines the need for an equation of state, which can take them into account properly. Near the edges of the quartzite sample, temperature and pressure peaks can be observed which are due to a focusing of the shock wave by the surrounding denser and less compressible materials, but at the resolution of our simulations and within the restrictions of the employed material model they do not reach values necessary for the formation of melt. The temperatures in the interior of the rock samples during shock passage remain well below the melting point of any of the materials. At the interface between rock sample and iron container a reflected wave can be observed.

According to data obtained from vertical sections through the centers of both rock samples, the phase transitions on top of the quartzite block begin about 3 microseconds (low pressure phase - high pressure phase) after the impact of the flyer plate. Pressure release and reverse transformation (high pressure phase - low pressure phase) start at 4 microseconds after impact. In the density profile the phase changes are visible by a step-like increase or decrease in density. Due to the thin flyer plate the shock extends only over a short vertical distance, and the quartzite section is never completely in its high-pressure phase. The phase changes are reversible under unloading, but affect strongly the propagation of the shock wave.

Fig. 5. (The following 3 pages) Horizontal sections across the middle of the rock sample, 18 mm below top of the iron container and 5 mm below top of the sample in uncompressed state (22 mm below top of the impacting flyer plate), showing variations of density, pressure and temperature across the probe at different times and illustrating phase transitions in quartzite (note the density changes in the quartzite section). Next to the interface to the denser and less compressible materials, pressure and temperature in quartzite show notable peaks in compression at t=3.5 microseconds.

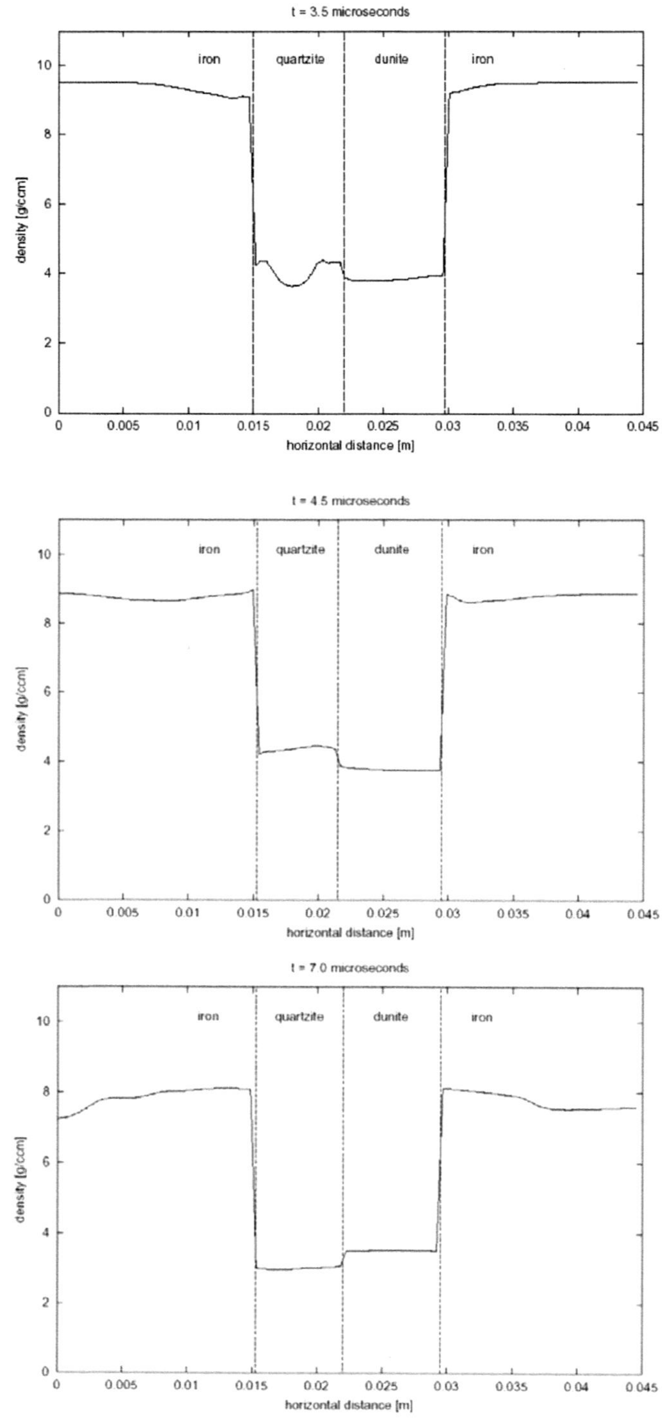
t = 3.5 microseconds
iron
quartzite
dunite
iron
density [g/ccm]
horizontal distance [m]
t = 4.5 microseconds
iron
quartzite
dunite
iron
density [g/ccm]
horizontal distance [m]
t = 7.0 microseconds
iron
quartzite
dunite
iron
density [g/ccm]
horizontal distance [m]

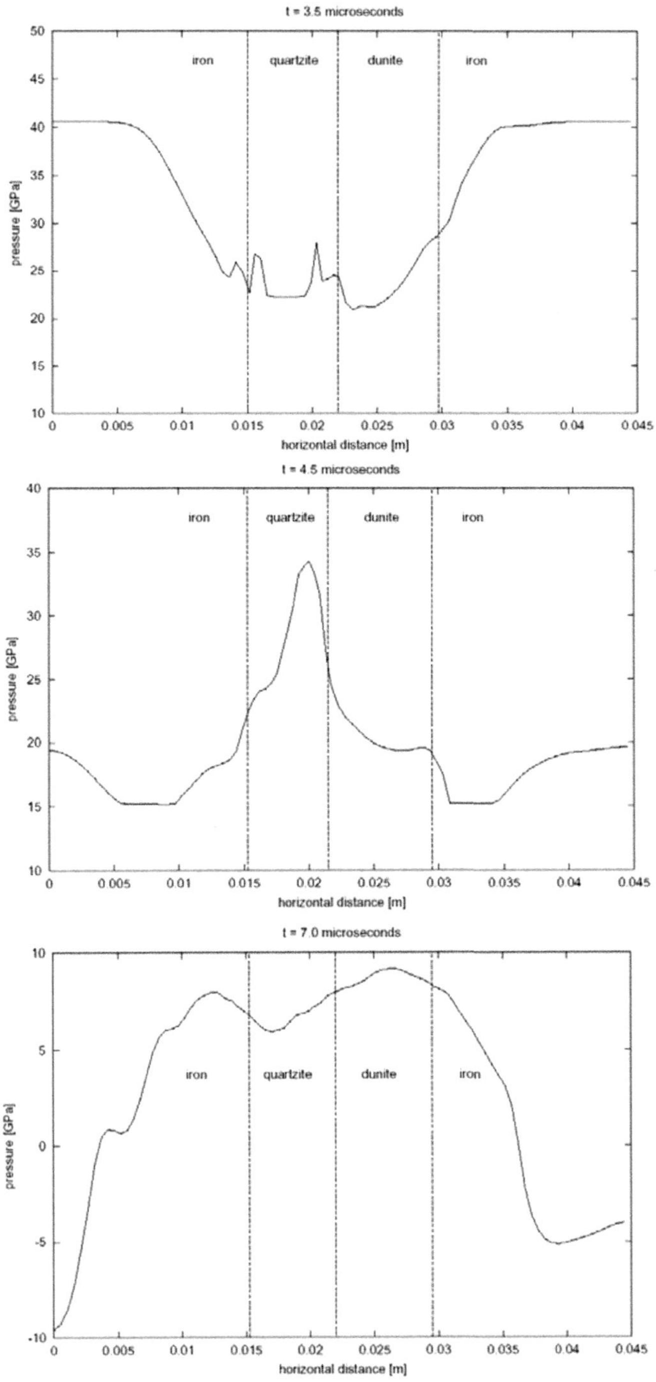
t = 3.5 microseconds
iron
quartzite
dunite
iron
pressure [GPa]
50
45
40
35
30
25
20
15
10
0
0.005
0.01
0.015
0.02
0.025
0.03
0.035
0.04
0.045
horizontal distance [m]
t = 4.5 microseconds
iron
quartzite
dunite
iron
pressure [GPa]
40
35
30
25
20
15
10
0
0.005
0.01
0.015
0.02
0.025
0.03
0.035
0.04
0.045
horizontal distance [m]
t = 7.0 microseconds
iron
quartzite
dunite
iron
pressure [GPa]
10
5
0
-5
-10
0
0.005
0.01
0.015
0.02
0.025
0.03
0.035
0.04
0.045
horizontal distance [m]

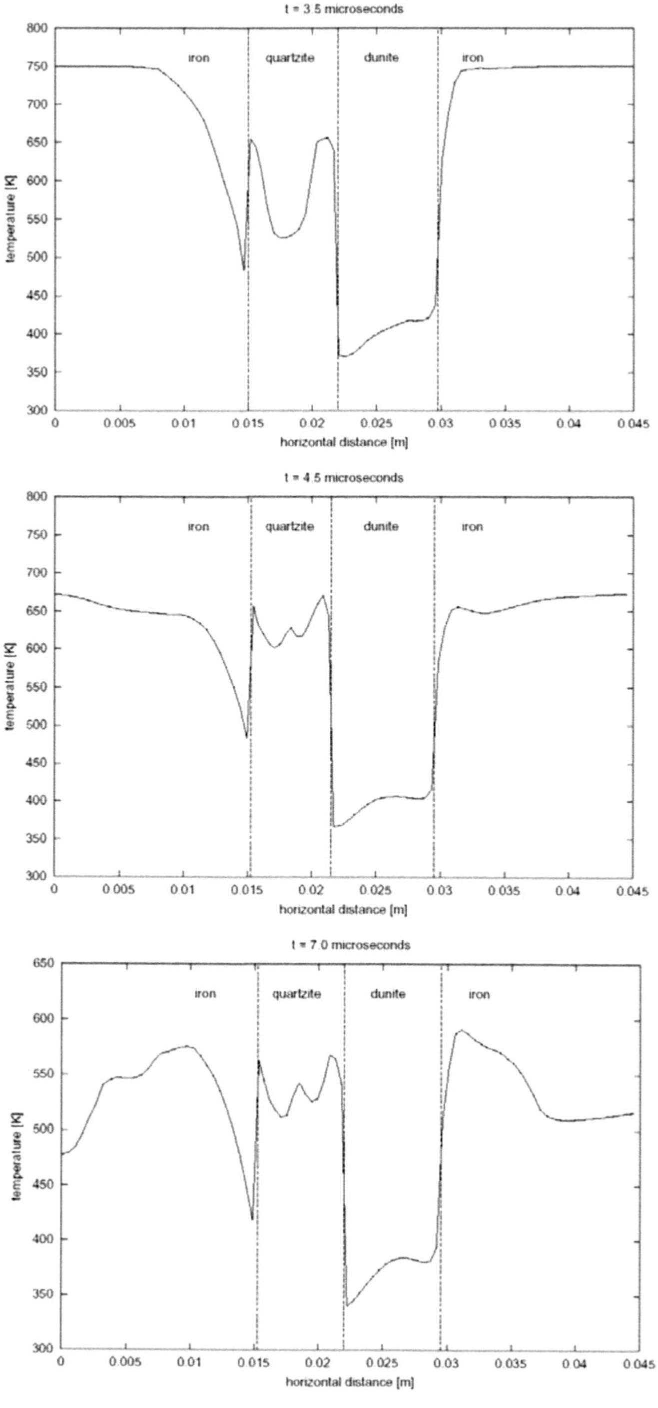
t = 3.5 microseconds
iron
quartzite
dunite
iron
temperature [K]
horizontal distance [m]
t = 4.5 microseconds
iron
quartzite
dunite
iron
temperature [K]
horizontal distance [m]
t = 7.0 microseconds
iron
quartzite
dunite
iron
temperature [K]
horizontal distance [m]

5 Simulations in Eulerian Mode

5.1 Geometry

In addition to the computations described in the previous section, we have investigated the behavior of an inclined material boundary under shock loading. The material interface cross-cutting the grid cells requires the use of the Eulerian mode. Here, too, the ANEOS equation of state has been employed. Because the original Eulerian version of SALE can not deal with more than one material, it has been modified to be able to reflect the behavior of mixed cells containing two materials (Ivanov et al. 1997) following techniques used in more recently developed hydrocodes (Shuvalov 1999, Shuvalov et al. 1999). However, even after this extension the possibilities of our code are limited, and we have therefore only worked with two mineralic components and have simulated a dunite flyer plate impacting a dunite sample with wedge-shaped quartzite inclusion. The geometry of this setup is given in Table 2. The shock wave passes first through a horizontal dunite-quartzite interface and then through a quartzite-dunite interface inclined 45 degrees to the shock wave plane.

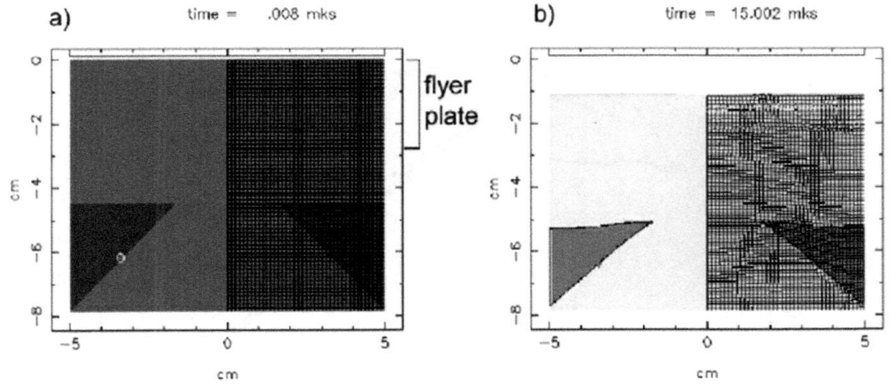

Fig. 6. Eulerian simulations of a shock wave passing an inclusion with an interface inclined 45 degrees to the shock wave plane. Width of sample: 5 cm. Height of sample: 8 cm. Width of quartzite inclusion: 4 cm. Height of inclusion: 4 cm. Flyer plate thickness: 2.7 cm. **a**). Displayed at t = 0 (Impact), **b**). At t = 15 microseconds. Left panels (plastic work field) are mirror images of right panels (temperature field). Resolution: 320 x 320 cells. The occurrence of a zone of concentrated shear with correspondingly elevated temperature near the material interface is reproduced by these simulations. Melt production is not observed because of the employed material model. Various gray shadings correspond to zero (t=0) and total (t=15 microseconds) damage state of minerals. The location of tracers adjusted to the quartz/dunite interface is shown with a white circle.

With this approach, we can at least qualitatively demonstrate the principal effects of an inclined material boundary. In Fig. 6 we show two snapshots of the shock wave compressing the sample, displaying the changes in plastic work and temperature by means of gray-scale plots, Fig. 7 shows a logarithmic intensity plot of the accumulated strain in the sample after passage of the shock, and in Fig. 8 the variations of pressure and temperature for two Lagrangian tracers (located close to the quartz/dunite interface) across a horizontal section through the sample are displayed. The peaks of those quantities in the region of the quartzite section bordering the dunite part deserve particular attention because they are signs of localised shear.

5.2 Results

In accordance with the experimental observations (Kenkmann et al. 2000) we could reproduce a concentration of plastic work in an oblique zone

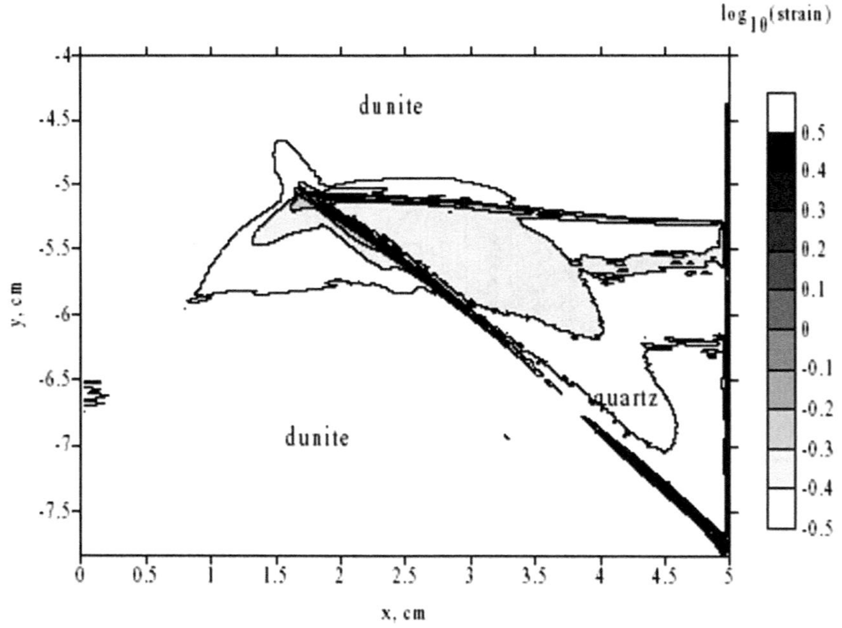

Fig. 7. Accumulated plastic strain field at the end of the model run. Logarithm of strain is displayed. The occurrence of a zone of concentrated shear strain with correspondingly elevated temperature near the material interface is reproduced by these simulations. Melt production is not observed because of the employed material model.

along the inclined interface, which leads to an additional temperature rise of 400 K above the average temperature in shock-compressed quartzite. This is due to the different compressibility of the materials which causes considerable localised shear at their interface. This temperature rise does not change any more for a finer mesh resolution, thus it reflects no numerical artifact. Shear heating of the material has occurred in its compressed state so that the differences between equilibrium release path (which is assumed in the employed equation of state) and experimentally observed release path (e. g. Ahrens and Rosenberg 1968, Swegle 1990) are not critical for the outcome of the simulations.

In recent additional simulations (Hertzsch 2003) with a slightly changed geometry (in particular with a material interface extending all the way across the sample width) the angle between material interface and shock wave plane has been varied in order to determine its effect on the magnitude of the shear heating. The dependence of the maximum temperature on the angle is shown in Fig. 9. The degree of shear heating is found to depend on the angle between incoming shock wave plane and

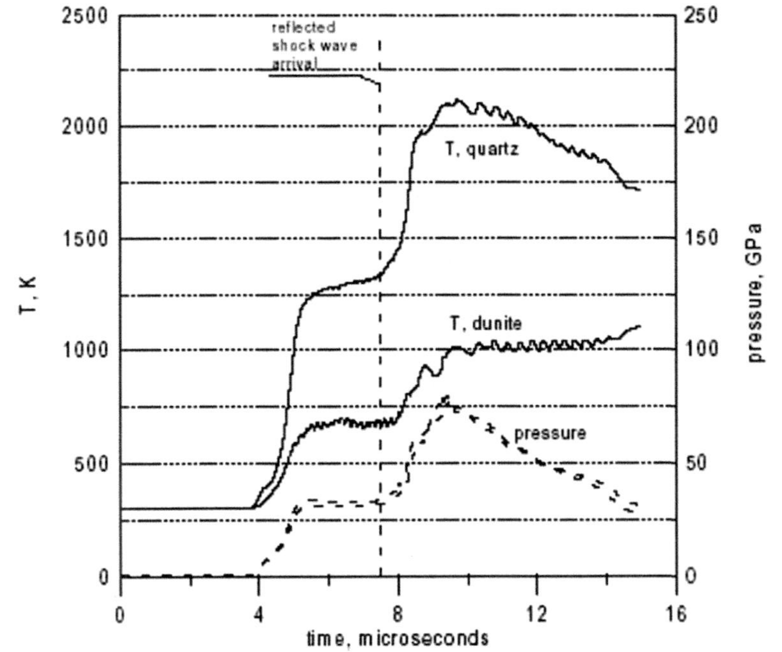

Fig. 8. Pressure and temperature recorded by two tracers at one cell distance from the interface (see tracer location in Fig. 6). The arrival of the shock wave reflected from the bottom boundary is shown as a vertical dashed line.

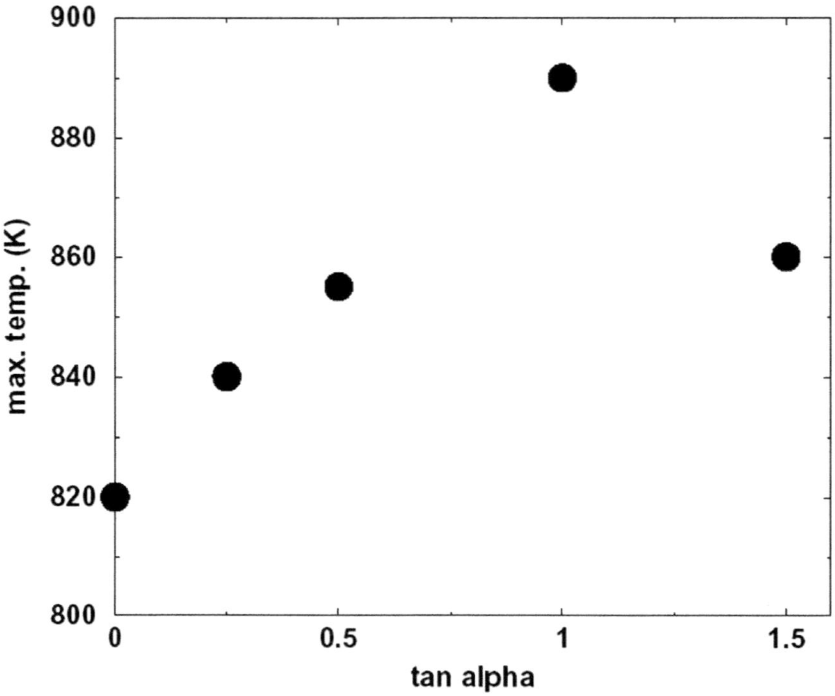

Fig. 9. Maximum temperature during passage of shock front depending on the angle of the material interface to the horizontal (initial shock wave) plane. The most pronounced temperature elevation is found for an angle of 45°.

material interface, and the temperature assumes its highest value for an angle of 45°. However, even with an improved material model, which would allow for melting, it is not possible to judge the amount of possible melt production from such information alone. A thorough comparison with the experimental results of Kenkmann et al. (2000) must be subject of more detailed investigations.

Although we observe temperatures at the material interface which are above those reached due to shock compression of the bulk of the material, the formation of melt could not be reproduced in the present simulations. This is due to the simple material model, which we have used, in which friction decreases with rising temperature and reaches zero at melt temperature. This results in a decrease of the heat production and a limited temperature rise. Consequently, the material cannot reach its melting point in this model. An additional mechanism of energy dissipation like viscosity of the material close to the melting point is needed to explain the experimentally observed melting. This is planned as a next step to improve our numerical model.

In addition to the model improvement the possibility of a thermo-chemical reaction at the quartz/forsterite interface may play a role in the onset of boundary melting observed in experiments by Kenkmann et al. (2000). The Mg-rich melt observed in these experiments may result from the reaction

$$Mg_2SiO_4 + SiO_2 \rightarrow 2MgSiO_3$$

In this case low eutectic melting is taking place. Comparing melting temperatures for SiO_2, forsterite, enstatite and forsterite/enstatite eutectic (Fig. 10, compiled according to Chen and Presnall (1975) and Presnall et al. (1998)) one can see that up to ~30 GPa the eutectic melting occurs ~1000 K below pure forsterite melting and ~2000 K below stishovite melting. Addition a of small amount of water from the sample preparation or from accessory minerals decomposed in the shock wave also may decrease the melting temperature at the mineral interface.

Finally, a slightly imperfect fitting of the two rock half cylinders along their polished interface may have produced gaps of a few micrometer width in the experiments. The closure of such a fissure during shock compression causes an additional heat pulse (Heider and Kenkmann 2003) that is not considered here.

6
Implications, Outlook, and Some Precautionary Notes

At the conditions used in the simulations, the materials do not melt by shock heating alone. However, experiments and Eulerian simulations suggest considerable shear at the interface of different materials and the possibility of massive thermal damage of material near interfaces under shock loading. If organic material were present in such regions (as it is usually expected to be found in cracks and near interfaces of rocks), it would very probably suffer destruction when its host rock is subjected to impacts with high shock pressures. Different compressibility of materials in contact can lead to a concentration of plastic work connected with a significant rise in temperature which makes localised melting possible. However, the material model, which was employed, was not fully adequate to describe this effect. Nevertheless we conclude that the presence of material interfaces can lead to increased thermal effects in shocked geomaterials, and should be considered if impact conditions are to be derived from material changes.

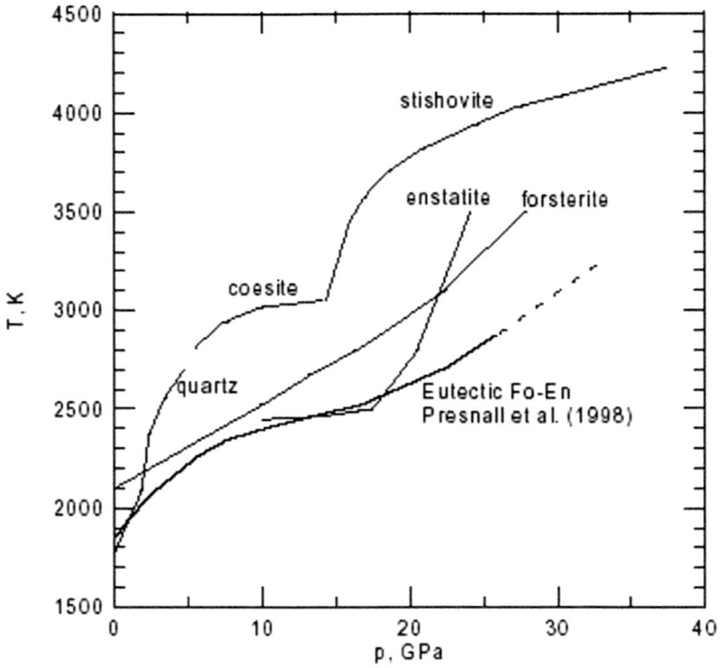

Fig. 10. Pressure dependence of melting temperatures for SiO_2, enstatite, forsterite, and forsterite/enstatite eutectic mixture after Chen and Presnall (1975) and Presnall et al. (1998).

The results of simulations like those presented in this work depend not only on the material models used, but also on details of the equation of state. Since ANEOS is mainly based on equilibrium thermodynamics, its usage for materials exhibiting complex unloading behavior and irreversible phase transitions is complicated. Their occurrence requires the separate treatment of low-pressure and high-pressure phases (Ivanov 2003) and the inclusion of transition kinetics. However, the differences in release paths were not of significant influence on the outcome of the simulations described above because the principal changes occurred in the compressed state of the material.

Future simulations must take into consideration effects of yield strength and brittle fragmentation of the material, and for a proper description of the melting process which is not possible using purely deviatoric stresses it is necessary to include the viscoelastic behavior of the material (in particular near the melting point) as a further sink of dissipated energy. The temperature dependence of the maximum shear stress, the kinetics of solid-solid phase transitions and melt production, and the viscous relaxation times of the materials need also to be taken into consideration.

Three-dimensional simulations are necessary for a full description of experiments. It is planned to investigate the effects of the cylindrical shape of the samples. The Eulerian version of the program needs to be extended further in order to be able to simulate systems where three materials (e. g. iron and two kinds of rock) may be in mutual contact. Alternatively, the input parameters may be adjusted such that the velocity of the shock wave and the pressure on top of the inclusion reach the values they would have for an iron container and an iron flyer plate so that the experimental situation is approximated. Studies of this problem and systematic simulations of the effect of the inclination of the material boundary on the temperature rise (Hertzsch 2003) have to be subject of future work.

Acknowledgements

The authors thank Thomas J. Ahrens and Paul S. DeCarli for their comments and suggestions.

References

Ahrens TJ, Rosenberg JT (1968) Shock metamorphism: Experiments on quartz and plagioclase. In: French BM, Short NM (eds.) Shock Metamorphism of Natural Materials, Mono Book Corp. Baltimore, pp 59-81

Amsden AA, Ruppel HM, Hirt CW (1980) SALE: A Simplified ALE Computer Program for Fluid Flow at All Speeds, Los Alamos Scientific Laboratory Report LA-8095

Benz W, Cameron AGW, Melosh HJ (1989) The origin of the moon and the single impact hypothesis III. Icarus 81: 113-131

Chen CH, Presnall DC (1975) The system Mg_2SiO_4-SiO_2 at pressures up to 25 kilobars. American Mineralogist 60: 398-406

Crawford DA, Barnouin-Jha OS, Cintala MJ (2003) Mesoscale computational investigation of shocked heterogeneous materials with application to large impact craters. Third International Conference on Large Meteorite Impacts, Lunar and Planetary Institute, Houston, LPI Contribution 1167, abs. #4119 (CD-ROM)

Heider N, Kenkmann T (2003) Numerical simulation of temperature effects at fissures due to shock loading. Meteoritics and Planetary Science 38: 14511-1460

Hertzsch JM (2003) Shock effects at inclined material interfaces - numerical simulations [abs.]. Third International Conference on Large Meteorite Impacts, Lunar and Planetary Institute, Houston, LPI Contribution 1167, abs. #4031 (CD-ROM)

Ivanov BA (2003) Modification of ANEOS for rocks in compression [abs.]. Workshop on Impact Cratering: abs. #8030

Ivanov BA (2004) Numerical modelling of cratering. In: Masaitis VL, Pevzner LA, Deutsch A, Ivanov BA (eds) Deep drilling in the Puchezh-Katunki impact structure. Impact Studies, Springer Verlag, Berlin, Heidelberg, in press.

Ivanov BA, DeNiem D, Neukum G (1997) Implementation of dynamic strength models into 2D hydrocodes: applications for atmospheric breakup and impact cratering. International Journal of Impact Engineering 20: 411-430

Johnson GR, Cook WC (1983) A constitutive model and data for metals subjected to large strains, high strain rates and high temperatures. In: Proceedings of the Seventh International Symposium on Ballistics, The Hague, The Netherlands, pp 1-7

Kenkmann T, Hornemann U, Stöffler D (2000) Experimental generation of pseudotachylites along lithological interfaces. Meteoritics and Planetary Science 35: 1275-1290

Kenkmann T, Walther K, Frischbutter A, Scheffzük C, Eichhorn F, Daymond MR (2003) Strain scanning across a shock-deformed quartzite/dunite interface using neutron and synchrotron radiation [abs.]. Lunar and Planetary Science 34: abs. #1374 (CD-ROM)

Langenhorst F, Deutsch A (1994) Shock experiments on pre-heated alpha- and beta-quartz. 1. Optical and density data. Earth and Planetary Science Letters 125: 407-420

Lundborg N (1968) Strength of rock-like materials. International Journal of Rock Mechanics and Mining Science 5: 427-454.

Luo SN, Cagin T, Strachan A, Goddard WA, Ahrens TA (2002) Molecular dynamics modeling of stishovite. Earth and Planetary Science Letters 202: 147-157

Melosh HJ (1989) Impact cratering. A geologic process, Oxford University Press, New York, 245 pp

Ohnaka M (1995) A shear failure strength law of rock in the brittle-plastic transition regime. Geophysical Research Letters 22: 25-28

Pierazzo E, Vickery AM, Melosh HJ (1997) A reevaluation of impact melt production. Icarus 127: 408-423

Presnall DC, Weng Y-H, Milholland CS, Walter MJ (1998) Liquidus phase relations in the system $MgO-MgSiO_3$ at pressures 3 up to 25 GPa-constraints on crystallization of a molten Hadean mantle. Physics of the Earth and Planetary Interiors 107: 83-95

Reimold WU, Stöffler D (1978) Experimental shock metamorphism of dunite. Proceedings of the 9th Lunar and Planetary Science Conference, pp 2805-2824

Schmitt RT (2000) Shock experiments with the H6 chondrite Kernouvé: Pressure calibration of microscopic shock effects. Meteoritics and Planetary Science 35: 545-560

Shuvalov VV (1999) 3D hydrodynamic code SOVA for interfacial flows, application to thermal layer effect. Shock Waves 9: 381-390

Shuvalov VV, Artemieva NA, Kosarev IB (1999) 3D hydrodynamic code SOVA for multimaterial flows - application to Shoemaker-Levy 9 comet impact problem. International Journal of Impact Engineering 23: 847-858

Stevenson DJ (1987) Origin of the moon - the collision hypothesis. Annual Review of Astronomy and Astrophysics 15:271-315

Stöffler D (1972) Deformation and transformation of rock-forming minerals by natural and experimental shock processes. Fortschritte der Mineralogie 49: 50-113

Stöffler D, Langenhorst F (1994) Shock metamorphism of quartz in nature and experiment: I. Basic observation and theory. Meteoritics 29: 155-181

Stöffler D, Keil K, Scott ERD (1991) Shock metamorphism of ordinary chondrites. Geochimica and Cosmochimica Acta 50: 889-903.

Swegle JW (1990) Irreversible phase transitions and wave propagation in silicate geologic materials. Journal of Applied Physics 68: 1563-1579

Thompson SL, Lauson HS (1972) Improvements in the Chart-D Radiation-Hydrodynamic CODE III: Revised Analytic Equations of State, Sandia National Laboratory Report SC-RR-71 0714

Tillotson JH (1962) Metallic equations of state for hypervelocity impact, General Atomic Report GA-3216

Walther K, Scheffzük C, Frischbutter A, Kenkmann T, Daymond MR, Eichhorn F (2002) Strain scanning across a shock-deformed quartzite/dunite compound. 2. MEACA-SENS Conference, Coimbra, Portugal, Juli 2002, Abstract.

The Kentland Impact Crater, Indiana (USA): An Apatite Fission-Track Age Determination Attempt

John C. Weber[1], Christina Poulos[1], Raymond A. Donelick[2], Michael C. Pope[3], and Nicole Heller[1]

[1]Department of Geology, 1 Campus Drive, Grand Valley State University, Allendale, MI 49401, USA (weber@gvsu.edu)

[2]Apatite to Zircon, Inc., 1075 Matson Road, Viola, ID 83872-9709, USA

[3]Department of Geology, Washington State University, Webster Physical Science Building 1228, Pullman, WA 99164-2812, USA

Abstract. We attempt to determine the poorly constrained age of the ≤13 km diameter Kentland impact crater in Indiana, USA (40°45'N, 87° 24'W) through a three-stage fission-track study of the apatite-bearing St. Peter Sandstone (Ordovician) using: 1) three outcrop samples from the Newton County stone quarry in Kentland, in the center of the crater, 2) three subsurface core samples from neighboring Indiana counties, 31-53 km away, and 3) twenty far-field subsurface and outcrop samples from Illinois and Indiana, 138-302 km away. All of the samples studied have been thermally reset and are significantly younger than the ~460 Ma St. Peter Sandstone depositional age. Modeling fission-track age and track length distributions indicates that the St. Peter Sandstone in the crater was heated to ≥135°C at some unknown time after deposition, and then, 184±13 m.y. ago in the Jurassic, cooled rapidly through 135°C.

The Jurassic cooling age is consistent with Mississippian-Pleistocene stratigraphic constraints on the age of the crater, and is significantly older than a Late Cretaceous (< 97±10 Ma) paleomagnetically determined age. We test a working hypothesis that if the reset fission-track ages in the crater are related to impact, uplift, excavation, and exhumation, then additional fission-track ages from more distant samples should decrease to some background age level within a few crater diameter distances away. Three subsurface samples 2 to 4 crater diameter distances away fail the hypothesis test; they yielded a model composite cooling age of 185±11

Ma, statistically identical to that obtained from the crater samples, and also a similar ≥130°C peak burial temperature. Seventeen of the twenty far-field samples, tens of crater diameters (237-251 km) from the crater, yielded apatite grains; ten yielded >5 grains each: eight gave fission-track ages that overlap at 2σ with those from the crater center, and fail the test of our working hypothesis; one gave an apparent outlier age; and only one outcrop sample gave a possible significantly older pooled fission-track age. We conclude that we have not dated the age of the crater, but rather some regional-scale cooling and exhumation event that either predates or postdates impact. This work provides no new constraints on the age of the crater, but illustrates some of the limitations related to using fission tracks to date deeply eroded craters. (U-Th)/He dating of the crater apatite samples, and a more robust paleomagnetic fold test, might help further constrain the crater age.

1 Introduction

The Kentland dome in east-central Indiana (40°45'N, 87° 24'W) exposes intensely faulted and folded, steeply dipping, Ordovician strata within the interior of the stable North American platform, a region otherwise covered by relatively flat-lying Late Paleozoic sedimentary rocks and Pleistocene drift (Fig. 1 – 3). Multiple lines of evidence suggest that the Kentland dome is a crater that was formed by meteorite impact (see below). A 13 km estimate is given for the crater diameter (ImpactDatabase), but the crater is buried and has no geomorphic expression. Diameter estimates can probably be improved with more and better subsurface geologic mapping. A major question remaining is when did the crater form? Unfortunately, glassy material, which is relatively straightforward to date, is not available from this deeply eroded crater. We attempt to determine the poorly constrained age of the Kentland crater through a three-stage fission-track study of the apatite-bearing St. Peter Sandstone.

Gutschick (1961, 1972, 1983, 1987) provided complete and detailed bibliographies of the history of the geologic investigation and quarrying at Kentland; both date back to the 1800s. Briefly, the dome was initially thought to be an organic reef, then a tectonic feature of unknown origin, and later a "cryptoexplosion" structure. The abundance of shatter cones, many of which are oriented normal to the tilted carbonate beds exposed and quarried near Kentland, inspired Dietz (1947) to propose that the dome was formed by meteorite impact. Shatter cones are now commonly regarded as key indicators of impact-related deformation (e.g., Dietz 1960,

1968; Gibson and Spray 1998; Sharpton et al. 1996; Nicolaysen and Reimold 1999; Baratoux and Melosh 2003); however, the physics of shatter cone development is still poorly understood (e.g., see discussion in Sharpton et al. 1996). Polymict breccia dikes occur near major faults in the Newton County stone quarry at Kentland (Gutschick 1983, 1987; Bjornerod 1998), and are also common at other proven or suspected impact sites (Dietz 1964; Wilshire et al. 1971; Degenhardt et al. 1994; Sharpton et al. 1996; Sharpton and Dressler 1997). Such breccia dikes, like shatter cones, are unusual geologic features that probably form only under rare geologic conditions (i.e., Spray 1997). Coesite (Cohen et al. 1961), pervasive brittle quartz microstructures (Laney and Van Schmus 1978), and quartz planar deformation features (PDFs) (Gutschick 1987; Huss 1996) indicative of shock metamorphism (e.g., Stöffler and Langenhorst 1994; French 1998; Alexopoulos et al. 1988), also occur in the St. Peter Sandstone in the quarry at Kentland. Thus, multiple lines of evidence suggest that the Kentland structure was formed by meteorite impact.

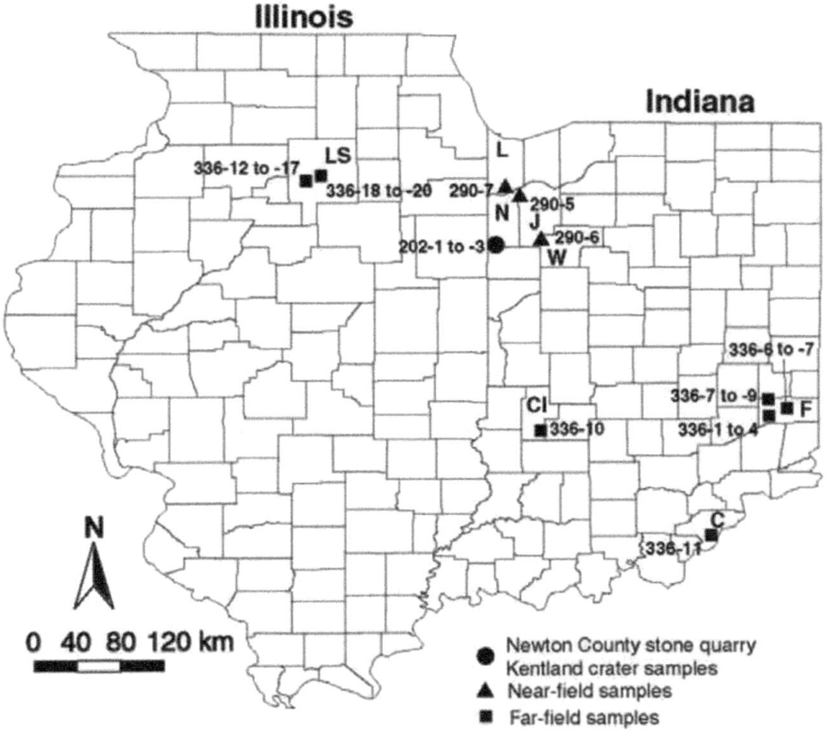

Fig. 1. Map showing fission-track sample locations, location of the Newton County stone quarry in Kentland, Indiana, and sampled counties in Illinois (LS-La Salle) and Indiana (C-Clark, Cl-Clay, F-Franklin, J-Jackson, L-Lake, N-Newton, and W-White). All samples collected are shown; only those listed in Tables 1-5 yielded enough apatite grains (> 5) to be analyzed and used in this study.

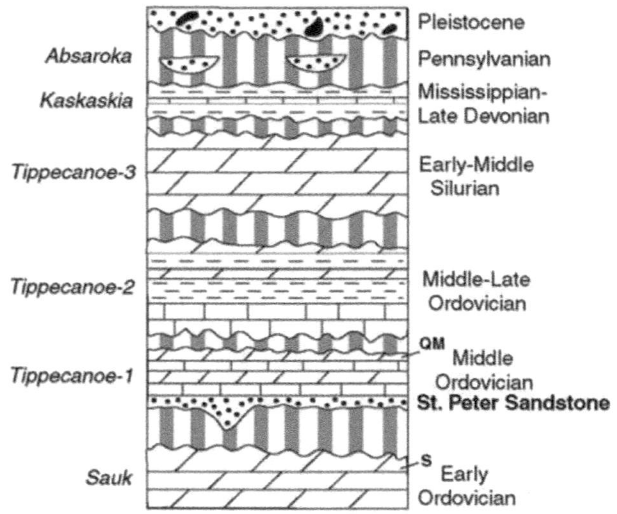

Fig. 2. Schematic stratigraphic column showing rock types and units exposed and drilled in and around the Newton County stone quarry, Kentland, Indiana, major unconformity-bounded transgressive-regressive Paleozoic depositional sequences, after Sloss (1963), and Pleistocene drift. Sequence names given on left are discussed in text; ages and some specific rock unit names discussed in text (S-Shakopee Formation; QM-Quimby's Mill Formation; St. Peter sandstone) are given on right. Rock types are represented by: brick pattern-limestone, diagonal brick pattern-dolostone, horizontal dashes-shale; small dots-sandstone; dots and blobs-glacial drift; the gray vertically ruled lines represent unconformities, gaps in the rock record.

Stratigraphic relations provide broad constraints on the timing of deformation at Kentland (Fig. 2, 3). Silurian carbonates, now oriented subvertically in the quarry, and tilted Mississippian strata mapped nearby in the subsurface (Gutschick 1987), are unconformably overlain by flat-lying, undeformed Pleistocene glacial deposits. Thus, the Kentland event occurred sometime between ~300 and 1 m.y. ago.

A previous paleomagnetic study determined a post-Late Cretaceous (< 97±10 Ma) age for the Kentland deformation event (Jackson and Van der Voo 1986). This determination was based on a paleomagnetic pole position obtained from 25 remagnetized Middle Ordovician Quimby's Mill Limestone samples collected in the quarry at Kentland (Fig. 2). The remagnetized samples failed a fold test (more accurately a tilt test, because all but two of the samples dipped to the northwest), and were thus interpreted to have acquired their magnetic signatures before Kentland doming; the ~97±10 paleomagnetic Ma age was, therefore, considered by Jackson and Van der Voo (1986) to be a maximum age for the deformation.

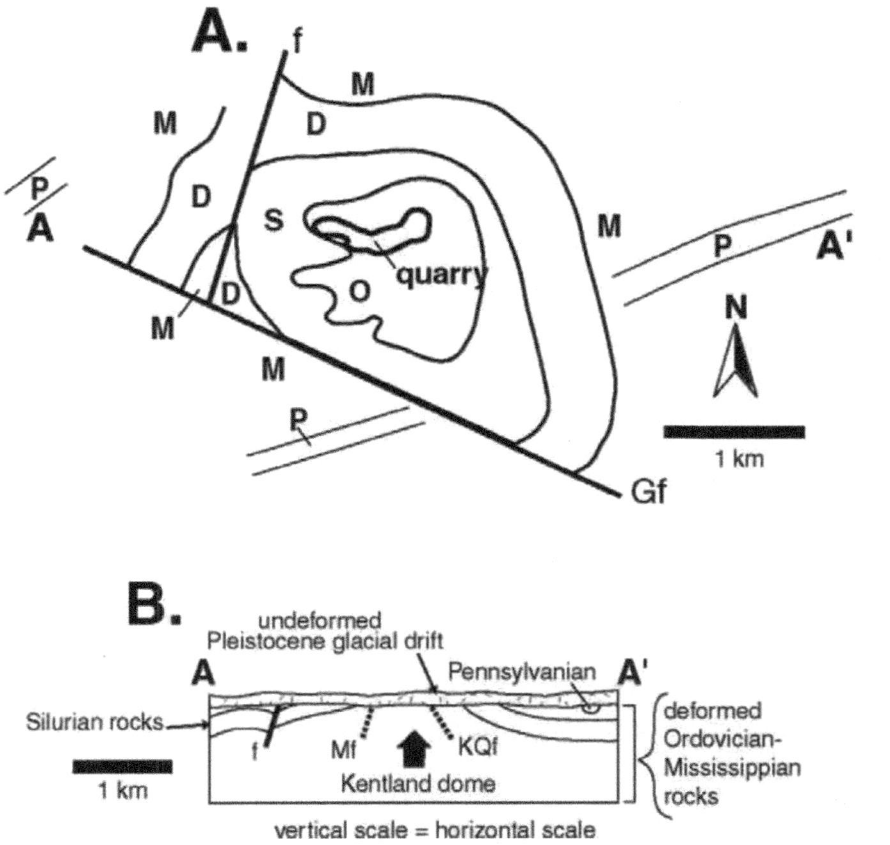

Fig. 3. A. Bedrock geologic map for the Kentland, Indiana area, after Gutschick (1987). Map is based on Newton County stone quarry (marked quarry on map) exposures and on > 100 drill cores that reached bedrock. No bedrock exposures exist outside of quarry for the area shown, which is covered by glacial drift. The Kentland crater is thus mostly buried, deeply eroded and highly exhumed, and has only a geologic expression and no geomorphic expression. Rock units mapped are shown by age: O-Ordovician, S-Silurian, D-Devonian, M-Mississippian, P-Pennsylvanian paleostream channel deposits; Gf-Gutschick Fault; f-unnamed fault. A-A' shows line of cross-section. **B.** Geologic cross-section, after Gutschick (1987), showing undeformed Pleistocene glacial drift cover (thickness exaggerated) over domed Ordovician-Mississippian strata. Means fault (Mf) and Kentland quarry fault (KQf) are well exposed and mapped in quarry but not shown in Fig. **2A**. Thickness of Pennsylvanian channel shown schematically.

This study outlines our attempt to use apatite fission tracks to better determine the age of the Kentland crater. Samples were collected and analyzed in three stages: 1) from the Newton County stone quarry in Kentland near the center of the crater, 2) tens of km from the crater, using subsurface cores from three surrounding counties, and 3) from distant subsurface cores and outcrops, hundreds of km from the crater in Indiana and Illinois. In stage 1, we observed apatite grains with reset fission tracks, and by modeling track length distribution, obtained a 184±13 Ma Jurassic rapid cooling age, consistent with the Mississippian-Pleistocene stratigraphic constraints on the age of the crater. We developed the working hypothesis that if our reset fission-track ages obtained from within the crater are related to impact, uplift, and exhumation, additional samples from outside the crater should yield fission-track ages that decrease to some background age within a few crater diameters away. The stage 2 subsurface samples, several crater diameter distances away, gave statistically identical cooling ages, failing our test. All but one stage 3 sample, tens of crater diameter distances from the crater center, gave pooled fission track ages statistically identical to the earlier, closer samples, again failing our test. We conclude that we did not date the age of the Kentland crater, but rather some regional-scale cooling and exhumation event that could either predate or postdate impact. This study illustrates some limitations related to the use of fission tracks to date deeply eroded craters.

2 Newton County Stone Quarry (Kentland) Geology

Gutschick's (1961, 1972, 1983, 1987) detailed, high-quality maps and published descriptions show that bedding and the major faults in the Kentland quarry generally dip steeply to the north and west of the apex of the Kentland dome (Fig. 3). The major fault in the quarry, the Kentland Quarry Fault, is a folded, bedding-parallel, reverse fault between the St. Peter Sandstone and overlying Platteville Group (Fig. 3 and below). The Kentland Quarry Fault and surrounding strata are folded into a steeply northward-plunging (~58°) macroscopic-scale fold train, which controls the principal outcrop pattern of rocks in the quarry. The Shakopee Formation, the oldest unit exposed in the quarry, has been uplifted more than 600 m above its regionally consistent subsurface position (Fig. 2, 3; Gutschick (1972, 1987)).

The > 300 m of Lower-Middle Paleozoic stratigraphy exposed in and drilled near the Kentland quarry (Gutschick, 1972, 1983, 1987) can be subdivided into seven unconformity-bounded depositional sequences (Fig. 2) (Sloss 1963). The Lower Ordovician Shakopee Formation is the oldest unit exposed and represents the top of the Sauk sequence. The overlying Tippecanoe-1 sequence consists of the basal, transgressive St. Peter Sandstone, a medium- to fine-grained, well-sorted, quartz sandstone containing the detrital apatite grains used in our fission-track analysis, and the overlying Middle Ordovician Platteville Group carbonates. The age of the St. Peter Sandstone is known from regional stratigraphic considerations to be approximately 460 Ma. Platteville conodonts in the quarry yielded a Conodont Alteration Index (CAI) of 1.5, indicating that these rocks were heated to < 90°C (Votaw 1980). The Quimby's Mill samples used for the paleomagnetic study of Jackson and Van der Voo (1986) were taken from near the top of the Platteville Group. The overlying Tippecanoe-2 sequence consists of the Late Middle to Late Ordovician Galena and Maquoketa Groups. The youngest strata exposed in the quarry are limestone, dolostone, and shale of the Lower to Middle Silurian Sexton Creek and Salamonie Formations, which constitute the Tippecanoe-3 depositional sequence.

Late Devonian shale occurs as rare fault breccia fragments in the quarry, and Mississippian limestone and shale were mapped in the subsurface surrounding the quarry (Fig. 2, 3; Gutschick 1987); these two units comprise the Kaskaskia sequence. Pennsylvanian channel sandstones of the Absaroka sequence cut into older strata outside the quarry, and 0-25 m of Pleistocene glacial drift unconformably cap the Paleozoic strata throughout the region (Fig. 2, 3; Gutschick (1983, 1987)).

3 Working Hypothesis

Fission tracks are linear damage zones caused by the spontaneous fission of uranium atoms in U-bearing minerals such as apatite, followed by passage of the expelled heavy fission products through a mineral's crystal lattice. Fission-track ages can be measured because ^{238}U decay occurs at a constant rate and the number of tracks per unit area, in conjunction with uranium concentration, gives a measure of the time since track accumulation began. Above a certain narrow range of annealing temperatures, fission tracks anneal and disappear. Fission-track ages are therefore "set" when U-bearing minerals cool through the annealing zone; track length distributions are related mainly to the rate of cooling. These basic principals and additional theory and details are well laid out in Crowley et al. (1989).

During impact cratering, fission tracks could be set: 1) directly by cooling following a pulse of impact-related heating and melting (e.g., Koeberl et al. 1993), or 2) indirectly by uplift, evacuation and erosion, and exhumation (or, more simply, unburial). Rocks are poor heat conductors and the rocks that are direct effects of impact-related heating are generally limited to fairly shallow depths in impact craters. General crater models (e.g., French 1998) indicate that in a ~13 km diameter crater, like that at Kentland, impact melting was probably limited to a ~2 km thick breccia lens on the crater floor; ~1 km deeper, shatter cones formed in a much cooler zone that probably experienced post-impact temperatures of < 100°C, just below the temperature required to reset apatite fission tracks.

Given the presence of shatter cones and the absence of a breccia lens and impact melt rocks at Kentland, we interpret that this crater was deeply eroded before it was buried by Pleistocene glacial drift, and that apatite grains now exposed in the crater were probably not affected by direct impact-related heating and cooling. However, it is possible that apatite grains were affected indirectly by cooling. According to empirical relations derived geologically given by Melosh and Ivanov (1999), 1.0-1.3 km of differential uplift should occur in the center of a 13 km diameter crater. Differential uplift tapers off radially outward, past the uplifted crater rims and rim flanks, to zero a few crater diameter distances away. Removal of such uplifted material by crater excavation, and erosion of the uplifted rims and rim flanks, would bring formerly buried rocks closer to Earth's surface, exhuming and cooling them. In the stable mid-continent, this type of indirect cooling signature could potentially remain undisturbed after impact. Therefore, we set out to test the hypothesis that fission-track ages set by impact-related uplift and exhumation at Kentland, if observed, should decay outward away from the crater center to some background age within a few crater diameter distances away; we also hypothesize that cooling rates related in this way to impact, particularly in the crater center, would likely have been rapid.

4 Samples and Fission-Track Methods

4.1 Center of Kentland Crater

Outcrops of the Middle Ordovician (~460 Ma) St. Peter Sandstone, which have a distinctive chalky white color and powdery texture, were identified using Gutschick's (1972, 1983, 1987) detailed maps of the quarry. Three ~2 kg samples of St. Peter Sandstone were collected, then broken, pulverized, sieved through a 300 μm screen, and separated using standard heavy liquid and magnetic techniques and analyzed separately. Two polished apatite grain mounts were made from each sample. Fission tracks were counted at 1562.5 times magnification using a 100X dry objective, 1.25X drawing tube, and 12.5X oculars, under unpolarized transmitted and reflected light.

Apatite fission-track ages were determined from the first set of grain mounts using the external detector method (Crowley et al. 1989, and references therein). These apatite grains were etched for 20 s in 5.5 M HNO_3 at 21°C to reveal the natural fission tracks crossing the etched grain surfaces. Low-uranium muscovite detector sheets were attached to the grain mounts to a sample of ^{235}U-doped glass (CN-1 glass), and to the Durango and Fish Canyon tuff standards of known ages. All samples were irradiated simultaneously for 60 min in position 'A-2' offset of the Texas A & M nuclear reactor, yielding a neutron fluence of approximately 1 x 10^{16} neutrons per cm^2. After irradiation, the muscovite detector sheets were etched for 14 min at 24°C to reveal the induced fission tracks. The muscovite sheets were then removed to reveal induced ^{235}U fission tracks in the apatite grains and CN-1 glass. The resultant samples are: (1) standard of known age, (2) glass of a known uranium concentration, and (3) three grain mounts of unknown age and unknown uranium concentration. By comparing the induced fission track densities in these samples, the ^{235}U concentration in the natural samples was determined. By calculating the $^{235}U/^{238}U$ ratio, the concentration of ^{238}U in the natural sample grains was determined. Fission-track ages were calculated using a zeta factor of 113.8±2.9 relative to the CN-1 dosimeter glass (Steiger and Jäger 1977; Hurford and Green 1983).

Fission-track length distributions in reset grains are related to the cooling rate in addition to the age of the cooling event. Apatite fission-track length distributions were measured on the second set of grain mounts. After polishing, these grain mounts were irradiated with approximately 1 x 10^7 ^{252}Cf-derived fission fragments per cm^2 to enhance the horizontal, confined fission tracks (Donelick and Miller 1991). Track lengths (± 20 μm) and track angles (with respect to the crystallographic c-axes; ± 2°) were measured using the microscope described above, coupled with a digitizing tablet interfaced with a personal computer, which projects an LED image from the tablet cursor onto the field of view via a drawing tube.

The fission-track length and age data were then modeled using the AFTSolve computer program (Ketcham et al. 2000), which is based on the general principle that fission-track length distribution and age are related to cooling history; specific details come from the experimental data of Carlson et al. (1999), and the kinetic models of Donelick et al. (1999) and Ketcham et al. (1999), which account for kinetic (observable) factors, such as the fission-track orientations and fluorine and chlorine contents. The kinetic parameter D_{par}, estimated from mean fission-track etch pit diameters on etched grain surfaces (Donelick 1993, 1995; Burtner et al. 1994), was also measured for all samples and used as input. In addition, the 460 Ma St. Peter Sandstone regional depositional age was used as a t_0 constraint.

4.2 Several Crater Diameter Distances From Crater Center

To further test the hypothesis that the rapid fission-track cooling age obtained in stage 1 (see Fission-Track Results below) represents the true age of the Kentland crater, we collected and analyzed three additional subsurface samples of St. Peter Sandstone (Table 4). One sample was obtained from each of the three Indiana counties, Jasper, White, and Lake, neighboring Newton County and Kentland, where the quarry is located (Fig. 1). The St. Peter Sandstone does not outcrop in north-central Indiana outside of Kentland, so samples from three drill cores were obtained from the Indiana Geological Survey core library. Radial distances range from 2 to 4 crater diameter distances (31-53 km) from sample sites to the crater center, outside of our hypothetical zone of indirect-impact-related cooling and exhumation. From core 271-Nipsco in Jasper County we sampled 47.3 km from the crater center at depths of 386-404 m (sample 290-5); from 689-Indiana Gas/Water core in White County we sampled 30.5 km from the crater center from depths of 433-447 m (sample 290-6); and from 826-St. Joe Lead core in Lake County we sampled 53.4 km from the crater center from depths of 373-389 m (sample 290-7). These samples were treated and analyzed identically to the stage 1 samples described above.

4.3 Tens of Crater Diameter Distances From Crater Center

To further test whether our stage 1 and 2 reset fission-track ages (see Fission-Track Results below) represent an age for the Kentland crater, we initiated a stage 3 study, collecting and analyzing twenty additional St. Peter Sandstone samples, including subsurface core samples from Franklin, Clark, and Clay Counties, Indiana, obtained from the Indiana Geological Survey core library, and outcrop samples from Mattheson and Buffalo Rock State Parks, in La Salle County, Illinois (Fig. 1). Radial distances range from 11 to 23 crater diameter distances (138-302 km) from sample sites to the crater center, clearly outside our hypothetical zone of

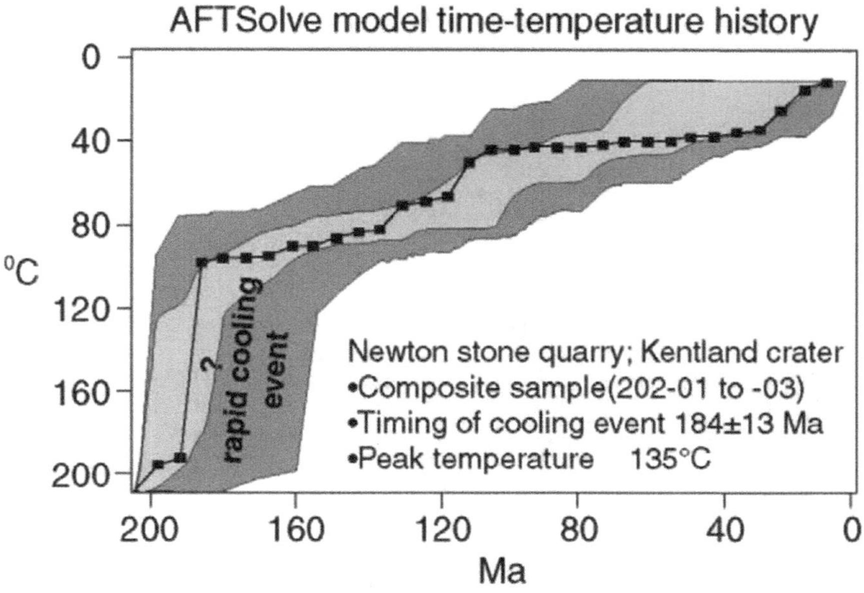

Fig. 4. Model (AFTSolve) time-temperature history for composite of St. Peter Sandstone apatite fission-track samples 202-01, 202-02, and 202-03 from Newton County stone quarry, crater center, Kentland, IN. Shaded regions show all fits generated by model simulation (outer shaded region-acceptable fits; inner shaded region-good fits); solid line represents best-fit model. Models are poorly constrained at temperatures above 150°C. See text for discussion.

indirect-impact-related cooling and exhumation. The stage 3 samples were treated and analyzed similarly to the stage 1 samples described above, with one exception; no modeling of track lengths was done.

From cores 307-B7-Gulf Oil and 307-B4-Gulf Oil in Franklin County we sampled 244.3 km from the crater center at depths of 383-387 m (samples 336-1 to 336-4); we sampled 250.8 km from the crater center from cores 305-B5-Gulf Oil and 305-B6-Gulf Oil in Franklin County at depths of 357-358 m (samples 336-5, 336-6); from cores 309-B2-Gulf Oil, 309-B3-Gulf Oil, and 309-B4-Gulf Oil in Franklin County we sampled 236.6 km from the crater center at depths of 449-452 m (samples 336-7 to 336-9); we sampled 170.3 km from the crater center at depths of 448-449 m from core 332-B3-Indiana Gas/Water in Clay County (sample 336-10); and from core 332-B1-Indiana Gas/Water in Clark County we sampled 302.2 km from the crater center at a depth of 449 m (sample 336-11).

Outcrop samples 336-12 to 336-17 were collected from the scenic gorge cut into the St. Peter Sandstone at Mattheson State Park in La Salle County, Illinois, 145 km from the crater center. Outcrop samples 336-18 to

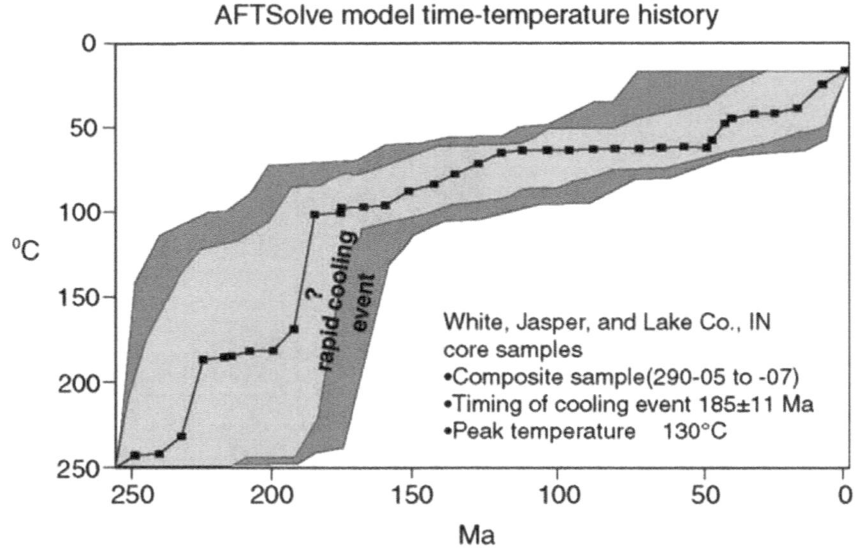

Fig. 5. Model (AFTSolve) time-temperature history for composite of St. Peter Sandstone apatite fission-track samples 290-5, 290-6, and 290-7, 31-53 km from crater center. Shaded regions show all fits generated by model simulation (outer shaded region-acceptable fits; inner shaded region-good fits); solid line represents best-fit model. Models are poorly constrained at temperatures above 150°C. See text for discussion.

336-20 were collected nearby at Buffalo Rock State Park, also in La Salle County, Illinois, 138 km from center of the crater.

5 Fission-Track Results

5.1 Center of Kentland Crater

Fission-track ages, oldest fission tracks retained, and mean fission-track lengths are statistically identical (within erro)r for the three crater samples (Tables 1, 2). This internal consistency allowed us to treat the three individual samples as a single composite sample. The estimated fission-track ages are significantly younger than the ~460 Ma depositional age of the St. Peter Sandstone. In addition, thermal history data for the St. Peter

Sandstone in the surrounding Illinois-Indiana (Illinois Basin) region (e.g., Pitman and Spötl 1996, and references therein) suggest that this unit was probably at peak burial depths and temperatures in the Pennsylvanian-Permian. The fission-track ages are reset and thus can be used to estimate the age of the exhumation and cooling event, perhaps related to Kentland cratering. Our AFTSolve modeling indicates that the samples studied cooled rapidly through 135°C, the apatite fission-track closure temperature determined from the observed D_{par} kinetic values, 184±13 m.y. ago (Fig. 4). Although consistent with the Mississippian-Pleistocene stratigraphic constraints, it is possible that the 184±13 Ma Jurassic rapid cooling event recorded at the crater is not related to local impact-related deformation and exhumation, but rather to regional-scale exhumation and cooling related to some other process. Additional St. Peter Sandstone outcrop samples are not available from the immediately surroundings, so to further test this possibility we used subsurface samples from the three neighboring counties in stage 2 of our study.

Table 1. Summary of apatite fission-track (FT) data for three individual samples (202-01, 202-02, 202-03) and composite from Newton County stone quarry in center of crater in Kentland, IN: Q-(dimensionless) probability of greater χ^2, after Galbraith (1981); D_{par}-mean etch pit diameter parallel to c-axis, see, e.g., Burtner et al. (1994); Pooled FT age-weighted average fission-track age; Track lengths-number of track lengths measured.

Sample	Number of grains	Q	D_{par} (μm)	Pooled FT age (Ma)	Track lengths	Mean length (μm)	Standard deviation (μm)
202-01	9	0	2.15	158±21	41	13.29±0.38	2.34
202-02	11	0.585	2.10	143±15	59	13.04±0.28	2.16
202-03	9	0.624	1.91	158±19	64	13.00±0.31	2.49
composite	29	0.001	2.06	152±11	164	13.09±0.18	2.35

Table 2. Summary of results obtained using AFTSolve (see Ketcham et al. 2000) for three individual samples (202-01, 202-02, 202-03) and composite sample from Newton County stone quarry, center of crater, Kentland, IN. Also see Fig. 4.

Sample	Peak burial temp (°C)	Oldest fission track retained (Ma)	Age of cooling event (Ma)
202-01	≥123°C	184±24	184±24
202-02	≥116°C	177±19	177±19
202-03	≥127°C	182±22	182±22
composite	≥135°C	184±13	184±13

5.2
Several Crater Diameter Distances From Crater Center

The subsurface St. Peter Sandstone samples used in this stage of the study came from cores in the neighboring three counties, where the St. Peter Sandstone is present at depths of ~396-457 m. All of these samples yielded reset apatite fission-track ages significantly younger than the ~460 Ma St. Peter Sandstone depositional age (Tables 3, 4). Combining the data from these three apparently thermally related samples yields a composite fission-track age of 175±14 Ma, a model composite cooling age of 185±11 Ma, and a composite peak burial temperature of ≥130°C (Fig. 5). Within 1σ error, there is no significant difference between these ages and those obtained from within the crater in stage 1 of the study. In addition, the inner "good-fit" envelope of the stage 2 model cooling history curve is similar to that of the stage 1 crater samples (Fig. 4, 5). These results failed the test of our working hypothesis; nonetheless, as a further check, we expanded the study to stage 3.

Table 3. Summary of apatite fission-track (FT) data for three individual samples (290-5, 290-6, 290-7) and composite from cores in Jasper, Lake, and White Counties, IN: Q-(dimensionless) probability of greater χ^2, after Galbraith (1981); D_{par}-mean etch pit diameter parallel to c-axis, see, e.g., Burtner et al. (1994); Pooled FT age-weighted average fission-track age; Track lengths-number of track lengths measured.

Sample	Number of grains	Q	D_{par} (μm)	Pooled FT age (Ma)	Track lengths	Mean length (μm)	Standard deviation (μm)
290-5	7	0.029	1.73	121±20	21	12.39±0.44	1.97
290-6	10	0.011	1.97	157±13	22	13.41±0.34	1.55
290-7	15	0	1.99	165±14	36	12.88±0.26	1.51
composite	32	0	1.92	157±9	79	12.89±0.19	1.68

Table 4. Summary of results obtained using AFTSolve (see Ketcham et al. 2000) for three individual samples (290-5, 290-6, 290-7) and composite from cores in Jasper, Lake, and White Counties, IN. Also see Fig. 5.

Sample	Peak burial temp (°C)	Oldest fission track retained (Ma)	Age of cooling event (Ma)
290-5	≥120°C	147 ± 24	147 ± 24
290-6	≥130°C	175 ± 14	175 ± 14
290-7	≥130°C	204 ± 17	204 ± 17
composite	≥130°C	185 ± 11	185 ± 11

5.3 Tens of Crater Diameter Distances From Crater Center

Of the twenty stage 3 samples that we collected 138-302 km from the crater, seventeen yielded apatite grains. Of these, ten yielded a sufficient number of apatite grains (e.g., > 5 apatite grains per sample) that we could calculate and measure robust fission-track ages and track lengths; no track length modeling was done on these samples (Table 5). Aside from sample 336-3, which is an apparent outlier, the Franklin County, Indiana core samples (336-1 to 336-9) give pooled fission-track ages that are, within 1σ, the same as the composite pooled age from the crater and the stage 2 composite pooled age. Mean track lengths, however, tend to be shorter with increasing distance from the crater, giving a hint of a possible radial decay in cooling rate. Outcrop sample 336-14 from Mattheson Park, Illinois, collected 145.2 km from the crater, gives a slightly older Permian 264±33 Ma pooled age, which at 2σ, is perhaps slightly different than the composite pooled age from the crater. This age may match what is known about the tectonics. The St. Peter Sandstone crops out in the asymmetric, fault-cored, La Salle anticline that formed in the late Paleozoic (Nelson 1990). The Permian fission-track age could reflect modest thickening of the crust and associated uplift and erosion during late Paleozoic low-magnitude crustal shortening.

Table 5. Summary of apatite fission-track (FT) age data for subsurface core samples from Franklin (336-1 to 336-9) County, IN, and outcrop sample 336-14 from Mattheson State Park, IL: Q-(dimensionless) probability of greater χ^2, after Galbraith (1981); D_{par}-mean etch pit diameter parallel to c-axis, see, e.g., Burtner et al. (1994); Pooled FT age-weighted average fission-track age; Track lengths-number of track lengths measured.

Sample	Number of grains	Q	D_{par} (μm)	Pooled FT age (Ma)	Tracks	Mean length (μm)	Standard Deviation (μm)
336-1	5	0.520	1.85	226±45	12	11.81±0.77	2.56
336-2	10	0	1.79	137±19	16	11.93±0.33	1.26
336-3	5	0	1.89	35.4±5.8	15	12.32±0.67	2.51
336-4	16	0	1.87	193±18	35	12.10±0.36	2.08
336-5	19	0.009	2.02	172±13	100	12.93±0.18	1.81
336-6	17	0.063	1.91	188±17	101	12.80±0.19	1.88
336-7	19	0.027	1.85	184±17	66	12.16±0.25	2.05
336-8	18	0	1.87	157±13	69	12.13±0.26	2.17
336-9	11	0.288	1.75	146±18	14	13.49±0.41	1.46
336-14	9	0.157	1.88	264±33	33	11.19±0.46	2.58

6 Discussion

In all the samples we studied from the Kentland crater, apatite grains have reset fission-track ages, thus were heated to at least > 110°C, and according to the observed D_{par} values, likely to ≥ 135°C (Fig. 4). The 184±13 Ma Jurassic cooling age we obtained is consistent with the available stratigraphic constraints on the age of the crater. It is significantly different from the paleomagnetically determined post-Late Cretaceous (< 97±10 Ma) crater age of Jackson and Van der Voo (1986). Without the hypothesis testing we did, and the additional stage 2 and 3 data we collected, the stage 1 Jurassic cooling age might have been interpreted as the true crater age.

Our three stage 2 subsurface core samples, collected 31-53 km from the crater at depths of 373-447 m, all have reset fission tracks and yielded fission-track ages and model cooling ages that statistically overlap with those obtained from the crater in stage 1, e.g., the model composite cooling age of 185±11 Ma from a peak burial temperature of ~130°C is, within 1σ error, identical to the 184±13 Ma cooling age we obtained from the crater. In addition, the cooling histories we model from the stage 1 and 2 data are quite similar (Fig. 4, 5). These observations failed the test of our working hypothesis. The stage 3 far-field samples were taken tens of crater diameter distances from the center of the Kentland crater. All but one of the stage 3 fission-track ages, sample 336-14, also failed the working hypothesis test. We conclude that we have not dated the age of the Kentland crater, but rather a regional-scale "rapid" cooling and exhumation event that could either predate or postdate impact at Kentland. Unfortunately, this study provides no new constraints on the age of the crater. Given the Jurassic age and regional nature of the cooling event, we speculate that it could be related to increased precipitation and accelerated erosion associated with the break-up of Pangea.

Our ≥ 135°C maximum heating temperature estimate differs significantly from the previous < 90°C estimate Votaw (1980) obtained for Platteville carbonates in the Kentland quarry using the Conodont Alteration Index (CAI). The lower temperature estimate of Votaw (1980) stood out as anomalously low for an impact site.

The empirical relations in Melosh and Ivanov (1999) suggest that 1.0-1.3 km of differential uplift should occur in the center of a 13 km diameter crater. Gutschick's (1972, 1987) mapping in and around the quarry at Kentland, however, shows that the oldest unit exposed, the Shakopee Formation, has been uplifted by only about 600 m above its regionally consistent subsurface position. That the observed stratigraphic uplift is smaller than the uplift empirically predicted may indicate that the diameter of the Kentland crater is overestimated, perhaps by as much as by a factor of 2. Allowing this possibility would place our stage 2 samples 4 to 8 crater diameters away, even further outside of the hypothesized zone of thermal influence, strengthening our results and conclusions. Diameter estimates for the Kentland crater will probably be improved with more and better subsurface geologic mapping.

Precise absolute age constraints are badly needed for this and other deeply eroded craters in the mid-continent, e.g., to help resolve controversies such as the debated timing and origin of the eight circular structures in the central U.S. "chain" (Rampino and Volk 1996; Luczaj 1998; Rampino 1999; Koeberl and Reimold 1999; Glikson 1999). Fission-track dating, unfortunately, is limited for this purpose and requires special circumstances to yield useful results, erosion down to and exhumation of the domed partial annealing zone formerly beneath a crater and no later resetting.

How can we better constrain the Mississippian-Pleistocene, possibly Late Cretaceous, age of the Kentland crater? Although Jackson and Van der Voo's (1986) tilt test provided a conclusive formal statistical result that the Late Cretaceous magnetic signature observed pre-dated doming, all but two of the samples studied had dips to the northwest. Additional paleomagnetic samples with a greater variety of dip directions, critically toward the southern quadrants, could be used to construct a more robust fold test and to better assess the possibility that the Late Cretaceous pole position actually post-dates rather than pre-dates doming. Because it provides information on low temperature (<100°C) cooling history additional to that obtained from fission-track analysis, (U-Th)/He dating (Reiners et al. 2002) of our crater apatite samples might also provide further insights on the relative timing of thermal events at Kentland.

Acknowledgements

We thank Dr. Ray Gutschick for providing the basic geologic data that made this study possible, and for his encouragement. Tom Goyette made our work in the Newton County stone quarry at Kentland possible. Dr. Jan Schreus of Corning Glass Works provided the CN-1 dosimeter glass. Grand Valley State University and the Michigan Space Grant Consortium supported some of the fieldwork and laboratory analyses. We appreciate the helpful and constructive reviews we got from David King and W. U. Reimold on this paper, and from Christian Koeberl and John Spray on an earlier, less comprehensive, manuscript. Finally, we thank the Indiana

Geological Survey, and specifically John Rupp and Sherry Cazee, for providing access to and use of their well-organized core library, and the Illinois Department of Natural Resources for permitting us to sample in Mattheson and Buffalo Rock State Parks.

References

Alexopolous JS, Grieve, RAF, Robertson PB (1988) Microscopic lamellar deformation features in quartz: Discriminative characteristics of shock generated varieties. Geology 16: 796-799

Baratoux D, Melosh HJ (2003) The formation of shatter cones by shock wave interference during impacting. Earth and Planetary Science Letters 216: 43-54

Bjornerud MG (1998) Superimposed deformation in seconds: breccias from the impact structure at Kentland, Indiana (USA). Tectonophysics 290: 259-269

Burtner RL, Nigrini A, Donelick RA (1994) Thermochronology of Lower Cretaceous source rocks in the Idaho-Wyoming thrust belt. American Association of Petroleum Geologists Bulletin 78: 1613-1636

Carlson WD, Donelick RA, Ketcham RA (1999) Variability of apatite fission-track annealing kinetics I: Experimental results. American Mineralogist 84: 1213-1223

Cohen AJ, Bunch TE, Reid AM (1961) Coesite discoveries establish cryptovolcanics as fossil meteorite craters. Science 134: 1624-1625

Crowley KD, Naeser CW, Naeser ND (1989) Fission track analysis: Theory and applications. Short course manual, Geological Society of America Annual Meeting, November 1989, St. Louis, Missouri, 296 pp

Degenhardt JJ Jr, Buchanan PC, Reid AM, Miller RMcG (1994) Breccia veins and dykes associated with the Roter Kamm Crater, Namibia. In: Dressler BO, Grieve RAF, Sharpton VL (eds) Large Meteorite Impacts and Planetary Evolution, Geological Society of America Special Paper 293: 197-208

Dietz RS (1947) Meteorite impact suggested by the orientation of shatter-cones at the Kentland, Indiana disturbance. Science 105: 42

Dietz RS (1960) Meteorite impact suggested by shatter cones in rock. Science 131: 1781-1784

Dietz RS (1964) Sudbury structure as an astrobleme. Journal of Geology 72: 412-434

Dietz RS (1968) Shatter cones in cryptoexplosion structures. In: French BM, Short NM (eds) Shock Metamorphism of Natural Materials, Mono Book Corp, Baltimore, pp 267-285

Donelick RA (1993) A method of fission track analysis utilizing bulk chemical etching of apatite. U. S. Patent Number 5, 267, 274

Donelick RA (1995) A method of fission track analysis utilizing bulk chemical etching of apatite. Australian Patent Number 658,800

Donelick RA, Miller DS (1991) Enhanced TINT fission track densities in low spontaneous track density apatites using ^{252}Cf-derived fission fragment tracks: A model and experimental observations. Nuclear Tracks and Radiation Measurements 18: 301-307

Donelick RA, Ketcham RA, Carlson WD (1999) Variability of apatite fission-track annealing kinetics II: Crystallographic orientation effects. American Mineralogist 84: 1224-1234

French BM (1998) Traces of a catastrophe: A Handbook of Shock-Metamorphic Effects in Terrestrial Meteorite Impact Structures. LPI Contribution No. 954, Lunar and Planetary Institute, Houston, Texas, USA, 120 pp

Galbraith RF (1981) On statistical models for fission-track counts. Journal of the International Association for Mathematical Geology 13: 471-478

Gibson HM, Spray JG (1998) Shock-induced melting and vaporization of shatter cone surfaces: Evidence from the Sudbury impact structure. Meteoritics and Planetary Science 33: 329-336

Glikson A (1999) Argument supporting explosive igneous activity for the origin of "cryptoexplosion" structures in the midcontinent, United States: Discussions and Reply. Geology 27: 279-285

Gutschick RC (1961) The Kentland structural anomaly, northwestern Indiana. Guidebook for field trips 1961 Cincinnati meeting, Geological Society of America Guidebook, pp 12-17

Gutschick RC (1972) Geology of the Kentland Structural Anomaly, Northwestern Indiana. Field Guide for the 35th Annual Meeting of the Meteoritical Society, pp 1-35

Gutschick RC (1983) Geology of the Kentland Dome structurally complex anomaly, northwest Indiana (Field Trip 5). In: Shaver RH, Sunderman JA (eds) Field trips in midwestern geology: Bloomington, Indiana, Geological Society of America, Indiana Geological Survey and Indiana University Department of Geology Guidebook 1, 105-138

Gutschick RC (1987) The Kentland Dome, Indiana: A structural anomaly. Geological Society of America Centennial Field Guide, North-Central Section, pp 337-342

Huss QS (1996) Shocked quartz of the St. Peter Sandstone from the Kentland structural anomaly, Indiana, USA [abs.]. Geological Society of America Abstracts with Programs 28(6): 46

Hurford AJ, Green PF (1983) The zeta age calibration of fission-track age dating. Isotope Geoscience 1: 285-317

Impact Data Base, (http://www.unb.ca/passc/ImpactDatabase/index.html)

Jackson M, van der Voo R (1986) A paleomagnetic estimate of the age and thermal history of the Kentland, Indiana cryptoexplosion structure. Journal of Geology 94: 713-723

Ketcham RA, Donelick RA, Carlson WD (1999) Variability of apatite fission-track annealing kinetics III: Extrapolated to geological time scales. American Mineralogist 84: 1235-1255

Ketcham RA, Donelick RA, Donelick MB (2000) AFTSolve: A program for multi-kinetic modeling of apatite fission-track data. Geological Materials Research 2(1): 1-32

Koeberl C, Hartung JB, Kunk MJ, Klein J, Matsuda J, Nagao K, Reimold WU, Storzer D (1993) The age of the Roter Kamm impact crater, Namibia: Constraints from ^{40}Ar-^{39}Ar, K-Ar, Rb-Sr, fission track, and ^{10}Be-^{26}Al studies. Meteoritics 28: 204-212

Koeberl C, Reimold WU (1999) Argument supporting explosive igneous activity for the origin of "cryptoexplosion" structures in the midcontinent, United States: Discussions and Reply. Geology 27: 279-285

Laney RT, van Schmus WR (1978) A structural study of the Kentland, Indiana impact site. Proceedings of the 9th Lunar and Planetary Science Conference, pp 2609-2632

Luczaj J (1998) Argument supporting explosive igneous activity for the origin of "cryptoexplosion" structures in the midcontinent, United States. Geology, 26, 295-298

Melosh HJ, Ivanov BA (1999) Impact crater collapse. Annual Reviews of Earth and Planetary Sciences 27: 385-415

Nelson W J (1990) Structural Styles of the Illinois Basin. In: Leighton MW, Kolata DR, Oltz DF, Eidel JJ (eds) Interior Cratonic Basins, American Association of Petroleum Geologists Memoir 51: 209-243

Nicolaysen LO, Reimold WU (1999) Vredefort shatter cones revisited. Journal of Geophysical Research 104: 4911-4930

Pitman JK, Spötl C (1996) Origin and timing of carbonate cements in the St. Peter Sandstone, Illinois Basin: Evidence for a genetic link to Mississippi Valley-type mineralization. Society of Economic Paleontologists and Mineralogists Special Publication 55: 187-203

Rampino MR (1999) Argument supporting explosive igneous activity for the origin of "cryptoexplosion" structures in the midcontinent, United States: Discussions and Reply. Geology 27: 279-285

Rampino MR, Volk T (1996) Multiple impact event in the Paleozoic: Collision with a string of comets or asteroids? Geophysical Research Letters 23: 49-52

Reiners PW (2002) (U-Th)/He chronometry experiences a renaissance, EOS Transactions of the American Geophysical Union 83 (3): 21, 26-27

Sharpton VL, Dressler BO (1997) Reply: New constraints on the Slate Islands impact structure, Ontario, Canada. Geology 25: 667-669

Sharpton VL, Dressler BO, Herrick RS, Schnieders B, Scott J (1996) New constraints on the Slate Islands impact structure, Ontario, Canada. Geology 24: 851-845

Sloss LL (1963) Stratigraphic sequences of the North American craton. Geological Society of America Bulletin 74: 93-114

Spray JG (1997) Superfaults. Geology 25: 579-582

Steiger RH, Jäger E (1977) Subcommission on geochronology: Convention on the use of decay constants in geo- and cosmo-chronology. Earth and Planetary Science Letters 36: 359-362

Stöffler D, Langenhorst F (1994) Shock metamorphism of quartz in nature and experiment: I. Basic observation and theory. Meteoritics 29: 155-181

Votaw RB (1980) Middle Ordovician conodonts from the Kentland structure, Indiana [abs.]. Geological Society of America Abstracts with Programs 12(5): 259

Wilshire HG, Howard KA, Offield TW (1971) Impact breccias in carbonate rocks, Sierra Madera, Texas. Geological Society of America Bulletin 82: 1009-1018

The Combined Petrographic and Chemical Analysis of end-Permian Kerogens

M.A. Sephton[1,3,*], C.V. Looy[2], H. Visscher[2], H. Brinkhuis[2] and J.W. de Leeuw[1,3]

[1]Department of Geochemistry, Institute of Earth Sciences, Utrecht University, PO Box 80.021, NL-3508 TA, Utrecht, The Netherlands

[2]Laboratory of Palaeobotany and Palynology, Utrecht University, Budapestlaan 4, NL-3584 CD, Utrecht, The Netherlands

[3]Department of Marine Biogeochemistry and Toxicology, Royal Netherlands Institute for Sea Research (NIOZ), PO Box 59, NL-1790 AB, Den Burg, Texel, The Netherlands

*Current address: Planetary & Space Science Research Institute, Open University, Milton Keynes MK7 6AA, UK (m.a.sephton@open.ac.uk)

Abstract. The end of the Permian was marked by one of the greatest mass extinctions of all time. A valuable record of life and death during this event is contained within sedimentary organic matter. The stable isotopic, molecular and morphological information contained within remains of end-Permian organisms represent an important resource for scientists attempting to produce paleoenvironment reconstructions. Most meaningful data derive from multidisciplinary analyses of the same samples. In these circumstances it is desirable that sample preparation for one approach does not hinder subsequent analysis by another. To ensure compatibility of sample processing procedures the petrographic and chemical consequences of two common kerogen preparation steps, demineralization and screening (sieving), were simultaneously monitored using transmitted light microscopy and flash pyrolysis. Two end-Permian sediments, whose organic content was predominated by land-plant debris, were chosen for this purpose. A limestone was used to assess the problem of fluoride production when dematerializing carbonates and a marl was used to investigate the possibility of introducing a sampling bias following kerogen screening. Flash pyrolysis results of demineralization residues indicate that neoformed fluorides can be effectively removed by repeated treatments with excess concentrated HCl. Flash pyrolysis of screened size fractions (<10 μm, 10-18 μm, 18-30 μm, 30-125 μm, 125-250 μm, >250 μm) suggest that, for the end-Permian kerogen used, the various fractions are qualitatively representative of the unscreened kerogen. In a paleoenvironmental context, the homogeneity of the land plant derived kerogen reflects a period of organic accumulation on land followed by rapid deposition and burial in a marine setting. These findings constitute a step forward in the quest for parity between petrographic and chemical analyses of the same kerogen samples.

1 Introduction

The end-Permian mass extinction, which occurred approximately 250 Myr ago, represents the most severe biotic crisis in the geological record (Raup and Sepkoski 1982). Around 90% of marine species disappeared (Raup, 1979), 70% of land invertebrate species became extinct (Maxwell 1982) and a worldwide destruction of standing woody biomass occurred (Visscher et al. 1996). The extinction appears to have been accompanied by changes in atmospheric and marine chemistry evidenced, in part, by dramatic global shifts in carbon isotopic compositions of marine and terrestrial carbonates and organic matter (Sephton et al. 2002 and references therein). A record of the struggle between life and death at the end of the Permian is entombed in the sedimentary record. Numerous end-Permian land and marine organisms are represented in sediments by their organic remains. Identifying these organic remains allows the reconstruction of ancient environments, enabling past variations in habitat, climate and biology to be understood. By far the majority of organic matter in sedimentary rocks is present as a high molecular weight kerogen.

Both petrographic and chemical approaches can be followed to understand kerogen composition. Petrographic analyses of kerogens provide information about the physical makeup of the organic matter in sediments whereas chemical analyses of kerogens reveal the molecular structure of the sedimentary organic components. Each type of analysis provides information that complements the other and, whenever possible, petrographic and chemical data should be combined to help determine the origin and diagenetic history of kerogens. Understanding the shared chemical and petrographic features of kerogens is especially important for end-Permian sedimentary rocks. Many end-Permian rocks are organic-poor and are characterized by unusual and often unique petrographic and chemical signals. When attempting to determine the paleoenvironmental significance of an end-Permian organic component, petrographic and chemical data can be mutually supportive.

Combined petrographic and chemical analyses are most meaningful when they are performed on kerogen samples that have been subjected to the same preparative procedures. Yet for this to be achieved a common method is required which is suitable for both analytical approaches. The preparative steps used for the petrographic analysis of kerogens are well established (e.g., Barss and Williams 1973; Vidal 1988), but if the same samples are to be used for chemical analysis certain requirements must be met. In particular, the preparative procedures must not 1) produce problematic by-products that hinder or preclude analysis by chemical techniques, or 2) introduce a sampling bias that results in a chemical signature that is not representative of the kerogen as a whole. These problems may be encountered following two common steps in the preparation of kerogens for petrographic analysis, i.e., demineralization and screening.

Demineralization isolates the kerogen by dissolving the inorganic matrix with hydrochloric (HCl) and hydrofluoric (HF) acids. A concern with this preparative step is that for the carbonate-rich samples often found in end-Permian sections the use of HF can produce fluoride gels. These fluorides are tolerable for petrographic analysis but can conceal the chemical signature of the organic kerogen. Previous work has used infrared spectroscopy to monitor changes in the functional group content of several kerogens and the introduction of analytical artefacts during the demineralization of rock samples (Durand and Nicaise 1980). No similar report exists that used flash pyrolysis as a method of detecting the chemical progress of kerogen isolation.

Petrographers screen kerogens in order to isolate organic components, the morphologies of which constitute a record of paleoenvironmental conditions. Screening separates the kerogen into size fractions by rinsing them with water over nesting screens of different mesh sizes. To justify the chemical analysis of screened kerogens, as representative aliquots of the whole kerogen, it is necessary to establish that the screening procedure does not introduce a sampling bias. Each of the selectively preserved biomacromolecules in Kerogen can differ in their chemical nature (Tegelaar et al. 1989; de Leeuw and Largeau 1993), Some consist of highly aliphatic networks, e.g., algaenans and cutans, while others are predominantly aromatic, e.g., lignins and tannins (Tegelaar et al. 1989). How the organic hosts of these various chemical structures are distributed within certain size fractions of kerogens is unknown, but it is conceivable that screening may concentrate or remove certain selectively preserved biomacromolecules depending on the mesh size of the screen used.

In this paper we report the effects of demineralization and screening on the petrographic and chemical characteristics of an end-Permian limestone and marl, respectively. The consequences of these preparative steps were monitored using transmitted light microscopy and flash pyrolysis. The significance of the data to the end-Permian paleoenvironment is also considered.

2 Experimental

Two different end-Permian sedimentary rocks were chosen for monitoring the chemical and petrographic consequences of sample preparation. To assess the adverse chemical effects of demineralization on carbonate-rich sediments we chose an end-Permian organic-rich limestone (SD55; TOC: 1.62 wt%) from the Bellerophon Formation at Val Badia northern Italy. To investigate the effects of screening we chose an organic-rich marl (S0; TOC: 2.56 wt%) from the same formation. S0 is the youngest Bellerophon marl and it lies 55 cm above SD55. Full location details for the section can be found in Cirilli et al. (1998). These sediments were deposited on a shallow shelf in the western Tethys Ocean (Bosellini and Hardie 1973), and their organic content is predominated by the debris of land plants (Cirilli et al. 1998).

To isolate the S0 and SD55 kerogens from their mineral matrix, the whole rock samples were subjected to cycles of hydrofluoric (HF) and hydrochloric (HCl) acids: 5 % HCl for 24 hours (x 2), 2 hours manual agitation in 40 % HF; 30 % HCl (x2), H_2O rinse. This procedure was performed twice and was later refined to include subsequent multiple treatments with excess 30 % HCl to remove neoformed fluorides. Successful fluoride removal was indicated by a clear supernatant. The isolated S0 kerogen was separated into six size fractions by screening using sieves of different sizes: <10 μm, 10-18 μm, 18-30 μm, 30-125 μm, 125-250 μm, >250 μm.

For petrographic analysis approximately 200 organic particles of each were counted for each sample using a transmitted light microscope and assigned to the following groups: 1) amorphous organic matter, 2) opaque lath shaped, 3) opaque equidimensional, 4) non-cellular sheet light brown, 5) irregular degraded yellow, 6) irregular degraded light brown, 7) irregular degraded dark brown, according to Tyson (1995).

In preparation for pyrolysis, the SD55 kerogen, S0 kerogen, S0 screened fractions were exhaustively extracted with methanol, dichloromethane/hexane (1:1) and hexane. The kerogens were analyzed by flash pyrolysis-gas chromatography (Py-GC) and flash pyrolysis-gas chromatography-mass spectrometry (Py-GC-MS). Samples were pressed onto a flattened ferromagnetic wire and pyrolyzed by inductive heating for 10 seconds (Curie temperature 610°C). The pyrolysis unit was directly coupled to a Hewlett Packard 5890 gas chromatograph equipped with a cryogenic unit and fitted with a CP-Sil 5 capillary column (25 m x 0.32 mm x 0.45 μm). The GC oven was held at 0°C for 5 min and subsequently programmed from 0°C to 320°C at 3°C min^{-1}. The final temperature was held for 10 min. Identification of the pyrolysis products was performed by Curie-point pyrolysis-gas chromatography-mass spectrometry (Py-GC-MS), where an identical GC was interfaced to a VG Autospec Ultima mass spectrometer operated at 70 eV with a mass range m/z 40-800 and a cycle time of 1.6 seconds (resolution 1000). GC conditions were as above.

3 Results and Discussion

3.1 Preparative Procedures for End-Permian Kerogens

Following the HF/HCl demineralization steps, the organic components in the carbonate-rich SD55 limestone could be identified by transmitted light microscopy. Irregular degraded light and dark brown plant tissue dominate the kerogen. Non-cellular sheets, which may represent plant cuticle, are also present. Figure 1a represents the Py-GC analysis of the HF/HCl residue of SD55 used to make the petrographic observations mentioned above. The organic pyrolysis products appear to be overwhelmed by the by-products of the demineralization procedure. These by-products are likely to be neoformed inorganic fluorides formed by reactions between the carbonate matrix and HF. It has been stated that in some cases it is impossible to avoid the formation of at least some fluorides during the HF/HCl demineralization procedure (Durand and Nicaise 1980). Hence, although appropriate for petrographic analysis the standard HF/HCl residue is unsuitable for chemical analysis. Figure 1b shows a Py-GC analysis of the same HF/HCl residue following repeated treatments with excess concentrated HCl. The organic pyrolysis products are now clearly recognizable as a mixture of *n*-alkene/*n*-alkane doublets produced from

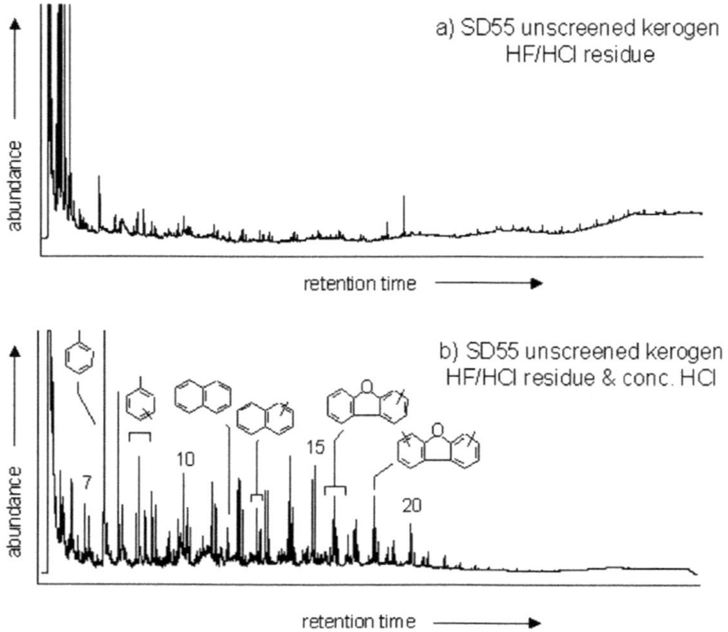

Fig. 1. Gas chromatograms of flash pyrolysates (Cu-temperature 610°C) of the residue from the HF/HCl demineralization procedure of the Val Badia organic-rich limestone (SD55) both a) before, and b) after multiple treatments with excess concentrated HCl to remove neoformed fluorides. Numbers indicate the retention times of *n*-alkene/*n*-alkane doublets. From left to right the other identified peaks are toluene (C_1 alkylbenzene), C_2 alkylbenzenes, naphthalene, C_1 alkylnaphthalene, C_1 alklydibenzofuran, C_2 alklydibenzofuran.

aliphatic biomacromolecules and aromatic hydrocarbons and furans from more condensed organic structures. Therefore, it appears that a refined demineralization procedure involving final multiple treatments of excess concentrated HCl, renders the HF/HCl residues of carbonates suitable for kerogen studies by combined petrographic and chemical techniques.

3.2 Organic Constitution of Kerogen Size Fractions

The general petrographic characteristics of the kerogen from the S0 organic-rich marl have been described previously (Sephton et al. 1999). Briefly, the kerogen consists of an unstructured, humic (i.e., predominated by plant tissues) cement with occasional woody fibers and particles of vitrinite and inertinite. The petrographic observations suggest a major input from land plants debris to produce a so-called type-III kerogen.

Published Rock Eval data (Sephton et al. 1999) confirm this assertion (hydrogen index = 145, oxygen index = 139). The results are consistent with previous work, which indicates that the debris of land plants becomes the major source of organic material at the top of the Bellerophon Formation in the southern Alps (Visscher and Brugman 1986).

Table 1. Organic constituents (%) in screened size fractions of isolated S0 kerogen.

Phytoclast	< 10 µm)	10-18 µm)	18-30 µm)	30-125 µm)	125-250 µm)	> 250 µm)	Unscreened S0
Irregular degraded dark brown	53.5	83.0	75.0	95.0	95.0	99.5	57.5
Irregular degraded light brown	43.0	14.5	22.5	4.0	4.0	0	40.5
Irregular degraded yellow	2.5	0	0	0	0	0	0
Non-cellular sheet light brown	0.5	0.5	1.0	0	0	0	0
Opaque equidimensional	0	0	0.5	0	0	0	0.5
Opaque lath shaped	0.5	2.0	1.0	1.0	1.0	0.5	0.5
Amorphous organic matter	0	0	0	0	0	0	1.0

Following the screening of S0 kerogen, petrographic analysis (Table 1) reveals that irregular degraded dark brown phytoclasts are the dominant component in all size fractions. Yet, there is some partitioning of organic matter types within the various fractions. Amorphous organic matter, present in small amounts in the unscreened sample, is absent in all of the screened fractions. Irregular degraded light brown phytoclasts appear to be more important in the smaller size fractions. Opaque lath shaped, opaque equidimensional and non-cellular sheet light brown elements appear to be concentrated in the medium size fractions (10 to 125 µm). Irregular degraded yellow phytoclasts appear to be concentrated exclusively into the <10 µm size fraction. Hence the various size fractions do show some slight dissimilarities in their organic constitution, particularly the loss of amorphous organic matter following screening.

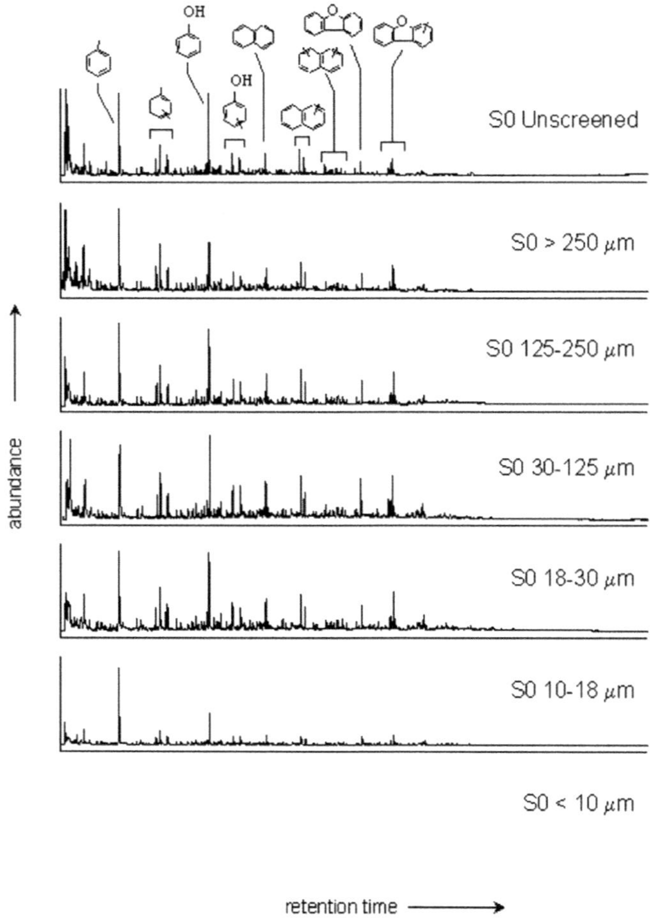

Fig. 2. Gas chromatograms of flash pyrolysates (Cu-temperature 610°C) of unscreened kerogen and six screened size fractions of the kerogen isolated from the Val Badia organic-rich marl (S0). From left to right, the identified peaks are: toluene (C_1 alkylbenzene), C_2 alkylbenzenes, phenol, C_1 alkylphenol, naphthalene, C_1 alkylnaphthalene, dibenzofuran, C_1 alklydibenzofuran.

The general chemical characteristics of the kerogen from the S0 organic-rich marl have been described previously (Sephton et al. 1999). In brief, Py-GC and Py-GC-MS analysis of the kerogen reveals a distribution predominated by peaks reflecting alkyl aromatic hydrocarbons, aromatic phenols and furans. Figure 2 shows the Py-GC results for the unscreened S0 kerogen and its various screened size fractions. The identifications for the major pyrolysis products, as determined by Py-GC-MS, are also indicated. Qualitatively the Py-GC results for the various size fractions appear very similar. The agreement between the unscreened material and the screened fractions suggests that the loss of amorphous organic matter revealed by petrographic analysis has no significant effect on the pyrolysates. It is likely that as amorphous organic matter is such a small component of the overall organic constitution of S0 kerogen its chemical signal is overwhelmed by the responses from the more abundant organic matter types. A similar argument may be invoked for the similarity of the Py-GC results from the various screened size fractions. In each size fraction the dominant irregular degraded dark brown components are contributing the majority of the pyrolysis products.

3.3 Paleoenvironmental Significance of Unstructured Type-III Kerogens

The lack of major chemical differences between the S0 size fractions may be partly a result of the depositional history of the organic matter. Usually, kerogens, which appear unstructured under the petrographic microscope, are assigned as type-II kerogens and are attributed to an algal or marine origin. Type III kerogens commonly exhibit a high level of structural integrity. However, the S0 kerogen appears to have originated from a terrestrial accumulation of decomposing land plant debris, specifically soil organic matter, which was eroded and redeposited in a shallow marine depositional environment (Sephton et al. 1999). Consequently the organic matter has undergone decomposition, mechanical breakage and compaction. The product of these events is likely to be a highly corrupted organic assemblage rather than a collection of coherent and recognizable plant fragments. The swift transportation and redeposition of soil organic matter is likely to produce a homogenized and unstructured mixture of organic constituents, which distributes itself throughout the various size fractions.

Hence, certain unstructured type-III kerogens may represent valuable records of paleoenvironmental conditions due to their particular depositional history. Following redeposition in the marine setting, the oxidation of large amounts of land-derived organic matter will quickly deplete oxygen supplies and aerobic microbial degradation will be halted. Consequently, easily degraded organic species may survive diagenesis and contribute to a more representative sedimentary record of the organic components that make up the terrestrial paleoenvironment. In this context it is interesting to note the presence of furans in the kerogen pyrolysates (Fig. 2). Furans are markers for polysaccharides that are often easily degraded by microbial action. Their presence in the S0 kerogen suggests relatively rapid transportation from land to sea and swift burial in the marine depositional setting. Once buried, the polysaccharide-related compounds were shielded from microbial degradation.

It is relevant to consider the causative mechanisms for the introduction of polysaccharide-rich soil organic matter into marine environments at the end of the Permian. At this time, the biosphere experienced a natural pollution event initiated by massive volcanism (Visscher et al. 1995). The Siberian Traps represent the largest of all flood basalt provinces and their emplacement would have involved the release of large amounts of acidifying gases and liquid aerosols to the atmosphere. Once deposited at the surface of the supercontinent Pangaea, chemical changes occurred which had a deleterious effect on ecosystems. It appears that eventually the terrestrial biosphere succumbed to the imposed stresses and a widespread loss of rooted vegetation preceded a dramatic episode of soil erosion. The influx of mobilized soil materials exacerbated stresses in the marine realm completing the global catastrophe.

4 Conclusions

Organic matter in an end-Permian marl (SD55) from Val Badia in northern Italy contains a mixture of aliphatic and aromatic biomacromolecules, which reflect the organic components present in the end-Permian terrestrial ecosystem adjacent to the continental margin. Treatment with HF/HCl acids followed by excess concentrated HCl allows the visualization of kerogen components by flash pyrolysis, even for the carbonate-rich sediments which are common in end-Permian sections.

Organic matter in the latest end-Permian marl (S0) is predominated by the organic remains of land plants. The land plant remains appear to have accumulated on land as soil organic matter prior to their rapid transportation and burial in the marine depositional setting to produce a type-III kerogen. Petrographic results of screened fractions of S0 illustrate that the various size fractions do show some slight dissimilarities in their organic constitution. However, chemical data in the form of flash pyrolysates fail to show significant differences. The data suggests that the S0 kerogen has been "naturally homogenized" by its depositional history and, as a result, the chemical composition of the kerogen is relatively unaffected by screening.

In a paleoenvironmental context, these data are consistent with the mobilization of soil organic matter, by volcanically-induced environmental degradation, and its redeposition in marine settings adjacent to the Pangaean land mass.

Acknowledgements

M. Dekker and W. Pool are thanked for analytical assistance. R. M. Hough and G. J. M. Versteegh are acknowledged for advice on demineralization and petrographic methods respectively. These investigations were, in part, supported by the Netherlands Geosciences Foundation (GOA) with financial aid from the Netherlands Organization for Scientific Research (NWO). This contribution is NSG number 990111 and NIOZ number 3347. Sylvie Derenne and an anonymous reviewer are thanked for their constructive comments.

References

Barss MS, Williams GL (1973) Palynology and nannofossil processing techniques. Geological Survey of Canada, Paper 73-26: 1-26

Bosellini A, Hardie AL (1973) Depositional theme of a marginal evaporite. Sedimentology 20: 5-27

Cirilli S, Pirini Radrizzani C, Ponton M, Radrizzani S (1998) Stratigraphical and palaeoenvironmental analysis of the Permian-Triassic transition in the Badia Valley (Southern Alps, Italy). Palaeogeography, Palaeoclimatology, Palaeoecology 138: 85-113

Durand B, Nicaise G (1980) Procedures for kerogen isolation, In: Durand B (ed) Kerogen - insoluble organic matter from sedimentary rocks. Editions Technip, Paris, pp 35-53

de Leeuw JW, Largeau C (1993) A review of macromolecular organic compounds that comprise living organisms and their role in kerogen, coal and petroleum formation. In: Engel MH, Macko SA (eds) Organic Geochemistry, Principles and Applications. Plenum Press, New York, 23-72

Maxwell WD (1992) Permian and Early Triassic extinction of non-marine tetrapods. Palaeontology 35: 571-583

Raup DM, Sepkoski JJ (1982) Mass extinctions in the marine fossil record. Science 215: 1501-1503

Sephton MA, Looy CV, Veefkind RJ, Visscher H, Brinkhuis H, de Leeuw JW (1999) Cyclic diaryl ethers in a Late Permian sediment. Organic Geochemistry 30: 267-273

Sephton MA, Looy CV, Veefkind RJ, Brinkhuis H, de Leeuw JW, Visscher H (2002) A synchronous record of $\delta^{13}C$ shifts in the oceans and atmosphere at the end of the Permian. In: Koeberl C, MacLeod K (eds) Catastrophic Events and Mass Extinctions: Impacts and Beyond: Boulder, Colorado. Geological Society of America Special Paper 356, 455-462

Tegelaar EW, de Leeuw JW, Derenne S, Largeau C (1989) A reappraisal of kerogen formation. Geochimica et Cosmochimica Acta 53: 3103-3106

Tyson RV (1995) Sedimentary organic matter: organic facies and palynofacies. Chapman & Hall, London

Vidal G (1988) A palynological preparation method: Palynology 12: 215-220

Visscher H, Brinkhuis H, Dilcher DL, Elsik WC, Looy CV, Rampino MR, Traverse A (1996) The terminal Paleozoic fungal event: Evidence of terrestrial ecosystem destabilization and collapse. Proceedings of the National Academy of Sciences USA. 93: 2155-2158

Economic Mineral Deposits in Impact Structures: A Review

Wolf Uwe Reimold[1], Christian Koeberl[2], Roger L. Gibson[1], and Burkhard O. Dressler[1,3]

[1]Impact Cratering Research Group, School of Geosciences, University of the Witwatersrand, Private Bag 3, P.O. Wits 2050, Johannesburg, South Africa (reimoldw@geosciences.wits.ac.za; gibsonr@geosciences.wits.ac.za)

[2]Department of Geological Sciences, University of Vienna, Althanstrasse. 14, A-1090 Vienna, Austria (christian.koeberl@univie.ac.at)

[3]185 Romfield Circuit, Thornhill, Ontario, Canada, L3T 3H7 (burkhard@attcanada.ca)

Abstract. Many large meteorite impact structures throughout the world host mineral resources that are either currently mined or have the potential to become important economic resources in the future. The giant Vredefort-Witwatersrand and Sudbury impact structures underline this statement, because of their enormous resources in gold and uranium, and nickel, copper, and PGEs, respectively. In relation to impact, three basic types of ore deposits in impact structure settings have been distinguished: (1) *progenetic* (i.e., pre-impact) deposits that already existed in the target regions prior to an impact event, but may have become accessible as a direct result of the impact; (2) *syngenetic* (syn-impact) deposits that owe their existence directly to the impact process, and (3) *epigenetic* (immediately post-impact) deposits that result from impact-induced thermal/hydrothermal activity. In addition to metalliferous ore deposits related to impact structures, impact structure-hosted epigenetic hydrocarbon deposits are reviewed and are shown to make a major contribution to the North American economies. Non-metallic resources, such as minerals derived from crater-lake deposits, dimension stone, and hydrological benefits, may also be derived from impact structures, and the educational and recreational value of many meteorite impact craters can be substantial.

Undoubtedly, impact structures - at least those in excess of 5-10 km diameter - represent potential exploration targets for ore resources of economic magnitude. This important conclusion must be communicated to exploration geologists and geophysicists. On the other hand, impact workers ought to be familiar with already established fact concerning ore deposits in impact environments and must strive towards further understanding of the ore generating processes and styles of emplacement in impact structures.

1 Introduction

Currently some 170 impact structures are known on Earth – presumably representing a mere fraction of the entire terrestrial cratering record for a meteorite impact structure list (e.g. Impact database) Other solid bodies of the Solar System display surfaces that have been thoroughly cratered, but have barely been accessible for detailed impact geological study. Only the Moon and Mars have been – and will in future be – targets of direct geological study, besides probing of large, impact-cratered asteroids. Future Space exploration, and perhaps habitation of other planetary bodies, will have to rely on natural resources obtained in Space. This also includes asteroids, the direct study of which has only been resumed in 2001 with the spectacular soft landing of the Shoemaker-NEAR spacecraft on the asteroid 433 Eros. The study of comets recently experienced a setback when NASA's Contour probe perished shortly after take-off, but several other projects (e.g., NASA's Stardust and ESA's Rosetta missions) currently attempt to provide new insight into the composition of cometary bodies. Mining of Lunar and Martian surfaces, as well as of asteroidal bodies, for the procurement of raw materials required in Space, has been the subject of discussions for years (e.g., Lewis 1997, and references therein).

Thus, a look at the economic potential of impact structures and impactites must be an integral part of any comprehensive treatise of impact phenomena. Grieve and Masaitis (1994), in their benchmark account of impact-related ore deposits, stated that "*impact is an extraordinary geologic process involving vast amounts of energy, resulting in near instantaneous rises in temperature and pressure, and in the structural redistribution of target materials*". In essence, impact is catastrophic and destructive, but it leads to the formation of specific rock units and may – directly or indirectly – trigger mineralization processes, both of which may have considerable economic significance. Here, we provide a review of the existing knowledge about ore-forming processes related to impact and describe the mineralization environments known from quite a number of terrestrial impact structures. Table 1 provides some pertinent detail about those impact structures refered to in the text.

Table 1. (continued on next two pages) Some pertinent information about those impact structures discussed in the text.

Crater Name	Long.	Lat.	Country	Diam. [km]	Age [Ma]	Economic Interest
Ames	36°15'N	98°12'W	Oklahoma USA	16	470±30	Hydrocarbons
Avak	71°15'N	156°38'W	Alaska USA	14	*ca.* 460	Hydrocarbons
Beyenchime Salaatin	71°00'N	121°40'E	Russia	8	40±20	Pyrite (minor)
Boltysh	48°45'N	32°10'E	Ukraine	24	65.2±0.6	Phosphorite; hydrocarbons
Bosumtwi	06°30'N	01°25'W	Ghana	10,5	1,07	Water reservoir; education /recreation; traces of agate; fishing
Brent Crater	46°05'N	78°29'W	Ontario Canada	3,8	396±20	Crater sediment
Carswell	58°27'N	109°30'W	Saskatch Canada	39	115±10	Uranium
Charlevoix	47°32'N	70°18'W	Quebec Canada	54	342±15	Ilmenite
Chesapeake Bay	37°17'N	76°01'W	Virginia USA	80	35.5±0.3	Water reservoir; education /recreation; traces of agate; fishing
Chicxulub	21°20'N	89°30'W	Mexico	180	65.00±0.05	Hydrocarbons; impact diamonds
Cloud Creek	43°10.6'N	106°42.5'W	Wyoming USA	*ca.* 7	*ca.* 190±20	Hydrocarbons
Crooked Creek	37°50'N	91°23'W	Missouri USA	7	320±80	Pb-Zn
Decaturville	37°54'N	92°43'W	Missouri USA	6	<300	Pb-Zn
Dellen	61°48'N	16°48'E	Sweden	19	89.0±2.7	Summer/winter sport; hydropower reservoir
Gardnos	60°39'N	09°00'E	Norway	5	500±10	Gardnos Breccia (decorative arts)
Houghton Dome	75°22'N	89°41'W	Nunavut Canada	24	23±1	Epigenetic overprint
Ilyenits	49°07'N	29°06'E	Ukraine	8,5	378±5	Agate (traces)
Kaluga	54°30'N	36°12'E	Russia	15	380±5	Water

Crater Name	Long.	Lat.	Country	Diam. [km]	Age [Ma]	Economic Interest
Kara	69°06'N	64°09'E	Russia	65	70.3±2.2	Impact diamonds, Pyrite (minor)
Karla	54°55'N	48°02'E	Russia	10	5±1	Mercury
Kentland	40°45'N	87°24'W	Indiana USA	13	97	Pb-Zn
Lake St. Martin	51°47'N	98°32'W	Manitoba Canada	40	220±32	Gypsum, anhydrite
Lappajärvi	63°12'N	23°42'E	Finland	23	73.3±5.3	Summer and winter sport; education/recreation; building stone
Logoisk	54°12'N	27°48'E	Belarus	15	42±1	Phosphorite; amber; groundwater recharge basin
Lonar	19°58'N	76°31'E	India	1,8	0.05±0.01	Trona; post impact hydrothermal alteration
Manicouagan	51°23'N	68°42'W	Quebec Canada	100	214±1	Water reservoir; hydro power
Manson	42°35'N	94°33'W	Iowa, USA	35	73.8±0.3	Epigenetic overprint
Marquez Dome	31°17'N	96°18'W	Texas, USA	12,7	58±2	Hydrocarbons
Meteor Crater	35°02'N	111°01'W	Arizona USA	1,2	0.049 ±0.003	Silica; museum
Morokweng	26°28'S	23°32'E	South Africa	70	145±1	None (suspected Ni/PGE mineralization)
Newporte	48°58'N	101°58'W	North Dakota USA	3,2	<500	Hydrocarbons
Obolon	49°35'N	32°55'E	Ukraine	20	169±7	Hydrocarbons (oil shale)
Popigai	71°39'N	111°11'E	Russia	100	35.7±0.2	Impact diamonds
Puchezh-Katunki	56°58'N	43°43'E	Russia	80	167±3	Impact diamonds, mercury; zeolite
Ragozinka	58°44'N	61°48'E	Russia	9	46±3	Diatomite
Red-Wing Creek	47°36'N	103°33'W	North Dakota USA	9	200±25	Hydrocarbons
Ries (Nördlinger Ries)	48°53'N	10°37'E	Germany	24	15.1±0.1	Impact diamonds; bentonite; lignite; building stone; museum; epigenetic overprint

Crater Name	Long.	Lat.	Country	Diam. [km]	Age [Ma]	Economic Interest
Rochechouart	45°50'N	00°56'E	France	23	214±8	Education/recreation/ Museum; building stone
Rotmistrovka	49°00'N	32°00'E	Ukraine	2.7	120±10	Hydrocarbons
Sääksjärvi	61°24'N	22°24'E	Finland	6	*ca.* 560	Agate (traces); recreation (summer/winter sport)
Serpent Mound	39°02'N	83°24'W	Ohio, USA	8	<320	Pb-Zn
Sierra Madera	30°36'N	102°55'	Texas USA	13	<100	Hydrocarbons
Siljan	61°02'N	14°52'E	Sweden	65	362±1	Pb-Zn; winter sport
Steen River	59°30'N	117°30'W	Alberta Canada	25	91±7	Hydrocarbons
Steinheim Basin	48°41'N	10°04'E	Germany	3,8	15±1	Museum
Sudbury	46°36'N	81°11'W	Ontario Canada	±250	1850±3	Ni, Cu, PGE; minor Cu-Pb-Zn; impact diamonds
Ternovka (Terny)	49°01'N	33°05'E	Ukraine	11	280±10	Iron ore; impact diamonds; uranium
Tookoonooka	27°07'S	142°50'E	Australia	55	128±5	Possible hydrocarbon target
Tswaing (Pretoria Saltpan)	25°24'S	28°05'E	South Africa	1.13	0.22±0.05	Trona; education/recreation/ Museum
Ust-Kara	69o18'N	65o18'E	Russia	25	70,3	Pyrite (minor)
Vepriai	55°05'N	24°35'E	Lithuania	8	160±10	Water reservoir
Viewfield	49°35'N	103°04'W	Sasketch Canada	2,5	190±20	Hydrocarbons
Vredefort-Witwatersrand	27°00'S	27°30'E	South Africa	250-300	2020±5	Gold, uranium; education/recreation; Kibaran bentonite
Zapadnaya	49°44'N	29°00'E	Ukraine	3,2	165±5	Impact diamonds
Zhamanshin	48°24'N	60°58'E	Kazakstan	14	0.9±0.1	Bauxite

(Unconfirmed impact structures mentioned in the text are Bangui (Central African Republic/DR Congo), Calvin (Michigan, USA), and Pechenga (northern Scandinavia)).

It was the interest in finding a potentially economic iron, nickel and platinum group element (PGE) deposit that, early in the last century, led Daniel Moreau Barringer to devote himself and his resources to the investigation of Meteor Crater in Arizona (Barringer 1906; Hoyt 1987). This exploration enterprise was doomed to fail, because, as we now understand, projectiles at hypervelocity speed do not survive the impact process intact. However, Barringer's persistence and continued efforts yielded an enormous amount of information about the structure of Meteor Crater and triggered much interest in the physical and geological aspects of impact cratering and impact structures, in general.

In recent decades, much geological and geophysical information about impact structures has been obtained from mineral and hydrocarbon exploration in impact structures. Two of the world's largest and richest mining camps, hosting the Ni-Cu-PGE deposits of the Sudbury Structure in Canada and the gold- and uranium-bearing sedimentary rocks of the South African Witwatersrand Basin, are located within the two largest known terrestrial impact structures (Sudbury and Vredefort, respectively). The third, truly large-scale impact structure known on Earth, Chicxulub in Mexico, was discovered as a direct consequence of extensive hydrocarbon exploration by geophysical methods and drilling in the Gulf of Mexico (Hildebrand et al. 1991, and references therein). Other large impact structures, such as the Chesapeake Bay structure thought to measure about 80 km in diameter (Poag et al. 2003, and references therein) off the east coast of North America and the 70-80 km Morokweng structure in South Africa (Reimold et al. 2002a), have been investigated - to a large degree - by geophysical methods and drilling because of commercial (e.g., base metals in the case of Morokweng – Maier et al. 2003) or strong environmental/hydrogeological (in the case of Chesapeake Bay – Poag et al. 2003, and references therein) interests.

Previous reviews of economic deposits (Grieve and Masaitis 1994; Grieve 2003) and hydrothermal mineralization in impact structures (Naumov 2002) have emphasized that a large proportion of the known terrestrial impact structures are either the locations of commercially exploited ore deposits or of potentially usable geological resources. Grieve (2003) estimated that 25% of all known terrestrial impact structures have economic ore resources associated with them and that these are currently exploited in about 12% of all known impact structures.

In this review we follow the approach taken by Grieve and Masaitis (1994), who distinguished three basic types of mineralization environments in terrestrial impact structures, which are based on the dominant parameters that govern their formation. The first type of mineralization is based on the metalliferous components and ore mineral enrichment already present in the target area (i.e., pre-impact mineralization) of a specific impact region. Grieve and Masaitis (1994) applied the term *progenetic deposit* for this case. Clearly, the most prominent example for this class of deposits is provided by the Archean gold-and-uranium ores of the Witwatersrand Basin, preserved in the mid-Proterozoic Vredefort impact structure. Second, there are distinct types of deposits that were formed during the impact event or as a direct, immediate consequence of the event. The term *syngenetic* with regard to an impact event applies to those deposits that are epitomized by the extraordinary mineralization of the Sudbury impact structure. Finally, largely hydrothermal-epithermal processes have led to the formation of a large number of so-called *epigenetic* mineralizations (i.e., mineralization that was formed as a direct consequence of the impact event but after it) and, in some cases, economic ore deposits in impact structures. Many hydrocarbon deposits, which have a particularly important economic role with regard to impact structures, belong into this third class of deposits.

2 Pre-Impact (Progenetic) Deposits in Large Impact Structures

2.1 Vredefort (South Africa)

The most prominent progenetic ore deposits on Earth related to impact are the gold deposits of the Vredefort-Witwatersrand Basin system. The Vredefort impact structure (Fig. 1) is centered on the Vredefort Dome at about latitude 27° S and longitude 27°30'E in north-central South Africa, in the heart of the Witwatersrand Basin.

The origin of the Vredefort Dome was controversial for most of the 20th century (see review by Gibson and Reimold 2001a). Since Du Toit (1954) published a tectonic model involving large-scale thrusting for the origin of the Vredefort Dome, ideas linking the origin of the Vredefort Dome with tectonic processes have been repeatedly proposed (e.g., Colliston 1990; Coward et al. 1995). Early workers, such as Shand (1916) and Hall and Molengraaff (1925), on the other hand, recognized the enigmatic rock deformations associated with the structure and already then related them to a 'catastrophic event'.

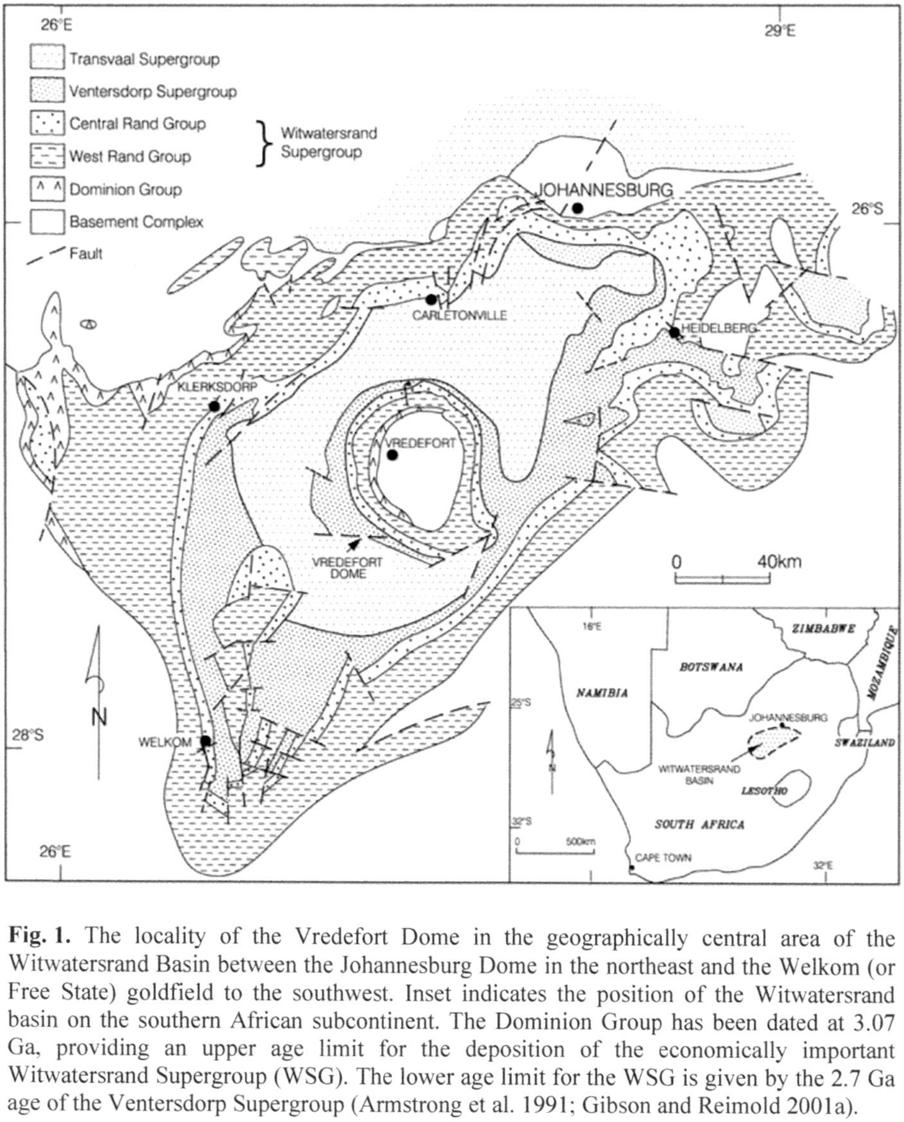

Fig. 1. The locality of the Vredefort Dome in the geographically central area of the Witwatersrand Basin between the Johannesburg Dome in the northeast and the Welkom (or Free State) goldfield to the southwest. Inset indicates the position of the Witwatersrand basin on the southern African subcontinent. The Dominion Group has been dated at 3.07 Ga, providing an upper age limit for the deposition of the economically important Witwatersrand Supergroup (WSG). The lower age limit for the WSG is given by the 2.7 Ga age of the Ventersdorp Supergroup (Armstrong et al. 1991; Gibson and Reimold 2001a).

First thoughts about a possible meteorite impact origin were published by Boon and Albritton (1937). A decade later, Daly (1947) concluded that, in the absence of any other conclusive hypothesis, the impact hypothesis

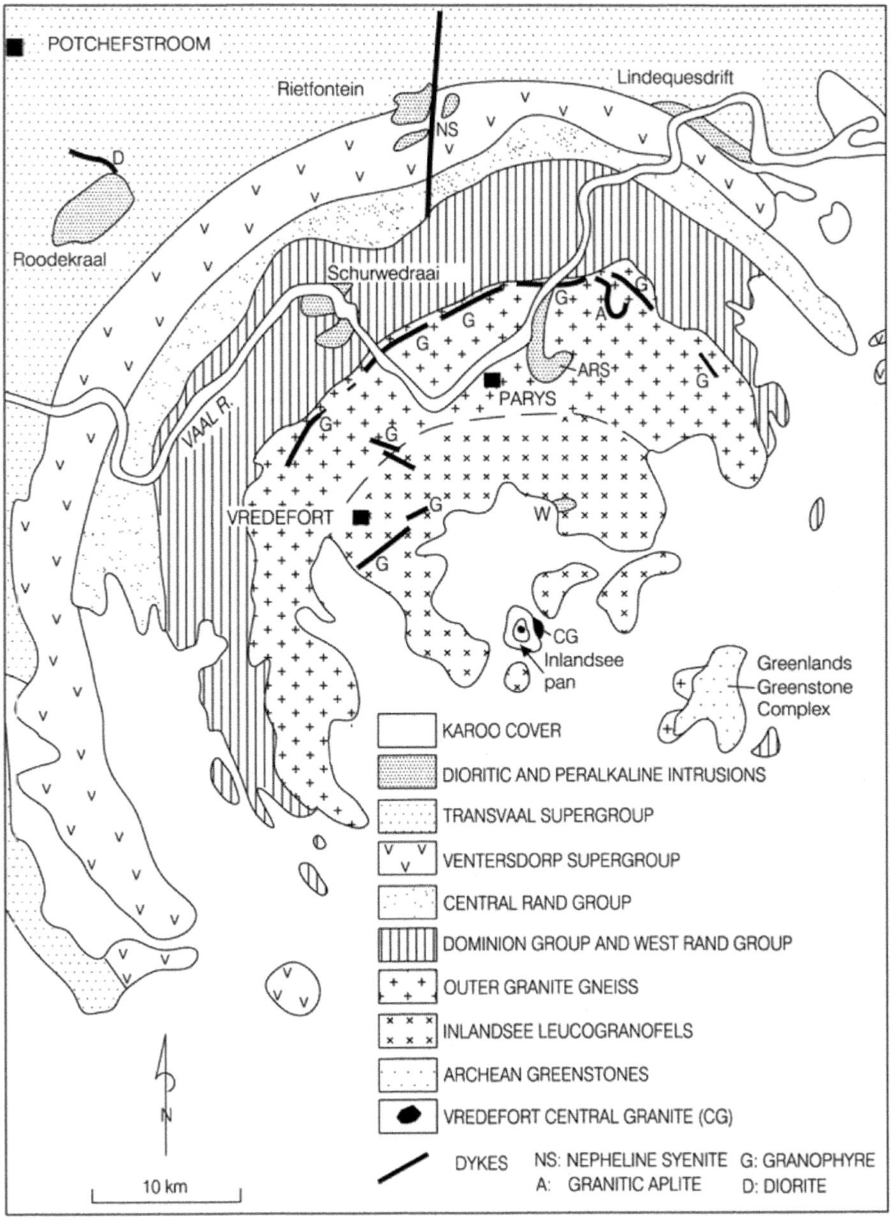

Fig. 2. Schematic geology of the Vredefort Dome (modified after Gibson and Reimold 2001b). CG – Central Granite, see text for further detail.

deserved to be further investigated. In 1961, Dietz discussed the possibility that the Vredefort Dome, which he compared with the Sudbury Structure, could also contain shatter cones, as identified a year earlier at Sudbury. Shortly thereafter, Hargraves (1961) recognized this conical fracture phenomenon in rocks of the Vredefort Dome, providing the first hard evidence for impact at Vredefort. Carter (1965, 1968) described planar deformation lamellae in quartz from Vredefort rocks, which at the time were investigated at a number of impact structures (see papers in French and Short 1968). The Vredefort 'lamellae', though, remained controversial for a long time (e.g. Alexopoulos et al. 1988; Grieve et al. 1990; Reimold 1990), until Leroux et al. (1994) showed conclusively that they indeed constituted *bona fide* shock deformation features (basal Brazil twin lamellae). Strong evidence in favour of an impact origin for the Vredefort Dome was also produced by Martini (1978), who discovered coesite and stishovite in the structure.

The Vredefort Structure is 2.02 Ga old (Kamo et al. 1996) and deeply eroded. Allogenic breccias and sheets of melt rocks, therefore, are absent. Furthermore, the Vredefort Dome has been subject to a widespread post-impact thermal overprint, which has obscured or modified many of the impact-diagnostic shock features. Pseudotachylitic breccias are common and locally abundant (Reimold and Colliston 1994). Impact melt breccias in the deeply eroded structure are only represented by a series of narrow dikes of the so-called Vredefort Granophyre (see review in Gibson and Reimold 2001a).

Koeberl et al. (1996a) applied the very sensitive Re-Os isotopic method to Vredefort Granophyre samples in comparison to various country rock types and established that the Granophyre contained a very small – only 0.2 % - meteoritic component. Shock deformation in zircon from rocks of the Vredefort Dome (Kamo et al. 1996; Gibson et al. 1997; Leroux et al. 1999; Reimold et al. 2002b) and in feldspar (Gibson et al. 2001) has been discovered, and just recently, Buchanan and Reimold (2002) identified - for the first time - shock deformation in lithic inclusions in the Granophyre. The Vredefort Dome is well known for another, regionally unique deformation phenomenon: ubiquitous veinlets and massive dikes or network breccias, for which Shand (1916) coined the term "pseudotachylyte" (modern spelling 'pseudotachylite'). In the past decades many such occurrences have been described from impact structures, whereby, however, different types of breccias have not been carefully distinguished. Reimold (1995, 1998) proposed to use the non-genetic term "pseudotachylitic breccia" where an origin by impact melting, friction melting, or combination of these processes can not be determined with certainty, and to reserve the expression "pseudotachylite" for *bona fide* friction melt rock. The largest occurrence of pseudotachylitic breccia observed by our group in the Vredefort Dome measures 2.6 km in length and 50 m in width (Dressler and Reimold 2004). Such breccias also occur in great abundance in the northern, well explored and heavily mined, part of the Witwatersrand Basin and have mostly been linked to the Vredefort impact event (Fletcher and Reimold 1989; Killick and Reimold 1990; Killick 1993; Reimold and Colliston 1994; Trieloff et al. 1994; Reimold et al. 1999b; Hayward et al. 2003), but some older occurrences have also been identified (Berlenbach and Roering 1992; Reimold and Colliston 1994).

2.1.1 Geological Setting

The 80-90-km-wide Vredefort Dome (Gibson and Reimold 2001a,b) is located in Archean and Paleoproterozoic rocks of the Kaapvaal craton of southern Africa (Fig. 1). The Dome comprises a central core of about 45-50 km diameter that is composed of poly-deformed, pre-3.1 Ga, Archean granitoid gneisses and granites, with fragments of upper amphibolite to granulite facies, mafic and felsic ortho- and paragneisses. The core is surrounded by a circa 20-km-wide collar of greenschist to amphibolite facies, metasedimentary and metavolcanic, Late Archean to Paleoproterozoic (ca. 3.1-2.15 Ga) strata of the Dominion Group and the Witwatersrand, Ventersdorp, and Transvaal supergroups (Gibson and Reimold 2001a). The southern parts of the Dome, as the southern part of the Witwatersrand Basin, are largely covered by Phanerozoic sedimentary rocks and dolerite intrusions of the 300-180 Ma Karoo Supergroup (Fig. 1, 2).

The Dome is surrounded by a 50-70 km wide rim syncline known as the Potchefstroom Synclinorium (Fig. 3), which is characterized by shallow dips of the Transvaal Supergroup and older strata. The Transvaal Supergroup strata are affected by kilometer-scale open folds that are tangentially arranged around the Dome (Simpson 1978). Ductile shear zones with associated meter-scale folds and a cleavage that displays centrifugal vergence in relation to the Dome are found in Transvaal Supergroup rocks along the northern margin of the Witwatersrand Basin, as far from the Dome as 150-200 km (McCarthy et al. 1986, 1990; Gibson et al. 1999). Intercalated with the collar strata of the Dome are intrusions that have been related to various magmatic events, including the emplacement of the 2.06 Ga Bushveld Complex and the 2.7 Ga Ventersdorp extrusive event.

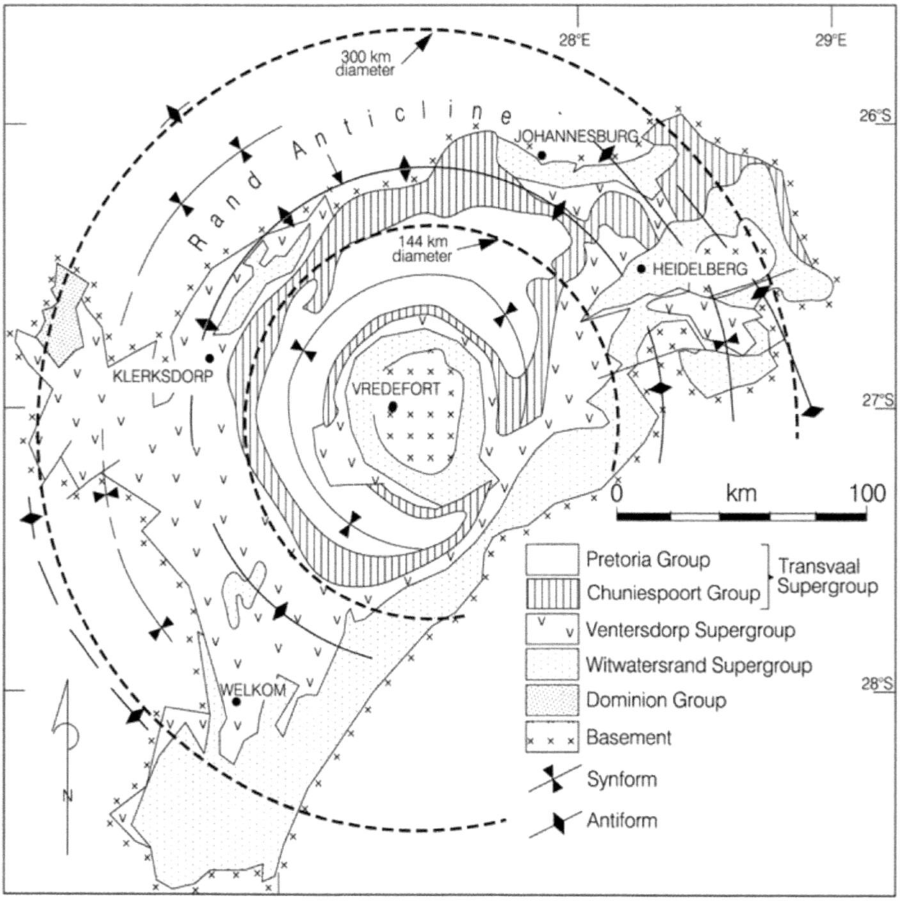

Fig. 3. Large-scale tectonic structure (anticlinal/synclinal structures) in the region of the Witwatersrand Basin and in the environs of the Vredefort Dome. The Rand Anticline is generally taken as the northern limit of the Witwatersrand Basin. It also represents the northernmost limit of known occurrence of pseudotachylitic breccia (personal observation, WUR). Modified after Therriault et al. (1997). The position of the Potchefstroom Synclinorium is marked by the wide occurrence of Chuniespoort strarta in the environs of the dome.

One controversial issue has been the question of whether strata exposed in the crystalline core of the Vredefort Dome are upturned, or even overturned, as observed in the adjacent supracrustal strata of the collar. Hart et al. (1981, 1990), Tredoux et al. (1999), and others have subscribed to this idea of a "crust-on-edge" geometry of the basement rocks. The geophysical modeling of Henkel and Reimold (1998) and detailed metamorphic and structural work by, for example, Stevens et al. (1997), Gibson and Reimold (2000, 2001b), Gibson et al. (1998, 1999), Lana (2004), and Lana et al. (2003a-d) has provided much evidence against the crust-on-edge hypothesis. This hypothesis postulates that in the center of the Dome rocks of the lower crust and, possibly, even of the upper mantle (Tredoux et al. 1999) are exposed. This is, however, not supported by the modeling of basement uplift based on geophysical data and the metamorphic grades determined for rocks of the central core. Recent detailed, small-scale mapping of the exposed basement in the core of the Vredefort Dome by Lana et al. (2003a, c, and d) also did not provide any structural geological support for the so-called 'crust-on-edge' model. Rotation of the subvertical Archean fabric by impact tectonics is restricted to a zone of a few kilometer width just inward from the core-collar contact. Henkel and Reimold (1998) suggested only gentle upward flexing of the lower crust, perhaps with a maximum gradient of about 4 km.

The best estimate for the age of the Vredefort impact event of 2020 ± 5 Ma (Kamo et al. 1996; Moser 1997; Gibson et al. 1997; Spray et al. 1995) clearly distinguishes this impact event from the about 50 Ma earlier (2.06 Ga) emplacement of the Bushveld Complex – a finding that with regard to thermal and hydrothermal overprint on the Witwatersrand ores (see below) is of significance.

2.1.2 Considerations and Evidence Regarding the Size of the Vredefort Structure - and its Spatial Relation to the Witwatersrand Basin

Originally, Vredefort researchers focused purely on the Vredefort Dome. Most early size estimates for the Vredefort Structure ranged from 70 (only the inner parts of the Vredefort Dome) to 140 (Dome plus surrounding Potchefstroom Synclinorium) kilometers (e.g., Bishopp 1941; Dietz 1961; Simpson 1978; Grieve and Pesonen 1992). Grieve and Masaitis (1994), however, estimated - on the basis of a map with concentric structural features published by McCarthy et al. (1990) - that the original diameter of the structure could have been as wide as 300 kilometers. Therriault et al. (1997) applied the spatial distribution information for various styles of deformation (occurrences of planar deformation features, shatter cones and impact-related brecciation) to empirically derived equations based on observations from other impact structures (Melosh 1989) to predict an original diameter for the Vredefort impact structure (Fig. 4), of similar magnitude. Henkel and Reimold (1996, 1998), who carried out integrated geophysical modeling of the whole Witwatersrand basin, derived at a

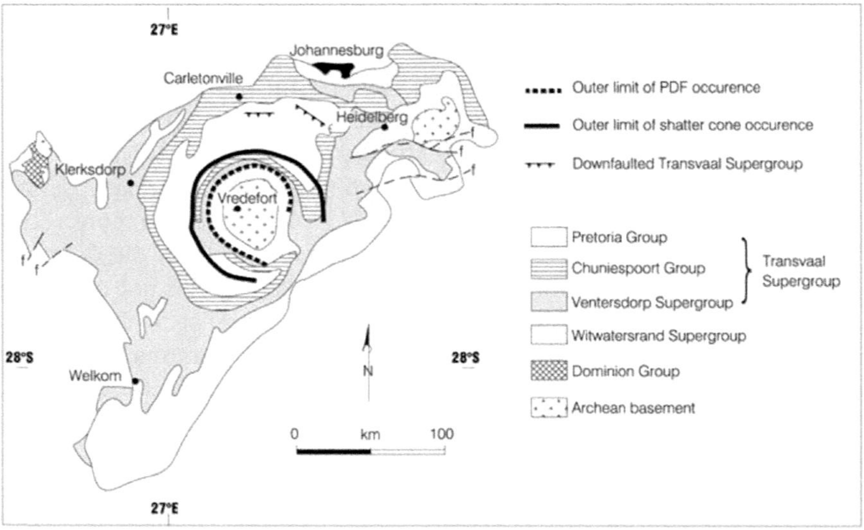

Fig. 4. Scaling of the spatial distribution of various deformation features caused by the Vredefort impact event in the environs of the Vredefort Dome (based on work by Therriault et al. 1997).

diameter in the order of 250-300 km, and remote sensing applications by Phillips et al. (1999) are in excellent agreement with these figures. Generally, most workers have accepted that Vredefort is the remnant of an originally ca. 300-km-diameter impact structure.

Vredefort, with regard to its size, falls into the same category as two other terrestrial impact structures, namely the Sudbury and Chicxulub structures (e.g., Grieve and Therriault 2000), for which original diameters of 200-250 (or even larger) and ca. 200 km are favored, respectively (also Deutsch et al. 1995; Sharpton et al. 1996; Morgan and Warner 1999; Snyder and Hobbs 1999). However, Vredefort differs from the other two structures in that it does not contain any evidence for a significant coherent impact melt body and impact breccia fill. The Sudbury Structure still includes a ca 3.5 km thick impact melt body and even overlying fallback breccia (see below), and Chicxulub is more or less fully preserved underneath Tertiary sediments. Based on what we know about the Sudbury Structure, the currently exposed erosion level at Vredefort is considerably lower than that at Sudbury. Published estimates of the depth of erosion of the Vredefort Dome range from a few hundred meters (Martini 1991) to nearly 18 kilometers (Schreyer and Abraham 1978). Figures of 7-10 km have been favored in recent years (e.g., Henkel and Reimold 1998; Gibson et al., 1998; Gibson and Reimold 2000, 2001b), as the absence of a coherent impact melt body and impact breccia crater fill make the lower estimates highly unlikely. In addition, as the stratigraphic and intrusive units that originally lay close to the surface (namely Transvaal Supergroup and Bushveld Complex strata) are still preserved with only gentle dips in the environs of the Vredefort Dome, the upper limit is unrealistic as well. The prefered values are also generally consistent with scaling equations (Grieve and Pilkington 1996) that relate the amount of structural uplift (SU) to crater (rim) diameter D as $SU = 0.086\ D^{1.03}$. Also, if the diameter of the central uplift (Vredefort Dome to axis of Potchefstroom Synclinorium) is taken at 80-100 km, the empirical relation between diameter of central uplift (D_{SU}) and total diameter of the impact structure (D) by Therriault et al. (1997), ($D_{SU} = 0.31\ D^{1.0}$), gives a 300 km diameter for the Vredefort impact structure.

In addition to the shock deformation features found in the rocks of the dome, several synclinal and anticlinal structures have been described from the region of the Witwatersrand Basin, and have been related to the Vredefort impact by McCarthy et al. (1986, 1990) and Brink et al. (1997, 1999, 2000). Beyond the Rand Anticline along the northern margin of the conventionally considered area of the Witwatersrand Basin (Pretorius et al. 1986), a series of smaller, and more periclinal, synclinal and anticlinal structures has been described (McCarthy et al 1990; Gibson et al. 1999), up to a distance of 150 km from the center of the Dome. Several workers (Spray 1998; Brink et al. 1997, 1999) have speculated that this series of morphological rings and/or alleged zones of enhanced development of pseudotachylitic breccia that generally have been associated with the Vredefort impact (Grieve and Therriault 2000) could be equivalent to the ring features of multi-ring basins (Spudis 1993). Reimold (1998) discussed the Vredefort case and emphasized that apparent enhancement of breccia development at specific distances from the center of the Vredefort dome could be a result of insufficient, discontinuous outcrop and localized underground observation that is restricted to the Witwatersrand goldfields (Fig. 5) where strata are accessible for geological analysis to a maximum depth of around 4 km.

2.1.3 Thermal Metamorphism

Geochronological evidence indicates that an early crustal component of the granitoids of the core of the Vredefort Dome was formed around 3.4 Ga ago (our group, unpublished SHRIMP U-Pb zircon data; Lana 2004). Major granitoid formation and high-grade (amphibolite or – in the innermost zone of the core – granulite facies) metamorphism occurred then between 3.2 and 3.07 Ga (Hart et al. 1999; Moser et al. 2001; Gibson and Reimold 2001b; Lana et al. 2003a-d). The next metamorphic stage in the evolution of this terrane was a thermal metamorphic overprint that has been described by a number of workers (e.g., Phillips and Law 2000, and references therein) from the Witwatersrand conglomerates. It is characterised by a mineral assemblage characteristic of lower greenschist facies metamorphism, with chloritoid as indicator mineral, corresponding to maximum thermal conditions of about 350 °C. Some researchers have speculated whether this event could be related to Ventersdorp magmatism at 2.7 Ga (Barnicoat et al. 1997), Transvaal basin development at 2.15-2.6 Ga (e.g., Robb et al. 1997), or to the Bushveld magmatic event at 2.06 Ga, prior to the impact event (Gibson and Wallmach 1995; Frimmel 1997a,b; Robb et al. 1997). Gibson and Wallmach (1995) and Stevens et al. (1997) discussed whether the pre-impact peak metamorphic grade was reached due to regional overprint related to the emplacement of the massive Bushveld magmas at 2.06 Ga ago, some 50 Ma prior to the impact event. That lower greenschist metamorphism regionally overprinted the impact-related pseudotachylitic breccias has been established through studies of such breccias from the outer collar of the Vredefort Dome and from the gold fields in the outer parts of the Potchefstroom Synclinorium (Foya 2002; Reimold et al. 1999b; see also section 4.3.).

The impact event at 2.02 Ga ago, whereby the rocks now exposed in the central core experienced local shock melting and widespread high to moderate degrees of shock metamorphism (Gibson and Reimold 1999, 2000, 2001b; Gibson 2001; Gibson et al. 2002), represents the next metamorphic phase for this region. Gibson (1996) and Gibson et al. (1997, 1998) established that the pseudotachylitic breccias in the Dome experienced post-impact, high-grade metamorphic overprint that marks a roughly radial temperature decrease from > 1000 °C in the center of the dome to 300 °C at its margin. In the wider region of the Witwatersrand Basin, two thermal metamorphic events can be recognized, being separated by the formation of pseudotachylitic breccia (e.g., Reimold et al. 1999b). However, instead of being represented by amphibolite facies grade, these two phases of metamorphism attained only lower greenschist facies grade. The pre-impact stage is characterised by chloritoid as the peak metamorphic (ca. 350 °C) indicator mineral, whereas the post-impact metamorphic assemblage is characterised by chlorite-dominated parageneses indicative of slightly lower metamorphic temperatures of about 300-330°C (Frimmel 1997a,b; Frimmel and Gartz 1997; Frimmel and Minter 2002; Gibson and Reimold 1999; Foya et al. 2000; Foya 2002). The post-impact metamorphism decreases in intensity outwards from the Vredefort Dome, from lower amphibolite (500-525 °C/0.3 GPa) grade in the inner parts of the collar to lower greenschist grade (300 °C/for an assumed pressure of 0.25 GPa – based on lithostratigraphic thicknesses) in the area of the gold fields.

2.1.4 Relation to Archean Witwatersrand Gold and Uranium Ore Deposits

The Vredefort impact structure encompasses the bulk of the Witwatersrand Basin (Fig. 3). The synclinal structure of the supracrustal sequence from the Witwatersrand to the Transvaal Supergroup in the environs of the Vredefort Dome represents effective downwarping of the stratigraphic units including those of the Witwatersrand Supergroup (McCarthy et al. 1990). This entire succession containing the gold- and uranium-rich conglomerate horizons (known locally as "reefs") of the Witwatersrand Supergroup could have been removed due to erosion since 2 billion years, if it had not been for the Vredefort impact event.

The Witwatersrand Basin represents the world's richest gold province. Some 40-50 % of all the gold ever mined on Earth has been produced from the basin (Robb and Robb 1998), which is roughly an amount of 45 000 to 50 000 tons, at an estimated value of more than 50 billion US$ at a gold price of US$ 310 per ounce. The remaining gold reserves of the Witwatersrand Basin are estimated at 45% of the world's known total reserves. Currently mined strata in the northern part of the Witwatersrand Basin generally dip between 15 and 20 degrees south, towards the Vredefort dome. With many mines already operating at mining depths between 2 500 and 4 000 m, it does seem unlikely, though, that much of the remaining resource will be accessible to mining. However, there are distinct extension possibilities to previously and currently mined areas, such as the so-called Argonaut region adjacent to and to the south/southwest of the Central Rand goldfield (Fig. 5), and the main mining-houses with large-scale operations in the Witwatersrand goldfields do not hesitate to emphasize that they foresee continued gold-mining in this region for many years to come. Although sustainability depends hugely on prevailing gold prices.

Most of the Witwatersrand gold has been mined from the Witwatersrand Supergroup, especially its upper succession, the Central Rand Group. In addition, subordinate amounts of gold have been obtained from Transvaal Supergroup strata, mainly in the region to the northwest of Johannesburg and along the base of the Transvaal Supergroup, the Black Reef Quartzite Formation, to the east of Johannesburg. Minor gold mining activity has also taken place in the 3.07 Ga Dominion Group strata that form the base onto which the Witwatersrand Supergroup was deposited, in an area to the west of the town of Klerksdorp (Fig. 5), along the western margin of the Witwatersrand Basin. In addition to these vast amounts of gold, some 150 000 tons of uranium have been mined, mainly as a by-product of gold mining, but also from several comparatively gold-poor reef horizons. The value of this mined product has been estimated at another 4 billion US$, and estimated remaining uranium reserves are of the order of 475 000 tons.

Mining in the Witwatersrand has been the mainstay of the South African economy for more than a century, and by supporting tens of thousands of migrant workers, has been beneficial for the economies of other southern African countries as well. It is estimated that several million people, directly or indirectly, still owe their livelihood to the Witwatersrand industry today.

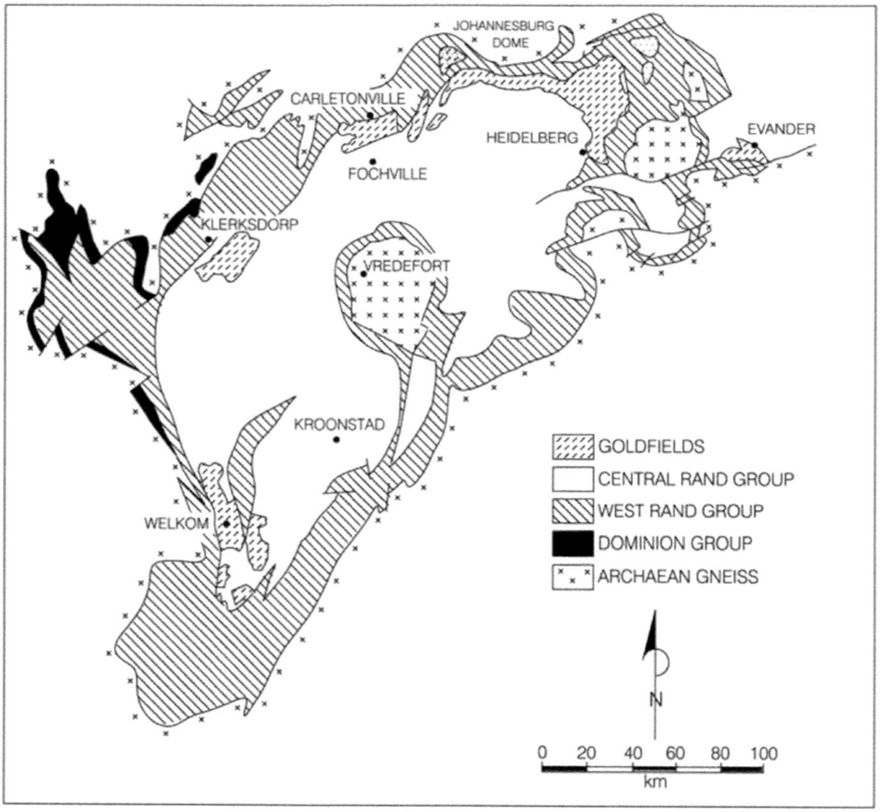

Fig. 5. The Golden Arc, the annular pattern of Witwatersrand goldfields around the Vredefort dome, extending from the Evander Basin (1) and East Rand (2), via the Central Rand (3) and the Carletonville goldfield ((4) - also known as West Wits Line), and the West Rand (5) goldfields to the southwestern Welkom (6) goldfield. The location of the hamlet of Venterskroon, the focus of a minor goldrush into the Vredefort Dome is denoted (7).

The major mining camps of the Witwatersrand Basin are located within a semi-circular pattern at a distance of approximately 30 to 70 km from the outer edge of the Vredefort Dome, known as the Golden Arc (Fig. 5). They are arranged at a distance from the Vredefort Dome that is determined by the depth to the major reefs that can be exploited with current technology.

A long-standing debate has centered on the origin of the Witwatersrand ores. As recently reviewed in great detail by, for example, Robb et al. (1997), Robb and Robb (1998), Phillips and Law (2000), and Frimmel and Minter (2002), one school of thought has favored an origin of the Witwatersrand gold by accumulation as purely detrital ore (e.g., Minter 1999; most recently, Kirk et al. 2002, and references therein; Jolley et al. 2004), whereas others (e.g., Barnicoat et al. 1998; Phillips and Law 2000, and other work comprehensively cited therein) have subscribed to a purely hydrothermalist view. In addition, a large number of Witwatersrand workers, especially those that in recent years carried out detailed petrographic studies on samples from the Witwatersrand reefs, have supported the so-called "modified placer theory" that depends on combined elements of the detrital ore hypothesis as well as the view that post-sedimentary thermal overprint affected the original placer deposits (e.g., Stevens et al., 1997; Robb et al. 1997; Frimmel 1997b; Frimmel and Gartz 1997; Frimmel and Minter 2002; Foya et al. 1999; Gibson and Reimold 1999; Reimold et al. 1999b; Foya 2002; Hayward et al. 2003). The timing of these growths of authigenic ore mineralization is, however, debated, with phases at 2.7, 2.5, 2.3-2.15, 2.06, and 2.02 Ga having been favored by different workers at different times. In addition, a number of workers have established that at post-Vredefort times thermal metamorphism and hydrothermal overprint affected the basin region at around 1.2-1.0 Ga (e.g., Reimold et al. 1995, 1999b; Friese et al. 1995, 2003). Hydrothermal overprint on the Ventersdorp Contact Reef and – to be precise – pseudotachylitic breccia along the reef was related by Reimold et al. (1999b) to autometasomatism of these materials as a direct and immediate consequence of the impact event. Hayward et al. (2003) reported results of a detailed petrographic study of numerous ore samples from all major gold-bearing Witwatersrand conglomerates and goldfields. These authors discussed basin-wide observed mineral deformation (cataclasis) and permeability increase, late in the evolution of the Witwatersrand ores, that was followed by an even later stage of authigenic sulfide-gold mineralization. Hayward et al. (2003) made a case for the most likely deformation event of basin scale having been the Vredefort impact at 2.02 Ga ago.

A number of workers have recognized a low-pressure hydrothermal, lower-greenschist metamorphic overprint in the gold ores throughout the expanse of the Witwatersrand Basin. This aspect of the Vredefort-Witwatersrand system is discussed in the section on epigenetic mineralization related to impact structures. There it is also emphasized that the lower-greenschist metamorphism and related mineralization can be related to the Vredefort event and do not have to be considered as of Ventersdorp (2.7 Ga) or Transvaal (2.6-2.15 Ga) age - as still favoured by many Witwatersrand workers.

In the Vredefort Dome itself, gold mining has taken place at various times in different geological settings. In the latter decades of the 19th century, shortly after the first white settlers had come to the region in the 1840s/1850s, small deposits of gold- and also silver-bearing vein quartz were exploited in the greenstone terrane of the southeastern part of the core (Nel 1927). Gold in strata of the Upper Witwatersrand Supergroup was mined or explored during two periods: In the years prior to the Anglo-Boer War (1899-1902), the Kimberley Reefs – locally known as the Amazon Reef – were mined in the Venterskroon gold field (also known as the Rooderand gold field) in the western collar of the Dome (Fig. 5). Mining was proclaimed in 1887, but, due to unsatisfactory results, operations were discontinued already in 1911 (Borchers 1964). Main mining operations in the early 1900s included the Amazon Mine which was opened in 1910 on farm Rooderand in the Transvaal Republic, and a mine on farm Elandslaagte No. 28 located in the then Orange Free State Republic. Resurgence in mining occurred in the 1930s when the Great Western Mining Company re-opened the mine on Elandslaagte and extended its operations into the Transvaal. In numerous places around the collar of the Dome can one still observe the remnants of gold exploration trenches and adits in the strata of the Kimberley Reefs (Turffontein Subgroup of the Upper Witwatersrand Supergroup, UWS) and also in conglomeratic units of the Johannesburg Subgroup of the UWS. Altogether, these operations never brought a thorough success. The Great Western Mining Company ceased production in 1937. The total amount of gold extracted from Vredefort sites is estimated at no more than some 130 kg – a trivial amount in comparison to the gold production from the entire Witwatersrand Basin.

In conclusion, the Vredefort impact event at 2.02 Ga secured one of the world's largest ore deposits, namely the Witwatersrand gold and uranium province. Based on the ample evidence for sedimentary structures related to gold mineralization, textural evidence that strongly supports a detrital origin of Witwatersrand gold, and, in essence, widespread occurrence of evidence of sedimentary controls on gold mineralization for all auriferous reefs in the basin (e.g., Robb and Robb 1998; Foya 2002; Minter 1999; Minter et al. 1993; Frimmel and Minter 2002), there can be no doubt that much of this deposit was in place prior to the impact event. This event itself caused preservation of these progenetic Archean deposits along the ring syncline of the impact structure, and the ejecta assumed to have covered the gold-bearing strata contributed to their protection from erosion since 2.02 Ga. Thus, it must be concluded that this ore deposit is primarily of a progenetic nature, according to the classification by Grieve and Masaitis (1994). By improving the understanding of the deformation imparted on the strata in the Witwatersrand region by the impact event, beneficial output will be obtained with regard to working effectivity and safety, besides the benefit that the gold-mining industry will reap from improved structural geological understanding of the Witwatersrand Basin.

2.2 Iron Ore at Ternovka (Ukraine)

The Krivoi Rog region of Ukraine is world-famous for its gigantic iron ore deposits. A 10-11 km diameter (or > 8 km, according to Krochuk and Sharpton 2002), perhaps originally 15-18 km large, now deeply eroded impact structure, known as Ternovka or Terny (Nikolskiy 1991; Nikolskiy et al. 1981, 1982; Val'ter 1988), is located in the Proterozoic fold belt of the Krivoi Rog basin, at 49°01'N/33°05'E. Ternovka was formed about 375 ± 25 Ma ago. Both iron and uranium ores have been mined here from several open pits and have also been known for decades from numerous underground workings and boreholes (Masaitis 1989). Mining of uranium ores ceased in 1967 (Grieve and Masaitis 1994). The ores occur in crater floor rocks as well as in impact breccias of this deeply eroded impact crater.

Ore formation was the result of Lower Proterozoic hydrothermal and metasomatic overprint on ferruginous quartzites, among other lithologies, which led to the formation of several mineralized zones that also involved uranium mineralization and that are now exposed in the crater floor. Post-impact hydrothermal activity caused uranium remobilization and the generation of secondary pitchblende veining. In 1994, Grieve and Masaitis estimated the remaining ore reserves for the Pervomaysk open pit mine at 74 million metric tons, but also mentioned additional lower-grade reserves of some 675 million tons. Due to macro-deformation as the result of the impact event, large blocks of iron ore have been displaced and rotated, and mixed in with barren blocks. This geological situation, a direct consequence of the impact event, causes some difficulties during mining operations. A positive result of the impact event is, however, the advantage of working deformed (highly fractured) ore, which proves beneficial

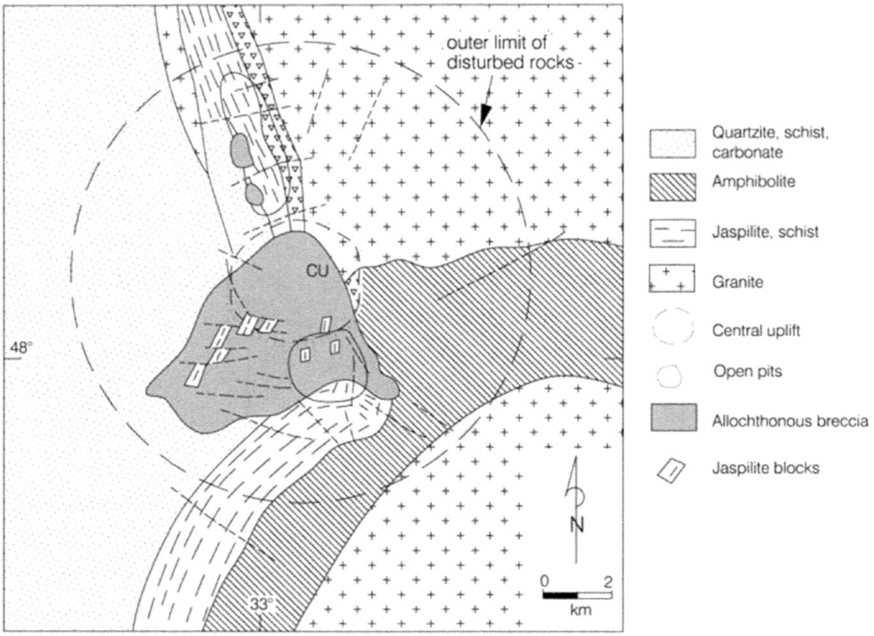

Fig. 6. Geology and structure of the Ternovka (Terny) impact structure, indicating the relationship between uplifted basement blocks and exploited ore deposits. After Grieve and Masaitis (1994).

during the extraction and processing stages (as it requires less mechanical work during the ore comminution stage).

2.3 Uranium at Carswell (Canada)

The outer limit of the Carswell impact structure in Sasketchewan (Canada; 58°27'N/109°30'W) is only vaguely defined, but the diameter of the exposed structure is generally estimated at 39 km, although Grieve and Masaitis (1994) suggested that the original diameter of the Carswell Structure could have been as large as 50-55 km. The structure comprises an annular, about 5 km wide trough around a 20-km-wide core terrane composed of metamorphic basement rocks (Fig. 6). Despite its relatively young age of 115 ± 10 Ma, the structure - similar to Vredefort- is deeply eroded to a level below the entire crater fill. Only dike breccias, as well as some other bodies of impact breccia and impact melt rock, remain in the exposed basement rocks.

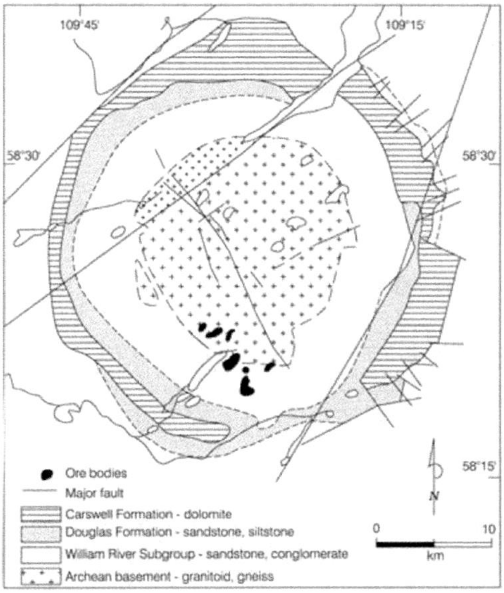

Fig. 7. Geology and structure of the Carswell impact structure. After Grieve and Masaitis (1994).

The general geology of the uplifted basement core and the surrounding younger formations has been reviewed by Grieve and Masaitis (1994) and in various papers in Lainé et al. (1985). The basement core comprises a mixture of Archean gneisses and granitoids that are surrounded by unmetamorphosed sedimentary lithologies including sandstones, siltstones, conglomerates, and carbonates of the Athabasca Group. Radial faulting has truncated and offset, in places, the contact between the basement rocks of the core and overlying Athabasca Group lithologies. Uranium mineralization is concentrated along the southern/southwestern contact between the uplifted core and the Williams River Subgroup of the Athabasca Group, and occurs in both these stratigraphic settings (Fig. 7). Harper (1983) estimated a reserve of 46,500 metric tons of uranium in the deposits known at that time. Ore formation is believed to be the result of regolith development due to lateritic weathering of the basement rocks under tropical climatic conditions, prior to the deposition of the Athabasca Group. The main mineralization apparently formed as a consequence of a hydrothermal event at approximately 1000 Ma ago that produced a uraninite-polymetallic sulfide assemblage. The ore evolution is complex and involves a series of remobilization events (further details are provided in Lainé et al. 1985).

Due to the impact event and associated kilometer-scale uplift of the basement core, the pre-impact ores were uplifted by about 2 km, in the formation of the central uplift structure, and brought into their present position where they can be exploited. Impact effects on these ores include some brecciation, as well as minor post-impact remobilization that produced a coffinite-sulfide paragenesis.

The Carswell uranium deposits are currently exploited. Grieve (2003) estimated that a reserve in excess of 45 000 tons remains.

2.4 Other Progenetic Ore Showings in Impact Structures

Several other pre-impact ore showings are known, as listed by Grieve and Masaitis (1994; their Table 2). Barringer (Meteor) Crater in Arizona has some reasonable grade silica. The 54-km-wide, 360 Ma old Charlevoix structure in Canada carries some ilmenite. And, finally, the 13.5 km, 0.9 Ma young Zhamanshin structure in Kazakhstan is known for some bauxite occurrences.

3 Syngenetic, Directly Impact-related Deposits

3.1 Diamonds

Impact diamonds are the result of shock transformation of graphite or coal that have been shocked to shock pressures in excess of 30 GPa (Masaitis 1993; Koeberl et al. 1997b; Gilmour 1998). Impact diamonds may appear as paramorphs of precursor crystals, often showing the crystallographic habits of these. Cubic and hexagonal impact diamond and lonsdaleite may occur as individual, nanometer to micrometer crystallites, but they can also occur in the form of large aggregates of up to 10 mm in size.

Such occurrences of impact diamond are known from a large number of impact structures (e.g., review by Gilmour 1998). They were probably first discovered when small diamonds, in association with lonsdaleite (a hexagonal high-pressure polymorph of carbon), were observed in the 1960s and 1970s in placer deposits, for example in the Ukraine; but at that time, their origin was unknown (e.g., Sokhor et al. 1973). For many decades enigmatic diamond finds in placer deposits in the Yakutia region in northeast Siberia had been known, but the source area for these so-called "yakutites" remained unknown. Only in the 1970s were diamonds

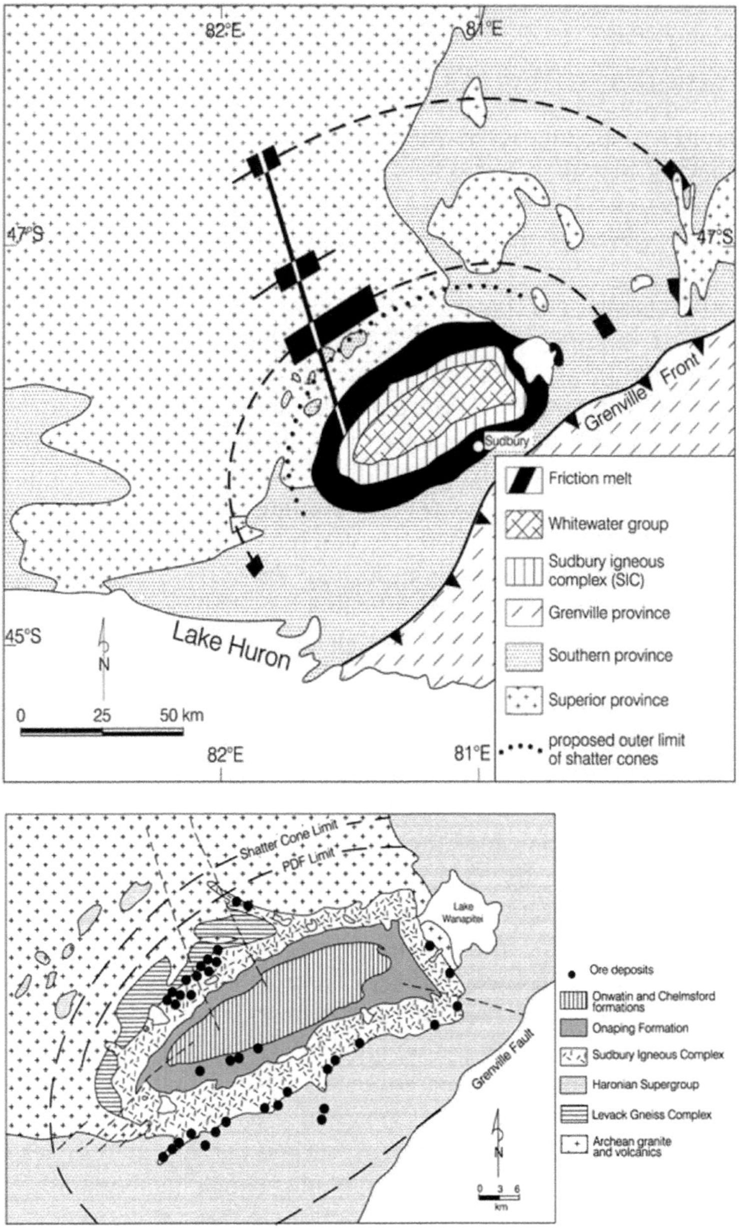

Fig. 8. Schematic geology of the Sudbury impact structure, indicating the occurrence of Sudbury Breccia and outer limit of shatter cone occurrence (after Spray and Thompson 1995). The diagram at the bottom (after a diagram by D.H. Rousell and co-workers) shows the distribution of major ore deposits in and around the Sudbury Structure, as well as the maximum limit of PDF and shatter cone development in the basement rocks to the structure. Lake Wanapitei is the site of a much younger (ca. 35 Ma) and 7.5 km diameter impact crater.

discovered in impactite of the Popigai Structure (e.g., Masaitis and Selivanovskaya 1972; Masaitis 1998), when the origin of the so-called yakutites was traced, through surface exploration, back to the Popigai area. Impact diamonds have since been discovered in a host of other impact structures, including Kara, Puchezh-Katunki, Ries, Ternovka, Zapadnaya, Sudbury, Chicxulub, and others (e.g., Masaitis 1993; Masaitis et al. 1999; Montanari and Koeberl 2000). Diamonds of impact (shock) origin have also been observed in a large number of meteorites (e.g., Rubin and Scott 1996).

Impact diamonds have been documented from a variety of impact breccias (Masaitis et al. 1999, and references therein; Montanari and Koeberl 2000), including impact melt breccia and suevite. At Zapadnaya, an impact structure of 4 km diameter and 115 Ma age in Ukraine, for example, impact diamonds have been observed in dikes of impact melt rock injected into the central uplift as well as in suevite from the trough surrounding the uplifted core. According to Gurov et al. (1985), the Zapadnaya crater was formed in a graphite-bearing Proterozoic granite. At Popigai, a stratigraphic section through the fill of the crater basin around the central uplift comprises allochthonous breccia that is overlain by diamond-bearing suevite and impact melt rock. In the case of this large impact structure, the source of carbon from which diamond was formed is Archean graphite-bearing gneiss. At the Kara impact structure in northern Russia, diamonds also occur in impact melt rock. This impact structure is deeply eroded; values for its diameter that have been cited range from 65 to 120 km; its age is 70.3 ± 2.2 Ma (Trieloff et al. 1998). Kara is located in Permian, coal-bearing sediments (Ezerskii 1982).

Diamonds generally occur in very small amounts in impact melt rocks. Grieve and Masaitis (1994) estimated an average concentration of about 10 ppb. Upon slow cooling of large melt bodies, diamond may become metastable and then revert back to the graphite structure. Masaitis (1993) discussed that the distribution of diamond within a particular impact structure, as well as within individual melt rock and suevite breccia bodies, may be very irregular. This author also stated that diamond occurrences can occur concentrated in rays or zones emanating from areas that had comprised prominent amounts of carbon-bearing lithologies. Near the center of an impact structure, high post-shock temperatures may cause rather rapid oxidation. In contrast, farther from the central area, shock pressures would be insufficient to allow the phase transformation to diamond to occur. Thus, the extent of diamond-bearing zones may have a finite radial extent.

To date, no impact diamonds have been exploited commercially. The vast resources of these microcrystalline diamonds at the Popigai structure are non-economic, due to the complete lack of infrastructure in this remote region of northeastern Siberia that would permit commercial mining of this industrial-grade diamond deposit.

3.2 The Carbonado Conundrum

Carbonados are polycrystalline diamond aggregates of generally irregular shapes that have been observed in placer deposits and low-grade metamorphic rocks of mainly Brazil, Russia, South Africa, Ukraine, Venezuela, and the Central African Republic. Some major occurrences have been exploited commercially for the production of industrial diamond (e.g., Trueb and de Wys 1971; Kaminskiy et al. 1978). Carbonados occur in crustal lithologies and have crustal signatures in terms of stable isotope characteristics, and can not be related to an origin from kimberlites (e.g., Ozima et al. 1991; Haggerty 1999). They do have variable $\delta^{13}C$ values that overlap those of diamond-lonsdaleite aggregates in Ukrainian placers (-15.8 to -20.5 ‰, Kaminskiy et al. 1977; -29.7 to 24.2 ‰, Kamioka et al. 1996; -23 to -30 ‰, De et al. 2001) and the diamond-lonsdaleite bearing aggregates in suevitic breccias from Popigai (-12.3 to -17.6 ‰, Galimov et al. 1980). Kaminskiy et al. (1978) reported the presence of lonsdaleite in some carbonados (an observation that has not been confirmed since) and that carbonados were generally of Precambrian age. Further evidence for a crustal source for the origin of carbonados includes their isotopically light character, noble gas contents interpreted as representing trapped atmospheric composition (Ozima et al. 1991), and rare earth element abundance patterns (Shibata et al. 1993; Kamioka et al. 1996) consistent with this interpretation. Also, the individual occurrences have widely different mineral parageneses and, thus, suggest a variety of different sources. Reviewing such evidence, Smith and Dawson (1985), consequently, suggested that carbonados could have been formed as a consequence of Precambrian impact events into carbon-bearing crustal rocks. All other traces of these impacts and the related impact structures apparently had been eroded, and only the carbonados had survived erosion and were then incorporated into sedimentary rocks.

No direct evidence for a shock origin of carbonados has ever been reported from any of these occurrences, and the geochemical evidence does not agree with this hypothesis either (Koeberl et al. 1997c). Shelkov et al. (1998) compared the ^{4}He signatures in diamond crystals from kimberlites with those in carbonado samples and concluded that both types of samples had very similar signatures. In general, currently favored

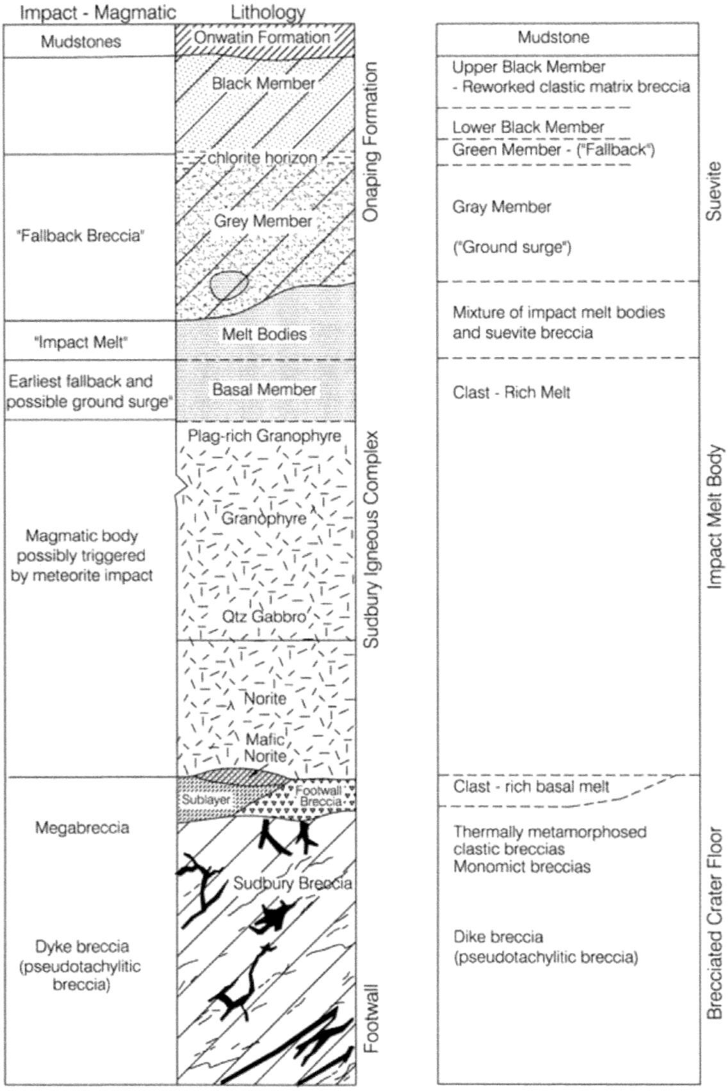

Fig. 9. A generalized stratigraphic column for the Sudbury Igneous Complex (SIC) and corresponding interpretation of the various stratigraphic intervals in terms of their impact generation. After Grieve et al. (1991).

hypotheses for the formation of carbonados include mutually exclusive preferences involving extraterrestrial, crustal, impact and mantle provenances (Kaminskiy 1987; Kagi et al. 1994; Kletetschka et al. 2000; De et al. 1998, 2001; Haggerty 1999).

The carbonado occurrence of the Central African Republic (and of Brazil) has repeatedly been related to the possible existence of a very large impact structure, termed the Bangui structure (Girdler et al. 1992; De et al. 1998) in the region of the Bangui basin in the northern parts of the Democratic Republic of Congo and southern Central African Republic. However, all evidence quoted in support of the existence of such a structure has been derived from regional geophysical data and does not constitute reliable first-order evidence for the existence of such a Bangui impact structure.

3.3 The Ni, Cu, PGE Deposits of the Sudbury Structure (Canada)

3.3.1 Geological Overview

The Sudbury Structure is located in the central part of the Province of Ontario, approximately 400 km north of Toronto in Canada. Its origin has been debated for more than 100 years since its discovery in the late 19th century. World-class Ni-Cu sulphide deposits, also containing noteworthy platinum group metal mineralization, are associated with the Sudbury Igneous Complex (SIC) in the central part of the Sudbury Structure (e.g., papers in Pye et al. 1984; Naldrett 1984a,b; papers in Lesher and Thurston 2002; Naldrett 2003). The SIC is ca. 2.5 km thick and forms an elliptical body of about 27 x 60 km extent (Fig. 8). The entire Sudbury Structure covers a much larger area, of some 15 000 km^2. This includes, besides the SIC, the fractured, brecciated and shocked footwall rocks of the structure, namely the Archean basement to the north and east of the SIC, and the supracrustal rocks of the Proterozoic Huronian Supergroup south of the SIC, and the impact breccias, mudstones and graywackes of the Whitewater Group in the Sudbury Basin, overlying the SIC (e.g., Giblin 1984; Dressler 1984b; Dressler et al. 1991; Rousell 1984a). For obvious economic reasons, the SIC has been the focus of much of the past investigations in the Sudbury Structure. The proceedings volumes of several recent conferences provide comprehensive information on the current state of Sudbury research (Dressler et al. 1994; Dressler and Sharpton 1999; Lesher and Thurston 2002).

The term Sudbury Structure collectively refers to the brecciated country rocks of the Superior and Southern provinces of the Canadian Shield in the environs of the Sudbury Igneous Complex, the SIC with its ore deposits, and the interior Sudbury Basin (Dressler 1984b; Giblin 1984). The Sudbury Structure straddles the present boundary of the Proterozoic

Huronian Supergroup metavolcanic and metasedimentary sequence in the south and the Archean basement rocks in the north. Shatter cones occur all around the SIC, up to a distance of about 17 km from the periphery of the SIC (Grieve and Therriault 2000). Pseudotachylitic breccias, the so-called "Sudbury Breccias", form numerous, often very large, bodies immediately around the SIC. There appear to occur zones of larger and more plentiful bodies at distinct distances around the SIC to the north (Dressler 1984a; Peredery and Morrison 1984; Fig. 8) that have been interpreted as being the likely expressions of the ring structures of a multi-ring Sudbury structure (e.g., Spray and Thompson 1995; Spray 1997). Near the lower contact of the SIC and mainly in the northern part of the structure, Footwall Breccia bodies up to 250 m thick occur, representing part of a strongly brecciated crater floor (Lakomy 1990). In several places, this Footwall Breccia hosts significant Cu-Ni-PGE mineralization.

3.3.2
The Sudbury Igneous Complex (SIC)

The various units of the SIC, from bottom to top, comprise the Sublayer, Norite, Quartz-Gabbro and Granophyre (Fig. 9). Apparently, there are several phases of Sublayer which, in general, consist of a noritic-quartz dioritic rock containing mafic and ultramafic inclusions as well as inclusions derived from the target rocks (Naldrett et al. 1982, 1984). Sulfide mineralization is associated mainly with the mafic and ultramafic inclusion rich Sublayer. The Sublayer is the main ore-bearing unit of the Sudbury Structure. Economic ore bodies also occur in the so-called Offset Dikes (Grant and Bite 1984) that are found around the SIC, both in the Proterozoic and Archean target rocks. According to Lightfoot and Farrow (2002), Offset Dikes host 50% of all the ores at Sudbury. The SIC and its associated Offset Dikes have been interpreted as a differentiated impact melt sheet (e.g., Grieve et al. 1991; Deutsch et al. 1995; Fig. 9) or as a combination of Granophyre impact melt rock with an underlying intrusive sequence of norite–quartz gabbro that was formed as a consequence of the impact event, possibly derived from lower crustal to upper mantle target rocks (Chai and Eckstrand 1994; Dressler and Sharpton 1999, and references therein). Presently, the "differentiated impact melt sheet" model is accepted by many, but not by all, Sudbury researchers.

The Whitewater Group, from bottom to top, consists of the impact melt breccias and suevitic breccias of the Onaping Formation, the mudstones of the Onwatin Formation, and the wackes of the Chelmsford Formation. Near the bottom of the Onwatin Formation is the stratigraphic position of the Vermilion Member. It consists of a distinct carbonate-chert unit with interbedded argillitic limestone and dolostone. Disseminated and massive pyrite, sphalerite, galena, markasite, and pyrrhotite occur in these rocks and have been the subject of mining efforts, without much success, before and after World War 2 (Rousell 1984b). Figure 9 illustrates the current interpretation of the various lithologies in terms of an impact model (after Grieve et al. 1991), whereby crater floor lithologies, impact melt body, and suevitic fall-back breccia are distinguished.

An impact origin for Sudbury was first proposed by Dietz (1962), who predicted that shock metamorphic evidence would be found at Sudbury, and from shatter cones recognized soon thereafter (Dietz 1964; Guy-Bray et al. 1966). Later, planar deformation features (PDFs) were reported from quartz in clasts of the Onaping Formation by French (1967; cf. also Joreau et al. 1996), and by Dence (1972) and Dressler (1984a) in the footwall rocks of the northern circumference of the SIC, the so-called North Range. Recently, PDFs have been found in a few inclusions in the SIC (Grieve and Therriault 2000).

Seismic reflection investigations (Milkereit et al. 1992, 1994; Lithoprobe reports in a dedicated issue of Geophysical Research Letters, volume 21, 1994) have shown that the Sudbury Structure is asymmetric at depth: the North Range dips at about 30° towards the south, and in the South Range a series of southward-dipping reflectors indicate thrust faulting. The Sudbury region was affected by the Penokean Orogeny, which was pene-contemporaneous with the formation of the structure. Some 800 million years after the Sudbury event at 1850 Ma ago (Krogh et al. 1984), the Grenville Thrust Front approached the structure from the southeast. Thompson et al. (1998) reported ^{40}Ar-^{39}Ar ages from pseudotachylitic breccias in the North Range that have been reset to 1 Ga; the Grenville structural event may, thus, have involved the entire North Range as well.

Although the geological community has generally agreed that a catastrophic "event" of sorts had taken place at Sudbury, several main arguments - including the elliptical shape of the SIC, the long-term focus on the central part of the structure (the SIC) only, and an origin of the SIC as a plutonic body thought to have been contaminated with crustal rocks - were widely accepted (e.g., Naldrett 1984a,b). Although impact was recognized long ago as an integral part of the formation of the structure, the vast volume of the SIC and its strong differentiation (cf below) posed serious problems for a simple impact model. Even strong proponents of an impact origin for Sudbury for long times favored some combined impact and magmatic process (e.g., French 1970; Dressler et al. 1987). It has only been during the last 15 years of the previous century that the impact origin of the Sudbury Structure has been generally accepted. The cause of the elliptical shape of the structure has been strongly debated as being related to either oblique impact, thrusting or faulted slices, or a combination of both, during either the Penokean or Grenville orogenies, or both (e.g., Wu et al. 1995; Fueten and Redmond 1997; Cowan and Schwerdtner 1994; Cowan 1999; and various papers in Dressler and Sharpton 1999).

The bulk SIC, as well as chill margins of the Offset Dikes and some glass fragments in the breccias of the Onaping Formation, are of granodioritic composition (Lightfoot et al. 1997a; Tuchscherer and Spray 2002; Ames et al. 2002). Isotopic and geochemical data are consistent with the derivation of the SIC from pre-existing crustal material. A significant crustal component (at least 50%) is indicated in the Main Mass Sm-Nd (Faggart et al. 1985; Dickin et al. 1992) and Re-Os (Walker et al. 1991) isotopic data, and by the results of trace-element geochemical studies (Lightfoot et al. 1997a). Studies of the ores and their host rocks have both consistently shown crustal signatures and have - equally significantly - failed to show any unambiguous evidence for a juvenile mantle component. Geochemical (Lightfoot et al. 1997b; Prevec et al. 2000) and combined Nd and Pb isotopic (Dickin et al. 1996) work on the Main Mass, Sublayer, and Offset Dikes indicates that initially similar rocks were affected by localized contamination from footwall (e.g., North versus South Range compositions). Prevec et al. (2000) further showed that the geochemical and Sm-Nd isotopic characteristics of the Sublayer and the Offset Dikes could be accounted for by a range of mixtures of Huronian basaltic rocks (comprising South Range footwall) and Archean gneissic rocks (of the North Range footwall).

Isotopic studies of the ores themselves also depict a mixture of crustal sources. Lead (Dickin et al. 1992) and Re-Os (Walker et al. 1991) isotope data are compatible with a mixture of evolved (crustal) and depleted (possible mantle or mantle-derived) materials. Pt-Os and Re-Os isotopic analysis on ore samples from a number of Sudbury mines by Morgan et al. (2002) showed that these ores could be obtained from binary mixtures of Superior Province and Huronian metasedimentary rocks, with some admixture of Archean to Proterozoic mafic rocks. Grieve et al. (1991) calculated that the bulk composition of the SIC was consistent with reasonable mixtures of the available target rock components and discussed the stratigraphic position and volume in terms of an impact melt model. The Offset Dikes that occur in the environs of the SIC at distances of up to 50 km are also regarded as impact melt from the Sudbury impact and have been shown (Ostermann et al. 1996; Wood and Spray 1998) to have an age equivalent to that of the SIC.

Overlapping ages have been obtained for several parts of the Sudbury Structure: The SIC has been dated at 1849.9 ± 3.6 Ma and the Offset Dikes at 1852 ± 4 Ma (Krogh et al. 1984; cf. also Ostermann et al. 1996). Similar ages were also obtained for the Sublayer (Krogh et al. 1984; Corfu and Lightfoot 1996) and inclusions in the Sublayer (Corfu and Lightfoot 1996), and also for some Sudbury Breccia (Thompson et al. 1998). Several workers proposed that the Offset Dikes were emplaced up to 300 000 years after the emplacement of the Main Mass of the SIC (Ostermann et al. 1996; Wood and Spray 1998), which is controversial in the light of (a) some field evidence suggesting that the Sublayer emplacement postdates that of the Main Mass and Offset Dikes, and (b) geochemical evidence closely relating the Main Mass and the Offsets (Lightfoot et al. 1997b; Tuchscherer and Spray 2002).

Whether all components of the SIC (i.e., the Main Mass, the Granophyre - disproportionately large from a magmatic perspective, and the Sublayer that contains ultramafic rocks and sulphide ores both traditionally associated with the mantle) represent impact melt has also remained a matter of debate (e.g., Dressler et al. 1996; Rousell et al. 1997). Although Chai and Eckstrand (1994) suggested distinct origins for the Main Mass and the Granophyre, a differentiated magma with a silicious parent, derived as suggested above, is generally favored (e.g., Therriault et al. 1996; Dickin and Crockett 1997). Similarly, derivation of the mafic to ultramafic inclusions in the Sublayer from SIC melts, rather than from a mantle-related source, has been successfully modeled (Prevec 2000).

Volume of impact melt is a function of size of an impact structure (Cintala and Grieve 1994, 1998). The size of the Sudbury Structure has been estimated on the basis of empirical relations, involving other impact structures, from the observed radial extent of shock metamorphic effects, such as shatter cones, planar deformation features, breccia distribution, as well as morphological and structural characteristics of impact structures. According to Grieve et al. (1991) and Deutsch et al. (1995), this translates to a size estimate in the range of 200-280 km for the original extent of the impact structure. Such a range is also consistent with estimates derived from the spatial distribution of Sudbury Breccia occurences (Spray and Thompson 1995). The results of a regional reflection seismic investigation (Milkereit et al. 1992, 1994) could be interpreted to indicate a transient cavity diameter of 100-200 km (Naldrett 2003), the upper limit of which would translate to an even larger (perhaps as much as 400 km) size of the original crater structure.

In contrast to the Vredefort Structure (see above), which is in its outer parts characterized by a series of concentric anticlinal and synclinal structures (Fig. 3), Sudbury does not reveal such alternating structural features, with the possible exception of a partial annulus of down-dropped Huronian cover rocks in the North Range. Grieve and Therriault (2000) proposed that this feature could be equivalent to one of the structural ring features around Vredefort. Results from a remote sensing investigation by Butler (1994) that suggested a series of ring features around Sudbury were challenged by Lowman (1999). A number of workers (Dressler 1984a; Müller-Mohr 1992; Spray and Thompson 1995; Spray 1997) have discussed apparent zones of relatively more intense development of Sudbury Breccia in the environs of the SIC. Spray (1997) equated some of these zones with discrete but very prominent zones of slip in basement rocks assumed to take place during the modification stage of the impact event ("superfaults"). Wood and Spray (1998) suggested that some of these zones were the locations for later emplacement of Offset Dikes.

As discussed by Spray (1997) and Grieve and Therriault (2000), it is possible that different zones of breccia development were formed at different stages in the development of the impact process, both at the time of shock compression and later modification – leading to the formation of so-called "A- and B-type pseudotachylite" (see also Lambert 1981 and Martini 1991) – corresponding in Spray's (1998) classification to so-called S- and E-type pseudotachylites, respectively. The validity of these classification schemes is discussed in detail in another paper in this volume (Reimold and Gibson, this volume).

According to Thompson et al. (1998), a thermal overprint related to the Penokean and Grenvillian orogenies and associated overthrusting can be observed in rocks in the environs of the SIC up to 50 km to the north of the North Range. Hydrothermal alteration and associated mineralization are also widespread in the Sudbury Structure and has been associated with the impact event (see section 4.1). Thermal metamorphism in the direct surrounds of the SIC is represented by a thermal aureole, within the confines of which quartz and feldspar are recrystallized and most PDFs have been annealed. In the North Range this zone of recrystallization extends for ca 500 m from the SIC, the thermal aureole to beyond 1 km. The metamorphic isograds evident in the contact aureole are also consistent with thermal modeling of a convecting, cooling impact melt sheet (Prevec and Cawthorn 2002).

Dressler (1984a) showed that at a distance of about 500 m from the SIC planar deformation features are recognizable again and can be observed to a distance of 8 km from the melt body (Dence 1972). The thermal aureole around the SIC is also observed in the South Range (Dressler 1984a). Fueten and Redmond (1997) and Molnár et al. (1999) have discussed metamorphic effects related to the Penokean Orogeny in this southern area. As much of the ore mineralization in the Sudbury Structure is located in the lower formations, it is mandatory to understand the thermal and deformation history of this important zone (see also Boast and Spray 2003).

3.3.3 Sudbury Ore Mineralization

The Cu-Ni and platinum group element ore bodies of the Sudbury Structure occur in the Sublayer, the Footwall Breccias, footwall rocks beneath the SIC, and the Offset Dikes. They are of extraordinary value. The total ore reserve is estimated at 1.65 x 10^9 metric tons at 1.2 % Ni and 1.05 % Cu. The cumulative value of ore extracted from the Sudbury Structure in more than 100 years of production has been estimated at over US$ 100 billion. The current production is valued at between US$1.5 and 2 billion per annum (Grieve and Masaitis 1994; Grieve and Therriault 2000). By far the majority of ore deposits in the Sudbury Structure occur at the base of the Sudbury Igneous Complex. Many of the ore bodies lie in what are locally known as "embayments", which have been interpreted (see Morrison 1984; Grieve and Masaitis 1994) as terraces in the original impact crater floor. Naldrett (1984b) distinguished five types of ore associations: (1) in the South Range where ores mostly occur as massive sulfide deposits at the base of the Sublayer; (2) in the North Range where ores occur as accumulations of sulfides (up to 60 %) in the Footwall Breccia, as well as in the form of dikes and stringers in the Footwall rocks; (3) the Offset Dikes where sulfide ores are mainly concentrated in breaks and constrictions in the dikes; (4) in fault settings where ores occur as remobilized masses; and (5) other associations.

Ore characteristics vary widely. However, all major ore occurences have in common that they occur within a broad zone at the base of the SIC and in the Offset Dikes. Naturally, this circumstance was previously interpreted by supporters of a magmatic ore genesis to result from segregation of sulfides from a silicate matrix as an immiscible liquid, triggered by assimilation of silicious crustal rock by a basaltic magma, gravitational settling out, and subsequent fractional crystallization and remobilization (e.g., Naldrett et al. 1982; Morrison et al. 1994). The main difference of the impact model in comparison to the magmatic model is that in the impact model the original source of metals is considered to have been entirely of crustal origin. In this scheme, sulfide immiscibility would be the result of melting together of diverse target rocks to form a massive volume of impact melt that from the onset was enriched in SiO_2 (ca. 64 wt%) in comparison to endogenic magmatic compositions (e.g., those of the Stillwater and Bushveld complexes).

The source of metals in the target lithologies has not been resolved completely. It is widely believed that mafic and ultramafic lithologies of the Proterozoic Nipissing Magmatic Belt and/or the East Bull Lake intrusive suite constituted a significant proportion of the basement in the impact area. The Nipissing Belt includes a volumetrically significant suite of leucogabbroic plutons and the gabbroic Nipissing rocks in the Sudbury area, all of which are locally PGE-Cu-Ni enriched, as well as Cu-Ni bearing tholeiitic basalts (Innes and Colvine 1984). The East Bull Lake intrusive suite comprises a discontinuous belt of bimodal volcanic rocks, felsic plutons, and the regionally extensive Heart and Matachewan dike swarms that are known to contain contact-type PGE-Cu-Ni mineralization (James et al. 2002). In addition, the granitoid basement rocks of the North Range also display ubiquitous showings of sulfide mineralization. The sulphide minerals associated with the Sudbury Structure, therefore, are possibly derived from pre-impact mineralizations that were incorporated in the impact melt and settled out in a process similar to segregation of immiscible sulfides in endogenic magmatic bodies.

It must be noted that the location of the ore bodies at the base of the Sudbury Igneous Complex corresponds to the base of the excavation flow (which was later flooded by impact melt from relatively near to the impact center). In this location along the final ejecta flow path, mixing of projectile matter into impact melt does constitute a possibility (H. Henkel, personal communication, 2003).

3.4 Pechenga – Another Sudbury-type Base Metal Deposit in Northern Scandinavia?

Recently, Jones et al. (2003) discussed the possibility that an alleged 80 km wide impact structure near the northern border between Norway and Russia – refered to as the Pechenga Structure – could represent another Sudbury-type, impact structure-hosted, layered intrusion of economic importance with regard to Ni and Cu sulphide mineralization. In this context it must be emphasized that the presence of a Pechenga impact structure has not been proven to date. Evidence quoted in favor of impact at Pechenga by Mutonen (2000) and Jones et al. (2003, their figure 2), allegedly representing impact diagnostic planar deformation features, can not be confirmed as such and, thus, do not represent acceptable shock metamorphic deformation. Thus, any reference to a Pechenga impact is not permissible, and further work is required to investigate a possible yet to date unsubstantiated link of this geological structure and its sulphide mineralization to impact processes.

3.5 Morokweng (South Africa)

The 70-80 km diameter (Reimold et al. 2002a) Morokweng impact structure in South Africa has been suspected for some years to host some significant Ni-PGE mineralization. The structure was initially explored because of distinct geophysical anomalies (e.g., Corner et al. 1997) occurring in its area. A thick (up to > 800 m in one borehole) body of impact melt rock occurs in the central part of this structure. Several exploration boreholes were sunk and investigated, with some workers hoping to find a melt body analogous to the Sudbury Igneous Complex (Andreoli et al. 1999; Reimold et al. 1999c; McDonald et al. 2000; Reimold et al. 2002a; Maier et al. 2003). The main aeromagnetic anomaly was shown to be caused by the thick body of impact melt rock. No progenetic base metal, or other, mineralization has been encountered. Only insignificant amounts of Fe sulfide and magnetite have been identified in drillcore. However, the impact melt rock is indeed enriched in Ni and PGE, but due to a significant admixture of up to 5% of the meteoritic projectile to the melt rock (Koeberl et al. 1997a; Hart et al. 2002; Maier et al. 2003). Projectile matter occurs at highly variable abundances throughout this body.

4 Hydrothermal Processes in Impact Structures

Impacts on land will involve volatiles and water of the lithosphere, and those in oceanic environments will directly affect water bodies as well as rock-bound volatiles. Even impact into arid environments will potentially result in release of volatiles through shock dissociation of hydrous minerals of the target. Interaction of aqueous solutions and other volatile components with hot, shocked rock volumes will result in hydrothermal activity, leading to effective alteration processes and potentially large-scale hydrothermal overprint on deformed and shock metamorphosed rocks of the crater floor and impact breccias, as well as deposition of secondary minerals from hydrothermal solutions. In favorable circumstances, these hydrothermal deposits could be of economic value. The size and duration of such a hydrothermal system will critically depend on the magnitude of the impact event, i.e., the amount of deposited kinetic energy, the rock volume affected, and the amount of volatiles released/set into circulation. Possible heat sources include shock heating, frictional melting/heating in the crater floor, the emplacement of a sizable impact melt sheet in the crater structure, and uplift of hot rocks from deeper crustal levels into the central uplift structure (see Fig. 10). The duration of the active hydrothermal process will be governed by the amount of heat available to drive a convective process. Fluid circulation is facilitated by the presence of large volumes of impact melt – as heat source - and shocked and fractured/brecciated target rocks, the unconsolidated nature of the crater fill, and enhanced permeability due to impact-induced fracturing and brecciation of target rocks.

Low-temperature hydrothermal processes may even provide for the

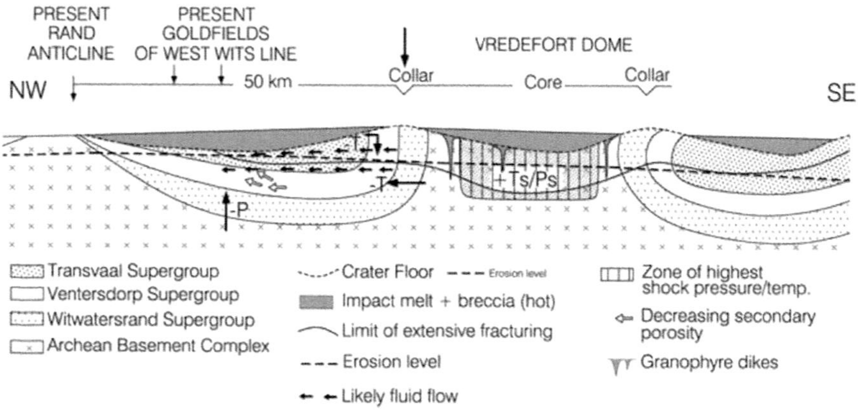

Fig. 10. Model for fluid flow in the Witwatersrand Basin generated by the Vredefort impact event. A schematic complex impact structure is shown, with a blanket of hot impact melt. The central uplift region comprises hot (high temperature = +T) mid-crustal rocks that are subjected to high shock pressures (+Ps). Temperature along a profile away from this hot central uplift decreases, and it also decreases away (downward) from an upper impact melt body (i.e., with depth). Outside of the central uplift, lithostatic pressure increases with depth (or, as indicated, P decreases upwards). In addition, one must assume that secondary (i.e., impact generated) porosity decreases away from the central part of the impact structure. The assumed combination of high T along the impact melt cover and in the volume of the central uplift, as well as decreasing porosity away from the center of the impact structure and decreasing lithostatic pressure with decreasing depth, lead necessarily to the conclusion that fluid flow will be largely laterally away from the central uplift, and enhanced at relatively shallow depths. Also shown is a schematic present depth of erosion, as it would apply to the Vredefort situation. This implies that the currently mined Witwatersrand strata just below this erosion depth would have likely been in the presumed flow channel.

generation of biologically active environments, perhaps creating unique niches for new development of life (e.g., Farmer 2000; Kring 2000; Cockell et al. 2002). Newsom et al. (1986) discussed the chemical effects that impact-hydrothermal alteration could have on regolith and soil development – obviously an issue of major interest in these days of preparation for surface sampling on Mars.

Naumov (2002) compiled a vast amount of mineralogical information from the impact crater literature and presented a detailed synthesis of investigations into low- to intermediate (ca. 50-350 °C) temperature hydrothermal processes as observed in the various lithologies in the Kara, Popigai, and Puchezh-Katunki impact structures. This work demonstrates the impact-triggered generation of hydrothermal cells, in the course of which extensive rock alteration, involving element leaching and redeposition, can take place. These three large impact structures, of 65, 80 and 100 km diameter, respectively, are all characterised by extensive alteration and secondary mineral formation. In his recent review, Naumov (2002) concludes:

- The kinetic energy of an impact event and the pre-impact target characteristics are the most important parameters for the development of a post-impact hydrothermal system.
- As a result of impact, a near-surface high-gradient zone of hot and high permeability rocks is generated. High pressure and temperature, as well as permeability gradients, in impact structures may lead to the formation of hot-water circulation that can last, in some cases, for thousands of years.
- The most extensive hydrothermal alteration has been observed in impact craters formed in shelf or intra-cratonic shallow basin environments.
- Mineral assemblages observed indicate post-impact hydrothermal alteration at 50-350 °C, pH of 6-8 (due to uptake of alkali elements and Ca from strongly deformed rocks), and E_h values of > -0.5.
- The composition of hydrothermal mineralization is determined primarily by the respective target rock composition found at a given structure (the more varied the spatially observed target composition is, the more variable may be the secondary mineral paragenesis), besides the physical properties of rocks after their deformation under shock-metamorphic conditions.
- Post-impact mineralization can be present in all parts of an impact structure and in all types of impact lithologies. But the crater fill (suevites, impact melt rock) is likely to be more affected than parautochthonous impactites of the central uplift and the crater rim.

- In accordance to the findings of fluid inclusion studies by Komor et al. (1987, 1988) on the Siljan structure, Naumov (2002) concludes that crystallization temperatures for secondary mineral parageneses decrease upward in an impact structure and outwards from the center of the structure (compare Fig. 10).
- Meteoric, groundwater, and surface water can be sources for the hot solutions in post-impact hydrothermal cells. The shocked target rocks are the sources for the adsorbed mineral constituents.
- A three-stage development of hydrothermal activity is envisaged: Stage 1, when isotherms are still in pre-impact configuration; Stage 2, after inversion of the thermal field (30 °C gradients in central part to 100 °C at the periphery), and Stage 3, with gradients of <10-30° C.
- Hot-water circulation affects only the upper parts of an impact structure (compare Fig. 10).

The proposed scheme involves that superficial aqueous fluids infiltrate hot rocks of the central uplift as well as impactites of the crater fill. Alkalinity rises due to uptake of K, Na and Ca. Silica is freely available. This represents a very favorable situation for the formation-deposition of Fe-smectites and zeolites, phases typically identified in impact crater settings (e.g., Stöffler et al. 1977). The alkaline components are largely deposited in the upper part of a crater fill. Upon ascent of a fluid, its temperature decreases, OH^- is taken up, and, especially in cases where ample carbonate is present, CO_2 contents of the fluids may increase. Consequently, solutions may become more acidic.

Clearly, the strong deformation, especially fracturing and brecciation, that affects huge rock volumes in large impact events, and the instantaneous increase in temperature over huge rock volumes provide ideal conditions for the initiation of hydrothermal systems.

So far, it does not seem possible to define parameters that could distinguish unambiguously between the results of an impact-triggered hydrothermal mineralization event and one that is the result of other geological processes such as volcanism or metamorphism. The geological context – impact or endogenic setting – will have to provide the vital clues to allow the determination of a cause of formation of a specific hydrothermal ore deposit.

4.1 Case Studies

4.1.1 Ries Crater (Germany)

So far, only a limited number of detailed studies of the products of impact-related hydrothermal alteration have been conducted. One of the first such investigations involved the mineralogical analysis of the complete impactite sequence of the interior of the Ries Crater, Germany, provided by the extensive analysis of the Nördlingen 1973 drill core (Stöffler et al. 1977). These authors concluded that it was possible that the crystallization of authigenic minerals observed in crater suevite could have commenced at temperatures as high as 400-500 °C, but that a zeolite paragenesis observed was indicative of temperatures well below 200 °C. They also found that authigenic mineralization in the upper suevite was similar to that observed in post-impact sediments (crater-lake deposits). Solutions would have been strongly alkalic, and rich in silica, alkali cations, and Ca.

Newsom and co-workers (Newsom 1980; Newsom et al. 1986, 1996, 1999) made extensive studies of the hydrothermal mineralization in Ries crater impactites, in particular the fall-out suevite deposits in the environs of the Ries. They related their findings to the potential of volatiles degassing from hot impact ejecta for the modification of existing soils and formation of soil constituents in extraterrestrial impact settings, especially on Mars where the presence of volatiles in crustal materials would provide a prerequisite for such processes.

Hagerty and Newsom (2003) investigated hydrothermal alteration associated with the basaltic target rock to the Lonar impact crater in India. They concluded that hydrothermal activity led to clay mineral formation at 130-200 °C temperatures. These authors also proposed that a ca. 2 km crater size could represent a minimum size for structures in which impact-induced hydrothermal activity could be noted.

4.1.2 Manson Impact Structure (USA)

A detailed mineralogical study of the products of hydrothermal alteration of various types of impactites, including fragmental and suevitic breccias, and of authigenic mineralization in cavities, fractures and clasts within impactites, from Manson impact crater (Iowa) was reported by McCarville and Crossey (1996). Further information was provided by Boer et al.

(1996). The mineralogical compositions of secondary parageneses directly related to the impact event were not found to be distinct from those of other hydrothermal provinces (see above, Naumov 2002). The results of McCarville and Crossey (1996) indicated that authigenic crystallization resulted from solutions and that the Manson hydrothermal system reached, at least, temperatures of 300 °C. They recorded the effects of cooling to approximately ambient temperature in the paragenetic sequence observed. A further important conclusion of this investigation was that, based on the observed lack of high-temperature minerals in the discontinuous impact melt sheet of this structure and the abundance of such minerals in rocks of deeper levels of the central uplift, heat from the central uplift and not from an overlying melt sheet powered the hydrothermal system (McCarville and Crossey 1996). This conclusion was further supported by the findings of Boer et al. (1996), who presented results of a first detailed fluid inclusion study of Manson impact crater samples.

4.1.3 Houghton Dome (Canada)

A detailed study of hydrothermal mineralization in the Houghton Dome impact structure (24 km diameter, 23 Ma age) was reported by Osinski et al. (2001). These authors found hydrothermal mineralization in localised pipes in impact-generated concentric fault systems and in cavity and fracture fillings within polymict impact breccias overlying the central part of the structure. Cavity and fracture fills involve sulfide-carbonate, sulfate, and carbonate parageneses. The pipes are subvertical, 1-7 m wide, and exposed at Houghton for lengths of up to 20 meters. The pipe fillings consist of quartz-carbonate breccia with Fe-hydroxide alteration, and associated marcasite, pyrite, and chalcopyrite. Osinski et al. (2001) proposed that the hydrothermal system comprised three stages: (1) an early stage characterised by processes involving a dominant vapor phase, resulting in production of quartz and taking place at temperatures between >600 and 200 °C; (2) a so-called main stage at 200-100 °C temperatures and involving a two-phase regime of vapor plus liquid leading to precipitation of calcite, celestite, barite, marcasite and fluorite; and (3) a late stage at temperatures below 100 °C, dominated by precipitation from liquid and producing selenite and fibroferrite deposits. It was estimated that the duration of cooling to temperatures below 50 °C lasted in the order of several tens of thousands of years – for a case of an impact structure of the size of Houghton Dome. Osinski et al. (2001) modeled that the Houghton crater during this time contained a 14 km wide crater lake and subsurface water system that provided a warm and wet environment relative to the environs of the crater structure. It is interesting to note that these authors, in contrast to others (e.g,. McCarville and Crossey 1996), believe that a heat source was not only provided by hot rocks of the central uplift, but by hot impact breccia fill of the crater interior as well.

4.1.4
Siljan Impact Structure (Sweden)

A prime example of an impact-related hydrothermal deposit with inherent economic importance are the large Pb-Zn deposits located in the 377 Ma old (Reimold et al. 2004), ca. 65-km-diameter (Kenkmann and von Dalwigk 2000; others favour ca. 75 km, Henkel and Aaro (this volume)). Siljan impact structure in Central Sweden. A major deposit, estimated at 0.3 million metric tons of ore at grades of 3 % Pb, 1.5 % Zn, and 70 ppm Ag, is located in Ordovician and Silurian rocks of the rim syncline around the central uplift of this complex impact structure and has been exploited near Boda town. According to Johansson (1984), this deposit is the largest one of an entire suite of similar Pb-Zn deposits in the Siljan structure (another notable deposit occurs at Söderön in the southwestern sector of the Siljan rim syncline). Mineralization occurs in the form of veins, as well as fracture and breccia fillings, in Ordovician limestone. It is thought that these traps for ore fluids formed as a consequence of tectonic disturbances resulting from the Siljan impact event. The mineralization at Boda, with regard to geological setting, mineralogy, and isotope geochemistry, seems to compare well with the general characteristics of Mississippi Valley-type deposits of these elements. A similar deposit (though of much smaller extent) is also present at the Kentland impact structure, Indiana, USA. The mineralization at Siljan comprises carbonate-hosted calcite-fluorite-barite-sphalerite-galena-pyrite-marcasite ores that are enriched in radiogenic Pb and heavy sulfide sulfur. Johansson (1984) presented Pb isotopic evidence that suggested that the source of the ore mineralization was the Proterozoic basement, either directly or indirectly, as well as redeposited Proterozoic lead in the Ordovocian limestone, where it was presumably located in the clastic silicate fraction. Textural evidence available suggests that rapid precipitation occurred – repeatedly - from supersaturated ore fluids. The Siljan impact event has been suggested as the cause of ground preparation with regard to leaching (production of hydrothermal fluids) and precipitation (providing suitable open space) of base metals. The Pb isotopes indicate an age for the mineralization similar to that of the impact itself.

4.2 Other Epigenetic Ore Mineralization in Impact Structures

Other epigenetic Pb-Zn (or Zn only) mineralization in impact structures has been reported (Grieve and Masaitis 1994) from the Crooked Creek and Decaturville structures of 7 and 6 kilometers diameter, respectively, and from Serpent Mound (8 km diameter) in the USA, as well as from Kara in Russia (65 km diameter). Small amounts of pyrite mineralization are known from the Beyenchime-Salaatin (8 km diameter) and the Kara and Ust-Kara (65 and 25 km, 73 Ma) structures in Russia. Mercury has been detected in the breccias of the Russian Puchezh-Katunki and Karla structures (80 km and 12 km diameters, respectively), and some agate has been described from impact breccias in Ilyenits (4.5 km) in the Ukraine (Gurov et al. 1998), from Sääksjärvi (6 km, ca. 560 Ma) in Finland (Kinnunen and Lindqvist 1998), and in suevite from Bosumtwi (10.5 km diameter, ca. 1 Ma) in Ghana (these authors, unpublished information).

Interesting mineral deposits have also formed from impact crater-lake deposits (Masaitis 1989; Grieve and Masaitis 1994). Crater-lake deposits from the Ries crater (Germany), Brent and Lake St. Martin (Canada), Lonar (India), and several other small structures contain small deposits of gypsum, anhydrite, and other evaporites related to temporary drying out or evaporation of the lakes. Since 1901, gypsum and anhydrite have been quarried in the Lake St. Martin structure (40 km diameter, 220 Ma). In the case of the small Lonar crater (1.8 km, 0.05 Ma, entirely formed in basalt), the evaporites are exploited for the production of trona (Nayak 1985). This author proposed that the high salinity of the lake water could be related to water having been heated by the impact event and subsequent leaching of alkali elements from the target rock. Masaitis (1989) prefers an evaporation-dominated process and believes that the small size of Lonar crater would not support the presence of a sufficiently large hydrothermal system that could account for the production of the total reserve of evaporites from the leaching process.

Another trona deposit that has been commercially worked in the past exists in the Tswaing Crater (formerly known as the Pretoria Saltpan) in South Africa (Reimold et al. 1999a). At this small (1.13 km diameter) and young (0.2 Ma) impact crater, trona deposits were worked between the late 19th century and mid-20th century. Mining was first started to provide a product needed in the gold mines of the Witwatersrand basin to neutralize acidic mine waters. In later decades, several companies attempted to produce a pure bicarbonate product for human consumption, but failed, which eventually led to the demise of mining at Tswaing around the mid-1950s. The alkali element concentrations are clearly derived, in this case, from meteoric water circulation and leaching of elements from the granites of the basement below the crater floor. Evaporation is responsible for concentration of lake waters and periodic deposition of salts on the lake floor. The impact event did not contribute the fluid volume, but provided the storage reservoir and low-density crater fill serving as the pathway for fluid release to the surface.

Deposits of diatomite occur at Ragozinka (9 km, 55 Ma, Russia) and phosphorite as well as amber at Logoisk (17 km, 40 Ma, Belarus). Phosphorite has also been reported (Masaitis 1989) from the Boltysh structure (24 km, 88 Ma) in Ukraine. Masaitis (ibid) also mentions the possibility that bentonite may occur in the Ries Crater.

Bentonite is also mined in the southeastern sector of the large Vredefort Structure in South Africa. Extensive deposits are mined along the contact between metasedimentary rocks belonging to the Ecca Formation of the Karoo Supergroup (300-180 Ma) and underlying greenstones of Archean age. Bentonite here occurs in the form of large lenses in shale of the Volksrust Formation. It is clear, in this case, that the bentonite formation cannot be related to post-impact effects, but rather to Mesozoic deposition and alteration of volcanic ash beds deposited in channel structures eroded into the Archean bedrock. The volcanic ash texture with typical glass shards is still preserved in the mined material (G. Henry, Council for Geoscience, Pretoria, personal communication 2001; Gibson and Reimold 2001b). Mining of these deposits was begun as early as the 1950s. Currently, open-cast operations exploit layers at 12 m depth below surface.

Another potentially economically interesting hydrothermal deposit occurs in the 80-km-diameter Puchezh-Katunki impact structure (Deutsch and Ivanov 2003). The post-impact sediments developed in a basin that extended over the central part of the central uplift structure. They not only contain gritstone and clays, but the gritstone is also altered to a product that contains up to 30% zeolite. Whilst this deposit is not yet exploited, it may become economic in the future.

4.3 Epigenetic Mineralization Related to the Sudbury Structure

Ames and Gibson (1995) and Ames et al. (1997, 1998) described extensive, regional hydrothermal alteration that pervasively affected the breccia accumulations in the Sudbury basin of the Sudbury impact structure, in particular the Onaping and lower Onwatin formations. They reported evidence for a regional sub-seafloor hydrothermal system that included vertically stacked, basin-wide, semi-conformable alteration zones, in which the rocks had undergone "from base to top, silicification, albitization, chloritization, calcitification, and complex feldspathification". Massive Zn-Cu-Pb sulfide deposits within the Vermilion Member of the Onwatin Formation are being mined at the Errington and Vermilion mines. Ames et al. (1998) obtained isotopic evidence that constrains this alteration phase to the immediate 4 Ma following the Sudbury impact event. These authors concluded that *"the short-lived impact-induced hydrothermal system was capable of producing large, pervasive, semiconformable alteration zones and Zn-Cu-Pb mineralization… "* – as indeed observed in the Vermilion Member. They, too, acknowledged the fact that this mineralization was similar to that in volcanic terranes.

The hydrothermal system generated by the SIC impact melt body may have been responsible for remobilization and redeposition of metals in fracture zones and breccia occurences, and produced a number of Cu and precious metal enriched ore mineralizations in the footwall to the SIC (Farrow and Watkinson 1997; Molnár et al. 1999).

4.4 Post-Impact Ore Mineralization in the Vredefort-Witwatersrand Structure

At Vredefort, a curiously complicated ore geological situation prevails. As explained earlier, the Archean Witwatersrand Supergroup strata of exceptional gold and uranium concentration were impacted by the Vredefort bolide some 700-900 Ma after their deposition. The current erosion level has been estimated for geological and metamorphic reasons (McCarthy et al. 1986; Gibson et al. 1998; Gibson and Reimold 2000) at ca. 7-10 km. Consequently, only a deep cross-section through the central uplift is currently accessible for investigation, and a relatively deep level through the surrounding ring basin (compare Fig. 10). Hydrothermal deposits that could be unequivocally related to the Vredefort impact event or a post-impact thermal/hydrothermal overprint have not been identified to date. However, isotopic work by a number of researchers (Trieloff et al. 1994; Reimold, 1994; Reimold et al. 1995; Friese et al. 1995, 2003) and mineralogical and chemical studies (Gibson and Wallmach 1995; Gibson et al. 1998; Gibson and Reimold 1999; Reimold et al. 1999b) have indicated that the whole area of the Vredefort Dome and environs has experienced late overprint at about impact times (2 Ga), as well as significantly later, during Kibaran (Grenvillian) times around 1 Ga ago.

The West Rand Group rocks exposed in the inner part of the collar of the Dome exhibit mid-amphibolite facies mineral assemblages, whereas the regional metamorphic grade in the surrounding basin is much lower at greenschist facies level. Detailed petrographic analysis of recent years (Gibson and Reimold 2000, 2001b; Gibson et al. 1998, 2001, 2002; Foya 2002) has shown that the Vredefort Dome experienced not one but two post-Archean metamorphic events. The first event is attributed to the 2.06 Ga (Walraven et al. 1990; R.E. Harmer, personal communication to RLG) Bushveld regional magmatic event, during which intraplating of lower- to mid-crustal mafic and ultramafic magmas raised the regional crustal geotherm to ca 40 °C/km. The lower greenschist facies metamorphism in the gold fields and in the Central Rand Group of the outer collar of the Dome is also related to this event. The close textural association between the silicate metamorphic minerals, authigenic pyrite and gold in the Witwatersrand reefs suggests that gold was mobile during this event. The second metamorphic event was centered on the Vredefort Dome and generated temperatures between 1000 and 1400 °C in the center of the Dome (Gibson 2002; Gibson et al. 2002), and between 300 and 500 °C in the Witwatersrand Supergroup strata in the Dome (Gibson et al. 1998; Foya 2002). Hydrothermal effects (T ~ 300 °C) related to this event extend to the gold fields along the northwestern margin of the Witwatersrand basin (Frimmel and Gartz 1997; Foya 2002). The extreme temperatures in the center of the Dome reflect a combination of uplift of hot rocks from mid-crustal levels and shock heating induced by release of elastic strain energy within mineral lattices following passage of the shock wave. Farther from the Dome, heat from a large impact melt sheet lying in the crater may have contributed, to some degree, to the generation and circulation of hydrothermal fluids.

Chemical alteration and isotopic resetting observed in Witwatersrand strata including gold mineralized reef horizons have both been related to the Vredefort event (e.g., Frimmel et al. 1993; Frimmel and Minter 2002; Reimold et al. 1995, 1999b; Foya et al. 1999; Foya 2002). In both the Vredefort Dome and surrounding Witwatersrand Basin, the abundant pseudotachylitic breccias (Reimold and Colliston 1994 and references therein; Gibson and Reimold 2001b) provide an excellent time marker, since it has been established that at least the majority of these occurences must be related to the impact event. Detailed ore mineralogical investigations, such as those by Foya (2002), Foya et al. (1999), Reimold et al. (1999b, 2002b), Frimmel and Gartz (1997), Gartz and Frimmel (1999) and Hayward et al. (2003), established that the present ore textures were developed at post-impact times (with the impact event characterised by basin-wide brittle deformation and pseudotachylitic breccia development), involving gold remobilization and redeposition throughout the basin.

Current thinking by some Witwatersrand workers (e.g., Gibson 2002; Gibson et al. 2002) involves a thermal/hydrothermal system that raised temperatures in the currently exposed inner parts of the central uplift to at least 1000 °C and in the surrounding collar of Witwatersrand Supergroup strata to between 300 and 500 °C. Associated hydrothermal effects at temperatures of about 300 °C extended throughout the Witwatersrand Supergroup in gold fields surrounding the Vredefort Dome at a distance of 40-60 km from the center of the impact structure (e.g., Gibson and Reimold 2001a,b). Under these conditions, much of the gold in important economic strata, such as the Ventersdorp Contact Reef or the Kimberley Reefs, has been redissolved and then redeposited, as a direct consequence of the impact event. It is not clear whether this event has, thus, caused local enrichment of the pre-existing ore, or whether it only resulted in redistribution and textural changes on small scales – but within a large ore province.

A conceptual model for hydrothermal processes in a Vredefort sized impact structure, in general agreement with many of the findings of Naumov (2002), is presented in Figure 10. A schematic complex impact structure is shown, with an original blanket of hot impact melt. An erosion level consistent with the Vredefort-Witwatersrand situation is also indicated. The central uplift region is made up of hot mid-crustal rocks that have been subjected to high shock pressures. Along a profile radially outward from this hot central uplift, temperature decreases continuously, and the same effect is noted along a profile downward from the upper impact melt layer. Outside of the central uplift region, lithostatic pressure increases with depth. One must also assume that impact-generated porosity (= permeability) decreases away from the central parts of the impact structure.

Because of elevated temperatures near the overlying impact melt body and in the central uplift region, in combination with a decrease of porosity away from the center of the impact structure and a decrease of lithostatic pressure in the uppermost strata, it can be concluded that fluid flow will be largely laterally away from the central uplift and will largely be confined to upper stratigraphic levels. As discussed, the Vredefort-Witwatersrand region has suffered extensive erosion since impact at about 2 Ga ago, with strata of several kilometer thickness having been removed. If erosion of 5-7 km is assumed for these outer parts of the Vredefort impact structure, it appears that the currently mined Witwatersrand strata would have been located in, or perhaps in the lower part, of the assumed flow regime. This implies that gold grade at even lower depths might be less for two reasons: (a) the strata would be more distal with regard to the assumed source regions for gold to the north/northwest of the Witwatersrand Basin (based on paleocurrent directions), and (b) less gold could have been remobilised from the Archean sediments below the favored flow regime for impact-mobilised fluids.

4.5 Hydrocarbons

A 5985 m deep borehole, Gravberg 1, was sunk in 1986-1987 into the 65 to 75 km diameter Siljan impact structure in Sweden. The project endeavoured to investigate the possibility that commercial volumes of mantle-derived abiogenic methane gas could be trapped in the fractured and brecciated lower parts of the impact structure and could provide a sustainable source of energy for Sweden (Castaño 1993). This country derives much of its energy supply from hydrotechnological facilities. It was hoped that impact-induced fracturing of the lower crust could tap a large gas potential (Gold 1988). The possibility of the presence of mature oil had also been discussed by Vlierboom et al. (1985). However, the project failed, as no mantle-derived methane could be detected. A further result was that Siljan does not possess a suitable, hydrocarbon-trapping caprock stratum. Pore-pressure at depth was determined to be largely hydrostatic in contrast to the a priori assumption that it would be lithostatic. No free gas was detected.

In contrast to the negative outcome of this project at Siljan, massive volumes of hydrocarbons have been confirmed or are already exploited in other impact structures. In fact, it has been estimated that total commercial hydrocarbon production from North American impact structures has been between 5 to 16 billion US$ per annum – and this since many years already (e.g., Grieve and Masaitis 1994; Donofrio 1981, 1997, 1998). Of 19 confirmed impact structures in 1997 in North America (Koeberl and Anderson 1996), 9 were exploited for oil or gas at the time (Donofrio 1981, 1997). Ages of these productive impact structures range from Cambrian/Ordovician to late Tertiary. Production comes from impact-affected basement granites, carbonate rocks, and sandstones. Donofrio (1997) estimated that productions ranged from 30 to 2 million barrels per day, plus more than 1.4 billion cubic feet of gas per day. Various hydrocarbon reservoirs may exist in all parts of an impact structure, including central uplifts, rim structures, slump terraces, and ejecta. In the case of very large impact structures, such as Chicxulub in Mexico, even disrupted and fractured rocks in the wider environs of an impact structure may be favorable exploration targets. Donofrio (1997) reported that approximately 50 % of confirmed impact structures as well as other only suspected (i.e., no definitive evidence for impact available yet) impact sites in petroleum provinces are commercial oil and gas fields.

The Ames impact structure in Oklahoma boasts a 7200 barrel oil per day well test, Sierra Madera in Texas a 4.3 billion cubic feet of gas per day calculated well test, and a well with a 2850 ft oil column is known from the Red Wing Creek impact structure. Other impressive reserves exist at the 25 km diameter Steen River impact structure (Alberta, Canada), with an estimated 3 million barrels of oil in structural traps in the rim strata of this structure (Grieve 2003), and in the Chicxulub region off the Yucatán peninsula (Mexico) with 30 billion barrels of oil and 15 trillion cubic feet of gas (see, for example, Grajales-Nishimura et al. 2000). Grieve (2003) emphasized that Steen River is located in a very remote part of northern Alberta Province, where lack of infrastructure inhibits exploitation. However, this author comments that this structure may be "a sleeping giant from a hydrocarbon perspective" (ibid).

Grieve and Masaitis (1994) discussed in detail hydrocarbon resources at the Ames, Red Wing Creek and Avak (Alaska) structures. In addition, an entire monograph (Johnson and Campbell 1997) has been devoted to the Ames Structure. Ames is an approximately 14 km wide complex impact structure, which comprises a central uplift surrounded by an annular graben, and an outer, slightly uplifted rim section. The structure is buried by several kilometers of Ordovician and post-Ordovician sediments. Particularly important horizons are the Arbuckle Dolomite Formation that occurs regionally and that is itself overlain by middle Ordovician Oil Creek Shale that forms an effective trap for hydrocarbons and is also regarded as the source for them (Kuykendall et al. 1997). The age of the structure has been estimated on stratigraphic grounds (as summarized in Koeberl et al. 2001) – namely the absence of the Arbuckle dolomite within the area of the structure, and complete cover with Oil Creek Shale - at approximately 460 Ma. The first hydrocarbon discoveries in this structure were made in 1990 within a 500 m thick section of Lower Ordovician Arbuckle dolomite of the rim section. The economic importance of this impact crater section of a regionally not very productive dolomite layer stems from the impact structure-specific amount of fracturing and associated karst formation. Wells into central granite breccia have also been very productive: for example, the famous Gregory 1-20 well has been appraised as representing the most productive oil well from a single pay zone in all of Oklahoma – a state with a more than 100 year hydrocarbon exploration history. In 1994, about 100 wells had been drilled into the Ames structure – with 52 of them producing oil, and another producing gas.

In this case, the impact event resulted in fracturing and brecciation, leading to enhanced porosity and permeability in rocks of all parts of the structure. It also led to significant topography in the crater area that could accelerate erosion of granite and development of karst topography in the crater rim section – resulting in further enhanced porosity of reservoir rocks. The source of the Ames oil is the Oil Creek Shale, which is unique to the structure and has not been recognized outside of it (Castaño et al. 1997). The Ames impact apparently produced a unique environment to deposit the post-impact oil shale and, in addition, allowed the formation of the structural traps for hydrocarbon accumulation.

The Ames situation, according to Grieve and Masaitis (1994), resembles that at Newporte, another oil-producing impact structure (Koeberl and Reimold 1995) in North Dakota. In contrast, at Red Wing Creek (Koeberl et al. 1996b), also located in North Dakota, and like Newporte in the Williston Basin, hydrocarbons are also recovered from the brecciated basement rocks of the central uplift, but the impact structure does only represent a structural trap and is not responsible for the accumulation of hydrocarbons. The Red Wing Creek structure was discovered when a pronounced seismic anomaly was drilled in 1965. After non-productive drilling results on the flank of the central uplift and in the annular trough, the central uplift eventually proved productive. It was estimated in 1994 that reserves within the ca 3 km wide central uplift were over 130 million barrels of oil, with up to 70 million barrels possibly recoverable (Donofrio 1981; Pickard 1994). The reserves of natural gas were estimated at that time at some 100 billion cubic feet. Red Wing Creek is considered the most productive oil play in the USA, with a cumulative production of 12.7 million barrels of oil and reserves of 20 million barrels of oil and 25 billion cubic feet of gas (Grieve 2003).

Another structure in the Williston Basin still needs to be confirmed as an impact structure: Viewfield, a small structure of 2.5 km diameter, has, at 20 million barrels, quite substantial oil reserves. Grieve (2003) proposed - on the basis of the terrestrial cratering rate - that there could be as many as 12 ± 6 impact structures > 10 km in diameter in the region of the Williston Basin alone, all of which could be viable exploration targets for hydrocarbon deposits. Grieve estimated that if only 50% of these impact structures had reserves similar to those associated with Red Wing Creek, the impact-related reserves in the Williston basin alone could amount to 1 billion barrels of oil and 600 billion cubic feet of gas.

Another structure, for which an impact origin was proposed but still remains to be confirmed, is the 7-8 km wide Calvin structure in Michigan (Milstein 1988). It is estimated that, by 1994, more than 500 000 barrels of oil had been produced from this structure.

The Avak structure, located in the Arctic coastal plain of Alaska, was shown to be of impact origin by Kirschner et al. (1992), who described shatter cones and planar deformation features in quartz. The age of this structure is given by Kirschner et al. (1992) as 100 ± 5 Ma, based on stratigraphic information. Avak, at about 12 km diameter, is a complex impact structure with an annular trough and central uplift. The central uplift has been drilled at the Avak well that penetrated the regional Lower Cretaceous to Ordovician successions. This well also showed some oil, but not of commercial amounts. However, in the immediate vicinity of the impact structure (Fig. 11), three major gas fields – Sikulik and East and South Barrow – occur and straddle annular structures that have been related to the impact event. The idea is, as explained by Grieve and Masaitis (1994), that listric faults of the crater rim, which truncated Lower Cretaceous Barrow sand and juxtaposed it against Lower Cretaceous Torok shale, created an effective gas seal. Both the South and East Barrow gas fields have been exploited. Lantz (1981) estimated a primary recoverable gas reserve of 37 billion cubic feet for this structure.

Hydrocarbon deposits are also known from crater sediments of the Boltysh and Rotmistrovka impact structures, in the Ries crater, and in several other structures. At Boltysh, a 25 km wide and 100 Ma old impact structure, the Eocene crater sedimentary sequence contains oil shales that are 400-500 m thick. According to Masaitis (1989), several tens, > 0.5 m thick, exploitable layers have been identified. Some of the most persistent and thickest layers have an average thickness of 4.4 m. The total resource reserves were estimated by Masaitis at 4.5 billion tons (see also Bass et al. 1967). Oil shales have also been reported from Rotmistrovka and Obolon craters (Masaitis et al. 1980; Gurov and Gurova 1991). In the Ries crater (24 km diameter, 15.1 Ma age), up to 1 m thick allochthonous layers of clayey lignite occur in the upper part of the sedimentary crater fill column (Wolf 1977). Some 600 barrels of oil were produced in 1994 from two wells on the northern rim of the 25-km-diameter Steen River structure in Canada. Gas is produced from the 22 km diameter Marquez Dome structure in Texas. And Gorter et al. (1989) postulated that the large, 55 km diameter, Tookoonooka crater structure of Australia could have potential to yield hydrocarbons because of its vicinity to the hydrocarbon-rich Eromanga Basin. Finally, a recently proposed new impact structure, the 7 km diameter Cloud Creek crater in Wyoming (USA) of about 190 ± 20 Ma age (Stone and Therriault 2003), also has several oil producing wells (in the so-called Lost Dome oil field) associated with the fault zone of the crater rim. It appears that several boreholes drilled into the central uplift area of this structure (Fig. 2, ibid) proved dry – however, it is not known how deep these holes extended.

Large amounts of oil and gas are also recovered from the Lomas Triste breccia (thought to represent brecciation related to seismic disturbances after the gigantic Chicxulub impact event) deposit of the Campeche oil field in the Gulf of Mexico 300 km from the Chicxulub impact structure (Camargo Zanoguera and Quezada Muñeton 1992; Limon et al. 1994). Production is estimated to be in excess of 2 million barrels of oil and 1.5 billion cubic feet of gas per day (Grieve 2003). The large Chicxulub impact structure that, with regard to the catastrophic events of global importance that took place at Cretaceous-Tertiary boundary times is most intriguing, was discovered and confirmed as the smoking gun for the K/T impact as a direct consequence of oil exploration by geophysical methods and drilling (e.g., Hildebrand et al. 1991; Grajales-Nishimura et al. 2000). These latter authors made a case for both the offshore oil-producing breccias and the sealing rocks from the oil fields (such as the Cantarell oil field) in the Campeche marine platform being probably related to the Chicxulub impact. Both the oil-producing carbonate breccias and the capping dolomitized layer contain impact products. Grajales-Nishimura et al. (2000) considered the dolomitized layer part of the impact ejecta layer. They emphasized that "the K-T breccia reservoir and seal ejecta layer of the Cantarell oil field, with a daily production of 1.3 million barrels of oil, are probably the most important known oil-producing units related to an impact event" (ibid). Grieve (2003) summarizes that the up to 300 m thick breccias of 10-20% porosity contain proven reserves of 30 billion barrels of oil and 15 trillion cubic feet of gas. These amounts, according to this author, exceed the entire on- and offshore reserves of the United States.

5 Other Deposits in, and Benefits from, Impact Structures

Besides metallic and mineral deposits, as discussed in detail in the above, impact structures may have much potential for other commercial or economic interests. Building materials have - for centuries - been obtained from impact structures, as, for example, witnessed by the century-old churches in Nördlingen in the Ries Crater of southern Germany and in Rochechouart in the meteorite crater of same name in the Haut-Limousin province of southwestern France. Like the medieval castle at Rochechouart, both historical churches have been erected, to a large degree, with blocks of suevite impact breccia. Impact melt rock has been utilised for the construction of a large hotel in the northern part of the Lappajärvi crater in Finland.

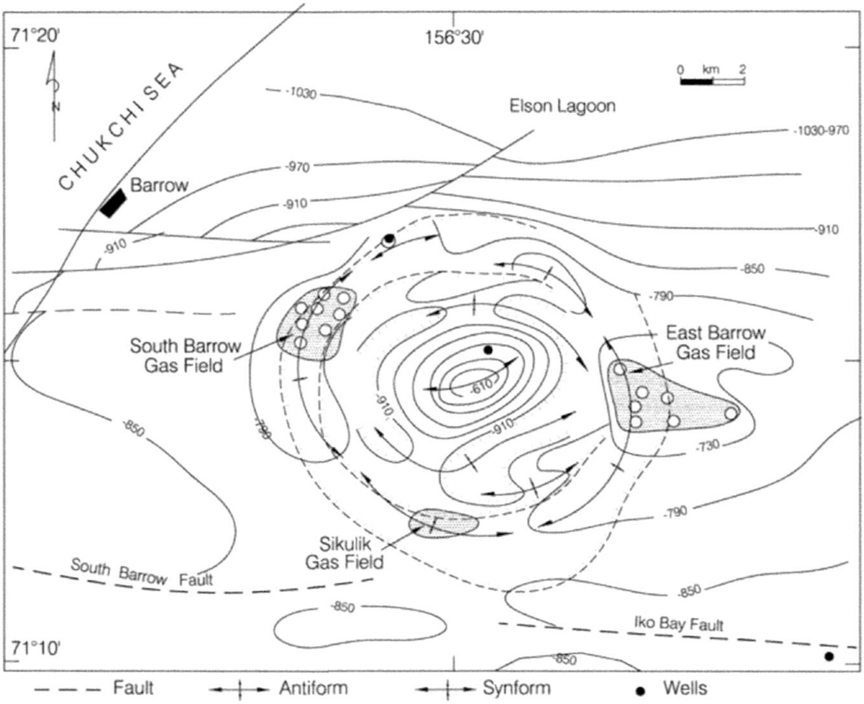

Fig. 11. Structure map for the region of the Avak impact structure, Alaska, with oil and gas fields superimposed (after Grieve and Masaitis 1994).

Dimension stone has also been quarried in the Vredefort dome, the central uplift region of the Vredefort impact structure (Reimold 1992). Prominent local applications of these granitic gneisses can be viewed in the main terminal building of Johannesburg International Airport, where many supporting columns also display highly instructive "exposures" of pseudotachylitic breccia veinlets, and in the entrance hall to the Council of Mineral Technology (MINTEK) in Randburg. Even the focal point of the Voortrekker historical monument near Pretoria, the sarcophagus, has been constructed from Vredefort granite gneiss. Dimension stone quarrying was abandoned in the Vredefort dome in the late 1990s, but the faces in the remaining quarries provide some of the best and most instructive exposures of Archean basement of the Kaapvaal craton and macro- and meso-deformation of the interior of a central uplift of a large impact structure.

In the Ries crater, suevite impact breccia has found a use in the manufacture of the so-called trass cement. This product has proven particularly useful for repair work on degraded stone monuments. Masaitis (1989) mentioned that some water-rich impact glasses could potentially become useful as swelling materials, along the lines of perlite applications. In Gardnos, a 5-km-diameter, ca. 500 Ma old, impact structure in Norway, a peculiar impact breccia known as Gardnos Breccia (French et al. 1998) is transformed into a variety of utility objects and jewellery. Jewellery has also been made, for several centuries, from tektites, especially Central European tektites (moldavites) from the Czech Republic, Germany, and Austria (see, e.g., papers in Konta 1988). Important sources for drinking or mineralized waters are the Manson structure (Iowa, USA), the Kaluga impact crater (15 km, 380 Ma, Russia), and the Vepriaj crater (8 km, 160 Ma, Lithuania), where high-porosity impact breccias are utilized as storage media. At Logoisk (17 km, 40 Ma, Belarus), thick layers of sand deposited in the crater lake form a useful local groundwater recharge basin and constitute an important source of water (Masaitis 1989). The Manicouagan reservoir in the 100 km wide Manicouagan impact structure in Quebec (Canada) is part of a large hydroelectric power development, and Lake Bosumtwi in Ghana, a 8.5 km diameter lake in the 10.5 km Bosumtwi impact crater, provides the livelihood for a large number of fishing communities along its shore.

Bosumtwi has additional potential in providing a unique recreational-educational area in a rather densely populated area of Ghana, near the major town of Kumasi. Already, a large lake in the Rochechouart structure of France that has been declared a protected national park (Geo-Parc) area is attracting thousands of tourists. The Tswaing meteorite crater in South Africa provides recreational and educational facilities and construction of a museum building is well under way. A number of meteorite craters in Scandinavia are recreational centers for summer as well as winter sports (the Dellen and Siljan areas in Sweden are favourite cross-country skiing areas).

A major tourist attraction is the museum at Meteor Crater in Arizona, and a thriving museum has also been developed in Nördlingen (Ries Crater). Museums and educational displays have either been established already at other impact structures (for example, at Lappajärvi one finds a jogging trail with a series of instructive plaques, and at Steinheim a very instructive and beautifully designed museum can be visited by appointment), or are being developed (e.g., at Tswaing in South Africa where hiking-trails have been laid out, game resettled, and a museum building is constructed; efforts are also made to establish a museum at Bosumtwi in Ghana). In this context, it is particularly evident that meteorite structures in developing countries, even if they are totally devoid of mineralization, may turn out to be highly profitable with regard to their potential as tourist attractions and suitable as outdoor classrooms – both of which are largely lacking in such countries. The large Popigai structure in Siberia has been suggested for World Heritage Site status, because of its unique combination of world-class geology and impact-diamond deposits, for which the term "national treasury" has been coined (Deutsch et al. 2000). However, it is understood that this proposal has not been taken further. In contrast, parts of the Vredefort Dome have been earmarked for dedicated eco-tourism development, have already been declared a national conservancy, and an application for World Heritage status – based on a unique combination of geological, cultural (iron age archaeology) and historic (early gold mining in the Ventersdorp gold field as well as Anglo-Boer War history) heritage - was submitted to UNESCO in January 2003.

The economic, touristic, and educational potential of meteorite impact structures in all parts of the world cannot be underestimated. Not only do many of the known structures have the potential to generate value in the future, but the commercial gain already derived from active mining of impact structures should make for powerful incentive to consider the prospective benefits of further exploration for still unknown impact structures.

6
Final Remarks

The extraordinary importance of ore resources in or associated with impact structures has been demonstrated: Vredefort-Witwatersrand, Sudbury and the North American impact-related hydrocarbon deposits are cases in point. Exploration and impact research will mutually benefit from better understanding of the economic potential of impact strcutures and improved understanding of impact crater geology. Already many impact structures have been identified as a direct consequence of remote sensing or geophysical investigations, many of which were undertaken as part of ore exploration.

The terrestrial impact record is far from complete, and many structures remain to be identified. Many of these are covered by post-impact cover strata and require geophysical methods and, ultimately, drilling to verify their existence. Many known impact structures remain to be investigated for their economic potential, and the economic value already established for a large number of impact structures ought to be strong incentive for continued "prospecting" for impact structures. Impact structures may have preserved roundish geometries and geophysical signatures that can be exploited for exploration strategies. Exploration geologists and geophysicists must be informed about impact structures and their potential. We hope that this contribution will go some way towards this goal.

Acknowledgements

We are most grateful for the opportunity to discuss, with many colleagues, especially Steve Prevec and Martin Tuchscherer, various types of ore deposits in impact structures. This work is a direct outcome of Uwe Reimold's Presidential Address 2002 (Reimold 2002) for the Geological Society of South Africa. CK is supported by the Austrian Science Foundation, project Y58-GEO. Detailed and constructive reviews by V. Puura and F. Tsikalas are much appreciated. This is University of the Witwatersrand Impact Cratering Research Group Contribution No. 66.

References

Alexopoulos JS, Grieve RAF, Robertson PB (1988) Microscopic lamellar deformation features in quartz: Discriminative characteristics of shock-generated varieties. Geology 16: 796-799

Ames DE, Gibson HL (1995) Controls on geological setting of regional hydrothermal alteration within the Onaping formation, footwall to the Errington and Vermilion base metal deposits, Sudbury Structure, Ontario. In: Current Research 1995-E. Ottawa, Ontario, Geological Survey of Canada, 161-173

Ames DE, Bleeker W, Heather KB, Wodicka N (eds) (1997) Timmins to Sudbury Transect: New Insights into the Regional Geology and the Setting of Ore Deposits. In: Joint Annual Meeting of the Geological Association of Canada - Mineralogical Association of Canada, Ottawa '97, Field Trip Guidebook B6, 133 pp

Ames DE, Watkinson DH, Parrish RR (1998) Dating of a regional hydrothermal system induced by the 1850 Ma Sudbury impact event. Geology 26: 447-450

Ames DE, Golightly JP, Lightfoot PC, Gibson HL (2002) Vitric compositions in the Onaping Formation and their relationship to the Sudbury Igneous Complex, Sudbury Structure. Economic Geology 97: 1541-1562

Andreoli MAG, Ashwal LD, Hart RJ, Huizenga JM (1999) A Ni- and PGE-enriched quartz norite melt complex in the Late Jurassic Morokweng impact structure, South Africa. In: Dressler BO, Sharpton VL (eds) Large Meteorite Impacts and Planetary Evolution II. Geological Society of America, Special Paper 339: 91-108

Armstrong RA, Compston W, Retief EA, Williams IS, Welke HJ (1991) Zircon ion-microprobe studies bearing on the age and evolution of the Witwatersrand Triad. Precambrian Research 53: 243-266

Barnicoat AC, Henderson IHC, Knipe RJ, Yardley BWD, Napier RW, Fox NPC, Kenyon AK, Muntingh DJ, Strydom D, Winkler KS, Lawrence SR, Cornford C (1997) Hydrothermal gold mineralisation in the Witwatersrand Basin. Nature 386: 820-824

Barringer DM (1906) Coon Mountain and its crater. Philadelphia Academy of Natural Sciences Proceedings 57: 861-886

Bass YB, Galaka VI, Grabovskiy VK (1967) The Boltysh oil shales. (in Russian) Razvedka I okhrana nedr 9: 11-15

Berlenbach JW, Roering C (1992) Sheath-fold-like structures in pseudotachylites. Journal of Structural Geology 14: 847-856

Bishopp DW (1941) The geodynamics of the Vredefort Dome. Transactions of the Geolological Society of South Africa 44: 1-18

Boast M, Spray JG (2003) Massive sulphide ore distribution at Sudbury revealed by the investigation of contact metamorphism and faulting [ext abs]. Applied Earth Science 112, No. 2: B144-B145

Boer RH, Reimold WU, Koeberl C, Kesler SE (1996) Fluid inclusion studies on drill core samples from the Manson impact crater: Evidence for post-impact hydrothermal activity. In: Koeberl C, Anderson RR (eds) The Manson Impact Structure: Anatomy of an Impact Crater. Geological Society of America, Special Paper 302: 377-382

Boon JD, Albritton CC (1937) Meteorite scars in ancient rocks. Field and Laboratory 5: 53-64

Borchers R (1964) Exploration of the Witwatersrand System and its extensions. In: Haughton SH (ed) The Geology of Some Ore Deposits in Southern Africa, vol. 1. Geological Society of South Africa, Johannesburg, 1-23

Brink MC, Waanders FB, Bisschoff AA (1997) Vredefort: a model for the anatomy of an astrobleme. Tectonophysics 270: 83-114

Brink MC, Waanders FB, Bisschoff AA (1999) Evolution of the ringed basin around Vredefort, South Africa [abs]. Meteoritics and Planetary Science 34: A19-A20

Brink MC, Waanders FB, Bisschoff AA (2000) The Foch Thrust-Potchefstroom Fault structural system, Vredefort, South Africa: a model for impact-related tectonic movement over a pre-existing barrier. Journal of African Earth Science 30: 99-117

Buchanan PC, Reimold WU (2002) Planar deformation features and impact glass in inclusions from the Vredefort Granophyre, South Africa. Meteoritics and Planetary Science 37: 807-822

Butler HR (1994) Lineament analysis of the Sudbury multiring impact structure. In: Dressler BO, Grieve RAF, Sharpton VL (eds) Large Meteorite Impact and Planetary Evolution, Geological Society of America, Special Paper 293: 319-329

Camargo Zanoguera A, Quezada Muñeton J (1992) Analysis of economic geology of areas of the Gulf of Mexico with hydrocarbon potential. Boletin Asociacion Geologos Petroleros XVI: 1-32

Carter NL (1965) Basal quartz deformation lamellae – a criterion for recognition of impactites. American Journal of Science 263: 786-806

Carter NL (1968) Dynamic deformation of quartz. In: French BM, Short NM (eds) Shock Metamorphism of Natural Materials, Mono Book Corp, Baltimore, 453-474

Castaño JR (1993) Prospects for commercial abiogenic gas production: Implications from the Siljan Ring area, Sweden. In: Howell DG (ed) The Future of Energy Gases. United States Geological Survey Professional Paper 1570, United States Government Printing Office, Washington, 133-154

Castaño JR, Clement JH, Kuykendall MD, Sharpton VL (1997) Source-rock potential of impact craters. In: Johnson KS, Campbell JA (eds) Ames Structure in Northwest Oklahoma and similar features: Origin and petroleum production (1995 Symposium). Oklahoma Geological Survey Circular 100: 100-103

Chai G, Eckstrand R (1994) Rare-earth element characteristics and origin of the Sudbury Igneous Complex, Ontario, Canada. Chemical Geology 113: 221-244.

Cintala MJ, Grieve RAF (1994) The effects of differential scaling of impact melt and crater dimensions on lunar and terrestrial craters: some brief examples. In: Dressler BO, Grieve RAF, Sharpton VL (eds) Large Meteorite Impacts and Planetary Evolution. Geological Society of America, Special Paper 293: 51-59

Cintala MJ, Grieve RAF (1998) Scaling impact melting and crater dimensions: Implications for the lunar cratering record. Meteoritics and Planetary Science 33: 889-912

Cockell CS, Lee P, Osinski G, Horneck G, Broady (2002) Impact-induced microbial endolithic habitats. Meteoritics and Planetary Science 37: 1287-1298

Colliston WP (1990) A model of compressional tectonics for the origin of the Vredefort structure. Tectonophysics 270: 83-114

Corfu F, Lightfoot PC (1996) U-Pb geochronology of the Sublayer environment, Sudbury Igneous Complex, Ontario. Economic Geology 91: 1263-1269

Corner B, Reimold WU, Brandt D, Koeberl C (1997) Morokweng impact structure, Northwest Province, South Africa: Geophysical imaging and shock petrographic studies. Earth and Planetary Science Letters 146: 351-364

Cowan EJ (1999) Magnetic fabric constraints on the initial geometry of the Sudbury Igneous Complex: A folded sheet or a basin-shaped igneous body? Tectonophysics 307: 135-162

Cowan EJ, Schwerdtner WM (1994) Fold origin of the Sudbury Basin. In: Lighfoot PC, Naldrett AJ (eds) Proceedings of the Sudbury-Noril'sk Symposium. Ontario Geological Survey, Special Volume 5: 45-55

Coward MP, Spencer RM, Spencer CE (1995) Development of the Witwatersrand Basin, South Africa. In: Coward MP, Ries AC (eds) Early Precambrian Processes. Geological Society of London, Special Publication 95: 243-269

Daly RA (1947) The Vredefort ring structure of South Africa. Journal of Geology 55: 125-145

De S, Heaney PJ, Hargraves RB, Vicenzi EP, Taylor PT (1998) Microstructural observations of polycrystalline diamond: a contribution to the carbonado conundrum. Earth and Planetary Science Letters 164: 421-433

De S, Heaney PJ, Vicenzi EP, Wang J (2001) Chemical heterogeneity in carbonado, an enigmatic polycrystalline diamond. Earth and Planetary Science Letters 185: 315-330

Dence MR (1972) Meteorite impact craters and the structure of the Sudbury Basin. In: Guy-Bray JV (ed) New Developments in Sudbury Geology. Geological Association of Canada Special Paper 10: 117-124

Deutsch A, Ivanov B (eds) (2004) Puchezh-Katunki deep borehole. Impact Studies series, Springer-Verlag Berlin-Heidelberg, in preparation

Deutsch A, Grieve RAF, Avermann M, Bischoff L, Brockmeyer P, Buhl D, Lakomy R, Müller-Mohr V, Ostermann M, Stöffler D (1995) The Sudbury Structure (Ontario, Canada): a tectonically deformed multi-ring basin. Geologische Rundschau 84: 697-709

Deutsch A, Masaitis VL, Langenhorst F, Grieve RAF (2000) Popigai, Siberia – well preserved giant impact structure, national treasury, and world's geological heritage. Episodes 23: 3-11

Dickin AP, Crockett JH (1997) Sudbury: geochemical evidence for a single-impact melting event [abs]. Lunar and Planetary Institute Contribution No 922, Houston, Texas, USA, pp 11-12

Dickin AP, Richardson JM, Crockett JH, McNutt RH, Peredery WV (1992) Osmium isotope evidence for a crustal origin of platinum group elements in the Sudbury nickel ore, Ontario, Canada. Geochimica et Cosmochimica Acta 56: 3531-3537

Dickin AP, Artan MA, Crockett JH (1996) Isotopic evidence for distinct crustal sources for North and South Range ores, Sudbury Igneous Complex. Geochimica et Cosmochimica Acta 60: 1605-1613

Dietz RS (1961) Vredefort ring structure: meteorite impact scar? Journal of Geology 69: 499-516

Dietz RS (1962) Sudbury Structure as an astrobleme [abs]. EOS, Transactions of the American Geophysical Union 43: 445-446

Dietz RS (1964) Sudbury Structure as an astrobleme. Journal of Geology 72: 412-434

Donofrio RR (1981) Impact craters: Implications for basement hydrocarbon production. Journal of Petroleum Geology 3: 279-302

Donofrio RR (1997) Survey of hydrocarbon-producing impact structures in North America: Exploration results to date and potential for discovery in Precambrian basement rock. In: Johnson KS, Campbell JA (eds) Ames Structure in Northwest Oklahoma and Similar Features: Origin and Petroleum Production (1995 Symposium). Oklahoma Geological Survey Circular 100: 17-29

Donofrio RR (1998) North American impact structures hold giant field potential. Oil & Gas Journal, May 18, 1998, pp 69-83

Dressler BO (1984a) The effects of the Sudbury Event and the intrusion of the Sudbury Igneous Complex on the footwall rocks of the Sudbury Structure. In: Pye E, Naldrett A, Giblin P (eds) The Geology and Ore Deposits of Sudbury. Ontario Geological Survey Special Volume 1: 97-136

Dressler BO (1984b) General geology of the Sudbury area. In: Pye EG, Naldrett AJ, Giblin PE (eds) The Geology and Ore Deposits of the Sudbury Structure. Ontario Geological Survey Special Volume 1: 57-82

Dressler BO, Sharpton VL (eds) (1999) Large Meteorite Impacts and Planetary Evolution II. Geological Society of America, Special Paper 339, 464 pp

Dressler BO, Morrison GG, Peredery WV, Rao BV (1987) The Sudbury Structure, Ontario, Canada – a review. In: Pohl J (ed) Research in Terrestrial Impact Structures. F Vieweg und Sohn, Braunschweig/Wiesbaden, 39-68

Dressler BO, Gupta VK, Muir TL (1991) The Sudbury Structure. In: Thurston P, Sutcliffe R, Stott G (eds) Geology of Ontario. Ontario Geological Survey, Special Volume 4, Part 1: 593-625

Dressler BO, Grieve RAF, Sharpton VL (eds) (1994) Large Meteorite Impacts and Planetary Evolution. Large Meteorite Impacts and Planetary Evolution. Geological Society of America, Special Paper 293, 348 pp

Dressler BO, Weiser T, Brockmeyer P (1996) Recrystallized impact glasses of the Onaping Formation and the Sudbury Igneous Complex, Sudbury Structure, Ontario, Canada. Geochimica et Cosmochimica Acta 60: 2019-2036

Du Toit AL (1954) The Geology of South Africa. Oliver & Boyd, Edinburgh, 3rd ed, 611 pp

Ezerskii VA (1982) Impact-metamorphosed carbonaceous matter in impactites. (in Russian) Meteoritika 41: 134-140

Faggart BE, Basu AR, Tatsumoto A (1985) Origin of the Sudbury complex by meteorite impact: Neodymium isotopic evidence. Science 230: 436-439

Farmer D (2000) Hydrothermal systems: Doorways to early biosphere evolution. GSA Today 10(7): 1-9

Farrow CEG, Watkinson DH (1997) Diversity of previous-metal mineralization in the footwall Cu-Ni-PGE deposits, Sudbury, Ontario: implications for hydrothermal models of formation. Canadian Mineralogist 35: 817-839

Fletcher P, Reimold WU (1989) Some notes and speculations on the pseudotachylites in the Witwatersrand basin and the Vredefort dome. South African Journal of Geology 92: 223-234

Foya SN (2002) Mineralogical and geochemical controls on gold distribution in the Kimberley Reefs, South Africa. University of the Witwatersrand, Johannesburg, PhD Thesis (unpublished), 383 pp

Foya SN, Reimold WU, Przybylowicz WJ, Gibson RL (1999) PIXE microanalysis of gold-pyrite associations from the Kimberley Reefs, Witwatersrand basin, South Africa. Nuclear Instruments and Methods in Physics Research B 158: 588-592

Foya SN, Reimold WU, Gibson RL (2000) Mineralogy of the Kimberley Reef, Witwatersrand Basin, South Africa: Implications for gold mineralization [abs]. Mineralogical Association of South Africa-Witwatersrand Interest Group and University of Natal Symposium on The Witwatersrand and its Correlatives: Implications for the Evolution and Mineralization of the Kaapvaal Craton, Durban, February 2000, University of Natal, Durban, 13-15

French BM (1967) Sudbury Structure, Ontario: some petrographic evidence for origin by meteorite impact. Science 156: 1094-1098

French BM (1970) Possible relations between meteorite impact and igneous petrogenesis, as indicated by the Sudbury Structure, Ontario, Canada. Bulletin Volcanologique 34: 466-517

French BM, Short NM (Eds) (1968) Shock Metamorphism of Natural Materials. Mono Book Corp, Baltimore, 644 pp

French BM, Koeberl C, Gilmour I, Shirey SB, Dons JA, Naterstad J (1998) The Gardnos impact structure, Norway: Petrology and geochemistry of target rocks and impactites. Geochimica et Cosmochimica Acta 61: 873-904

Friese AE, Charlesworth EG, McCarthy TS (1995) Tectonic processes within the Kaapvaal craton during the Kibaran (Grenville) orogeny: structural, geophysical and isotopic constraints from the Witwatersrand basin and environs. University of the Witwatersrand, Johannesburg, Economic Geology Research Unit Information Circular 292, 67 pp

Friese AEW, Reimold WU, Layer PW (2003) ^{40}Ar-^{39}Ar dating and structural information on tectonite-bearing faults in the Witwatersrand Basin: Evidence for multi-stage tectono-thermal activity in the Central Kaapvaal Craton. South African Journal of Geology 106: 41-70

Frimmel HE (1997a) Chlorite thermometry in the Witwatersrand basin: Constraints on the Paleoproterozoic geotherm in the Kaapvaal craton, South Africa. Journal of Geology 105: 601-615

Frimmel HE (1997b) Detrital origin of hydrothermal Witwatersrand gold: a review. Terra Nova 9: 192-197

Frimmel HE, Gartz VH (1997) Witwatersrand gold particle chemistry matches model of metamorophosed, hydrothermally altered placer deposits. Mineralium Depositum 32: 523-530

Frimmel HE, Minter WEL (2002) Recent developments concerning the geological history and genesis of the Witwatersrand gold deposits, South Africa. Society of Economic Geologists Special Publication 9: 17-45

Frimmel HE, Le Roux AP, Knight J, Minter WEL (1993) A case study of the post-depositional alteration of the Witwatersrand basal reef gold placer. Economic Geology 88: 249-265

Fueten F, Redmond DJ (1997) Documentation of a 1450 Ma contractional orogeny preserved between the 1850 Ma Sudbury impact structure and the 1 Ga old Grenville orogenic front, Ontario. Geological Society of America Bulletin 109: 268-279

Galimov EM, Ivanovskaya IN, Klyuyev YA, Nepsha VI, Yepishina NI, Plotnikova SP, Galyukina OI, Poberezhskiy VA, Gritsik VV, Koptil VA (1980) Isotopic composition and peculiarities of crystalline structure of natural diamond polycrystals with lonsdaleite. (in Russian) Geokhimiya 4: 533-539

Gartz VH, Frimmel HE (1999) Complex metasomatism of an Archaean placer in the Witwatersrand Basin, South Africa: The Ventersdorp Contact Reef – A hydrothermal aquifer? Economic Geology 94: 689-706

Giblin PE (1984) Glossary of Sudbury geology terms. In: Pye E, Naldrett A, Giblin PE (eds), The Geology and Ore Deposits of Sudbury. Ontario Geological Survey Special Volume 1: 571-574

Gibson RL (1996) The Vredefort hornfelses revisited.South African Journal of Geology 99: 93-96

Gibson RL (2002) Impact-induced melting in Archaean granulites in the Vredefort Dome, South Africa: I. Anatexis of metapelitic granulites. Journal of Metamorphic Geology 20: 57-70

Gibson RL, Reimold WU (1999) Significance of the Vredefort Dome for metamorphic-mineralization studies in the Witwatersrand Basin. Mineralogy and Petrology 66: 25-53

Gibson RL, Reimold WU (2000) Deeply exhumed impact structures: A case study of the Vredefort Structure, South Africa. In: Gilmour I, Koeberl C (eds), Impacts and the Early Earth. Lecture Notes in Earth Sciences 91: 249-277, Springer, Heidelberg

Gibson RL, Reimold WU (2001a) The Vredefort Impact Structure, South Africa: The scientific evidence and a two-day excursion guide. Council for Geoscience (Geological Survey), Pretoria, Memoir 92: 111 pp

Gibson RL, Reimold WU (2001b) Implications of the Vredefort meteorite event for the Witwatersrand basin. The Witwatersrand Goldfield [abs]. Mini-Symposium of the Geological Society of South Africa, Johannesburg, 17 May 2001, Randburg, 3 pp

Gibson RL, Wallmach T (1995) Low-pressure-high temperature metamorphism in the Vredefort Dome, South Africa: Anticlockwise pressure-temperature path followed by rapid decompression. Geological Journal 30: 319-331

Gibson RL, Armstrong RA, Reimold WU (1997) The age and thermal evolution of the Vredefort impact structure: a single-grain U-Pb zircon study. Geochimica et Cosmochimica Acta 61:1531-1540

Gibson RL, Reimold WU, Stevens G (1998) Thermal-metamorphic signature of an impact event in the Vredefort dome, South Africa. Geology 26: 787-790

Gibson RL, Courtnage PM, Charlesworth EG (1999) Bedding-parallel shearing and related deformation in the lower Transvaal Supergroup north of the Johannesburg Dome, South Africa. South African Journal of Geology 102: 99-108
Gibson RL, Reimold WU, Ashley AJ, Koeberl C (2001) Shock pressure distribution in the Vredefort impact structure, South Africa: Evidence from feldspar and ferromagnesian minerals [abs]. Lunar and Planetary Science 32, Lunar and Planetary Institute, Houston, abstract 1012: 2 p
Gibson RL, Reimold WU, Ashley AJ, Koeberl C (2002) Drilling the Vredefort central uplift, South Africa – petrographic, geochemical and structural results. Geological Society of America Annual Meeting, Denver, Abstracts with Programs 34: 466
Gilmour I (1998) Geochemistry of carbon in terrestrial impact processes. In: Grady MM, Hutchison R, McCall GJH, Rothery DA (eds) Meteorites: Flux with Time and Impact Effects. Geological Society of London, Special Publication 140: 205-216
Girdler RW, Taylor PT, Frawley JJ (1992) A possible impact origin for the Bangui magnetic anomaly (Central Africa). Tectonophysics 212: 45-58
Gold T (1988) The deep earth gas theory with respect to the results from the Gravberg-1 well. In: Bodén A, Eriksson KG (eds) Deep Drilling in Crystalline Bedrock. Springer, Berlin-Heidelberg, vol. 1: 18-27
Gorter JD, Gostin VA, Plummer PS (1989) The enigmatic sub-surface Tookoonooka complex in south-west Queensland: Its impact origin and implications for hydrocarbon accumulations. In: O'Neil BF (ed) The Cooper and Eromanga Basins, Australia. Petroleum Exploration Society of Australia, Society of Petroleum Engineers, Australian Society of Exploration Geophysicists, 441-456.
Grajales-Nishimura JM, Cadillo-Pardo E, Rosales-Domínguez, Moráan-Zenteno D, Alvarez W, Claeys P, Ruíz-Morales J, García-Hernández J, Padilla-Avila P, Sánchez-Ríos A (2000) Chicxulub impact: The origin of reservoir and seal facies in the southeastern Mexico oil fields. Geology 28: 307-310
Grant RW, Bite A (1984) Sudbury quartz diorite Offset Dikes. In: Pye EG, Naldrett AJ, Giblin PE (eds) The geology and ore deposits of the Sudbury Structure. Ontario Geological Survey Special Volume 1: 275-300
Grieve RAF (2003) Extraterrestrial triggers for resource deposits [ext abs]. Applied Earth Science 112, No. 2: B145-B147
Grieve RAF, Pesonen LJ (1992) The terrestrial impact record. Tectonophysics 216: 1-30
Grieve RAF, Masaitis VL (1994) The economic potential of terrestrial impact craters. International Geology Review 36: 105-151
Grieve RAF, Pilkington M (1996) The signature of terrestrial impacts. AGSO Journal of Australian Geology and Geophysics 16: 399-420
Grieve RAF, Therriault AM (2000) Vredefort, Sudbury, Chicxulub: Three of a kind? Annual Review of Earth and Planetary Science 28: 305-338
Grieve RAF, Coderre JM, Robertson PB, Alexopoulos JS (1990) Microscopic planar deformation features in quartz of the Vredefort structure: anomalous but still suggestive of an impact origin. Tectonophysics 171: 185-200
Grieve RAF, Stöffler D, Deutsch A (1991) The Sudbury Structure: controversial or misunderstood. Journal of Geophysical Research 96: 22 753-22 764
Gurov EP, Gurova EP (1991) Geological structure and composition of rocks in impact craters. Nauka Press, Kiev, 160 pp (in Russian)

Gurov EP, Melnychuk EV, Metalidi SV, Ryabenko VA, Gurova EP (1985) The characteristics of the geological structure of the eroded astrobleme in the western part of the Ukrainian Shield. (in Russian) Dopovidi Akademii Nauk Ukrains'koi RSR: 8-11

Gurov EP, Koeberl C, Reimold WU (1998) Petrography and geochemistry of target rocks and impactites from the Ilyenits Crater, Ukraine. Meteoritics and Planetary Science 33: 1317-1333

Guy-Bray JG, Geological Staff (1966) Shatter cones at Sudbury. Journal of Geology 74: 243-245

Hagerty JJ, Newsom HE (2003) Hydrothermal alteration at the Lonar Lake impact structure, India: Implications for impact cratering on Mars. Meteoritics and Planetary Science 38: 365-381

Haggerty SE (1999) A diamond trilogy: Superplumes, supercontinent, and supernovae. Science 185: 851-860

Hall AJ, Molengraaff GAF (1925) The Vredefort Mountain Land in the Southern Transvaal and the Northern Orange Free State. Verhandelingen der Koninklijke Akademie van Wetenschappen te Amsterdam, Sect 2, part 24: 183 pp

Hargraves RB (1961) Shatter cones in the rocks of the Vredefort Ring. Transactions of the Geological Society of South Africa 64: 147-154

Harper CT (1983) The Geology and Uranium Deposits of the Central Part of the Carswell Structure, Northern Sasketchewan, Canada. Unpublished PhD Thesis, Colorado School of Mines, Golden, CO, USA, 337 pp

Hart RJ, Welke HJ, Nicolaysen LO (1981) Geochronology of the deep profile through Archean basement at Vredefort, with implications for early crustal evolution. Journal of Geophysical Research 86: 10663-10680

Hart RJ, Andreoli MAG, Tredoux M, De Wit MJ (1990) Geochemistry across an exposed section of Archaean crust at Vredefort, South Africa: with implications for mid-crustal discontinuities. Chemical Geology 82: 21-50

Hart RJ, Moser D, Andreoli MAG (1999) Archaean age for the granulite metamorphism near the center of the Vredefort Structure, South Africa. Geology 27: 1091-1094

Hart RJ, Cloete M, McDonald I, Carlson RW, Andreoli MAG (2002) Siderophile-rich inclusions from the Morokweng impact melt sheet, South Africa: possible fragments of a chondritic meteorite. Earth and Planetary Science Letters 198: 49-62

Hayward CL, Reimold WU, Robb LJ, Gibson RL (2003) The Witwatersrand gold deposit, South Africa: an impact-modified metamorphosed placer [ext abs]. Applied Earth Science 112, No. 2: B147-B149

Henkel H, Reimold WU (1996) Integrated gravity and magnetic modeling of the Vredefort impact structure – Reinterpretation of the Witwatersrand Basin as the erosional remnant of an impact basin. University of the Witwatersrand, Johannesburg, Economical Geology Research Unit Information Circular 299: 89 pp

Henkel H, Reimold WU (1998) Integrated geophysical modeling of a giant, complex impact structure: anatomy of the Vredefort Structure, South Africa. Tectonophysics 287: 1-20

Hildebrand AR, Penfield GT, Kring DA, Pilkington M, Camargo AZ, Jacobsen SB, Boynton WV (1991) Chicxulub Crater: A possible Cretaceous/Tertiary boundary impact crater on the Yucatàn Peninsula, Mexico. Geology 19: 867-871

Hoyt WG (1987) Coon Mountain Controversies. University of Arizona Press, Tucson, 442 pp

Impact Database, http://www.unb.ca/passc/ImpactDatabase /index.html.

Innes DG, Colvine AC (1984) The regional metallogenetic setting of Sudbury. In: Pye EG, Naldrett AJ, Giblin PE (eds) The Geology and Ore Deposits of the Sudbury Structure, Ontario Geological Survey Special Volume 1: 45-56

James RS, Easton RM, Peck DC, Hrominchuk JL (2002) The East Bull Lake intrusive suite: Remnants of a ~2.48 Ga large igneous and metallogenic province in the Sudbury area of the Canadian Shield. Economic Geology 97: 1577-1606

Jolley SJ, Freeman SR, Barnicoat AC, Phillips GM, Knipe RJ, Pather A, Fox NPC, Strydom D, Birch MTG, Henderson IHC and Rowland TW (2004) Structural controls on Witwatersrand gold mineralisation. Journal of Structural Geology, doi 10.1016/j.jsg.2003.11.011

Johansson Å (1984) Geochemical studies on the Boda Pb-Zn deposit Siljan astrobleme, central Sweden. GFF 106: 15-25

Johnson KS, Campbell JA, eds (1997) Ames Structure in Northwest Oklahoma and Similar Features: Origin and Petroleum Production (1995 Symposium). Oklahoma Geological Survey Circular 100: 396

Jones AP, Mutanen T, Tuisku P, Hanski E, Price GD (2003) The Pechenga Structure, Russia: giant Ni-Cu mineralisation related to large meteorite impact?

Joreau P, French BM, Doukhan J-C (1996) A TEM investigation of shock metamorphism in quartz from the Sudbury impact structure (Canada). Earth and Planetary Science Letters 138: 137-143

Kagi H, Takahashi K, Hidaka H, Masuda A (1994) Chemical properties of Central African carbonado and its genetic implications. Geochimica et Cosmochimica Acta 58: 2629-2638

Kaminskiy FV (1987) Genesis of carbonado; crystalline aggregate of diamond. (in Russian) Dokladi Akademii Nauk SSSR, Earth Science Section 291: 439-440

Kaminskiy FV, Galimov EM, Ivanovskaya IN, Kirilitsa SI, Polkanov YA (1977) Isotopic distribution in carbon of small diamonds from Ukraine. (in Russian) Dokladi Akademii Nauk SSSR, Earth Science Section 236: 174-175

Kaminskiy FV, Klyuyev YA, Prokopchuk BI, Shckeka SA, Smirnow VI, Ivanovskaya IN (1978) First carbonado and new ballas find in the Soviet Union. (in Russian) Dokladi Akademii Nauk SSSR, Earth Science Section, 242: 152-155

Kamioka H, Shibata K, Kajizuka I, Tomoko O (1996) Rare-earth element patterns and carbon isotopic composition of carbonados: implications for their crustal origin. Geochemical Journal 30: 189-194

Kamo SL, Reimold WU, Krogh TE, Colliston WP (1996) A 2.023 Ga age for the Vredefort impact event and a first report of shock metamorphosed zircons in pseudotachylitic breccias and Granophyre. Earth and Planetary Science Letters 144: 369-387

Kenkmann T and von Dalwigk I (2000) Radial transpression ridges: A new structural feature of complex impact craters. Meteoritics and Planetary Science 35:1189-1202

Killick AM, Reimold WU (1990) Review of the pseudotachylites in and around the Vredefort "Dome", South Africa. South African Journal of Geology 93: 350-365

Kinnunen KA, Lindqvist K (1998) Agate as an indicator of impact structures: An example from Sääksjärvi, Finland. Meteoritics and Planetary Science 33: 7-12

Kirk J, Ruiz J, Chesley J, Titley S, Walshe J (2002) A detrital model for the origin of gold and sulfides in the Witwatersrand basin based on Re-Os isotopes. Geochimica et Cosmochimica Acta 65: 2149-2159

Kirschner CE, Grantz A, Mullen MW (1992) Impact origin of the Avak structure, Arctic Alaska, and genesis of the Barrow gas fields. American Association of Petroleum Geologists Bulletin 76: 651-679

Kletetschka G, Taylor PT, Wasilewski PJ, Hill HGW (2000) Magnetic properties of aggregate polycrystalline diamond: implications for carbonado history. Earth and Planetary Science Letters 181: 279-290

Koeberl C, Anderson RR (eds) (1996) The Manson Impact Structure, Iowa: Anatomy of an Impact Crater. Geological Society of America Special Paper 302: 468 pp

Koeberl C, Reimold WU (1995) The Newporte impact structure, North Dakota, USA. Geochimica et Cosmochimica Acta 59: 4747-4767

Koeberl C, Reimold WU, Shirey SB (1996a) A Re-Os isotope study of the Vredefort granophyre: Clues to the origin of the Vredefort structure, South Africa. Geology 24: 913-916

Koeberl C, Reimold WU, Brandt D (1996b) Red Wing Creek structure, North Dakota: Petrographical and geochemical studies, and confirmation of impact origin. Meteoritics and Planetary Science 31: 335-342

Koeberl C, Armstrong RA, Reimold WU (1997a) Morokweng, South Africa: A large impact structure of Jurassic-Cretaceous boundary age. Geology 25: 731-734

Koeberl C, Masaitis VL, Shafranovsky GI, Gilmour I, Langenhorst F, Schrauder M (1997b) Diamonds from the Popigai impact structure, Russia. Geology 25: 967-970

Koeberl C, Schrauder M, Andreoli MAG, Brandstätter F, Lengauer C, and Gilmou, I (1997c) Carbonados from Central Africa and Brazil: Are they related to impact events? A geochemical, spectroscopic, and X-ray study. Abstracts, 7th Annual Goldschmidt Conference, p. 114, LPI Contribution No. 921, Lunar and Planetary Institute, Houston.

Koeberl C, Reimold WU, Kelley SP (2001) Petrography, geochemistry, and argon-40/argon-39 ages of impact melt-rocks and breccias from the Ames impact structure, Oklahoma: The Nicor Chestnut 18-4 drill core. Meteoritics and Planetary Science 36: 651-669

Komor SC, Valley JW, Brown PE, Collini B (1987) Fluid inclusions in granite from the Siljan Ring impact structure and surrounding regions. In: Bodén A, Eriksson KG (eds) Deep Drilling in Crystalline Bedrock. Springer-Verlag, Heidelberg, 181-208

Komor SC, Valley JW, Brown PE (1988) Fluid-inclusion evidence for impact heating at the Siljan Ring, Sweden. Geology 16: 711-715

Konta J (ed) (1988) 2nd International Conference on Natural Glasses, Prague. Charles University, Prague, 423 pp

Kring DA (2000) Impact events and their effect on the origin, evolution, and distribution of life. GSA Today 10(8): 1-7

Krochuk RV, Sharpton VL (2002) Overview of Terny astrobleme (Ukrainian Shield) studies [abs]. Lunar and Planetary Science 33: abs #1832, CD-ROM

Krogh TE, Davis DW, Corfu F (1984) Precise U-Pb zircon and baddeleyite ages for the Sudbury area. In: Pye E, Naldrett AJ, Giblin PE, The Geology and Ore Deposits of the Sudbury Structure. Ontario Geological Survey Special Volume 1: 431-446

Kuykendall MD, Johnson CL, Carlson RA (1997) Reservoir characterization of a complex impact structure: Ames Impact Structure, Northern shelf, Anadarko Basin. In Johnson KS, Campbell JA (eds) Ames Structure in Northwest Oklahoma and Similar features: Origin and Petroleum Production (1995 Symposium), Oklahoma Geological Survey Circular 100: 199-206

Lainé R, Alonso D, Svab M (eds) (1985) The Carswell Structure Uranium Deposits. Geological Association of Canada, Special Paper 29: 230 pp

Lakomy R (1990) Implications for cratering mechanics from a study of the Footwall Brecciaof the Sudbury impact structure, Canada. Meteoritics 25: 195-207

Lambert P (1981) Breccia dikes: geological constraints on the formation of complex craters. In: Schultz PH, Merrill RB (eds) Multi-Ring Basins. Proceedings of the Lunar and Planetary Science Conference 12A: 59-78

Lana C (2004) Structural and petrogenetic studies related to the geological evolution of the Archean basement complex, Vredefort Dome. PhD Thesis, University of the Witwatersrand, Johannesburg

Lana C, Gibson RL, Reimold WU (2003a) Archean crustal structure of the Kaapvaal craton, South Africa – evidence from the Vredefort dome. Earth and Planetary Science Letters 206: 133-144

Lana C, Reimold WU, Gibson RL, Koeberl C (2003b) Nature of the Archean mid-crust in the core of the Vredefort Dome, central Kaapvaal craton, South Africa. Geochimica et Cosmochimica Acta 68: 623-641

Lana C, Gibson RL, Reimold WU (2003c) Impact tectonics in the core of the Vredefort Dome, South Africa: Implications for central uplift formation in very large impact structures. Meteoritics and Planetary Science 38: 1109-1130

Lana C, Gibson RL, Reimold WU (2003d) Geology and geochemistry of the Greenlands greenstone complex, Vredefort Dome: implications for the early and mid-Archean crustal evolution of the Kaapvaal craton. South African Journal of Geology 106: 291-314

Lantz R (1981) Barrow gas fields – N. Slope, Alaska. Oil and Gas Journal 79: 197-200

Leroux H, Reimold WU, Doukhan JC (1994) A TEM investigation of shock metamorphism in quartz from the Vredefort Dome, South Africa. Tectonophysics 230: 223-230

Leroux H, Reimold WU, Koeberl C, Hornmann U (1999) Experimental shock deformation in zircon: a transmission electron microscopic study. Earth and Planetary Science Letters 169: 291-301

Lesher CM, Thurston PC (eds) (2002) A special issue devoted to the mineral deposits of the Sudbury basin. Economic Geology 97: 1371-1375

Lewis JS (1997) Mining the Sky: Untold Riches from the Asteroids. Comets and Planets. Perseus Publishing, Reading, Massachussetts, 288 pp

Lightfoot PC, Farrow CEG (2002) Geology, geochemistry, and mineralogy of the Worthington Offset Dike: A genetic model for offset dike mineralization in the Sudbury Igneous Complex. Economic Geology 97: 1419-1446

Lightfoot PC, Keays RR, Morrison GG, Bite A, Farrell KP (1997a) Geochemical relationships in the Sudbury Igneous Complex: origin of the main mass and offset dikes. Economic Geology 92: 289-307

Lightfoot PC, Keays RR, Morrison GG, Bite A, Farrell KP (1997b) Geologic and geochemical relationships between the contact sublayer, inclusions, and the main mass of the Sudbury Igneous Complex: a case study of the Whistle Mine embayment. Economic Geology 92: 647-673

Limon M, Cedillo E, Quezada JM, Grajales JM, Alvarez W, Hildebrand AR, Sanchez MA, Rosales C, and Gonzalez V (1994) Cretaceous-Tertiary boundary sedimentary breccias from southern Mexico: Normal sedimentary deposits or impact-related breccias? [abs] American Association of Petroleum Geologists Annual Convention, p 199

Lowman PD (1999) Original size of the Sudbury Structure: evidence from field observations and imaging radar [abs]. Joint Conference of the Geological Association of Canada - Mineralogical Association of Canada 24, p 74

Magloughlin JF, Spray JG (1992) Frictional melting processes and products in geological materials: Introduction and discussion. Tectonophysics 204: 197-206

Maier WD, Andreoli MAG, McDonald I, Irvine GJ, Ashwal LD, Prevec S (2003) The Morokweng impact melt sheet, South Africa: a reconnaissance study with implications for Ni-Cu-PGE sulphide mineralisation [ext abs]. Applied Earth Science 112, No. 2: B150-B152

Martini JEJ (1978) Coesite and stishovite in the Vredefort Dome, South Africa. Nature 272: 715-717

Martini JEJ (1991) The nature, distribution and genesis of the coesite and stishovite associated with pseudotachylite of the Vredefort Dome, South Africa. Earth and Planetary Science Letters 103: 285-300

Masaitis VL (1989) The economic geology of impact craters. International Geology Review 31: 922-933

Masaitis VL (1998) Diamond-bearing impactites of the Popigai astrobleme. (in Russian) VSEGEI Press, 178

Masaitis VL (1993) Diamondiferous impactites, their distribution and petrogenesis. Reg Geol Metall 1: 121-134 (in Russian)

Masaitis VL, Selivanovskaya TV (1972) Shock-metamorphosed rocks and impactites of the Popigay meteor crater. (in Russian) Zapiski Vsesoyuznogo Mineralogicheskogo Obshchestva 100, Issue 4: 385-393

Masaitis VL, Danilin AI, Mashchak MS, Raikhlin, AI, Selivanovskaya TV, Shadenkov EM (1980) The Geology of Astroblemes. (in Russian) Nedra Press, Leningrad, 231

Masaitis VL, Shafranovsky GI, Grieve RAF, Langenhorst F, Peredery WV, Therriault AM, Balmasov EL, Fedorova IG (1999) Impact diamonds in the suevitic breccias of the Black Onaping Formation, Sudbury Structure, Ontario, Canada. In: Dressler BO, Sharpton VL (eds) Large Meteorite Impacts and Planetary Evolution II, Geological Society of America, Special Paper 339: 317-321

McCarthy TS, Charlesworth EG, Stanistreet IG (1986) Post-Transvaal structural features of the northern portion of the Witwatersrand Basin, South Africa. Transactions of the Geological Society of South Africa 89: 311-324

McCarthy TS, Stanistreet IG, Robb LJ (1990) Geological studies related to the origin of the Witwatersrand Basin and its mineralization – an introduction and a strategy for research and exploration. South African Journal of Geology 93: 1-4

McCarville P, Crossey LJ (1996) Post-impact hydrothermal alteratuion of the Manson impact structure. In: Koeberl C, Anderson RR (eds) The Manson Impact Structure: Anatomy of an Impact Crater. Geological Society of America, Special Paper 302: 347-376

McDonald I, Andreoli MAG, Hart RJ, Tredoux M (2000) Platinum-group elements in the Morokweng impact structure, South Africa: Evidence for the impact of a large ordinary chondrite projectile at the Jurassic-Cretaceous boundary. Geochimica et Cosmochimica Acta 65: 299-309

Melosh HJ (1989) Impact Cratering: A Geologic Process. Oxford Univ Press, New York, 245 pp

Milkereit B, Green A, Sudbury Working Group (1992) The geometry of the Sudbury Structure from high-resolution seismic reflection profiling. Geology 20: 807-811

Milkereit B, Green A, Wu J, White D, Adam E (1994) Integrated seismic and borehole geophysical study of the Sudbury Igneous Complex. Geophysical Research Letters 21: 931-934

Milstein RL (1988) The Calvin 28 structure: Evidence for impact origin. Canadian Journal of Earth Sciences 25: 1524-1530

Minter WEL (1999) Irrefutable detrital origin of Witwatersrand gold and evidence of eolian signatures. Economic Geology 94: 665-670

Minter WEL, Goedhart M, Knight J, Frimmel HE (1993) Morphology of Witwatersrand gold grains from the Basal reef: Evidence for their detrital origin. Economic Geology 88: 237-248

Molnár F, Watkinson DH, Everest JO (1999) Fluid-inclusion characteristics of hydrothermal Cu-Ni-PGE veins in granitic and metavolcanic rocks at the contact of the Little Stibie deposit, Sudbury, Canada. Chemical Geology 154: 279-301

Montanari A, Koeberl C (2000) Impact Stratigraphy. Lecture Notes in Earth Sciences 93, Springer-Verlag, Berlin-Heidelberg, 364 pp

Morgan J, Warner M (1999) Chicxulub: the third dimension of a multi-ring impact basin. Geology 27: 407-410

Morgan JW, Walker RJ, Horan MF, Beary ES, Naldrett AJ (2002) ^{190}Pt-^{186}Os and ^{187}Re-^{187}Os systematics of the Sudbury Igneous Complex, Ontario. Geochimica et Cosmochimica Acta 66: 273-290

Morrison GG (1984) Morphological features of the Sudbury Structure in relation to an impact origin. In Pye EG, Naldrett AJ, Giblin PE (eds) The Geology and Ore Deposits of the Sudbury Structure. Ontario Geological Survey Special Volume 1: 513-520

Morrison GG, Jago BC, White TL (1994) Footwall mineralization of the Sudbury Igneous Complex. In: Lightfoot PC, Naldrett AJ (eds), Footwall mineralization of the Sudbury Igneous Complex. Proceedings of the Sudbury-Noril'sk Symposium, Ontario Geological Survey Special Volume 5: 57-64

Moser DE (1997) Dating the shock wave and thermal imprint of the giant Vredefort impact, South Africa. Geology 25: 7-10

Moser D, Flowers RM, Hart RJ (2001) Birth of the Kaapvaal tectosphere 3.08 billion years ago. Science 291: 465-468

Müller-Mohr V (1992) Breccias in the basement of a deeply eroded impact structure, Sudbury, Canada. Tectonophysics 216: 219-226

Mutonen T (2000) The big Pechenga bang (abs). In Plado J and Pesonen L (eds), Meteorite Impact in Precambrian Shields. Program and Abstracts, 4th Workshop of the European Science Foundation, Lappajärvi, Finland. Geological Survey of Finland, Espoo, p 43

Naldrett AJ (1984a) Summary, discussion, and synthesis. In: Pye E, Naldrett A, Giblin PE (eds), The Geology and Ore Deposits of Sudbury, Ontario Geological Survey Special Volume 1: 533-570

Naldrett AJ (1984b) Ni-Cu ores of the Sudbury Igneous Complex – Introduction. In: Pye E, Naldrett AJ, Giblin PE (eds), The Geology and Ore Deposits of Sudbury. Ontario Geological Survey Special Volume 1: 302-307

Naldrett AJ (2003) From impact to riches: Evolution of geological understanding as seen at Sudbury, Canada. GSA Today, February 2003: 4-9

Naldrett AJ, Innes DG, Soea J, Gorton M (1982) Compositional variation within and between five Sudbury ore deposits. Economic Geology 77: 1519-1534

Naldrett AJ, Hewins RH, Dressler BO, Rao BV (1984) The Contact Sublayer of the Sudbury Igneous Complex. In: Pye EG, Naldrett AJ, Giblin PE (eds), The Geology and Ore Deposits of the Sudbury Structure. Ontario Geological Survey Special Volume 1, 253-274

Naumov MV (2002) Impact-generated hydrothermal systems: Data from Popigai, Kara, and Puchezh-Katunki impact structures. In: Plado J, Pesonen LJ (eds), Impacts in Precambrian Shields, Impact Studies Series, Springer-Verlag, Berlin-Heidelberg, 117-171

Nayak VK (1985) Trona in evaporite from the Lonar impact crater, Mahrashtra. Journal of Indian Earth Sciences: 221-252

Nel LT (1927) The Geology of the Country around Vredefort. An explanation of the geological map. Geological Survey of South Africa, Special Publication 6: 134 pp

Newsom HE (1980) Hydrothermal alteration of impact melt sheets with implications for Mars. Icarus 44: 207-216

Newsom HE, Graup G, Sewards T, Keil K (1986) Fluidization and hydrothermal alteration of the suevite deposit at the Ries Crater, West Germany, and implications for Mars. Proceedings of the 17th Lunar and Planetary Science Conference, Journal of Geophysical Research 91: E239-E251

Newsom HE, Britelle GE, Hibbitts CA, Crossey LJ, Kudo AM (1996) Impact cratering and the formation of crater lakes on Mars. Journal of Geophysical Research 101: 14951-14955

Newsom HE, Hagerty JJ, Goff F (1999) Mixed hydrothermal fluids and the origin of Martian soil. Journal of Geophysical Research 104: 8717-8728.

Nikolskiy AP (1991) Geological structure of iron ore deposit Pervomayskoe and its transformation caused by meteorite impact. Nedra Press, Moscow, 71 pp (in Russian)

Nikolskiy AP, Naumov VP, Korobko NI (1981) Pervomaysk iron deposit, Krivoi Rog, and its transformations by shock metamorphism. (in Russian) Geologia rudnykh mestorozhdeniy 5: 92-105

Nikolskiy AP, Naumov VP, Mashchak MS, Masaitis VL (1982) Shock metamorphosed rocks and impactites of Ternovka astrobleme. (in Russian) Transactions VSEGEI 238: 132-142

Osinski GR, Spray JG, Lee P (2001) Impact-induced hydrothermal activity within the Haughton impact structure, arctic Canada: Generation of a transient, warm, wet oasis. Meteoritics and Planetary Science 36: 731-745

Ostermann M, Schärer U, Deutsch A (1996) Impact melt dikes in the Sudbury multi-ring basin (Canada): implications from uranium-lead geochronology on the Foy Offset Dike. Meteoritics and Planetary Science 31: 494-501

Ozima M, Zashu S, Tomura K, Matsuhisa Y (1991) Constraints from noble-gas contents on the origin of carbonado diamonds. Nature 351: 472-474

Peredery WV, Morrison GG (1984) Discussion of the origin of the Sudbury Structure. In Pye EG, Naldrett AJ, Giblin PE (eds) The Geology and Ore Deposits of the Sudbury Structure. Ontario Geological Survey Special Volume 1: 491-511

Phillips GN, Law JDM (2000) Witwatersrand gold fields: Geology, genesis and exploration. Society of Economic Geologists Reviews 13: 1-31

Phillips ME, Bussell MA, McDonald I, Hart RJ, Andreoli MAG (1999) A remote sensing and geological investigation of the Vredefort impact structure (South Africa) using LANDSAT TM imagery [abs]. Meteoritics and Planetary Science 34: A92-A93

Pickard CF (1994) Twenty years of production from an impact structure, Red Wing Creek field, McKenzie County, North Dakota [abs.], American Association of Petroleum Geologists, Annual Convention, p 234

Poag CW, Koeberl C, Reimold WU (2004) The Chesapeake Bay impact structure. Impact Studies Series vol. 4, Springer-Verlag Berlin-Heidelberg, 522 pp

Pretorius DA, Brink WCJ, Fouche J (1986) Geological map of the Witwatersrand Basin. In: Anhaeusser CR, Maske S (eds.) Mineral Deposits of Southern Africa. Geological Society of South Africa, Johannesburg, scale 1: 250 000

Prevec SA (2000) An examination of modal variation mechanisms in the contact sublayer of the Sudbury Igneous Complex, Canada. Mineralogy and Petrology 68: 141-157

Prevec SA, Cawthorn RG (2002) Thermal evolution and interaction between impact melt sheet and footwall: a genetic model for the Contact Sublayer of the Sudbury Igneous Complex, Canada. Journal of Geophysical Research 107, B8: 10.1029/2001JB000525

Prevec SA, Lightfoot PC, Keays RR (2000) Evolution of the sublayer of the Sudbury Igneous Complex: geochemical, Sm-Nd isotopic and petrologic evidence. Lithos 51: 271-292

Pye E, Naldrett A, Giblin P (eds) (1984) The Geology and Ore Deposits of Sudbury. Ontario Geological Survey, Special Volume 1: 604 pp

Reimold WU (1990) The controversial microdeformations in quartz from the Vredefort structure, South Africa. South African Journal of Geology 93: 645-663

Reimold WU (1992) Dimension stone quarrying and the environment (A geologist's view of the case of the Vredefort Structure, south-west of Johannesburg). Environmental Planning and Management 3: 4-11

Reimold WU (1994) Hydrothermal Witwatersrand gold mineralisation caused by the Vredefort mega-impact event [abs]. In: New Developments Regarding the KT event and Other Catatsrophes in Earth History (Snowbird III Conf, Houston, Febr 1994). Lunar and Planetary Institute Contributions 825: 92-93

Reimold WU (1995) Pseudotachylite – Generation by friction melting and shock brecciation? – A review and discussion. Earth-Science Reviews 39: 247-364

Reimold WU (1998) Exogenic and endogenic breccias: a discussion of major problematics. Earth-Science Reviews 43: 25-47

Reimold WU (2002) Mineralization associated with impact structures, with special reference to the Vredefort-Witwatersrand system [abs.]. Presidential Address, Geocongress 2002 and 11th Quadriennial Meeting of the International Association for the Genesis of Ore Deposits, Windhoek, Namibia, July 2002, CD-ROM, 3 p

Reimold WU, Colliston WP (1994) Pseudotachylites of the Vredefort Dome and the surrounding Witwatersrand Basin, South Africa. In: Dressler BO, Grieve RAF, Sharpton VL (eds) Large Meteorite Impacts and Planetary Evolution. Geological Society of America, Special Paper 293: 177-196

Reimold WU, Gibson RL, Friese A, Layer PW (1995) A chronological framework for the Witwatersrand Basin – with emphasis on metamorphic/hydrothermal events [abs.]. In: Reimold WU (comp), The Economic Significance of Metamorphism and Fluid Movement Within the Witwatersrand Basin [abs], Geological Society of South Africa Symposium, 26 October 1995, Western Deep Levels Gold Mine, 21-25

Reimold WU, Brandt D, De Jong R, Hancox PJ (1999a) Tswaing Meteorite Crater – A Natural and Cultural History of the Tswaing Crater Region including a description of the Hiking Trail. Popular Geoscience Series, Council for Geoscience, Pretoria, vol 1: 171

Reimold WU, Koeberl C, Fletcher P, Killick AM, Wilson JD (1999b) Pseudotachylitic breccias from fault zones in the Witwatersrand Basin, South Africa: evidence of autometasomatism and post-brecciation alteration processes. Mineralogy and Petrology 66: 25-53

Reimold WU, Koeberl C, Brandstätter F, Kruger FJ, Armstrong RA, Bootsman C (1999c) Morokweng impact structure, South Africa: Geologic, petrographic, and isotopic results, and implications for the size of the structure. In: Dressler BO, Sharpton VL (eds) Large Meteorite Impacts and Planetary Evolution II. Geological Society of America Special Paper 339: 61-90

Reimold WU, Armstrong RA, Koeberl C (2002a) A deep drillcore from the Morokweng impact structure, South Africa: petrography, geochemistry and constraints on the crater size. Earth and Planetary Science Letters 201: 221-232

Reimold WU, Leroux H, Gibson RL (2002b) Shocked and thermally metamorphosed zircon from the Vredefort impact structure, South Africa: a transmission electron microscopic study. European Journal of Mineralogy 14: 859-868

Reimold WU, Kelley SP, Sherlock S, Henkel H., Koeberl C (2004) Laser argon dating of melt breccias from the Siljan Impact Structure, Sweden – Implications for possible relationship to Late Devonian extinction events [abs] 35th Lunar and Planetary Science Conference, Houston, CD-ROM, abstract No 1480, 2 pp

Robb LJ, Robb VM (1998) Gold in the Witwatersrand Basin. In: Wilson MGC, Anhaeusser CR (eds) The Mineral resources of South Africa, Council for Geoscience, Pretoria, Handbook 16, pp 294-349

Robb LJ, Charlesworth EG, Drennan GR, Gibson RL, Tongu EL (1997) Tectono-metamorphic setting and paragenetic sequence of Au-U mineralization in the Archaean Witwatersrand Basin, South Africa. Australian Journal of Earth Sciences 44: 353-371

Rondot J (1994) Recognition of eroded astroblemes. Earth-Science Reviews 35: 331-365

Rousell DH (1984a) Structural geology of the Sudbury basin. In Pye EG, Naldrett AJ, Giblin PE (eds) The Geology and Ore Deposits of the Sudbury Structure. Ontario Geological Survey Special Volume 1: 83-95

Rousell DH (1984b) Mineralization in the Whitewater Group. In: Pye EG, Naldrett AJ, Giblin PE (eds) The Geology and Ore Deposits of the Sudbury Structure. Ontario Geological Survey Special Volume 1: 219-232

Rousell DH, Gibson HC, Jonasson IR (1997) The tectonic, magmatic and mineralization history of the Sudbury Structure. Exploration and Mining Geology 6, No1: 1-22

Rubin AE, Scott ERD (1996) Abee and related EH chondrite impact-melt breccias. Geochimica et Cosmochimica Acta 61: 425-435

Schreyer W, Abraham K (1978) Symplectitic cordierite-orthopyroxene-garnet assemblages as products of contact metamorphism of pre-existing basement granulites in the Vredefort structure, South Africa. Contributions to Mineralogy and Petrology 68: 53-62

Shand EJ (1916) The pseudotachylite of Parys. Quarterly Journal of the Geological Society, London 62: 198-221

Sharpton VL, Marin LE, Carney JL, Lee S, Ryder G, Schuraytz BC, Sikora P, Spudis PD(1996) A model for the Chicxulub impact basin based on the evolution of geophysical data, well logs and drill core samples. In: Ryder G, Fastovsky D, Gartner S (eds), The Cretaceous-Tertiary event and other catastrophes in Earth history, Geological Society of America, Special Paper 307: 819-821

Shelkov DA, Verchovsky AB, Milledge HJ, Pillinger CT (1998) The radial distribution of implanted and trapped ^{4}He in single diamond crystals and implications for the origin of carbonado. Chemical Geology 149: 109-116

Shibata K, Kamioka H, Kaminsky KF, Koptil V, Svisero D (1993) Rare earth patterns of carbonado and yakutite: Evidence for their crustal origin. Mineralogical Magazine 57: 607-611

Simpson C (1978) The structure of the rim synclinorium of the Vredefort Dome. Transactions of the Geological Society of South Africa 81: 115-121

Smith JV, Dawson JB (1985) Carbonado: Diamond aggregates from early impacts of crustal rocks? Geology 13: 342-343

Snyder DB, Hobbs RW, the Chicxulub Working Group (1999) Ringed structural zones with deep roots formed by the Chicxulub impact. Journal of Geophysical Research 104: 10743-10755

Sokhor MI, Polkanov YA, Yeremenko GK (1973) Find of the hexagonal polymorphic modification of diamond (lonsdaleite) in placers. (in Russian) Dokladi Akademii Nauk SSSR, Earth Science Section 209: 933-936

Speers EC (1957) The age relationship and origin of common Sudbury Breccia. Journal of Geology 65: 497-514

Spray JG (1997) Superfaults. Geology 25: 627-630

Spray JG (1998) Localized shock- and friction-induced melting in response to hypervelocity impact. In: Grady MM, Hutchison R, McCall GJH, Rothery DA (eds) Meteorites: Flux with Time and Impact effects. Geological Society of London, Special Publication 140: 195-204

Spray JG and Thompson LM (1995) Friction melt distribution in a multi-ring impact basin. Nature 373: 130-132

Spray JG, Kelley SP, Reimold WU (1995) Laser probe argon-40/argon-39 dating of coesite- and stishovite-bearing pseudotachylytes and the age of the Vredefort impact event. Meteoritics 30: 335-343

Spudis PD (1993) The Geology of Multi-Ring Impact basins. Cambridge University Press, New York, 263 pp

Stevens G, Gibson RL, Droop GTR (1997) Mid-crustal granulite facies metamorphism in the central Kaapvaal Craton: the Bushveld Complex connection. Precambrian Research 82: 113-132

Stone D, Therriault AM (2003) Cloud Creek structure, central Wyoming, USA: Impact origin confirmed. Meteoritics and Planetary Science 38: 445-455

Stöffler D, Ewald U, Ostertag R, Reimold WU (1977) Research drilling Nördlingen 1973 (Ries): composition and texture of polymict impact breccias. Geologica Bavarica 75: 163-189

Therriault AM, Grieve RAF, Ostermann M, Deutsch A (1996) Sudbury Igneous Complex: how many melt systems? [abs]. Meteoritics and Planetary Science 31: A141-A142

Therriault AM, Grieve RAF, Reimold WU (1997) Original size of the Vredefort Structure: implications for the geological evolution of the Witwatersrand Basin. Meteoritics and Planetary Science 32: 71-77

Thompson LM, Spray JG, Kelley SP (1998) Laser probe argon-40/argon-39 dating of pseudotachylyte from the Sudbury Structure: evidence for postimpact thermal overprinting in the North Range. Meteoritics and Planetary Science 33: 1259-1269

Tredoux M, Hart RJ, Carlson RW, Shirey SB (1999) Ultramafic rocks at the center of the Vredefort structure: Further evidence for the crust-on-edge model. Geology 27: 923-926

Trieloff M, Reimold WU, Kunz J, Boer RH, Jessberger EK (1994) ^{40}Ar-^{39}Ar thermochronology of pseudotachylites at the Ventersdorp Contact Reef, Witwatersrand Basin. South African Journal of Geology 97: 365-384

Trieloff M, Deutsch A, Jessberger EK (1998) The age of the Kara impact structure, Russia. Meteoritics and Planetary Science 33: 361-372

Trueb LF, de Wys EC (1971) Carbon from Ubangi – A microstructural study. American Mineralogist 56: 1252-1268

Tuchscherer MG, Spray JG (2002) Geology, mineralization, and emplacement of the Foy Offset Dike, Sudbury Impact Structure. Economic Geology 97: 1377-1397

Val'ter AA (1988) Geochemical features of the meteoritic material in the impactites of the Terny astrobleme. (in Russian) In: The Composition and Origin of Meteorites, 85-92

Vlierboom FW, Collini B, Zumberge JE (1985) The occurrence of petroleum in sedimentary rocks of the meteor impact crater at Lake Siljan, Sweden. Advances in Organic Geochemistry 10: 153-161

Walker RJ, Morgan JW, Naldrett AJ, Li C, Fasett JD (1991) Re-Os isotope systematics of Ni-Cu sulfide ores, Sudbury Igneous Complex, Ontario; evidence for a major crustal component. Earth and Planetary Science Letters 105: 416-429

Walraven F, Armstrong RA, Kruger FJ (1990) A chronostratigraphic framework for the north-central Kaapvaal Craton, the Bushveld Complex and the Vredefort structure. Tectonophysics 171: 23-48

Wood CR, Spray JG (1998) Origin and emplacement of Offset Dikes in the Sudbury impact structure: constraints from Hess. Meteoritics and Planetary Science 33: 337-347

Wolf M (1977) Kohlenpetrographische Untersuchung der See-Sedimente der Forschungsbohrung Nördlingen 1973 und Vergleich mit anderen Untersuchungsergebnissen aus dem Ries. Geologica Bavarica 75: 127-138

Wu J, Milkereit B, Boerner D (1995) Seismic imaging of the enigmatic Sudbury Structure. Journal of Geophysical Research 100: 4117-4130

GPSR Compliance

The European Union's (EU) General Product Safety Regulation (GPSR) is a set of rules that requires consumer products to be safe and our obligations to ensure this.

If you have any concerns about our products, you can contact us on ProductSafety@springernature.com

In case Publisher is established outside the EU, the EU authorized representative is:

Springer Nature Customer Service Center GmbH
Europaplatz 3
69115 Heidelberg, Germany

Batch number: 10755324

Printed by Printforce, the Netherlands